Springer Proceedings in Physics 51

Springer Proceedings in Physics

Managing Editor: H. K. V. Lotsch

30 *Short-Wavelength Lasers and Their Applications* Editor: C. Yamanaka

31 *Quantum String Theory*
Editors: N. Kawamoto and T. Kugo

32 *Universalities in Condensed Matter*
Editors: R. Jullien, L. Peliti, R. Rammal, and N. Boccara

33 *Computer Simulation Studies in Condensed Matter Physics: Recent Developments*
Editors: D. P. Landau, K. K. Mon, and H.-B. Schüttler

34 *Amorphous and Crystalline Silicon Carbide and Related Materials*
Editors: G. L. Harris and C. Y.-W. Yang

35 *Polycrystalline Semiconductors: Grain Boundaries and Interfaces*
Editors: H. J. Möller, H. P. Strunk, and J. H. Werner

36 *Nonlinear Optics of Organics and Semiconductors*
Editor: T. Kobayashi

37 *Dynamics of Disordered Materials*
Editors: D. Richter, A. J. Dianoux, W. Petry, and J. Teixeira

38 *Electroluminescence*
Editors: S. Shionoya and H. Kobayashi

39 *Disorder and Nonlinearity*
Editors: A. R. Bishop, D. K. Campbell, and S. Pnevmatikos

40 *Static and Dynamic Properties of Liquids*
Editors: M. Davidović and A. K. Soper

41 *Quantum Optics V*
Editors: J. D. Harvey and D. F. Walls

42 *Molecular Basis of Polymer Networks*
Editors: A. Baumgärtner and C. E. Picot

43 *Amorphous and Crystalline Silicon Carbide II: Recent Developments*
Editors: M. M. Rahman, C. Y.-W. Yang, and G. L. Harris

44 *Optical Fiber Sensors*
Editors: H. J. Arditty, J. P. Dakin, and R. Th. Kersten

45 *Computer Simulation Studies in Condensed Matter Physics II: New Directions*
Editors: D. P. Landau, K. K. Mon, and H.-B. Schüttler

46 *Cellular Automata and Modeling of Complex Physical Systems*
Editors: P. Manneville, N. Boccara, G. Y. Vichniac, and R. Bidaux

47 *Number Theory and Physics*
Editors: J.-M. Luck, P. Moussa, and M. Waldschmidt

48 *Many-Atom Interactions in Solids*
Editors: R. Nieminen, M. Puska, and M. Manninen

49 *Ultrafast Phenomena in Spectroscopy*
Editors: E. Klose and B. Wilhelmi

50 *Magnetic Properties of Low-Dimensional Systems II: New Developments*
Editors: L. M. Falicov, F. Mejía-Lira, and J. L. Morán-López

51 *The Physics and Chemistry of Organic Superconductors*
Editor: G. Saito and S. Kagoshima

Volumes 1–29 are listed on the back inside cover

The Physics and Chemistry of Organic Superconductors

Proceedings of the ISSP International Symposium,
Tokyo, Japan, August 28–30, 1989

Editors: G. Saito and S. Kagoshima

With 300 Figures

Springer-Verlag Berlin Heidelberg New York
London Paris Tokyo Hong Kong

Professor Gunzi Saito
Department of Chemistry, Kyoto University, Sakyo–ku, Kyoto 606, Japan

Professor Seiichi Kagoshima
Department of Pure and Applied Sciences, University of Tokyo,
Komaba 3-8-1, Meguro, Tokyo 153, Japan

ISBN 3-540-52157-7 Springer-Verlag Berlin Heidelberg New York
ISBN 0-387-52157-7 Springer-Verlag New York Berlin Heidelberg

This work is subject to copyright. All rights are reserved, whether the whole or part of the material is concerned, specifically the rights of translation, reprinting, reuse of illustrations, recitation, broadcasting, reproduction on microfilms or in other ways, and storage in data banks. Duplication of this publication or parts thereof is only permitted under the provisions of the German Copyright Law of September 9, 1965, in its current version, and a copyright fee must always be paid. Violations fall under the prosecution act of the German Copyright Law.

© Springer-Verlag Berlin Heidelberg 1990
Printed in Germany

The use of registered names, trademarks, etc. in this publication does not imply, even in the absence of a specific statement, that such names are exempt from the relevant protective laws and regulations and therefore free for general use.

2154/3150(3011)-543210 – Printed on acid-free paper

Preface

This volume contains the proceedings of the first ISSP International Symposium on the Physics and Chemistry of Organic Superconductors, which was held at the Komaba Eminence Hotel in Tokyo, August 28–30, 1989.

This symposium was attended by 205 scientists from 12 countries. In total 106 papers were presented: 61 as posters, and 39 original papers and 6 review papers in oral sessions. Of these, 102 papers are included in these proceedings.

These contributions cover the interdisciplinary field of physics and chemistry of organic superconductors with particular emphasis on the following subjects and materials:

– superconducting properties,
– spin density waves,
– electronic and structural properties,
– TMTSF salts and their derivatives,
– BEDT-TTF salts and their derivatives,
– metal coordinated organic conductors.

The contributions to this volume are arranged in 11 categories.

The Organizing Committee would like to acknowledge all participants, who contributed to the great success of this symposium on a growing field in both physics and chemistry. The editors express their gratitude to the members of the Organizing and Executive Committees for their cooperation. We also wish to thank Dr. H. Lotsch of Springer-Verlag for his management of the publication and Miss S. Shibata for her assistance in editing this volume.

Tokyo
December 1989

G. Saito
S. Kagoshima

Opening Address

Ladies and Gentlemen,

On behalf of the Organizing Committee I have the privilege of opening this first ISSP International Symposium on the Physics and Chemistry of Organic Superconductors. We deeply appreciate your interest and efforts in helping us make this symposium possible, especially those who have traveled great distances and taken valuable time from their very busy schedules to attend.

This symposium is designed to integrate the recent fundamental advances in the physics and chemistry of organic superconductors. This field is one of the most rapidly expanding and interdisciplinary areas, ranging from synthesis, crystal growth and structural analysis to condensed matter physics and theory.

It is a great honor and pleasure for us to have the first International Symposium of the Institute for Solid State Physics, ISSP, on the subject of organic superconductors in this year. The timing of this symposium coincides with the tenth anniversary of the discovery of the organic superconductor $(TMTSF)_2PF_6$ in 1979.

So far, the TMTSF family has produced 7 superconducting members with the highest T_c around 3 K at 5 kbar. The BEDT-TTF family followed, and has about 13 members with T_c higher than 11 K. Then DMET has 7 members with T_c of 1.9 K and MDT-TTF has one member with T_c of 5 K. All four families are based on the molecule TTF, which was synthesized almost 20 years ago. Another family, the dmit system, has about 4 members with T_c of 6 K at 19 kbar. This family is an anion-based superconductor. Thus, at present we have five superconducting families with more than 30 members.

In the field of experimental and theoretical condensed matter physics, dimensionality, the spin density wave state, the anion order-disorder transition and, more recently, field induced spin density waves have been extensively studied. Furthermore, precise pictures of the electronic structures of organic metals and superconductors have become available from detailed measurements of NMR, ESR, thermal, optical, transport and structural properties, and especially from the recent success of the experimental and theoretical investigations of Fermi surfaces.

In order to vitalize the field of organic superconductors, it is essential to develop new materials, new synthetic methods, new molecular and crystal designs, and to understand the physical and structural aspects of organic superconductors. I am very happy that the key people from both the chemistry and physics sides are attending this symposium to give us reviews and up-to-date reports on their exciting experiments and theories.

The history of the highest T_c of organic superconductors tells us that the organics will reach 25 K at the end of this century, i.e. about ten years from now.

Then the organics will meet the oxides at the end of the third quarter of the next century, or about 85 years from now, if the oxides stay unchanged. But who can wait 85 years? I hope that within this century a significant breakthrough in organic superconductors will take place and I also hope that this symposium will provide many new ideas in physics and chemistry to catalyze such research for the next decade.

This symposium is almost twice as large as what I had originally anticipated. The total number of registered participants is almost 200, including 33 distinguished scientists from abroad. A total of 12 countries are represented, namely, Canada, Denmark, Federal Republic of Germany, France, Greece, Israel, Mexico, United Kingdom, USA, USSR, Yugoslavia and Japan. Forty students have registered for this symposium. It would be extremely valuable for young students and newcomers to this field to be presented with an overview of where we are today as well as with up-to-date work.

I would like to thank the following organizations for the cooperation and financial support we received in organizing this symposium: Ministry of Education, Science and Culture, Japan; Japan Society for the Promotion of Science; Nishina Memorial Foundation; Chemical Society of Japan; Physical Society of Japan; International Superconductivity Technology Center; and Commemorative Association for the Japan World Exposition. I would also like to thank the organizations and companies listed separately, especially Fujitsu, Hitachi, IBM Japan, NEC, NTT, Sony and Toshiba, for financial support which allowed us to hold this symposium.

Finally, I would like to emphasize the goals of this symposium as follows: First, it should provide an opportunity to establish and renew personal relationships between participants. Second, it should provide a forum for the exchange of information between participants in this interdisciplinary meeting. And finally it should stimulate the interest and ambition of participants to develop new physics and chemistry of organic superconductors.

Thank you very much for your attention.

Gunzi Saito, Chairman

Welcome Address

Ladies and Gentlemen,

It is my great pleasure to welcome you all to the first ISSP International Symposium on the Physics and Chemistry of Organic Superconductors.

The Institute for Solid State Physics was established in 1957. Since then one of the most important functions assigned to this institute has been to promote basic research activities in condensed matter physics and its related fields, in close collaboration with scientists throughout this country and possibly abroad. As a program in this direction we have been organizing several domestic symposia every year for 30 years.

In view of the rapidly growing frontiers of science and the increasing importance of international cooperation, we have decided to host a series of international symposia on various topics in the field of condensed matter physics and its related interdisciplinary fields of science. The present symposium is the first of this series. Although this particular symposium is rather large in scale, we do not necessarily expect that every symposium will be like this. The size and format of each symposium will be chosen flexibly considering the nature of its topic. However, in addition to promoting exchange of expertise we would like to encourage young participants, including students, in each symposium to present papers on their new results from the frontiers of science and to get an overview of the field they have been and will be involved in.

The topic of the present symposium is organic superconductors. Although the concept of organic superconductors seems to be nearly as old as the BCS theory and as old as our institute, discoveries of organic superconducting materials have started only in this decade. This research area, I believe, is still young and full of promise. I do not know how high we can go with the T_c of organic superconductors, but I certainly expect substantial developments in this area of condensed matter science in coming years.

Finally, I would like to thank the various organizations listed separately for their cosponsorship of this symposium. A number of industrial firms have also given us generous financial support, for which I would like to express my deep appreciation. My special gratitude goes to Professor Saito and other members of the Organizing Committee for their efforts in preparing this symposium.

I would like to close my remarks by expressing my sincere wish for the success of this symposium. Thank you.

Toru Moriya
Director of the Institute
for Solid State Physics,
University of Tokyo

Symposium Committees

Organizing Committee

G. Saito, Chairman (Univ. Tokyo)
H. Fukuyama (Univ. Tokyo)
I. Ikemoto (Tokyo Metropolitan Univ.)
H. Inokuchi (Inst. Molecular Science)
T. Ishiguro (Kyoto Univ.)
S. Kagoshima (Univ. Tokyo)
H. Kobayashi (Toho Univ.)
N. Miura (Univ. Tokyo)
S. Tanaka (Tokai Univ. & ISTEC)

Executive Committee

T. Enoki (Tokyo Inst. Technology)
T. Mori (Inst. Molecular Science)
K. Nakasuji (Inst. Molecular Science)
T. Osada (Univ. Tokyo)
T. Takahashi (Gakushuin Univ.)
M. Tokumoto (Electrotechnical Lab.)
K. Yakushi (Inst. Molecular Science)
H. Yamochi (Univ. Tokyo)

ISSP Executive Committee: H. Fukuyama, N. Miura, G. Saito, H. Shiba

International Advisory Committee

K. Andres (FRG)
K. Bechgaard (Denmark)
P. Cassoux (France)
P.M. Chaikin (USA)
P. Day (UK)
D. Jérome (France)
I.F. Schegolev (USSR)
J.M. Williams (USA)
F. Wudl (USA)

In Cooperation with

Ministry of Education, Science and Culture, Japan
Japan Society for the Promotion of Science
Nishina Memorial Foundation
International Superconductivity Technology Center
Commemorative Association for the Japan World Exposition
Chemical Society of Japan
Physical Society of Japan

Sponsors

FUJITSU LTD.
Hitachi, Ltd.
(Advance & Central Research Lab.)
IBM JAPAN, Ltd.

NEC Corp.
NIPPON TELEGRAPH AND TELEPHONE
 CORP.
SONY CORP.
TOSHIBA CORP.

Ajinomoto Co., Inc.
BRIDGESTONE CORP.
CHISSO CORP.
Ciba-Geigy (Japan) Ltd.
HONDA R & D CO., Ltd.
HOYA CORP.
Idemitsu Kosan Co., Ltd.
Japan Analytical Industry Co., Ltd.
THE JAPAN CARLIT CO., LTD.
KANEBO, LTD.
Kao Corp.
Matsushita Electric Industrial Co., Ltd.
Matsushita Research Institute Tokyo, Inc.
MITSUBISHI GAS CHEMICAL
 COMPANY, INC.
Mitsui Petrochemical Industries, Ltd.
Mitui Toatsu Chemicals, Inc.

THE NIKKAN KOGYO SHIMBUN, Ltd.
(The Business & Technology Daily News)
NIPPON CHEMI-CON CORP.
NIPPON KOKAN K.K.
NIPPON STEEL CORP.
The Ogasawara Foundation for the
 Promotion of Science and Engineering
Osaka Organic Chemical Ind., Ltd.
Sanyo Electric Co., Ltd.
Showa Denko K.K.
SUMITOMO CHEMICAL COMPANY, LTD.
Sumitomo Electric Industries, Ltd.
Tokyo Kasei Kogyo Co., Ltd.
TORAY INDUSTRIES, INC.
TOYO INK MFG. CO., LTD.
WAKO PURE CHEMICAL INDUSTRY, Ltd.
YAZAWA Scientific Inc.

Contents

Part I	Organic Superconductors: Overview and Comparison with Oxide Superconductors

Recent Developments in Organic Superconductors
By D. Jérome, P. Auban, W. Kang, and J.R. Cooper (With 4 Figures) ..   2

Trends in Structures and Properties of Organic and Inorganic Superconductors
By P. Day (With 4 Figures)   8

On Some Organic Conductors in the Light of Oxide Superconductors
By H. Fukuyama (With 3 Figures)   15

Part II	Metal Coordinated Organic Conductors and Related Materials

Conductive and Superconductive Coordination Compounds
By P. Cassoux, L. Valade, J.-P. Legros, C. Tejel, J.-P. Ulmet, and L. Brossard (With 3 Figures)   22

Synthesis and Properties of the New Molecular Metals $Na[Ni(dmit)_2]_2$ and $NH_4[Ni(dmit)_2]_2$
By R.A. Clark, A.E. Underhill, R. Friend, M. Allen, I. Marsden, A. Kobayashi, and H. Kobayashi (With 3 Figures)   28

Effect of Counter Ions on Alkali Metal Salts of $M_x[Ni(dmit)_2]$ (M=Na, NH_4...)
By A. Izuoka, A. Miyazaki, N. Sato, T. Sugawara, and T. Enoki (With 1 Figure)   32

The Organic π-Electron Metal System Interacting with Mixed-Valence Copper Ions $(R_1,R_2\text{-DCNQI})_2\text{Cu}$ (DCNQI=N,N'-dicyanoquinonediimine; $R_1, R_2 = CH_3, CH_3O, Cl, Br$)
By R. Kato, H. Kobayashi, and A. Kobayashi (With 4 Figures)   36

Metallic DCNQI Salts: Influence of Oxygen, Alloying and Dimensionality
By M. Bair, U. Langohr, J.U. von Schütz, H.C. Wolf, P. Erk, H. Meixner, and S. Hünig (With 2 Figures)   41

Anomalous Temperature Dependence of the Resistivity of (DMeO-DCNQI)$_2$Cu at High Pressure
By H. Kobayashi, A. Miyamoto, R. Kato, Y. Nishio, K. Kajita, W. Sasaki, and A. Kobayashi (With 3 Figures) 45

Reflectance Spectra of DCNQI Salts
By H. Tajima, G. Ojima, T. Ida, H. Kuroda, A. Kobayashi, R. Kato, H. Kobayashi, A. Ugawa, and K. Yakushi (With 5 Figures) 49

Optical Spectra of Highly Conducting Phthalocyanine Salts
By K. Yakushi, H. Yamakado, T. Ida, A. Ugawa, H. Masuda, and H. Kuroda (With 3 Figures) 54

Preparation of Electroconducting Materials Containing Copper Compounds: A Copper Oxide with 1,3,4,6-Tetrathiapentalene-2,5-dione
By M. Inoue, C. Cruz-Vázquez, M.B. Inoue, K.W. Nebesny, and Q. Fernando (With 3 Figures) 58

Part III TMTSF Family: Superconductivity and Spin Density Waves

A Hidden Low-Temperature Phase in the Organic Conductor (TMTSF)$_2$ReO$_4$
By S. Tomić and D. Jérome (With 3 Figures) 64

NMR Evidence for the Existence of 1D Paramagnons in Organic Conductors
By P. Wzietek, F. Creuzet, C. Bourbonnais, D. Jérome, P. Batail, and K. Bechgaard (With 2 Figures) 68

Long-Range Spin-Fluctuations and Superconductivity in the Quasi-One-Dimensional Hubbard Model
By H. Shimahara (With 2 Figures) 73

Transport Properties of Impure Anisotropic Quasi-One-Dimensional Superconductors
By Y. Suzumura (With 3 Figures) 77

Some Recent Experiments on the Field Induced Spin Density Wave States in the Bechgaard Salts
By P.M. Chaikin, J.S. Brooks, S.T. Hannahs, W. Kang, G. Montambaux, and L.Y. Chiang (With 2 Figures) 81

Phase Diagram of the Spin Density Waves Induced by the Magnetic Field in Organic Metals
By F. Pesty, P. Garoche, and M. Héritier (With 2 Figures) 87

Spin Density Wave and Field Induced Spin Density Wave Transport
By K. Maki (With 1 Figure) 91

Magnetothermodynamics and Magnetotransport in (TMTSF)$_2$ClO$_4$
By G. Montambaux (With 3 Figures) 97

Quantized Hall Effect in Spin-Wave Phases of Two-Dimensional Conductors
By M. Kohmoto 102

^{1}H Spin-Lattice Relaxation in the SDW State of $(TMTSF)_2PF_6$ Under Pressure
By T. Takahashi, T. Ohyama, T. Harada, K. Kanoda, K. Murata, and G. Saito (With 3 Figures) 107

Non-ohmic Electrical Transport in the Spin-Density Wave State of Organic Conductors
By S. Tomić, J.R. Cooper, W. Kang, and D. Jérome (With 3 Figures) .. 111

Part IV	**BEDT-TTF Family: Superconductivity**

Superconductivity in BEDT-TTF Based Organic Metals: An Overview
By M. Tokumoto (With 1 Figure) 116

T-P Phase Diagram of β-$(ET)_2I_3$
By V.N. Laukhin, V.B. Ginodman, A.V. Gudenko, P.A. Kononovich, and I.F. Schegolev (With 3 Figures) 122

"2K-Superconducting State" in the Organic Superconductor β-$(BEDT$-$TTF)_2I_3$
By S. Kagoshima, M. Hasumi, Y. Nogami, N. Kinoshita, H. Anzai, M. Tokumoto, and G. Saito (With 3 Figures) 126

A Change of the Incommensurate Superstructure in the Organic Superconductor β-$(BEDT$-$TTF)_2I_3$
By Y. Nogami, S. Kagoshima, H. Anzai, M. Tokumoto, G. Saito, and N. Mori (With 3 Figures) 130

Effect of Annealing on the Superconductivity of β-$(BEDT$-$TTF)_2I_3$
By K. Kanoda, K. Akiba, K. Suzuki, T. Takahashi, and G. Saito (With 2 Figures) 134

Evolution of the "High-T_c" States at Ambient Pressure in β-$(BEDT$-$TTF)_2I_3$
By M. Tokumoto, Y. Yamaguchi, N. Kinoshita, and H. Anzai (With 3 Figures) 138

Electrical Resistance and Upper Critical Field in the "2K-Superconducting State" of β-$(BEDT$-$TTF)_2I_3$
By T. Sasaki, N. Toyota, M. Hasumi, T. Osada, S. Kagoshima, M. Tokumoto, N. Kinoshita, and H. Anzai (With 2 Figures) 142

Bulk Superconductivity at Ambient Pressure in Polycrystalline Pressed Samples of Organic Metals
By D. Schweitzer, S. Kahlich, S. Gärtner, E. Gogu, H. Grimm, R. Zamboni, and H.J. Keller (With 2 Figures) 146

An Ambient Pressure Organic Superconductor κ-(BEDT-TTF-h$_8$ and -d$_8$)$_2$Cu(NCS)$_2$ with T$_c$ Higher than 10K
By H. Mori, S. Tanaka, H. Yamochi, G. Saito, and K. Oshima
(With 5 Figures) 150

Nuclear Spin-Lattice Relaxation in the Organic Superconductor (BEDT-TTF)$_2$Cu(NCS)$_2$: Measurements by the Field Cycling Technique
By T. Takahashi, K. Kanoda, K. Sakao, M. Watabe, H. Mori, and G. Saito
(With 3 Figures) 155

Magnetic-Field Penetration Depth of (BEDT-TTF)$_2$Cu(NCS)$_2$ Determined by Complex Susceptibility
By K. Kanoda, K. Akiba, T. Takahashi, and G. Saito (With 1 Figure) .. 159

Tunneling Spectroscopic Study of the Superconducting Gap of (BEDT-TTF)$_2$Cu(NCS)$_2$ Crystals
By Y. Maruyama, T. Inabe, H. Mori, H. Yamochi, and G. Saito
(With 4 Figures) 163

STM Measurements of Superconducting Properties in κ-(BEDT-TTF)$_2$Cu(NCS)$_2$
By H. Bando, S. Kashiwaya, T. Tokumoto, H. Anzai, N. Kinoshita, M. Tokumoto, K. Murata, and K. Kajimura (With 2 Figures) 167

Effect of Tensile Stress on the Superconducting Transition Temperature in (BEDT-TTF)$_2$Cu(NCS)$_2$
By H. Kusuhara, Y. Sakata, Y. Ueba, K. Tada, M. Kaji, and T. Ishiguro
(With 4 Figures) 171

Highly Correlated Fermi Liquids in the High-T$_c$ Organic Conductor κ-(BEDT-TTF)$_2$Cu(NCS)$_2$
By N. Toyota, E.W. Fenton, T. Sasaki, and M. Tachiki 177

Electronic Properties of (BEDT-TTF)$_3$Cl$_2\cdot$2H$_2$O
By S.D. Obertelli, I.R. Marsden, R.H. Friend, M. Kurmoo, M.J. Rosseinsky, P. Day, F.L. Pratt, and W. Hayes (With 3 Figures) ... 181

Part V BEDT-TTF Family: Fermiology and Related Subjects

Galvanomagnetic Properties of the Organic Metals β-(ET)$_2$X: Magnetoresistance and Shubnikov–de Haas Oscillations
By V.N. Laukhin, M.V. Kartsovnik, S.I. Pesotskii, I.F. Schegolev, and P.A. Kononovich (With 5 Figures) 186

The Fermi Surface in the Organic Superconductor β-(BEDT-TTF)$_2$IBr$_2$
By T. Sasaki, N. Toyota, T. Fukase, K. Murata, M. Tokumoto, and H. Anzai (With 4 Figures) 191

On the Electronic Properties of ET$_2$Cu(NCS)$_2$ as well as of Some New Organic Salts
By H. Müller, C.-P. Heidmann, A. Lerf, W. Biberacher, R. Sieburger, and K. Andres (With 4 Figures) 195

Fermi Surface and Band Structure of κ-(BEDT-TTF)$_2$Cu(NCS)$_2$
By F.L. Pratt, J. Singleton, M. Kurmoo, S.J.R.M. Spermon, W. Hayes,
and P. Day (With 1 Figure) 200

Fermi Surface and Thermoelectric Power of Two-Dimensional Organic
Conductors
By T. Mori and H. Inokuchi (With 1 Figure) 204

Self-Consistent Band Structure and Fermi Surface for β-(BEDT-TTF)$_2$I$_3$
By J. Kübler and C.B. Sommers (With 7 Figures) 208

Anomalous Magneto-oscillation in θ-Type Crystals of (BEDT-TTF)$_2$I$_3$
By K. Kajita, Y. Nishio, T. Takahashi, W. Sasaki, R. Kato, H. Kobayashi,
A. Kobayashi, and Y. Iye (With 6 Figures) 212

Nearly Complete Quantization in Quasi-Two-Dimensional Organic
Superconductors
By K. Yamaji (With 4 Figures) 216

High-Field Magnetotransport in the Organic Conductor (BEDT-
TTF)$_2$KHg(SCN)$_4$
By T. Osada, R. Yagi, S. Kagoshima, N. Miura, M. Oshima, and G. Saito
(With 4 Figures) 220

Electronic Properties in (BEDT-TTF)$_2$X: Magnetoresistance and Hall
Effect
By K. Murata, M. Ishibashi, Y. Honda, T. Komazaki, M. Tokumoto,
N. Kinoshita, and H. Anzai (With 3 Figures) 224

Part VI DMET Salts and Their Families

Physical Properties and Crystal Structures of DMET Superconductors and
Conductors
By K. Kikuchi, K. Murata, K. Saito, K. Kobayashi, and I. Ikemoto
(With 4 Figures) 230

Superconductivity and Spin Density Waves in (DMET)$_2$Au(CN)$_2$
By K. Murata, K. Kikuchi, Y. Honda, T. Komazaki, K. Saito,
K. Kobayashi, and I. Ikemoto (With 3 Figures) 234

Phase Transition of the Organic Metal (DMET)$_2$Au(CN)$_2$ at 180K
By K. Saito, K. Kikuchi, K. Kobayashi, and I. Ikemoto (With 4 Figures) 238

Antiferromagnetic Transitions in (DMET)$_2$X and (DMPT)$_2$X
By K. Kanoda, S. Okui, T. Takahashi, K. Kikuchi, K. Saito, I. Ikemoto,
and K. Kobayashi (With 2 Figures) 242

Conducting and Superconducting Salts Based on MDTTTF, EDTTTF,
VDTTTF, EDTDSDTF, MDSTTF, BMDTTTF, Pd(dmit)$_2$, and Ni(dcit)$_2$
By G.C. Papavassiliou, G. Mousdis, V. Kakoussis, A. Terzis, A. Hountas,
B. Hilti, C.W. Mayer, and J.S. Zambounis (With 2 Figures) 247

Part VII Crystal and Electronic Structures

Structural Instabilities and Electronic Structures of Some Organic
Conductors and Superconductors
By S. Ravy, E. Canadell, and J.P. Pouget (With 5 Figures) 252

Structural and Physical Properties of (BEDT-TTF)$_2$[KHg(SCN)$_4$]
By M. Oshima, H. Mori, G. Saito, and K. Oshima (With 9 Figures) 257

Importance of Weak Hydrogen Bonding C-H···Donor and C-H···Anion
Interactions in Governing the Structural Properties of Organic Donors
BEDT-TTF and BEDO-TTF and Their Charge-Transfer Salts
By M.-H. Whangbo, D. Jung, J. Ren, M. Evain, J.J. Novoa, F. Mota,
S. Alvarez, J.M. Williams, M.A. Beno, A.M. Kini, H.H. Wang,
and J.R. Ferraro 262

Electronic Structure of β-(BEDT-TTF)$_2$I$_3$ Studied by Positron
Annihilation
By S. Tanigawa, P.K. Tseng, T. Kurihara, K.Y. Chang, K. Watanabe,
T. Kubota, M. Tokumoto, N. Kinoshita, and H. Anzai (With 6 Figures) . 267

Pressure Dependence of the Transport Properties of κ-(BEDT-
TTF)$_2$Cu(NCS)$_2$
By I.D. Parker, R.H. Friend, M. Kurmoo, P. Day, C. Lenoir, and P. Batail
(With 2 Figures) 272

Anomalous Transport Behavior in κ-(BEDT-TTF)$_2$Cu(NCS)$_2$
By K. Oshima, R.C. Yu, P.M. Chaikin, H. Urayama, H. Yamochi,
and G. Saito (With 2 Figures) 276

STM Study of (BEDT-TTF)$_2$Cu(NCS)$_2$ Surface
By M. Yoshimura, N. Ara, M. Kageshima, R. Shioda, A. Kawazu,
H. Shigekawa, H. Mori, M. Oshima, H. Yamochi, and G. Saito
(With 5 Figures) 280

Crystal Growth and Properties of (BEDT-TTF)$_2$Cu(NCS)$_2$
By Y. Ueba, T. Mishima, H. Kusuhara, and K. Tada (With 4 Figures) .. 284

Synthesis, Crystal Structure and Properties of (BEDT-TTF)$_3$CuCl$_4$·H$_2$O
By M. Kurmoo, T. Mallah, P. Day, I. Marsden, M. Allan, R.H. Friend,
F.L. Pratt, W. Hayes, D. Chasseau, J. Gaultier, and G. Bravic
(With 2 Figures) 290

Electronic Properties of Charge Transfer Complexes of BEDT-TTF and
Related Donors with Transition Metal Halides
By T. Enoki, I. Tomomatsu, Y. Nakano, K. Suzuki, and G. Saito
(With 4 Figures) 294

Structural Electrical and Magnetic Properties of the (BEDT-TTF)$_4$Ni(CN)$_4$
Complex
By M. Tanaka, H. Takeuchi, A. Kawamoto, J. Tanaka, T. Enoki,
K. Suzuki, K. Imaeda, and H. Inokuchi (With 4 Figures) 298

Crystal Structures and Electrical Conductivities of EDT-TTF Salts
with TaF_6^-, AsF_6^-, PF_6^-, ReO_4^-, ClO_4^-, BF_4^-, $Au(CN)_2^-$ and
$Ni(dmit)_2^{n-}$
By A. Kobayashi, R. Kato, and H. Kobayashi (With 2 Figures) 302

A New Transformable Cation-Radical Salt $(EPT)_2I_7$
By V.E. Korotkov, R.P. Shibaeva, N.D. Kushch, and E.B. Yagubskii
(With 1 Figure) 306

Microwave Conductivity of the Phthalocyanine and
Dicyanoquinonediimine Salts
By H. Yamakado, A. Ugawa, T. Ida, and K. Yakushi (With 2 Figures) .. 311

Electron–Molecular Vibration Coupling in Organic Superconductors
By T. Sugano and M. Kinoshita (With 2 Figures) 315

Dynamics of Charged Domain Walls in Semiconducting Charge Transfer
Compounds
By Y. Iwasa, N. Watanabe, T. Koda, S. Koshihara, Y. Tokura,
N. Iwasawa, and G. Saito (With 4 Figures) 319

The Effect of Pressure on the High Magnetic Field Electronic Phase
Transition in Graphite
By Y. Iye, C. Murayama, N. Mori, S. Yomo, J.T. Nicholls,
and G. Dresselhaus (With 4 Figures) 324

Ferro- and Antiferromagnetic Intermolecular Interactions of Organic
Radicals, α-Nitronyl Nitroxides
By K. Awaga, T. Inabe, U. Nagashima, and Y. Maruyama
(With 3 Figures) 329

Part VIII Structural Design of Organic Superconductors

Structure–Property Correlations in the Design of Organic Metals and
Superconductors: An Overview
By A.M. Kini, M.A. Beno, K.D. Carlson, J.R. Ferraro, U. Geiser,
A.J. Schultz, H.H. Wang, J.M. Williams, and M.-H. Whangbo
(With 4 Figures) 334

Organic Conductors and Superconductors Based on (BEDT-TTF)-
Polyiodides
By R.P. Shibaeva, E.B. Yagubskii, E.E. Laukhina, and V.N. Laukhin
(With 2 Figures) 342

Stoichiometry Control in Organic Metals
By K. Bechgaard, K. Lerstrup, M. Jørgensen, I. Johannsen,
and J. Christensen (With 8 Figures) 349

Unusual Molecular Systems of Organic-Inorganic Character and Increased
Architectural Complexity
By P. Batail, K. Boubekeur, A. Davidson, M. Fourmigué, C. Lenoir,
C. Livage, and A. Pénicaud (With 4 Figures) 353

Part IX New Molecules and Materials

Salts Derived from Bis(ethylenedioxa)tetrathiafulvalene ("BO")
By F. Wudl, H. Yamochi, T. Suzuki, H. Isotalo, C. Fite, K. Liou,
H. Kasmai, and G. Srdanov (With 7 Figures) 358

New Cation-Radical Salts $(ET)_3CuCl_4 \cdot H_2O$ and $(ET)_2CuCl_4$ with Metallic
and Semiconducting Properties
By A.V. Gudenko, V.B. Ginodman, V.E. Korotkov, A.V. Koshelap,
N.D. Kushch, V.N. Laukhin, L.P. Rozenberg, A.G. Khomenko,
R.P. Shibaeva, and E.B. Yagubskii (With 5 Figures) 364

New Organic Synthetic Metals Derived from BEDT-TTF, $Ni(dsit)_2$ and
BEDO-TTF
By M.A. Beno, A.M. Kini, U. Geiser, H.H. Wang, K.D. Carlson,
and J.M. Williams (With 3 Figures) 369

Synthesis and Crystal Structures of Multi-Chalcogen TTF Derivatives and
Conducting Organic Salts
By T. Nogami, H. Nakano, S. Ikegawa, K. Miyawaki, Y. Shirota,
S. Harada, and N. Kasai (With 4 Figures) 373

Preparation of Methylated BEDT-TTFs for Controlling Intermolecular S-S
Contacts
By A. Izuoka, S. Matsumiya, and T. Sugawara (With 3 Figures) 379

Preparation and Properties of Dimeric TTFs
By K. Lerstrup, M. Jørgensen, I. Johannsen, and K. Bechgaard
(With 1 Figure) 383

Preparation and Properties of p-Quinodimethane Analogues of
Tetrathiafulvalene
By Y. Yamashita, Y. Kobayashi, and T. Miyashi 387

Syntheses and Physical Properties of Oligothiophene Charge-Transfer
Complexes
By S. Hotta and K. Waragai (With 4 Figures) 391

$3,3':4,4'$-Bis(thieno[2,3-b]thiophene) with an Isoelectronic Structure of
Perylene
By T. Otsubo, Y. Kono, H. Miyamoto, Y. Aso, F. Ogura, T. Tanaka,
and M. Sawada (With 1 Figure) 395

Design of Organic Molecular Metals Based on New Multi-Stage Redox
Systems in the Non-TTF Family: Peri-Condensed Weitz-Type Donors
By K. Nakasuji (With 2 Figures) 399

Conjugated Heteroquinonoid Isologues of TCNQ as Novel Electron
Acceptors
By F. Ogura, K. Yui, H. Ishida, Y. Aso, and T. Otsubo (With 1 Figure) . 403

Design of Two-Dimensional Stacking Structures: Twin-Type Molecules
and Steric Interaction of Axial Substituents
By T. Inabe, T. Mitsuhashi, and Y. Maruyama (With 3 Figures) 408

New Molecular Conductors Based on Metal Complex Anions
By A.E. Underhill, K.S. Varma, R.A. Clark, and C.E. Wainwright 412

Conducting Evaporated Film of $Pt_2(CH_3CS_2)_4I$
By I. Shirotani, Y. Inagaki, and M. Yamashita (With 4 Figures) 416

Electroactive Langmuir Blodgett Films of Tetrathiafulvalene Derivatives
By A.S. Dhindsa, M.R. Bryce, and M.C. Petty (With 2 Figures) 420

Physical Properties of Conductive Langmuir-Blodgett Films of
Tridecylmethylammonium-Au(dmit)$_2$ and Its Derivatives
By T. Nakamura, Y. Miura, M. Matsumoto, H. Tachibana, M. Tanaka,
and Y. Kawabata (With 2 Figures) 424

Polymerization of Diacetylenes in Liquid Crystal Phases and Its
Application to the Preparation of High Spin Polydiacetylenes
By A. Izuoka, T. Okuno, and T. Sugawara (With 3 Figures) 428

Part X	Theory

Novel Superconductivity from an Insulator
By Y. Takada and M. Kohmoto (With 1 Figure) 434

Bethe Ansatz Wavefunction, Momentum Distribution and Spin Correlation
in the One-Dimensional Strongly Correlated Hubbard Model
By M. Ogata and H. Shiba (With 4 Figures) 438

New High-Temperature Cooper-Pairing Phase for Vibronic
Superconductivity
By A. Tachibana, S. Ishikawa, T. Tada, and T. Yamabe (With 4 Figures) 444

Examination of Pairing via Effective van der Waals Interaction in High-
Temperature Superconductivity
By K. Tanaka, Y. Yamaguchi, and T. Yamabe (With 1 Figure) 449

Reassessment of the Excitonic Mechanism of Little's Superconductivity
By K. Tanaka, M. Okada, Y. Huang, and T. Yamabe (With 2 Figures) .. 453

Multi-Valence Resonance-Condensation Model: A Possible Novel and
Universal Origin of Superconductivity
By A. Nakamura (With 2 Figures) 457

Density Functional Theory for the New High-Temperature
Superconducting Phase Transition
By A. Tachibana (With 4 Figures) 461

Possible Role of Two-Dimensionality for the Enhancement of
Superconducting T_c
By K. Fukushima and H. Sato (With 1 Figure) 465

Part XI	Summary

New Developments That Emerged in the ISSP Symposium
By T. Ishiguro 470

Index of Contributors................................ 475

Part I

**Organic Superconductors:
Overview and Comparison with
Oxide Superconductors**

Recent Developments in Organic Superconductors

D. Jérome, P. Auban, W. Kang, and J.R. Cooper*

Laboratoire de Physique des Solides (associé au CNRS),
Université Paris Sud, F-91405 Orsay, France
*Permanent address: Institute of Physics of the University, Zagreb, Yugoslavia

Abstract. Superconductivity of radical ion salts is now well established in two families of organic conductors: the quasi-one-dimensional family to which the prototype material $(TMTSF)_2PF_6$ belongs with T_c in the one Kelvin range and the two-dimensional series with T_c up to 10 K which is illustrated by salts such as $(BEDT-TTF)_2X$ i.e $(ET)_2X$ with $X= I_3$, AuI_2, $Cu(NSC)_2$, etc. We first recall the instabilities which are inherent to the $(TMTTF-TMTSF)_2X$ series giving as examples the behavior under pressure of the newly synthesized mixed sulfur-selenium compound $(TMDTDSF)_2PF_6$ and the discovery of a hidden spin density wave phase in $(TMTSF)_2ReO_4$. Secondly, we present recent experimental results about the quantization of the Hall constant in the field-induced semimetallic states of $(TMTSF)_2PF_6$ under pressure and the observation of giant magnetoresistance oscillations in the high T_c phase of β-$(BEDT-TTF)_2I_3$ illustrating the two dimensional character of the latter superconductor.

I-Instabilities in the $(TMTSF)_2X$ series

Superconductivity has always been a strong motivation for the research in organic conductors since the original suggestion by Little [1] that organic matter could provide critical temperatures near ambient conditions. This driving force has proved to be justified since superconductivity was indeed discovered in the 1K range in 1980 [2] and significant improvements have been achieved in other materials discovered more recently [3][4]. However, the nature of the superconducting pairing in organic materials remains (as it is for high T_c superconductors) an open question: phonon or non-phonon mediated, singlet versus triplet pairing, etc.). It becomes clear now that a better understanding of the context in which the superconducting instability sets in is a major step towards understanding organic superconductivity.

Superconductivity is only one among the various instabilities which are observed at low temperature in the isostructural series to which $(TMTSF)_2PF_6$ belongs. As shown in Figure 1 the electronic properties of the sulfur or selenium series are illustrated by a generalized phase diagram. The all-sulfur compound $(TMTTF)_2PF_6$ exhibits a marked charge localization below $T_\rho \approx 200K$ which is attributed to strong on-site Coulomb repulsions [5] (1-D quantum antiferromagnet) and the onset of a 3-D spin singlet state (spin-Peierls) accompanied by a lattice distortion below 19 K[6]. However, the Br salt of the same sulfur molecule exhibits a weaker charge localization below 100 K or so and the onset of a spin modulated state below 15-19K without any visible lattice distortion[7]. Quite a different behavior is observed instead with the all-selenium compound (TMTSF)$_2$ClO$_4$ as no charge localization is visible above the superconducting transition at 1.2 K under ambient pressure.

The generalized phase diagram reveals two competitions between ground states: SP versus SDW on the one hand and SDW versus superconductivity on the other.

The charge localization and the SP-SDW competition are governed by the amplitude of the Umklapp scattering term g3 which decreases under pressure and also when moving from PF$_6$ to Br compounds. The SDW-SC competition is very likely governed by the interplay between the SDW phase transition temperature of the Q-1-D conductor and the small energy ($\approx t^2/t_{//}$) which characterizes the deviation of the Q-1-D Fermi surface from perfect nesting[8]. A recently synthesized S-Se hybrid molecule TMDTDSF has enabled a more thorough study of the SP-SDW competition[9]. As shown on the phase diagram of (TMDTDSF)$_2$PF$_6$, Figure 1, a spin-Peierls ordering is detected at 19 K by its signature in the behavior of the ESR spin susceptibility. However, the nature of the ground state which is stabilized at 7 K is clearly magnetic according to a divergence of the nuclear spin relaxation rate and the observation of antiferromagnetic resonance modes instead of a Zeeman resonance below 7K. The conducting ground state becomes stable at low temperature above 20 kbar but no superconductivity has been detected so far down to 0.4 K. At present we can think of two possible interpretations for the absence of superconductivity: the pair-breaking effect of the disorder introduced by the non-symmetric organic molecule or the weakness of the interchain coupling.

The intrinsic existence of the SDW-SC competition in the TMTSF series is also revealed by a recent investigation of the phase diagram of the organic superconductor (TMTSF)$_2$ReO$_4$. This latter system undergoes a superconducting transition at $\approx$10 bar which is necessary to stabilize the ordering of the non-centrosymmetric ReO$_4$ anions with the wave vector (0, 1/2, 1/2). Below the critical pressure the stable (1/2, 1/2, 1/2) ordering

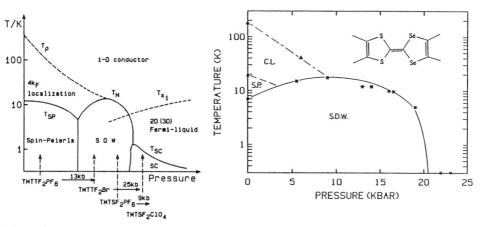

Fig.1: Generalized phase diagram of (TMTCF)$_2$X conductors (left) and phase diagram of (TMDTDSF)$_2$PF$_6$ showing the reentrance of SDW below the spin-Peierls state at 7 K and the stabilization of a conducting ground state above 20 kbar (without superconductivity down to 0.4K).

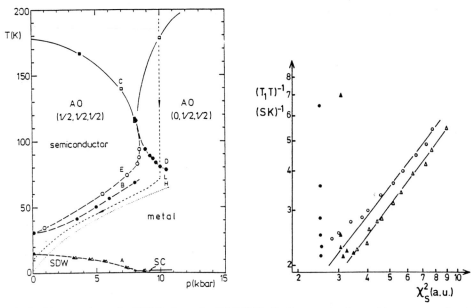

Fig.2: Observation of a SDW phase in (TMTSF)$_2$ReO$_4$ competing with superconductivity when a particular cooling procedure under pressure is followed (dotted line with arrows). Below 8 kbar the anions can be frozen at low temperature in the (0, 1/2, 1/2) configuration, allowing the conducting state to be stable down to the SDW or SC temperature (left). Temperature dependence of $(T_1T)^{-1}$ from ^{77}Se data plotted versus χ^2_s for (TMTSF)$_2$PF$_6$ (circles) and (TMTSF)$_2$ClO$_4$ (triangles). The departure from the straight line marks the onset of 2k$_F$ low temperature spin fluctuations at 100 and 30 K respectively in PF$_6$ and ClO$_4$ salts.

doubles the lattice periodicity along the chain direction and opens a gap at the Fermi level. The occurrence of this anion-ordered semiconducting state prevents us from studying the genuine instability of the Q-1-D organic stacks at low temperature. However, following a proper cooling procedure under pressure, the (0, 1/2, 1/2) configuration of the anions can be preserved at low temperature even under ambient pressure[10]. Thus, another insulating state is revealed, Figure 2, below 15 K. It is assumed that this new insulating state is analogous to the SDW phase which is stabilized in (TMTSF)$_2$PF$_6$ below the critical pressure.

In conclusion, the study of the (TMTSF)$_2$PF$_6$ and (TMTSF)$_2$ReO$_4$ phase diagrams has shown that the competition between antiferromagnetism and superconductivity is a firmly established character of Q-1-D superconductors, irrespective of the anion symmetry. The stabilization of superconductivity at a temperature higher than the 1 to 2 K range seems to be forbidden in Q-1-D conductors such as (TMTSF)$_2$X by the intrinsic competition with magnetic ordering. However, the border line between SDW and SC phases represents the optimum situation for superconductivity at variance with high T$_c$ cuprate materials.

Another facet of this competition is revealed by the contribution of the low frequency antiferromagnetic fluctuations to the nuclear spin-lattice relaxation as shown in Figure 2.

As far as $(TMTSF)_2ClO_4$ is concerned the temperature dependence of $(T_1T)^{-1}$ scales with χ^2_s in the domain where uniform $q = 0$ spin fluctuations are responsible for the temperature dependence of the spin susceptibility (T>30K)[11]. Below 30 K, $2k_F$ spin correlations contribute predominantly to the relaxation rate which then becomes nearly temperature independent down to the dimensionality cross-over at $\approx$ 8 K where a renormalized Fermi liquid description is recovered. As shown in Figure 2, the effect of $2k_F$ spin correlations is even more pronounced in the case of $(TMTSF)_2PF_6$ (below 100 K) at ambient pressure in agreement with the stronger tendency of that compound towards antiferromagnetism.

In conclusion, the proximity between magnetism and superconductivity is a remarkable feature of $(TMTSF)_2X$-like materials. In addition, NMR studies indicate that the long range superconductivity order develops in a superconductivity background of strong antiferromagnetic fluctuations. Some theoretical approaches have used these experimental facts to propose a pairing mechanism in $(TMTSF)_2X$ based on the interchain exchange of antiferromagnetic spin fluctuations[12].

II - Fermiology in organic conductors

The open and quasi planar nature of the Fermi surface of $(TMTSF)_2X$ materials is illustrated by the influence of the magnetic field on the stability of the conducting ground state. Figure 3 reports recent data of Hall effect and magnetoresistance obtained in $(TMTSF)_2PF_6$ under pressure (when the material undergoes a superconducting transition at 1.2 K[13]. The paramagnetic conducting state becomes unstable above 5T (T$\approx$0.5 K) against the formation of an antiferromagnetic semimetal displaying a sequence of subphases. Each of these subphases is characterized by a field independent Hall voltage. Furthermore, the Hall voltage is quantized according to the law $R_H=h/Ne^2$ (N=integer). As shown in

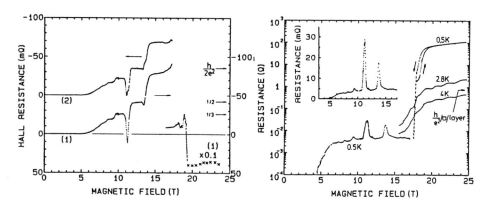

Fig.3: Hall resistance versus magnetic field along the c*axis for two $(TMTSF)_2PF_6$ samples (P $\approx$9 kbar, T = 0.5 K). The quantized values $h/2ne^2$=12.9/n kΩ per molecular layer are marked on the right for sample #1. The n = 0 phase is reached above 18 T (left). Magnetoresistance of the same sample up to 25T (linear scale in the inset).

Figure 3, not only are the ratios of the plateaus given by successive integers but the magnitude of the highest plateau between 14 and 18 T corresponds rather well to $h/2e^2$ = 12.9 kΩ/layer, the value expected for the quantum Hall effect (N=1) in the presence of spin degeneracy. In addition, magnetoresistance data show well-defined peaks at the fields where the Hall voltage jumps to the next plateau. Above 18 T, the dramatic increase of the magnetoresistance by about four orders of magnitude suggests that the N = 0 state is attained. This state is likely to be similar to the SDW ground state observed under ambient pressure.

The ClO$_4$ compound behaves differently at very high fields. It exhibits a reentrance of the non-ordered phase. The ultra one dimensionalization of the electron motion under large magnetic fields may be responsible for the reentrance phenomenon of (TMTSF)$_2$ClO$_4$.

Other spectacular Fermi surface effects related to low-dimensionality have also been observed in the (ET)$_2$X series. In the (ET)$_2$I$_3$ material superconductivity can be stabilized at T_c= 8.1 K in the β_H phase provided the cooling procedure avoids crossing a transition line below which an incommensurate lattice modulation develops[14].

Figure 4 displays giant oscillations of the magnetoresistance which exhibit a perfect (1/H) periodicity with the fundamental field H_0 = 3730 T and a beating phenomenon characterized by the much smaller field H_1 = 36.8 T[15]. The large amplitude and the beating phenomenon can both be understood in terms of a tube-like Fermi surface in a 2-D conductor. For such a situation of an anisotropic surface the amplitude of the oscillations is enhanced by the factor $(m_c/m_a)^{1/2}$, where m_c and m_a are respectively the effective mass perpendicular and parallel to the conducting planes. The beating frequency is related to the warping of the tube and thus to the interplanar coupling, namely $H_1/H_0 \propto t_a/t_c$. For the data in figure 4 a ratio $t_a/t_c \approx 140$ is obtained. The giant magnetoresistance oscillations emphasize clearly the 2D nature of the FS of β-(ET)$_2$X superconductors. The two dimensionality is very likely responsible for both the absence of competition between superconductivity and antiferromagnetism (as the latter instability is favoured by the nesting of quasi planar surfaces) and for the enhancement of T_c above 8K as the negative effect of

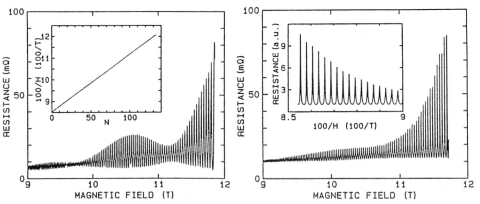

Fig.4: Magnetoresistance of β_H-(ET)$_2$I$_3$ between 9 and 12 T at 0.38 K and 1/H plot of the peak positions versus integer numbers (left inset). The oscillations become strongly anharmonic at highfields (right inset).

fluctuations is less severe for 2-D than for Q-1-D ordering. Furthermore, band structure calculations point out that the Fermi energy of these 2-D conductors might be located close to a van-Hove singularity in the density of states. We believe that such a situation should be considered for the interpretation of the high value of T_c and for its very strong pressure dependence.

III - Conclusion

Organic superconductivity is now been found in two families of organic compounds. In the first series, $(TMTSF)_2X$, the quasi one dimensionality manifests itself first in the competition between various ground states, which is governed by parameters such as the efficiency of Coulomb interactions, the amplitude of the interchain coupling, etc., and secondly in the remarkable field-induced spin density wave states.

The β and κ-phases of $(ET)_2X$ have proved to be two dimensional conductors after the observation of magnetoresistance oscillations which can exhibit remarkably large amplitudes in the case of β-$(ET)_2I_3$.

The nature of the pairing interaction remains an open question in organic superconductors. However, the existence of strongly developed antiferromagnetic fluctuations in $(TMTSF)_2X$ materials makes a non-phonon mediated pairing mechanism quite plausible.

[1] W.A.Little, Phys.Rev. 134 A, 1416 (1964)
[2] D.Jérome, A.Mazaud, M.Ribault and K.Bechgaard, J.Phys.Lett.Paris 41, L-95 (1980)
[3] V.N.Laukhin, E.E.Kostyuchenko, Y.V.Susko, I.F.Schegolev and E.B.Yagubski, JETP.Lett. 41,81(1985)
[4] H.Urayama, H.Yamochi, G.Saito, K.Nozawa, M.Kinoshita, S.Sato, K.Oshima, A.Kawamoto and J.Tanaka, Chem.Lett.55 (1988)
[5] V.J.Emery, R.Bruinsma, S.Barisic, Phys.Rev.Lett.48, 1039 (1982)
[6] J.P.Pouget in Low Dimensional Conductors and Superconductors, D.Jérome and L.G. Caron editors, p.17, NATO ASI Series, Plenum (1987)
[7] F.Creuzet, T.Takahashi, D.Jérome and J.M.Fabre, J.Phys.Lett.Paris 43, L-755 (1982)
[8] K.Yamaji, Mol.Cryst.Liq.Cryst.119, 105 (1985)
[9] P.Auban, D.Jérome, K.Lerstrup, I.Johansen, M.Jorgensen, and K.Bechgaard, J.Physique Paris 50, 2727 (1989)
[10] S.Tomic and D.Jérome, J.Phys.Cond.Matter 1, 4451 (1989)
[11] C.Bourbonnais, P.Wzietek, D.Jérome, F.Creuzet, P.Batail and K.Bechgaard, Phys.Rev.Lett. 62, 1532 (1989)
[12] C.Bourbonnais and L.Caron, Europhys.Lett.6, 177 (1988)
[13] J.R.Cooper, W.Kang, P.Auban, G.Montambaux, D.Jérome and K.Bechgaard, Phys.Rev.Lett.63, 1984 (1989)
[14] F.Creuzet, G.Creuzet, D.Jérome, D.Schweitzer and H.J.Keller, J.Phys.Lett.Paris 46, L-1079 (1985)
[15] W.Kang, G.Motambaux, J.R.Cooper, D.Jérome, P.Batail and C.Lenoir, Phys.Rev.Lett.62, 2559 (1989)

Trends in Structures and Properties of Organic and Inorganic Superconductors

P. Day

Institute Laue-Langevin, 156 X, F-38042 Grenoble Cedex, France

Abstract. In the past few years, unexpected and exciting discoveries of superconductivity have been made in two apparently quite different series of chemically-based materials: the organic molecular charge transfer salts and the inorganic ternary copper oxides. Several intriguing analogies exist between the structures and properties of these apparently disparate series, including the importance of low-dimensionality and the proximity of magnetic and superconducting states tunable by pressure or chemical composition. Similarities and differences between the organic and inorganic superconductors are surveyed.

1. Introduction

Through the long history of superconductivity, only two classes of materials have shown such a rapid advance that their critical temperatures have increased by nearly an order of magnitude in less than 10 years. These are the mixed valency copper oxides and the molecular charge transfer salts. In part, this must be because both families are extensive, and creative chemistry has uncovered a large number of variants. Such richness of chemical variety helps us to delineate the structural and electronic features needed to observe superconductivity (and optimize it) in each structure type. It also provides the foundation for a comparison between the significant features of both classes of material. The outlines of such a comparison are given in this note, in terms of three aspects: the dimensionality of the lattice, the influence of structural phase transformations and instabilities, and the presence either of localized magnetic moments or long-range magnetic order.

2. Dimensionality of Structure

Although the detailed structural chemistry of the molecular and copper oxide superconductors is quite different, there are broader features that they have in common, sufficient to make a comparison worthwhile. The most obvious of these is dimensionality.

Given that the dominant structural feature in conducting molecular crystals is the columns formed by plane to plane stacking of flat, or nearly flat, conjugated molecules, the physics of the earlier examples of the type were dominated by the properties of a near one-dimensional Fermi surface, and in particular by the various instabilities to which such a surface is subject.[1]

The notable feature of the earliest examples of charge transfer salts is the existence of insulating rather than metallic behaviour at low temperature. As is well known, the one-dimensional electron gas is unstable to either a charge density wave (CDW) or a spin density wave (SDW) state, the former via the Peierls distortion.[2] In the $(TMTSF)_2X$ series, one might have thought that there would be a Peierls distortion because there is 1/2 an unpaired electron per molecule and the resulting CDW would be commensurate. However, the structures of the $(TMTSF)_2X$ salts are unusual in that they are triclinic ($P\bar{1}$) with each unit cell containing two TMTSF molecules.[3] The TMTSF stacks are zig-zag and very slightly dimerized, with S...S intermolecular spacings of (3.964, 4.021) versus (3.874, 3.934)Å. However, although the valence band would be 3/4 filled, on the assumption of uniform stacking, the triclinic space group and slight dimerization finally lead to a half-filled band. The final ambient pressure ground state at low temperature is now not a CDW but an SDW where the X^- ion is octahedral (PF_6, AsF_6),[4] but eventually super-

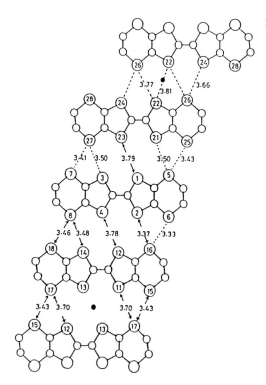

Fig. 1. Interstack S...S interactions in $(BEDT-TTF)_3Cl_2 \cdot 2H_2O$[10]

conducting in the ClO_4^- salt.[5] Thus $(TMTSF)_2ClO_4$ was the first ambient pressure organic molecular superconductor, although the SDW state in $(TMTSF)_2PF_6$ is suppressed by pressure to give a superconducting state (T_c 1.0K) at 8 kbars.[6]

An important consideration in deciding what compounds to synthesise in order to stabilise the superconducting state versus the CDW or SDW in molecular lattices is the effective dimensionality. A first priority is to make the lattice less purely one-dimensional so as to make it stable against Peierls distortion. In that context, the compound (HMTSF)(TCNQ) was a significant early example.[7] HMTSF (hexa-methylene-tetraselenofulvalene) was synthesised to explore the consequence of introducing more bulky substitution of the TSF moieties with the expectation that the system would be rendered more purely one-dimensional. Yet it was found that the hexamethylene-substituted complex retains metallic conductivity down to very low temperatures. Far from decreasing the interaction between stacks, the bulky substituents induce a change in the packing such that a terminal N of a TCNQ is brought within van der Waals contact (3.18Å) of an Se from TMTSF. A suitable strategy for maintaining metallic conduction down to temperatures where superconductivity has a chance to take over is thus to promote such close non-bonded contacts. In this context, one should compare the transfer integrals along and between stacks in (TTF)(TCNQ) and superconducting $(TMTSF)_2ClO_4$. In the former, the transfer integrals between TTF molecules along and between stacks ($t_{FF} : t_{FF'}$) are in the ratio 500:1, and between TCNQ's ($t_{QQ} : t_{QQ'}$) are 25:1.[8] On the other hand, the corresponding ratio ($t_{SS} : t_{SS'}$) in the superconductor is only 10:1.[9] In both, the intrastack transfer integrals are of the same order of magnitude (0.1 eV).

Evidence that chalcogen-chalcogen interactions are effective in stabilising the superconducting state came with the discovery of the superconductivity in the BEDT-TTF salts, which contain four additional chalcogen atoms per molecule. The network of interstack interactions is very extensive. An example is shown in Figure 1.[10] The ratio of intra- to interstack transfer integrals is then still smaller (~5:1). In the so-called β-series of salts with linear mononegative triatomic anions (I_3^-, IBr_2^-, AuI_2^- etc.), the structure approximates much more closely to two-dimensional, since the anions are segregated quite clearly into layers, separated by 'blocks' of the organic material. The conductivity anisotropy parallel and per-

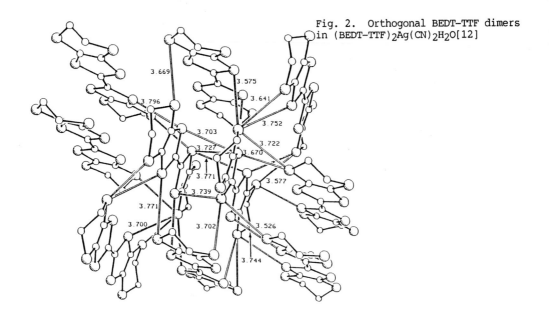

Fig. 2. Orthogonal BEDT-TTF dimers in $(BEDT-TTF)_2Ag(CN)_2H_2O$[12]

pendicular to these blocks is of the order of 500. However, the highest presently known T_c among the $(BEDT-TTF)_2X$ is not a β-salt, but belongs to the κ-series.[11] In the latter, there are no stacks of cations, but only face-to-face pairs with the planes of the nearest neighbour dimers orthogonal to each other. An example from our own group is shown in Figure 2.[12] With a segregation of the anions into layers, similar to that found in the β-series, the layers of organic cations are all but isotropic, as is the conductivity within that plane. Thus, overall, there is a clear evolution in the metallic charge transfer salts from the strongly one-dimensional TTF-TCNQ, with a transition at low temperatures to a CDW state, towards more and more two-dimensionally isotropic structures in the $(TMTSF)_2X$, β-$(BEDT-TTF)_2X$ and finally κ-$(BEDT-TTF)_2Cu(NCS)_2$, having superconducting ground states with higher and higher T_c.

Turning to the ternary and quaternary mixed valency copper oxide class of superconductors, the immediate feature of structural comparison with the charge transfer salts is their two dimensionalities. The only structural feature shared by all five of the copper oxide categories is the presence of square planar CuO_4 units linked through all four of their corners to form an infinite layer with the overall stoichiometry CuO_2. An example is shown in Figure 3. The oxygen coordination number of the Cu atoms in these layers is sometimes strictly fourfold, but it may also be five or six. However, even when one or two oxide ions are coordinated to the Cu perpendicular to the plane of the layers, as in Figure 3, the axial Cu-O bond distance is always much greater than the equatorial. Hence the physical properties are very strongly two-dimensional. Recall that in the 10.5K molecular superconductor κ-$(BEDT-TTF)_2Cu(NCS)_2$, the transport within the layers of BEDT-TTF is also almost isotropic but with an anisotropy of at least 100 parallel and perpendicular to the layers.[11]

In four of the five categories of copper oxide superconductors, there is also a small breaking of the fourfold symmetry of the CuO_2 layer, the structural mechanism being different in each case. Only in the electron superconductor $Nd_{1.85}Ce_{0.15}CuO_4$ is the tetragonal symmetry preserved. The separation of the CuO_2 layers is achieved either by large individual cations (e.g. Ln^{3+}, Ba^{2+}) with 8- or 9-fold oxygen coordination or by complete layers with 1:1 cation-anion stoichiometry, and average (though not local) NaCl structure. In the 1:2:3 series there is the further complication of the one-dimensional CuO_4 chain parallel to the b-axis which, in the early days of the high T_c superconductors, was thought to be the origin of the enhancement in T_c compared with $La_{1.85}Sr_{0.15}CuO_4$. However, the subsequent discovery of the Bi and Tl phases, with even higher T_c's and no Cu chains, indicated that this structural feature was not crucial to high T_c.

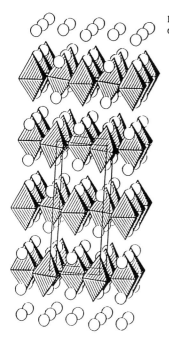

Fig. 3. The layer structure of $La_{2-x}Sr_xCuO_4$[16]

3. Structural Phase Transformations

Both the high T_C copper oxides and the molecular charge transfer salts give evidence of lattice instabilities which may have a bearing on the occurrence and mechanism of the superconductivity.

Particularly in the (BEDT-TTF)$_2$X series, polymorphism is rife, as witness the nomenclature established for the various phases ($\alpha, \alpha', \beta, \beta', \beta'', \kappa$ etc). The extreme sensitivity of the electronic properties of the ground state to the fine details of intermolecular packing can be exemplified by the fact that superconductivity has been found in the β and κ series while the α phases, which differ by the introduction of a zig-zag alternation in the molecular stacks, are Mott-Hubbard insulators that behave as low-dimensional antiferromagnets.[13]

A particularly interesting phenomenon occurs in the 10.5K superconductor κ-(BEDT-TTF)$_2$Cu(NCS)$_2$, whose resistivity first increases on cooling below room temperature, and passes through a maximum around 100K before becoming metallic in its temperature dependence (see Figure 4). Similar resistivity maxima have been seen in related compounds such as (DMET)$_2$AuBr$_2$, which becomes superconducting under pressure (1.5 kbar, 1.6K)[14]. In the Cu(NCS)$_2$ salt, application of a modest hydrostatic pressure (~ 1 kbar) sharpens the maximum, but at higher pressures it progressively diminishes till above 5 kbar it disappears, leaving a metallic temperature dependence from room temperature down to the superconducting T_C (Figure 4)[15]. The hypothesis is that there is a structural phase transformation, largely uncorrelated between the layers at ambient pressure, with the ambient pressure low temperature phase being stabilised over the whole temperature range above 5 kbar. High pressure crystallographic studies are in progress.

When superconductivity was found in the $La_{2-x}Ba_xCuO_4$ system, it was thought at first that the exceptionally high T_c was due to coupling of the conduction electrons with a high density of low energy phonons arising from a soft mode associated with a structural phase transition from orthorhombic to tetragonal. Thus, La_2CuO_4 itself is orthorhombic (Cmca)(O) at room temperature, with a transition to the tetragonal (T) (I4/mmm) K_2NiF_4 structure at 540K. In the orthorhombic phase the CuO$_2$ layers are puckered along [110] so that the Cu-O-Cu angle decreases from 180° to 173° (see Figure 3). On doping with M^{II}, the O → T transition temperature falls until above

11

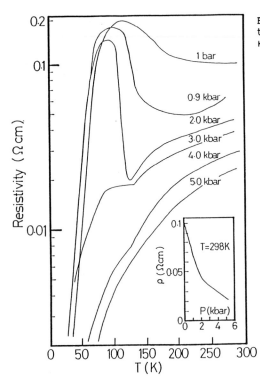

Fig. 4. Pressure dependence of the normal state resistivity of κ-(BEDT-TTF)$_2$Cu(NCS)$_2$[15]

x ~ 0.20, only the T phase occurs at all temperatures. It was originally thought to be significant that this is just the composition beyond which superconductivity is no longer found. However, careful studies of the temperature dependence of the structural parameters by high resolution powder neutron diffraction showed that, for the compositions that optimize the superconductivity (x ~ 0.15), both La:Sr[16] and La:Ba[17] compounds were already orthorhombic at T_c, e.g. in La$_{1.85}$Sr$_{0.15}$CuO$_4$, T_{OT} was 180K while T_c was 36K. An orthorhombic-tetragonal transition also occurs in the 1:2:3 phases, not as a function of temperature, but either oxygen deficiency or disorder.

4. Magnetic States

A striking similarity between the charge transfer salts and oxides is the importance of electron correlations and hence the proximity of superconducting and magnetic states in the phase diagrams as a function of band filling or structural modifications. This arises from the fact that both types of compound are narrow band systems.

The superconducting β-(BEDT-TTF)$_2$X are Pauli paramagnets in their normal states but the α-salts with the same X have the strongly enhanced and temperature-dependent susceptibility characteristic of a low-dimensional antiferromagnet[14]. The magnitude of the susceptibility is exactly what one would expect for one unpaired electron per (BEDT-TTF)$_2$ dimer[18]. An even closer concurrence of magnetism and superconductivity was found in the early work of Jérome and his colleagues on (TMTSF)$_2$PF$_6$[6]. At ambient pressures this compound is metallic down to 10K, where it undergoes a transition to a spin density wave (antiferromagnetic) state, and the resistivity increases again. If pressure is applied, though, the SDW state is suppressed, and the compound becomes superconducting.

Given that the molecular superconductors are salts, there is the further option of including magnetic ions into the lattice by incorporating them into the anion X$^-$. Thus TMTSF and BEDT-TTF salts of complex anions containing 3d elements have been prepared, such as $(BEDT-TTF)_2FeCl_4$[19] and $(BEDT-TTF)_3CuCl_4 \cdot H_2O$[20]. Unfortunately no superconducting examples have been found yet, so it has not been possible to test for the coexistence of superconductivity and magnetism in this class of compound.

The transformation from SDW to superconductivity occurs in the copper oxide series, not as a function of pressure like the TMTSF case, but as a function of band filling. Thus, La_2CuO_4, with a half-filled band based on Cu $3d(x^2-y^2)$ and O $2p$, is antiferromagnetic with $T_N \sim 270K$. On doping the La^{3+} site with Sr^{2+}, and so bringing about a corresponding decrease in the Cu-O band filling, T_N decreases very rapidly, so that for $x > \sim 0.06$ both T_N and the staggered moment tend to zero. It is significant that for $0 < x < 0.05$ there is no detectable variation in the periodicity of the SDW, which remains [1/2 1/2 1].[21] Hence one can talk of an antiferromagnetic state. Still, the limiting value of the moment for $x = 0$ ($\sim 0.5\mu_B$) is strongly reduced by comparison with what one would expect for a localized $Cu^{2+}(3d^9)$ configuration. This is because of the very extensive Cu-O covalent mixing: indeed, the electrons at the Fermi surface have more O(2p) than Cu(3d) character. Rather analogous results are also found in the 1:2:3 series, where the Cu-O band filling is systematically varied by changing the occupancy of the O between the Cu forming the chains parallel to the b-axis. The moments reside in the CuO_2 planes, and $YBa_2Cu_3O_6$ is a two-dimensional antiferromagnet[22]. With increasing x in $YBa_2Cu_3O_{6+x}$, the moment and T_N fall, and are replaced by superconductivity.

An alternative method for introducing localized magnetic moments into the Cu oxide superconductors is similar to the method used in the organics, though in this case by substituting the cations rather than anions. Thus in the 1:2:3 series, all Ln^{3+} except Pr and Ce yield superconductors but the Ln moments do not order until far below T_c. Then they order antiferro-magnetically with no great effect on the superconductivity. The Ln(4f) magnetic electrons have only negligible interaction with those of the Cu-O planar band. In contrast, substitution of Cu by elements of the 3d block depresses T_c markedly.

5. Conclusion

This brief overview of the molecular and oxide superconductors has emphasised the similarities between them, that may point the way to a further evolution of structures. Dimensionality provides a common theme, and the occurrence of antiferromagnetic phases differing only marginally in electronic and crystal structure from the superconducting ones is a further point of similarity. It is highly probable that neither group has yet reached the upper limit of T_c: a challenge for the future.

References

1. For reviews, see e.g. "The Physics and Chemistry of Low Dimensional Solids", L. Alcacer, ed., D. Reidel Publ. Co. (1980).
2. R. E. Peierls: "Quantum Theory of Solids", Clarendon Press, (1964).
3. N. Thorup, G. Rindorf, H. Soling, I. Johannsen, K. Mortensen and K. Bechgaard: J. Physique, Colloques C3, **44**, 1017 (1983).
4. D. Jérome and H. J. Schultz: Adv. Phys. **31**, 299 (1982).
5. K. Bechgaard, K. Carneiro, F. B. Rasmussen, M. Olsen, G. Rinsdorf, C. S. Jacobsen, H. J. Pedersen and J. C. Scott: J. Amer. Chem. Soc. **103**, 2440 (1981).
6. D. Jerome, A. Mazaud, M. Ribault and K. Bechgaard: J. Physique Lett. **41**, L95 (1980).
7. A. N. Bloch, D. O. Cowan, K. Bechgaard, R. E. Pyke and R. H. Banks: Phys. Rev. Lett. **34**, 1561 (1975).
8. V. K. S. Shante, A. N. Bloch, D. O. Cowan, W. M. Lee, S. Choi and M. H. Cohen: Bull. Amer. Phys. Soc. **21**, 287 (1976).
9. P. M. Grant: Phys. Rev. **B26**, 6888 (1982).
10. D. Chasseau, D. Watkin, M. J. Rosseinsky, M. Kurmoo and P. Day in 'Crystalline Low-dimensional Organic and Inorganic Solids', ed. P. Delhaes and M. Drillon, New York, Plenum Press, 1987, p.317.

11. H. Urayama, H. Yamochi, G. Saito, K. Nozawa, T. Sugamo, M. Kinoshita, S. Sato, K. Oshima, A. Kawamoto and J. Tanaka: Chem. Lett. **55** (1988).
12. M. Kurmoo, K. Pritchard, D. Talham, P. Day, A. Stringer and J. A. K. Howard: Synth. Met. **27**, A, 177 (1988); idem, Acta Cryst. B (in press).
13. D. Talham, M. Kurmoo, P. Day, D. Parker and R. H. Friend: J. Phys. Cond. Matter, **1**, 5671 (1989).
14. K. Kikuchi, K. Murata, Y. Houda, T. Namiki, K. Saito, T. Ishiguro, K. Kobayashi and I. Ikemoto: J. Phys. Soc. Jap. **56**, 3436 (1987).
15. I. D. Parker, R. H. Friend, M. Kurmoo and P. Day: J. Phys.: Cond. Matt. (in press).
16. P. Day, M. J. Rosseinsky, K. Prassides, W. I. F. David, O. Moze and A. Soper: J. Phys. C: Sol. St. Phys. **20**, L729 (1987).
17. D. McK. Paul, G. Balakrishnan, N. R. Bernhoeft, W. I. F. David and W. T. A. Harrison: Phys. Rev. Lett. **58**, 1976 (1987).
18. S. D. Obertelli, R. H. Friend, D. R. Talham, M. Kurmoo and P. Day: Synth. Met. **27**, A375 (1988).
19. T. Mallah, C. Hollis, S. Bott, P. Day and M. Kurmoo: Synth. Mat. **27**, 381 (1988); idem, J. Chem. Soc., Dalton Trans. (in press).
20. D. Chasseau, G. Bravic, J. Gaultier, M. Kurmoo, T. Mallah and P. Day: Acta Cryst. C (submitted).
21. M. J. Rosseinsky, K. Prassides and P. Day (to be published).
22. P. Burlet, C. Vettier, M. J. G. M. Jurgens, J. Y. Henny, J. Rossat-Mignod, H. Noel, M. Potel, P. Gougeon and J. C. Levat: Physica C **153**, 1115 (1988).

On Some Organic Conductors in the Light of Oxide Superconductors

H. Fukuyama

Institute for Solid State Physics, University of Tokyo,
7-22-1 Roppongi, Minato-ku, Tokyo 106, Japan

Abstract. The realization of high T_c Cu oxides may have consequences on the direction of material research in organic conductors. Here the present status of the basic understanding of the electronic state in Cu-oxides is explained, and then some organic conductors with possible Cu^{++} ions, which are considered as essential in oxides, are briefly analyzed.

1. Introduction

Recent discovery of high T_c Cu-oxides has clearly disclosed that the electronic states and then the transport properties in solids can have a large versatility. As regards this versatility organics are very unique, but the highest T_c there is admittedly very low in comparison to oxides, though the rate of realizing new systems is very encouraging. To achieve higher T_c in organics, it will be very important to understand Cu-oxides.

In this short article the present understanding of the electronic states in Cu-oxides is introduced. Though the mechanism of high T_c in oxides has not yet been clarified, it is generally believed that the high T_c superconductivity is closely related to the existence of strong Coulomb interaction since it exists in proximity to the antiferromagnetism associated with Cu^{++} ions [1].

From this understanding two cases of organic conductors with possible Cu^{++} ions, Cu(PC)·I and $(R_1, R_2\text{-DCNQI})_2 Cu$, are briefly analyzed.

2. Electronic States in Cu-Oxides

It is widely accepted that the essential structure in crystals of high T_c Cu-oxide is the Cu-O plane. Moreover the electronic transport is considered to be associated with atomic orbitals of $Cu:d_{x^2-y^2}$ and $O:p_\sigma$ in these planes as schematically shown in *Fig.1*, where the various microscopic parameters characterizing this system are also shown. The photoemission spectroscopy and the band theoretical investigation have yielded an estimate of these parameters, which indicate $U_d \simeq 8eV$, $U_p \simeq 5eV$, $V_0 \simeq 1eV$, $t_0 \sim 1.2eV$, $t_1 \sim 0.6eV$, $t_2 \sim 0.2eV$, $\Delta = \epsilon_p - \epsilon_d \sim 3eV$, respectively [2,3].

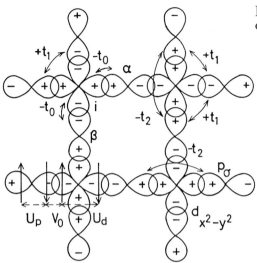

Fig.1 Various parameters of the model consisting of Cu:$d_{x^2-y^2}$ and O:p_σ.

From these, it is understood that these Cu-oxides are charge-transfer type insulators [4] in the undoped system, e.g. in La_2CuO_4, and that carriers introduced by doping, e.g. in $La_{2-x}Sr_xCuO_4$, occupy mainly O:p_σ orbitals [5,6]. The antiferromagnetism observed in the undoped case is due to Cu^{++} ions with spin S=1/2, and extra holes on oxygen sites introduced by doping couple to these Cu-spins, reducing the antiferromagnetic coupling between Cu-spins in the light-doping case and becoming mobile above some small critical rate of doping. The effective Hamiltonian describing the low lying excitation compatible with the estimate of the parameters given in *Fig. 1* for high energy region indicates that holes moving through oxygen sites coupled to the periodic array of spins on Cu-sites [7]. It turned out that the exchange coupling between Cu-spins and hole spins on oxygen, J_K, is very strong $J_K \simeq 1eV$, while the kinetic energy of these holes due to not only direct transfer integrals between oxygens, t_1 (nearest neighbors) and t_2 (next nearest neighbors across Cu-atoms), but also indirect transfer integral via Cu-sites, T_0, is not necessarily large compared to J_K, i.e. $t_1 \simeq 0.6eV$, $t_2 \simeq 0.3eV$, $T_0 \simeq 0.2eV$. These are schematically shown in *Fig.2*.

Although the exact nature of extra holes in such circumstances is not known yet, one plausible argument to this is that an extra hole on oxygen tightly couples to a Cu spin by occupying the symmetric orbital around that particular Cu-spin forming a singlet state [8]. This singlet state has a charge due to the fact that it is a composite state of a Cu-spin with an extra hole the latter of which has both spin and charge. This is the basic reasoning arguing for the validity of the *t-J* model, which Anderson [9] proposed to have the essence of high T_c Cu-oxides even at the beginning of high T_c studies. There are by now many theoretical investigations on this *t-J* model [1].

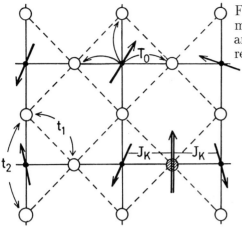

Fig.2 Schematic representation of the model derived from *Fig.1*, where dots and open circles indicate Cu and O sites, respectively.

3. Organic Conductors with Cu^{++}

There exist some experimental reports on organic conductors with Cu^{++}, though superconductivity is not yet observed.

a) $Cu(PC) \cdot I_x$

The mother material of this salt (*i.e.* $x = 0$) is the transition metal (M)–phtalocyanine (PC) complex salts M(PC). While the metallic state is stabilized for M=Ni, Cu(PC) is insulating and forms one-dimensional Heisenberg spin systems with S=1/2 due to Cu^{++} [10]. The introduction of iodine (I), possibly in the form of I_3^- in the crystal, acts like doping, introducing holes on PC–π bands, and make the system conductive [11]. As in the case of $La_{2-x}Sr_xCuO_4$ these dopants, I_3^-, are located in space apart from the conduction paths. This system then has a very unique feature of having one-dimensional periodic array of Cu-spins together with carriers on the π band of PC molecules surrounding each Cu-spin; a dense Kondo system with the one-dimensional carriers. In the case of $x=1$, which seems to be the only one among $x \neq 0$ so far realized experimentally, the conductivity is increased very much compared to that in Cu(PC) but is not truly metallic [12] on one hand and the spin susceptibility is modified but only slightly [13] on the other hand. Hence the realization of cleaner samples would be necessary before the interesting problem of the possible competition of the Kondo effect and one-dimensional singularities [14] can be understood.

b) $(R_1, R_2 - DCNQI)_2 Cu$

The crystal structure of this salt in the case of $R_1 = R_2 = DMeO$ viewed from the stacking axis is shown in *Fig.3* [15]. The first example of this family with

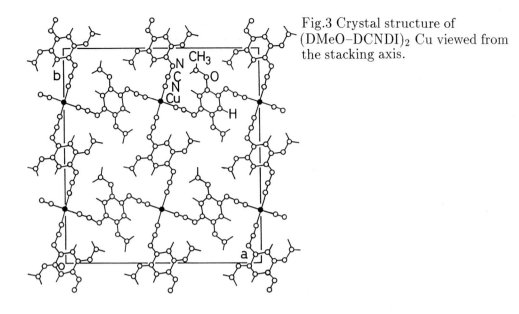

Fig.3 Crystal structure of (DMeO–DCNDI)$_2$ Cu viewed from the stacking axis.

$R_1 = R_2 =$ DMe which shows metallic behavior down to low temperature was discovered by Aumuller et al. [16,17] and the extreme sensitivity of this salt to pressure has been disclosed by Mori et al. [18] and Tomic et al. [19]. In the case of (DMeO–DCNQI)$_2$Cu, which has been explored by Kobayashi and collaborators [20,21], it has been reported that the temperature dependence of resistivity also shows anomalous behavior depending on applied pressure. Below the characteristic temperature, T*, of the sharp change of the resistivity it is claimed [15] that the Cu atoms are in the state of the mixed valence with Cu$^+$:Cu^{++} = 2 : 1 in view of the experimental observation of the superlattice with the period of 3c along the stacking axis in different but similar salts. If this is the case Cu spins are periodically arrayed along the conducting axis together with the charge density variation with the same periodicity. This periodicity turns out to be same as $2k_F$ periodicity of carriers on DCNQI molecules. Hence there is possible entanglement of various effects: dense Kondo, mixed valence and 1d instability. From these, it is noteworthy that for a finite range of the pressure the metallic state is stabilized down to low temperatures [20,21].

4. Discussions

We have seen that the crystal structures of organic conductors with Cu^{++} ions are very different from high T$_c$ Cu-oxides. In the case of (R$_1$,R$_2$-DCNQI)$_2$Cu the one-dimensional stacking is the main coupling even though the crystal structure projected onto the plane perpendicular to the stacking axis has a common feature as oxides as seen in *Fig.3*. This one-dimensional stacking will have appreciable mixing integral with Cu:3d states. The electronic states of this system have been discussed by Kobayashi et al. [15] based on extended Huckel approximation and by use of the tight-binding approximation. Such an investigation will give us information of overall features of electronic structures in the high temperature region with Pauli-like susceptibility. This will, however, break down below T^*, if Cu^{++} state is formed in this temperature range. (Formation of Cu^{++} state is not definite. The claimed spatial array of $-Cu^+-Cu^+-Cu^{++}-Cu^+-$ can be considered as the limiting case of spin density wave (SDW) and it will be possible that the amplitude of such SDWs is small.) In this context it is not obvious why average valence of Cu is fixed to $+4/3 = 1.33$ below T^* and whether this valence is the same above T^* or there exists a sharp change of the amount of charge transfer at T^*. Of particular interest is the nature of the metallic state below T^*. While the complete re-entrance of the resistivity has been observed in the case of $(DMe-DCNQI)_2 Cu$, the case of $(DMeO-DCNQI)_2 Cu$ is somewhat different and it appears that the metallic state stabilized at low temperatures in some particular range of pressure has very different values of resistivity extrapolated from above T^*. Physical properties, especially magnetic, have not been fully explored in this regime. If Cu^{++} states are really formed together with the presence of strong mixing between Cu:d state and holes in $P\pi-$states of DCNQI molecules, this system shares a common feature with high T_c Cu oxides, though the dimensionality of the hole band will be essentially different. This particular aspect is also interesting in comparison with heavy electrons realized in dense Kondo systems. In the latter case there are some systems which show superconductivity. Hence clarification of existence or absence of heavy electron states and search for superconductivity at lower temperatures in this case will be very important in understanding this possibly novel state. At the same time realization of similar salts but with the planar structure, e.g. a structure of *Fig.3* but layered with smaller transfer integrals between layers, will be very interesting in more direct comparison with oxides.

Acknowledgements

The author thanks M. Kinoshita, S. Kagoshima, Y, Murata, G. Saito, T. Takahashi, especially H. Kobayashi, for informative discussions.

References

[1] For example, Proc. of the Adriatico Research Conf., *Towards the Theoretical Understanding of High T_c Superconductors* (1988, Trieste), Int. J. of Mod. Phys. B **1**, ed. by S. Lundqvist, E. Tossatti, M.P. Tosi and Y. Yu (World Scientific, Singapore 1988).

[2] H. Eskes and G.A. Sawatzky, Phys. Rev. Lett. **61**, 1415 (1988); L.H. Tjeng, H. Eskes and G.A. Sawatzky, *Strong Correlation and Superconductivity* (Springer Verlag, 1989) ed. H. Fukuyama, S. Maekawa and A. Malozemoff, p.33.

[3] K.P. Park, K. Terakura, T. Oguchi, A. Yanase and M. Ikeda, J. Phys. Soc. Jpn. **57**, 3445 (1988).

[4] J. Zaanen G.A. Sawatzky and J.W. Allen, Phys. Rev. Lett. **55**, 418 (1985).

[5] A. Fujimori, Phys. Rev. B **39**, 793 (1989).

[6] V.J. Emery, Phys. Rev. Lett. **58**, 2794 (1987).

[7] H. Fukuyama, H. Matsukawa and Y. Hasegawa, J. Phys. Soc. Jpn. **58**, 364 (1989); H. Matsukawa and H, Fukuyama, ibid **58**, 2845 (1989), **58**, 3687 (1989).

[8] F.C. Zhang and T.M. Rice, Phys. Rev. B **37**, 3759 (1988).

[9] P.W. Anderson, Science **235**, 1196 (1987).

[10] S. Lee, M. Yudkowsky, W.P. Halperin, M.Y. Ogawa and B.M. Hoffman, Phys. Rev. B **35**, 5003 (1987); Phys. Rev. Lett. **57**, 1177 (1986).

[11] M.Y. Ogawa, B.M. Hoffman, S. Lee, M. Yudkowsky and W.P. Halperin, Phys. Rev. Lett. **57**, 1177 (1986).

[12] G. Quirion, M. Poirier, K.K. Liou, M.Y. Ogawa and B.M. Hoffman, Phys. Rev. B **37**, 4272 (1988).

[13] M.Y. Ogawa, S. M. Oalmer, K. Liou, G. Quirion, J.A. Thompson, M. Poirier and B.M. Hoffman, Phys. Rev. B **39**, 10682 (1989).

[14] B. Guay, L.G. Caron and C. Bourbonais, Synthetic Metals **29**, F557 (1989).

[15] A. Kobayashi, R. Kato, H. Kobayashi, T. Mori and H. Inokuchi, Solid State Commun. **64**, 45 (1987); Synthetic Metals **27**, B275 (1988).

[16] A. Aumuller, P. Erk, G. Klebe, S. Hunig, J.U. Schutz and H.P. Werner, Angew. Chem. Int. Ed. Engl. **25**, 740 (1986).

[17] H.P. Werner, J.U. von Schutz, H.C. Wolf, R. Kremer, M. Gehrke, A. Aumuller, P. Erk and S. Hunig, Solid State Commun. **65**, 809 (1988).

[18] T. Mori, K. Imaeda, R. Kato, A. Kobayashi, H. Kobayashi, and H. Inokuchi, J. Phys. Soc. Jpn. **56**, 3429 (1987).

[19] S. Tomic, D. Jerome, A.Aumuller, P. Erk, S. Hunig and J.U. von Schutz, J. Phys. C **21**, L203 (1988).

[20] R. Kato, H. Kobayashi, A. Kobayashi, T. Mori and H. Inokuchi, Chem. Lett., 1579 (1987); Synthetic Metals **27**, B263 (1988).

[21] H. Kobayashi, R. Kato, A. Kobayashi, T. Mori and H. Inokuchi, Solid State Commun. **65**, 1351 (1988).

Part II

**Metal Coordinated
Organic Conductors and
Related Materials**

Conductive and Superconductive Coordination Compounds

P. Cassoux [1], *L. Valade* [1], *J.-P. Legros* [1], *C. Tejel* [1], *J.-P. Ulmet* [2], *and L. Brossard* [3]

[1]Laboratoire de Chimie de Coordination du CNRS associé à l'Université Paul Sabatier et à l'INP de Toulouse, 205 Route de Narbonnne, F-31077 Toulouse Cedex, France

[2]Service des Champs Magnétiques Intenses, Laboratoire de Physique des Solides associé au CNRS, INSA, Avenue de Rangueil, F-31077 Toulouse Cedex, France

[3]Laboratoire de Physique des Solides associé au CNRS, Bât. 510, Université de Paris Sud, F-91405 Orsay Cedex, France

Abstract. A comprehensive review of the latest results obtained by several physical techniques on the molecular superconductors derived from $M(dmit)_2$ complexes is given. The phase diagrams of $TTF[Ni(dmit)_2]_2$ and α'-$TTF[Pd(dmit)_2]_2$ are discussed. Preliminary results on the chemistry, transport properties and magnetoresistance of $(NHMe_3)_{0.5}[M(dmit)_2]$ are reported.

1. Introduction

Most of the 40 or so known molecular superconductors are derived from purely organic molecules and originate from the same "founding father" TTF : TMTSF, BEDT-TTF, DMET or MDT-TTF are chemically closely related to TTF (Fig. 1). Moreover, a d^8 $metal^{2+}$ ion being isolobal to the C_2^{4+} ethylene fragment [1], the $[M(dmit)_2]$ complexes may also be considered as belonging to the same family of molecules. However, in the $[M(dmit)_2]$ systems the role of the central transition metal atom is crucial in determining the transport properties.

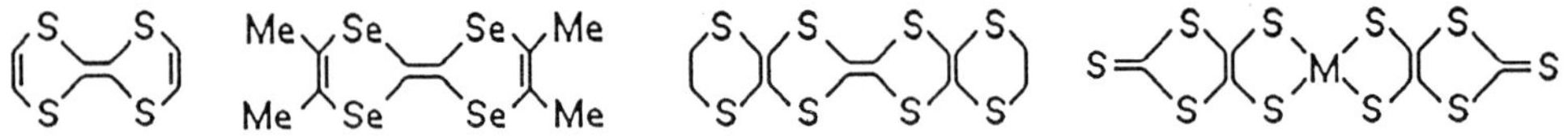

Fig. 1. The TTF, TMTSF, BEDT-TTF, and $[M(dmit)_2]$ molecules.

In our work, our main guideline was based on the observation of the number of sulfur or selenium atoms on the periphery of the molecules (4 for TTF or TMTSF, 8 for BEDT-TTF), and our basic strategy was the incorporation of additional sulfur or selenium atoms on the periphery of the molecule, in order to enhance interchain interactions. The $[M(dmit)_2]$ complexes have 10 sulfur atoms on their periphery. The dmit-ligand is obtained by reduction of CS_2 [2]. The divalent and monovalent salts, $C_n[M(dmit)_2]$ ($C = NR_4^+$, AsR_4^+,...; n = 2, 1; M = Ni, Pd, Pt) can be used for the preparation of either donor-acceptor complexes $D[M(dmit)_2]_y$ with different TTF-like donor molecules D [3], or of complexes with fractional oxidation state, $C_x[M(dmit)_2]$ [4]. Four superconductors have been characterized within this family : $TTF[Ni(dmit)_2]_2$ [5], α-$TTF[Pd(dmit)_2]_2$ [6], and α'-$TTF[Pd(dmit)_2]_2$ [7], have been prepared in our group; $(NMe_4)[Ni(dmit)_2]_2$ has been studied by Kobayashi et al. [8].

Springer Proceedings in Physics, Vol. 51
The Physics and Chemistry of Organic Superconductors
Editors: G. Saito · S. Kagoshima

We report here the latest results that we have obtained in the physics of two donor-acceptor compounds, TTF[Ni(dmit)$_2$]$_2$ and TTF[Pd(dmit)$_2$]$_2$, and in the chemistry (and physics) of a new series of "fractional-oxidation-state" metal-dmit complexes, (NH$_y$Me$_{4-y}$)$_x$ [M(dmit)$_2$].

2. The α'-TTF[Pd(dmit)$_2$]$_2$ phase

At room temperature, three phases of TTF[Pd(dmit)$_2$]$_2$ have been obtained : the α, α', and δ phases [9]. Apparently, the α and α' phases have the same crystal structure, similar to that of the nickel analogue compound (vide infra), but exhibit different conductivity behaviors. The structure of the δ phase consists of [Pd(dmit)$_2$]$_2$ stacked dimers, but the TTF molecules are not stacked, and yet this phase exibits a metal-like behavior down to 120 K. The α and α' phases are both superconducting under pressure [7].

As shown in the phase diagram of α'-TTF[Pd(dmit)$_2$]$_2$ (Fig. 2a), this compound undergoes a superconductive transition at high pressure.

At low pressure, the resistivity of α'-TTF[Pd(dmit)$_2$]$_2$ exhibits a broad minimum at a Tmin temperature, which decreases with increasing pressure (··· curve in Fig. 2a). No discontinuity is observed in the slope dLogR/d(1/T), which exhibits a maximum

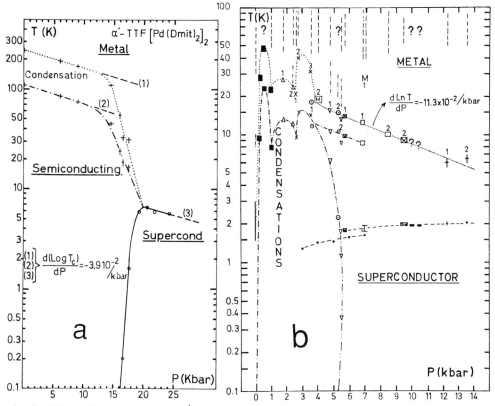

Fig. 2. Phase diagrams of (a) α'-TTF[Pd(dmit)$_2$]$_2$, and (b) TTF[Ni(dmit)$_2$]$_2$

at a Ts temperature (___ curve in Fig. 2a). Condensation of the carriers occurs between Tmin and Ts. There is a tricritical point at 6.5 K and 20 kbar. The same slope is observed for the variation of Log Tc, Log Tmin and Log Ts versus pressure : this means that the same exponential pressure-dependence governs the superconducting condensation as well as the carrier localization. Moreover, the slope Log Tc versus pressure is lower than that observed for $(TMTSF)_2PF_6$ in which a low temperature SDW ground state in the close vicinity of the superconductivity was observed [10]. This discrepancy and other differences in the transport properties and ESR spectra [7] suggest that in α'-$TTF[Pd(dmit)_2]_2$ the localization is not related to a similar low temperature SDW ground state. In fact, X-ray diffuse scattering experiments [11] show diffuse lines at $q_1 = 0.5$ b* and $q_2 = \pm 0.31$ b* in the high temperature range; these 1-dimensional fluctuations become correlated along the c axis below 200 K, and analysis of the temperature dependence of the intensities of the corresponding superlattice spots leads one to assign the CDW condensation at T1 = 150 K and T2 = 105 K. These results are developed in more detail in [12], but the basic result is that α'-$TTF[Pd(dmit)_2]_2$ is the first example of competition between CDW and superconductivity.

3. The $TTF[Ni(dmit)_2]_2$ compound.

We observed in 1984 that $TTF[Ni(dmit)_2]_2$ remains metallic at low temperatures and that its structure consists of segregated stacks of TTF and $[Ni(dmit)_2]$ molecules [13]. Moreover, lateral S...S interstack contacts shorter than the sum of the Van der Waals radii suggested at first that this compound is three-dimensional., at least from a structural point of view. These unusual features prompted us to carry out detailed studies of the transport properties of this "low-temperature metal".

At ambient pressure a minimum of the resistivity of $TTF[Ni(dmit)_2]_2$ was observed at ca. 3 K and this compound was not superconducting above 47 mK. A low anisotropy in the magnetoresistance was observed [14], but proton NMR experiments [15] indicated that the TTF stacks are 1-dimensional. This 1-D character was confirmed by band structure calculations [16]. Finally, thermopower experiments [17] showed an inversion in the sign of the carriers at 70 K: electrons from the $Ni(dmit)_2$ chains dominate the conductivity above 70 K whereas the TTF holes, not only from the TTF but also from the $Ni(dmit)_2$ chains, dominate at low temperatures. Under pressure the compound undergoes a superconductive transition [5]. Finally, Radio-Frequency penetration-depth measurements [18] showed that $TTF[Ni(dmit)_2]_2$ is the first superconductor in which the transition temperature Tc increases with increasing pressure (Fig. 2b).

This unusual increase of Tc (• dots in Fig. 2b) has been confirmed by resistivity measurements [19] with a slight shift due to either differences in pressure measurements and/or differences in Tc measurements either by low field ESR or by resistivity measurements. In the same high pressure range, the resistivity versus temperature curves show a minimum at a temperature Tmin. The slope of the variation of Log Tmin versus pressure is twice that observed for the α'-$TTF[Pd(dmit)_2]_2$ phase. This minimum reflects a carrier condensation (high-pressure condensation of type II). However, in $TTF[Ni(dmit)_2]_2$ this instability and superconductivity co-exist, or do not strongly compete as in α'-$TTF[Pd(dmit)_2]_2$.

Superconductivity stumbles upon a low-pressure condensation mode of type I : thus, there is a boundary between two modes of metal-to-insulator transition, type I and type II condensation. In the 5-6 kbar pressure range, superconductivity and type-I low-pressure condensation are opposed. However, superconductivity is re-entrant at lower temperatures. Extrapolation for very high pressure should lead to a tri-critical point. Consequently, the diagram of TTF[Ni(dmit)$_2$]$_2$ exhibits three tri-critical points.

Ambient pressure X-ray diffuse scattering experiments [11-12] have detected 1D-fluctuations as diffuse lines on the X-ray patterns at room temperature. This type of scattering condenses below 40 K into satellite reflections, which provides evidence for a CDW instability in TTF[Ni(dmit)$_2$]$_2$, also. At low pressure, there is a dramatic increase of Tmin versus pressure, followed by a dispersion of the experimental Tmin values which is not related to crystal quality or induced by thermal cycling. These features may be due to the complexity of the multiple Fermi surface sheets. Recent calculations of band electronic structure [20] shed some light on the phase diagram of TTF[Ni(dmit)$_2$]$_2$ (and of α'-TTF[Pd(dmit)$_2$]$_2$ as well). These calculations include the HOMO-based bands as well as the LUMO-based bands, and therefore differ from the previous calculations [16] in which only the LUMO-based bands were considered. In fact, the LUMO-based bands overlap appreciably with the HOMO-based bands in the band electronic structure of TTF[Ni(dmit)$_2$]$_2$. If the charge transfer can be reasonably estimated between 1 and 0.5, then there is a 2-dimensional character at the Γ center of the Brillouin zone. The effect of pressure will be an increase of the energy dispersion of the bands and the Fermi level will go below the lowest HOMO, as it is observed in the band structure of α'-TTF[Pd(dmit)$_2$]$_2$. Thus, in TTF[Ni(dmit)$_2$]$_2$ the 1-dimensionality is re-established by applying the pressure.

In conclusion, additional work is clearly needed in order to elucidate the intricate phase diagram of TTF[Ni(dmit)$_2$]$_2$, especially the peculiar behavior observed in the low pressure range, i.e. the dramatic increase of Tmin with increasing pressure and its anomalous dispersion.

4. The (NH$_y$Me$_{4-y}$)$_x$[M(dmit)$_2$] series

From the previous results, it is clear that small changes in the chemistry may lead to new compounds exhibiting quite different properties. (NMe$_4$)$_{0.5}$[Ni(dmit)$_2$] is a superconductor under pressure [8]. We were tempted to check whether substituting hydrogen atoms for methyl groups in [NH$_y$Me$_{4-y}$)$_x$[M(dmit)$_2$] would result in dramatic changes in the properties.

The standard procedure [2] had to be modified in order to obtain the (NH$_y$Me$_{4-y}$)$_2$[M(dmit)$_2$] precursor complexes [21]. Iodine oxidation of these complexes gives the corresponding monovalent (NH$_y$Me$_{4-y}$)$_1$[M(dmit)$_2$] complex. Finally, electrochemical oxidation of the monovalent salt gives the corresponding mixed valence salt. A number of these mixed valence salts have been obtained for different metals (M = Ni, Pd, Pt), and for different values of y (y = 1, 2, 3).

The stoichiometry of (NHMe$_3$)$_{0.5}$[Ni(dmit)$_2$] has been determined by X-ray structure resolution [21]. One interesting feature of this structure is the "dimerization" within the stacks of Ni(dmit)$_2$ with alternating interplanar distances of 3.51 and 3.60 Å. The room temperature conductivity of this compound is ca. 300

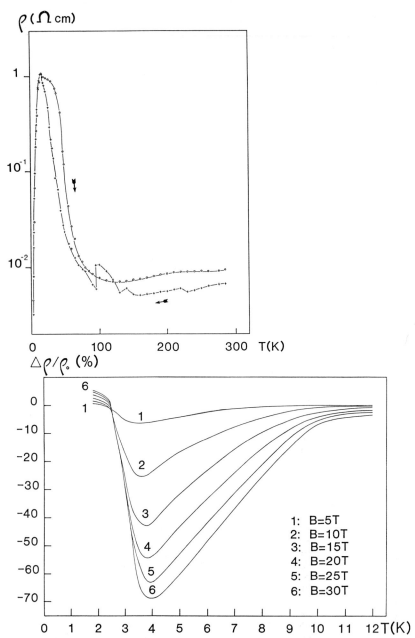

Fig. 3. Temperature dependent resistivity and magnetoresistance of $(NMe_4)_{0.5}[Ni(dmit)_2]$

S/cm [22]. At ambient pressure, a metal-like behavior is observed down to 120 K (Fig. 3).

At this temperature the compound undergoes a metal-to-insulator transition accompanied by several, sample dependent, resistance jumps. This behavior is similar to that observed for $(NMe_4)_{0.5}[Ni(dmit)_2]$ [8]. However, in the present case, a striking feature is observed at lower temperature: all studied samples, but one, exhibit a maximum in the resistance at ca. 15 K. Below this temperature a sharp decrease of the resistance is observed down to 2 K. A negative magnetoresistance appears below 20 K (Fig. 3) and decreases with decreasing temperature down to 3.8 K. Surprisingly, at lower temperatures the trend is reversed and the magnetoresistance increases again and even becomes slightly positive below 2.5 K [22]. In conclusion, these preliminary results open a vast new area of extensive work on these quite original systems. The dmit-systems remain a "gold mine" and we will be digging in it for a while.

References

1. S. Alvarez, R. Vicente, R. J. Hoffmann: J. Am. Chem. Soc. **107**, 6253-6277 (1985).
2. G. Steimecke, H. J. Sieler, R. Kirmse, E. Hoyer:. Phosphorus and Sulfur **7**, 49-55 (1979).
3. M. Bousseau, L. Valade, J.-P. Legros, P. Cassoux, M. Garbauskas, L. V. Interrante: J. Am. Chem. Soc. **108**, 1908-1916 (1986).
4. L. Valade, J.-P. Legros, M. Bousseau, P. Cassoux, M. Garbauskas, L. V. Interrante: J. Chem. Soc. Dalton Trans. 783-794 (1985)
5. L. Brossard, M. Ribault, M. Bousseau, L. Valade, P. Cassoux: C. R. Acad. Sc. Paris, Série II **302**, 205-210 (1986).
6. L. Brossard, H. Hurdequint, M. Ribault, L. Valade, J.-P. Legros, P. Cassoux: Synth. Met. **27**, B157-B162 (1988).
7. L. Brossard, M. Ribault, L. Valade, P. Cassoux: J. Phys. France **50**, 1521-1534 (1989).
8. A. Kobayashi, H. Kim, Y. Sasaki, R. Kato, H. Kobayashi, S. Moriyama, Y. Nishio, K. Kajita, W. Sasaki: Chem. Lett. 1819 (1987).
9. J.-P. Legros, L. Valade: Solid State Commun. **68**, 559-604 (1988).
10. D. Jérome, H. J. Schulz: Adv. Phys., **31**, 299 (1982).
11. S. Ravy, J.-P. Pouget, L. Valade, J.-P. Legros: Europhys. Lett. **9**, 391-396 (1989).
12. S. Ravy, J.-P. Pouget, R. Moret: this volume.
13. M. Bousseau, L. Valade, M.-F. Bruniquel, P. Cassoux, M. Garbauskas, L. V. Interrante, K. Kasper: Nouv. J. Chim. **8**, 653-658 (1984).
14. J.-P. Ulmet, P. Auban, A. Khmou, L. Valade, P. Cassoux: Phys. Lett. **113A**, 217-219 (1985).
15. C. Bourbonnais, P. Wzietck, D. Jérome, F. Creuzet, L. Valade, P. Cassoux: Europhys. Lett. **6**, 177-182 (1988).
16. A. Kobayashi, H. Kim, R. Kato, H. Kobayashi: Solid State Commun. **62**, 57-60 (1987).
17. W. Kang: PhD. Thesis, Université de Paris-Sud, Orsay, France (1988).
18. J. E. Schirber, D. L. Overmyer, J. M. Williams, H. H. Wang, L. Valade, P. Cassoux: Phys. Lett. **120**, 87-88 (1987).
19. L. Brossard, M. Ribault, L. Valade, P. Cassoux: C. R. Acad. Sc. Paris, in press.
20. E. Canadell, I. E. Rachidi, S. Ravy, J.-P. Pouget, L. Brossard, J.-P. Legros: J. Phys. France, in press.
21. C. Tejel, B. Pomarede, J.-P. Legros, L. Valade, P. Cassoux: Chem. Mat., in press.
22. J.-P. Ulmet, M. Mazzaschi, C. Tejel, P. Cassoux, L. Brossard: Solid State Commun., in press.

Synthesis and Properties of the New Molecular Metals Na[Ni(dmit)$_2$]$_2$ and NH$_4$[Ni(dmit)$_2$]$_2$

R.A. Clark[1], *A.E. Underhill*[1], *R. Friend*[2], *M. Allen*[2], *I. Marsden*[2], *A. Kobayashi*[3], *and H. Kobayashi*[4]

[1] Department of Chemistry and Institute of Molecular and Biomolecular Electronics, University of Wales, Bangor, UK
[2] Cavendish Laboratory, University of Cambridge, Madingley Road, Cambridge, UK
[3] Department of Chemistry, Faculty of Science, University of Tokyo, Bunkyo-ku, Tokyo 113, Japan
[4] Department of Chemistry, Faculty of Science, Toho University, Miyama 2-2-1, Funabashi, Chiba 274, Japan

Abstract. Electrocrystallisation of solutions of (TBA)$_2$[M(dmit)$_2$] with excess NaClO$_4$ or NH$_4$ClO$_4$ produced Na[Ni(dmit)$_2$]$_2$ and NH$_4$[Ni(dmit)$_2$]$_2$. Single crystals of the sodium salt exhibited metallic properties down to 25 mK whilst a compressed pellet of the microcrystalline ammonium salt behaved as a semiconductor. Studies of the magnetic susceptibility and conductivity under pressure of the sodium salt are reported.

Introduction

Recently a number of superconductors have been reported based on the [M(dmit)$_2$] anion [1,2,3]. Superconductivity in all cases was observed at high pressure and low temperatures. Although, on the basis of the crystal structure, the TTF[Ni(dmit)$_2$]$_2$ salt was initially considered to be essentially 3-dimensional with S...S distances shorter than the Van der Waals radii, more recently, tight binding band calculations have shown the system to be essentially a 1-dimensional metal [4].

The discovery that the salt TMA[Ni(dmit)$_2$]$_2$, in which the TMA$^+$ acts as a closed shell spectator cation is a superconductor, has shown that superconductivity can arise purely through the presence of [M(dmit)$_2$] anions. Salts of the [M(dmit)$_2$] anion containing various small cations such as group I metals may give rise to materials which show increased inter-stack interaction and consequently higher room temperature conductivities and the maintenance of the metallic state to much lower temperatures. It may not be unreasonable to expect such materials to show precursor or superconducting properties under pressure.

As part of this strategy we have studied the synthesis and properties of Na[Ni(dmit)$_2$]$_2$ and NH$_4$[Ni(dmit)$_2$]$_2$.

Experimental

(a) Na[Ni(dmit)$_2$]$_2$. Electrochemical oxidation of a solution of (TBA)$_2$[Ni(dmit)$_2$] (0.2 g) containing a 30-fold molar excess of NaClO$_4$ (~1.00 g) in acetonitrile (~85 cm^3) with a constant current of 2 µA and a current density of ~6 µA cm^2, gave thin black needle like crystals after 5-10 days. (Found: C, 15.75; H, 0.0; N, 0.0; Na, 3.0. Calc. for C$_{12}$S$_{20}$Ni$_2$Na; C, 15.56; H, 0.0; N, 0.0; Na, 2.5%).

(b) NH$_4$[Ni(dmit)$_2$]$_2$. (NH$_4$)$_2$[Ni(dmit)$_2$] was prepared as described by Steimecke [5]. Electrochemical oxidation of a solution of (NH$_4$)$_2$[Ni(dmit)$_2$] (0.2 g) containing a 15 x molar excess of NH$_4$ClO$_4$ (~0.5 g) in acetonitrile with a constant current of 1 µA and a current density of ~3 µA cm^2 produced small black plate-shaped crystals after 5-10 days. (Found: C, 14.88; H, 0.32; N, 1.40. Calc. for C$_{12}$H$_4$N$_1$S$_{20}$Ni$_2$; C, 15.65; H, 0.43; N, 1.50%).

Results and Discussion

(a) Na[Ni(dmit)$_2$]$_2$. Although black needle-shaped crystals were obtained of this salt, the crystal quality was not sufficient for a full X-ray structural analysis. However, Weissenberg photographs suggest that the compound possesses an equidistant stack structure with an intra-stack spacing of the anions of about 3.6 Å.

The room temperature conductivities of three crystals were determined and were found to lie between 1 and 100 S cm^{-1} with an average of 50 S cm^{-1}. The temperature dependence of the conductivity was examined and a typical example is shown in Figure 1.

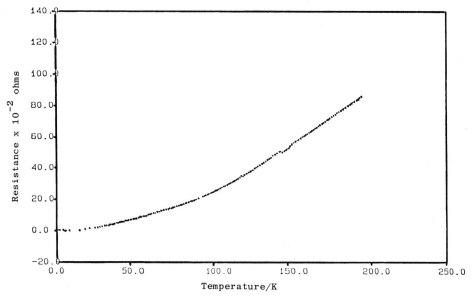

Figure 1. The temperature dependence of the conductivity of a crystal of Na[Ni(dmit)$_2$]$_2$

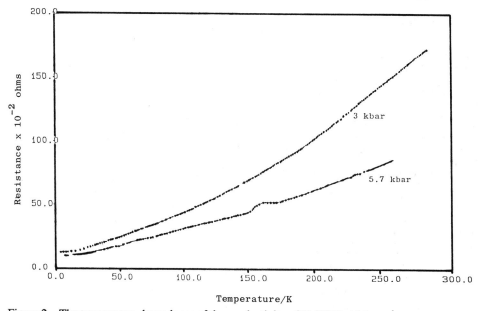

Figure 2. The temperature dependence of the conductivity of Na[Ni(dmit)$_2$]$_2$ under pressure

All the crystals studied showed simple metallic behaviour down to at least 2 K. One crystal was examined further and showed metal-like conductivity down to 25 mK at which point the resistance had dropped by a factor of 10 compared with room temperature. The conductivity was also studied under pressures of up to 6 kbar and whilst resistivity shows the expected decrease with increasing pressure, we saw no evidence for superconductivity at temperatures down to 2 K (see Figure 2).

The variation of thermopower with temperature was also studied. The thermopower is negative with a value of -15 μV/K at 300 K. It falls on cooling towards zero, with no suggestion of a sign

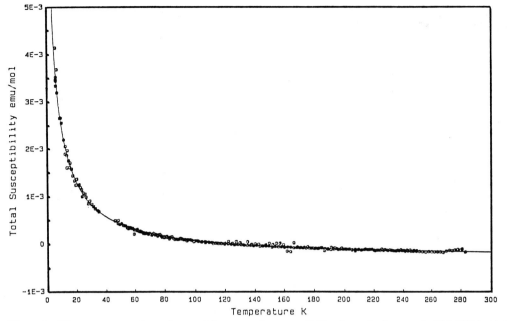

Figure 3. The temperature dependence of the magnetic susceptibility for a single crystal of Na[Ni(dmit)$_2$]$_2$

change above 30 K. This behaviour is characteristic of a simple metal. The sign of the thermopower indicating electrons as the charge carriers.

The magnetic susceptibility of the salt as a function of temperature is illustrated in Figure 3. The rise at low temperatures is well fitted by a Curie-Weiss law, and the solid line shows the fit obtained as

$$\chi = \chi_o + \frac{C}{T-\Theta}$$

with $\chi_o = -2.96 \pm 0.06 \times 10^{-4}$ emu/mole, $C = 3.76 \pm 0.05 \times 10^{-2}$ emu K/mole and $\Theta = -3.6 \pm 0.2$ K.

Since the salt shows metallic properties, we would expect to see a temperature independent Pauli contribution from the conduction electrons. The large Curie-Weiss term is, we consider, due to magnetic impurities, which, if spin 1/2, are present in a concentration of 10% per formula unit. The value of the temperature independent term is consistent with contributions from both core diamagnetic and Pauli paramagnetic terms.

(b) (NH$_4$)[Ni(dmit)$_2$]$_2$. This material was obtained as a black microcrystalline product on the electrodes. Since the crystals were not of sufficient size for electrical measurements, a compressed disc of the powdered material was examined. The room temperature conductivity of three such samples were found to lie between 1×10^{-3} and 1×10^{-1} S cm^{-1} with an average of 1×10^{-2} S cm^{-1}. The temperature dependence of the conductivity was measured down to 60 K for all three samples with consistent results. The material behaves as a semiconductor with an activation energy for conduction of 83 meV.

Discussion

The vast majority of room temperature molecular metals based on metal complex anions exhibit a metal to semiconductor transition at temperatures above 20 K. The behaviour of Na[Ni(dmit)$_2$]$_2$ is most unusual in retaining its metallic properties down to such a low temperature as 25 mK. Unfortunately a full crystal structure for the compound cannot be determined as yet and so the reason for this unusual behaviour cannot be ascertained. However, the preliminary evidence for an equidistant stack structure may be significant. As pointed out earlier, superconductivity in metal complexes has only been observed at low temperatures under high pressure. Up to the present time, the conduction studies have been restricted to pressures up to 6 kbar and no evidence of superconductivity has been obtained. Furthermore, these studies give no indication that superconductivity could be expected at higher pressures. Nevertheless, the synthesis of a material of this type which retains its metallic properties down to such low temperatures is an important development in the field of molecular metals and superconductors.

References

[1] Cassoux, P., Valade, L., Legros, J.P., Interrante, L., Rocs, C. (1986) *Physica B & C*, **143**, 313;
Brossard, L., Hurdequint, H., Ribault, M., Valade, L., Legros, J.P., Cassoux, P. (1988) *Synthetic Metals*, **27**, B157.

[2] Brossard, L., Ribault, M., Valade, L., Cassoux, P. (1986) *Physica B & C (Amsterdam)*, **143** (1-3) 378;
Schirber, J.E., Overmyer, D.L., Williams, J.M., Wang, H.H., Valade, L., Cassoux, P. (1987) *Phys. Lett. A*, **120** (2) 87.

[3] Kim, H., Kobayashi, A., Sasaki, Y., Kato, R., Kobayashi, H. (1987) *Chem. Lett.*, 1799;
Kobayashi, A., Kim, H., Sasaki, Y., Kato, R., Kobayashi, H., Moriyama,S., Nishio, Y., Kajita, K., Sasaki, W. (1987) *Chem. Lett.*, 1819.

[4] Kobayashi, A., Kim, H., Sasaki, Y., Kato, R., Kobayashi, H. (1987) *Solid State Commun.*, **62**, 57.

[5] Steimecke, G., Kirmse, R., Hoyer, E. (1975) *Z. Chem.*, **15** (1) 28.
Steimecke, G., Sieler, H.J., Kirmse, R., Hoyer, E. (1979) *Phosphorus and Sulphur,* **7** (1) 49.

[6] P.I. Clemenson, PhD Thesis, University of Wales 1987.

Effect of Counter Ions on Alkali Metal Salts of $M_x[Ni(dmit)_2]$ ($M = Na, NH_4 \ldots$)

A. Izuoka[1], A. Miyazaki[2], N. Sato[2], T. Sugawara[1], and T. Enoki[3]

[1]Department of Pure and Applied Sciences, University of Tokyo,
Komaba 3-8-1, Meguro, Tokyo 153, Japan
[2]Department of Chemistry, University of Tokyo, Komaba 3-8-1,
Meguro, Tokyo 153, Japan
[3]Department of Chemistry, Tokyo Institute of Technology, Ookayama 2-12-1,
Meguro, Tokyo 152, Japan

Abstract. Semiconducting $Ni(dmit)_2$ salt (1) with onium ions as a counter cation was obtained by the anodic oxidation of $n\text{-}Bu_4N[Ni(dmit)_2]$ in a $MeNO_2$-acetone solution using NH_4BPh_4 as an electrolyte. The $Ni(dmit)_2$ anions are "trimerized" and stacked along the c-axis in the crystal. On the other hand, metallic salt (2a) was obtained from an acetonitrile solution. Pauli paramagnetism of the salt was observed together with Curie paramagnetism by magnetic susceptibility measurement.

1. Introduction

Systematic investigation of counter ions in $Ni(dmit)_2$ salts is expected to elucidate the relation between conductivity and crystal structure in this unique molecular metal system. In a series of quarternary ammonium cations as counter ions, the conductivity is found to be increasing when the size of ammonium ions becomes smaller [1,2]. Especially, $Me_4N[Ni(dmit)_2]_2$ salt has a superconducting phase under high pressures[2]. Thus it may be of great interest to examine smaller counter ions such as NH_4^+.

It is also pointed out that tetrahedral counter ions can cause order-disorder transitions (e.g. $(TMTSF)_2ClO_4$[3]). In this respect alkali metal cations, which have spherical symmetry, may be informative as reference compounds.

Therefore we have synthesized $Ni(dmit)_2$ salts involving onium ions or alkali metal cations. We are also intrigued by the structural dependence on conductivities of these salts.

2. Experimental

2.1 Electrocrystallization

$(NH_4)_x[Ni(dmit)_2]$. The galvanostatic ($I=1\mu A$) anodic oxidation of $n\text{-}Bu_4N[Ni(dmit)_2]$ (2.5mM) was carried out using NH_4BPh_4 (30mM) as a supporting electrolyte. Black platelets(1) grew on the platinum anode when a mixture of nitromethane and acetone (1:1) was used as a solvent. Black needles(2) were obtained from an acetonitrile solvent.
$Na_x[Ni(dmit)_2]$. Using $NaClO_4$ as an electrolyte and acetonitrile as a solution, black needles (3) were obtained.
Other alkali metal salts. Potassium and rubidium salts were grown as black needles in acetonitrile using $KBPh_4$, $RbBPh_4$ as electrolytes.

2.2 Crystal Structure Determination

The crystal structure of **1** was determined by X-ray diffraction method. Crystal data: Monoclinic, Space group C2/c, a=37.940(3)Å, b=11.186(1), c=11.840(1), β=104.93°, V=4854.6(6)Å^3, Z=12. Reflection data (2θ<45°) were recorded on an automated four-circle diffractometer (Rigaku AFC-5) using MoKα, radiation. The structure was solved by the heavy atom method, refined by the block-diagonal least-squares method using the program UNICS-III[4]. Anisotropic thermal parameters were adopted for Ni and S atoms, and isotropic ones for other atoms. The final reliable factor was R=0.093 for 1599 independent reflections (F>3σ(F)).

No single crystals were obtained for the other salts.

2.3 Physical Measurements

DC conductivity. A four-probe technique was used. Contacts of gold wires (diameter 0.025mm) were made by gold paint.

Magnetic Susceptibility. Temperature dependence of susceptibility for the salt **2** was measured down to 2K, using a Faraday balance at the main magnetic field of 1 Tesla(Instrument center, Institute for Molecular Science).

3. Results and Discussion

3.1 Composition

No characteristic absorption around 3220cm^{-1} assignable to a stretching of N-H bond was observed in IR Spectra of the salts **1** and **2**. Elemental analyses of **1** and **2** also show that there is very little or no nitrogen in these salts. Presumably hydroxonium cation was incorporated instead of ammonium ions.

3.2 Crystal Structure of 1

In the crystal of the salt **1**, Ni(dmit)$_2$ molecules are stacked along the c-axis, and three molecules make a unit (Fig. 1). The molecule A is positioned at an inversion center and the molecules B and B' are related by the inversion symmetry. The distance between interstack S-S contacts are approximately the same as other Ni(dmit)$_2$ salts[1,2].

Counter cations and/or water molecules are represented by circles in Fig. 1(a). For each position, occupancy probability is 0.5. But the occupying species cannot be specified at the present stage.

3.3 Conductivity

The conductivities of various Ni(dmit)$_2$ salts are summarized in Table 1. The salt **1** showed a semiconductive behavior, which is consistent with the trimeric crystalline structure. On the other hand, salts **2** obtained from an acetonitrile solution showed a sample dependency. Most samples (**2a**) were metallic down to 70-75K. Semiconducting samples (**2b**) were also found within the same batch. The sodium salt **3** and the other alkali metal salts [5] show metallic behavior down to ca. 100K.

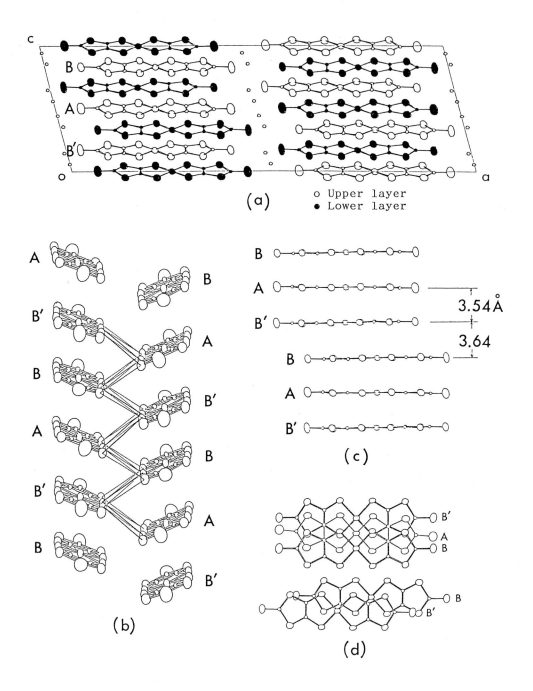

Fig. 1. The crystal structure of **1**. (a) Projection along the b-axis. Two layers are observed. (b) Seen along the a axis tilted by 5°. Interstack S-S contacts shorter than 3.70Å are shown. (c) Ni(dmit)$_2$ molecules from upper layer viewed along the molecular plane. (d) Two modes of overlap of Ni(dmit)$_2$ molecules.

Table 1. Conductivities of $M_x[Ni(dmit)_2]$ salts

	Electrolyte / Solvent	σ_{rt}/Scm^{-1}	E_a/eV
1	NH_4BPh_4 / $MeNO_2$-acetone	0.3	0.11
2a	NH_4BPh_4 / MeCN	50	Metallic (> 70-75K)
2b	NH_4BPh_4 / MeCN	0.15	0.06
3	$NaClO_4$ / MeCN	50-300	Metallic (> 60K)
	$KBPh_4$ / MeCN	200	Metallic (> 100K)
	$RbBPh_4$ / MeCN	300	Metallic (> 110K)

3.4 Magnetic Susceptibility of 2

Magnetic susceptibility of the gross sample 2 was measured. The temperature dependent term of the observed susceptibility follows Curie-Weiss law with a small negative Weiss temperature (θ = -2.3K). The observed paramagnetism may be ascribed to defects and/or impurities, due to the imperfect quality of the sample.

It is to be noticed that the temperature independent term $\chi_{dia}+\chi_{Pauli}$ has a positive value, which means the absolute value of Pauli paramagnetism in 2 is larger than that of diamagnetism. The value of the Pauli paramagnetism was constant down to 2K. These behaviors can be ascribed mainly to the metallic sample 2a. The result suggests that conduction electrons exist in a short range, although the bulk conductivity is not necessarily metallic at lower temperatures.

4. Conclusion

The crystal structure of the semiconducting salt 1 is quite different from the structure of the Me_4N salt[2], and is characterized by the "trimerized" structure.

Although position and nature of counter ions in salts 1 and 2 are not fully determined, the hydroxonium ion is the most probable candidate as in the case of $Li_{0.8}H_3O_{0.33}[Pt(mnt)_2]\cdot 1.67H_2O$ [6]. By controlling the amount of H_3O^+ and H_2O in the crystal, a hydrogen-bonded network may be built up. In such a case, phonon modes in the network will perturb transport property of conduction electrons, which may lead to a novel property of great interest.

References.

[1] L.Valade, J.-P.Legros, M.Bousseau, P.Cassoux, M.Garbauskas and L.V.Interrante, J. Chem. Soc., Dalton Trans., 783(1985).
[2] H.Kim, A.Kobayashi, Y.Sasaki, R.Kato and H.Kobayashi, Chem. Lett., 1799(1987); A.Kobayashi, H.Kim, Y.Sasaki, R.Kato, H.Kobayashi, S.Moriyama, Y.Nishio, K.Kajita and W.Sasaki, Chem. Lett., 1819(1987).
[3] P.Garoche, R.Brusetti and K.Bechgaard, Phys. Rev. Lett., **49**, 1346(1982).
[4] T.Sakurai and K.Kobayashi, Rep. Inst. Phy. Chem. Res., **55**, 69(1979).
[5] Potassium salts were prepared in a different way. A.Clark, A.E.Underhill, I.D.Parker and R.H.Friend, J. Chem. Soc., Chem. Commun., 228(1989).
[6] A.Kobayashi, T.Mori, Y.Sasaki, H.Kobayashi, M.M.Ahmad and A.E.Underhill, Bull. Chem. Soc. Jpn, **57**, 3262(1984).

The Organic π-Electron Metal System Interacting with Mixed-Valence Copper Ions $(R_1,R_2\text{-DCNQI})_2Cu$ (DCNQI = N,N'-dicyanoquinonediimine; R_1, R_2 = CH_3, CH_3O, Cl, Br)

R. Kato[1], H. Kobayashi[1], and A. Kobayashi[2]

[1]Department of Chemistry, Toho University, Funabashi, Chiba 274, Japan
[2]Department of Chemistry, The University of Tokyo, Hongo, Bunkyo-ku, Tokyo 113, Japan

Abstract. A solid-state chemistry of highly conducting anion-radical salts $(R_1,R_2\text{-DCNQI})_2Cu$ is described based on the systematic crystal structure analyses and the tight-binding band calculations. The essence of this system is the mixed valency of Cu which brings about the mixing of the organic $p\pi$ orbitals and metallic d orbitals.

1. Introduction

Since the discovery of $(DMe\text{-DCNQI})_2Cu$ (DMe-DCNQI = 2,5-dimethyl-N,N'-dicyanoquinonediimine) by Aumüller et al. in 1986 [1], increasing attention has been devoted to a new class of molecular materials, $(R_1,R_2\text{-DCNQI})_2Cu$. This paper will give an outline of this exotic molecular system according to our "multi-Fermi surface" model.

2. Preparation and Structure

The copper salts of DCNQI can be prepared by various methods. In our study, the single crystals were obtained electrochemically using $[Cu(CH_3CN)_4]ClO_4$ as the supporting electrolyte and acetonitrile as solvent. The large crystals were prepared by the diffusion method using the chemical reaction with CuI in acetonitrile. The single crystals were grown under nitrogen, and studied in the air. All these crystals prepared from two different routes gave the same physical properties.
All these copper salts are isomorphous [2]. The planar DCNQI molecules are uniformly stacked and form a one-dimensional column. The most important feature of this system is that the Cu cation is coordinated to the nitrogen atoms of the DCNQI molecule in a D_{2d} distorted tetrahedral fashion.

3. Temperature Dependence of Resistivity (at Ambient Pressure)

Temperature dependence of the electrical conductivity at ambient pressure classifies these Cu salts into two groups (Table 1). The group I compounds are metallic down to about 0.5 K. The group II compounds show a sharp metal-insulator (M-I) transition at rather high temperatures. Although all these salts have the same structure, they are quite different in their electrical behavior.

Table 1. Metal-insulator transition temperature (T_{M-I}) of (R_1,R_2-DCNQI)$_2$Cu

	DCNQI	R_1	R_2	T_{M-I}/K
Group I	DMe-DCNQI	CH_3	CH_3	metal
	DMeO-DCNQI	CH_3O	CH_3O	metal
	MeI-DCNQI	CH_3	I	metal [3]
Group II	MeBr-DCNQI	CH_3	Br	152
	DBr-DCNQI	Br	Br	161
	MeCl-DCNQI	CH_3	Cl	210
	BrCl-DCNQI	Br	Cl	213
	DCl-DCNQI	Cl	Cl	230

4. Nature of the M-I Transition of the Group II

This M-I transition is a cooperative structural phase transition induced by the CDW formation on the DCNQI column and the intensive distortion of the coordination tetrahedron around Cu [2]. At the transition temperature (T_{M-I}), the satellite reflections indicating the superstructure wavevector $(0,0,c*/3)$ develop discontinuously. These spots come from the CDW on the DCNQI column and the superstructure wavevector indicates that the formal charge of Cu is +4/3, in other words, the mixed valence state of Cu. The XPS measurements also indicated the mixed valency of Cu, and suggested that the formal charge of Cu in the group I is also +4/3 [4]. This transition is not a simple CDW transition of the one-dimensional system because there is no one-dimensional precursor diffuse scattering, and this transition is accompanied by an abrupt distortion of the coordination tetrahedron of Cu.

The static magnetic susceptibility jumps at T_{M-I}, indicating that the localized magnetic moment appears on the Cu^{2+} cation when the conduction electrons on the DCNQI column disappear. The g-value of ESR signal and the Curie-Weiss behavior of its intensity below the transition temperature indicate an antiferromagnetic ordering of the spins on Cu^{2+} [5].

5. Mechanism of the M-I Transition

We have proposed a mechanism of this transition, based on the extended Hückel MO calculation and tight-binding band calculation [2]. The essential point is that the mixed valence state of Cu brings about the mixing of the organic $p\pi$ orbitals and d orbitals and stabilizes the metallic state. At first we consider the system with no $p\pi$-d mixing. The DCNQI column forms a one-dimensional metallic band. Since the Cu cation is coordinated in D_{2d} distorted tetrahedral fashion, the highest d orbital is d_{xy} (Fig. 1). The energy splitting (ΔE) depends on the degree of the distortion and has an important effect on the electronic structure. When there is no $p\pi$-d mixing, the band structure consists of a one-dimensional $p\pi$ band and d bands with constant energy level. It should be noted that the one-dimensional $p\pi$ band is four-fold degenerate because the unit cell contains four columns and the mixed valency of Cu locates the highest d level near the Fermi level of the organic one-dimensional band. The introduction of the $p\pi$-d mixing removes the degeneracy of the $p\pi$ bands and generates a multiple Fermi surface (Fig .2a). The DCNQI bands give three pairs of waving planes. The closed Fermi surface represented by a dotted line in Fig. 2 arises mainly from

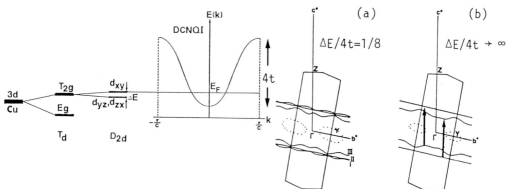

Fig. 1. Schematic interaction diagram in $(R_1,R_2\text{-DCNQI})_2\text{Cu}$.

Fig. 2. Multiple Fermi surface of $(R_1,R_2\text{-DCNQI})_2\text{Cu}$.

the narrow d bands. The electrons in the narrow d bands will tend to localize. The characteristic feature is that there is no single modulation wave vector which can nest all the Fermi surface. In the simple electronic structure examination of the dmit (dmit = 4,5-dimercapto-1,3-dithiole-2-thione) compounds, we have noticed that all the superconducting salts have such a multiple Fermi surface [6]. We propose that such a multiple Fermi surface will be related to the stable metallic state and the distortion of the coordination tetrahedron makes the Fermi surface simpler and brings about the one-dimensional instability. The distortion will stabilize the lower d orbitals and the total energy of Cu cation will be lowered, like the case of the Jahn-Teller distortion. When the lower d level is located far below the Fermi level, the planes I and II are merged into one and there appears a single nesting vector (Fig. 2b). This feature well explains the CDW formation coupled with the intensive distortion of the coordination tetrahedron at T_{M-I}.

6. Nature of the Group I

In the light of our model, the group I is considered to be a system with smaller ΔE value. The multiple Fermi surface in the group I remains down to lower temperature. The transition temperature T_{M-I} and the energy splitting ΔE calculated from the crystal structure data show an excellent correlation [2]. Application of pressure will cause the distortion of the coordination tetrahedron and the group I will turn to the group II at higher pressure. This is a possible mechanism of the pressure-induced M-I transition of the group I compounds. On the way from the group I to the group II, the system shows a very interesting behavior. Tomic et al. first reported a reentrant behavior in the temperature dependence of the resistivity of $(DMe\text{-}DCNQI)_2\text{Cu}$ under pressure [7]. We have found the same behavior in another group I compound, $(DMeO\text{-}DCNQI)_2\text{Cu}$ (Fig. 3) [8]. Such a reentrant behavior is also observed when the group I compound is alloyed with small amounts of the group II compound. For example, in the alloy $[(DMe\text{-}DCNQI)_{1-x}(MeBr\text{-}DCNQI)_x]_2\text{Cu}$ ($x < 0.1$), reentrant behavior is observed (Fig. 4) [9]. Temperature-dependence of the resistivity suggests an interaction of the conduction electrons and the magnetic ions (Cu^{2+}). In the case of $(DMeO\text{-}DCNQI)_2\text{Cu}$ under pressure, between T_{min} and T_{max}, the resistivity (ρ) linearly depends on $\log T$, indicating the

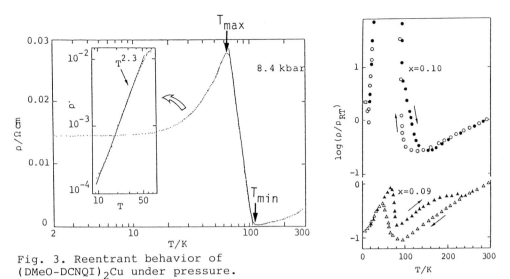

Fig. 3. Reentrant behavior of (DMeO-DCNQI)$_2$Cu under pressure.

Fig. 4. Reentrant behavior of [(DMe-DCNQI)$_{1-x}$(MeBr-DCNQI)$_x$]$_2$Cu.

scattering of the conduction electrons by the magnetic ions (Fig. 3). The lower-temperature resistivity shows a power law ($\rho = \rho_0 + AT^n$; $n \simeq 2.4$) similar to that of the dense-Kondo material. Of course, the situation is not so simple because the electron-lattice coupling, which is important in the M-I transition of the group II, will also play an important role. But it should be noted that the M-I transition of the group II is coupled with the appearance of the magnetic moment. Therefore, it is quite natural that the "partial insulating" state in the reentrant region is accompanied by the "partial appearance" of the magnetic moment. In order to understand the reentrant behavior, there remain the following questions; (1) Is the CDW formed or not? (2) Are the electrons heavy or not? (3) Does the oxidation state of Cu change or not? Anyway, we feel that the existence of the Cu^{2+} cation will play a central role in the transport properties at the reentrant region.

7. Conclusion

The Cu-DCNQI salts contain fertile solid state chemistry and physics. They come from the mixing of the organic pπ orbitals and the metallic d orbitals. In this sense, this system is a new material situated between the organic system and the coordination system. Further study will open the way for new molecular materials.

Acknowledgement. The authors are grateful to Dr. T. Mori (IMS), Mr. A. Miyamoto, Dr. Y. Nishio, Prof. K. Kajita, and Prof. W. Sasaki (Toho University) for valuable discussions.

References

1. A. Aumüller, P. Erk, G. Klebe, S. Hünig, J. U. von Schütz, and H. -P. Werner, Angew. Chem., Int. Ed. Engl., 25, 740 (1986).
2. R. Kato, H. Kobayashi, and A. Kobayashi, J. Am. Chem. Soc., 111, 5224 (1989).
3. P. Erk, S. Hünig, J. U. von Schütz, H. -P. Werner, and H. C. Wolf, Angew. Chem., 100, 286 (1988).
4. A. Kobayashi, R. Kato, H. Kobayashi, T. Mori, and H. Inokuchi, Solid State Commun., 64, 45 (1987).
5. T. Mori, H. Inokuchi, A. Kobayashi, R. Kato, and H. Kobayashi, Phys. Rev. B, B38, 5913 (1988).
6. A. Kobayashi, H. Kim, Y. Sasaki, R. Kato, H. Kobayashi, Solid State Commun., 62, 57 (1987).
7. S. Tomic, D. Jerome, A. Aumüller, P. Erk, S. Hünig, J. U. von Schütz, J. Phys., C 21, L203 (1988).
8. H. Kobayashi, A. Miyamoto, R. Kato, A. Kobayashi, Y. Nishio, K. Kajita, and W. Sasaki, Solid State Commun., in press.
9. A. Kobayashi, R. Kato, and H. Kobayashi, Chem. Lett., in press.

Metallic DCNQI Salts: Influence of Oxygen, Alloying and Dimensionality

M. Bair[1], U. Langohr[1], J.U. von Schütz[1], H.C. Wolf[1], P. Erk[2], H. Meixner[2], and S. Hünig[2]

[1]3. Physikalisches Institut, Universität Stuttgart, Pfaffenwaldring 57,
 D-7000 Stuttgart 80, Fed. Rep. of Germany
[2]Institut für Organische Chemie, Universität Würzburg, Am Hubland,
 D-8700 Würzburg, Fed. Rep. of Germany

Abstract. Measurements of conductivity and of proton spin lattice relaxation in different DCNQI salts with metallic character as a function of temperature and experiments with anaerobically grown crystals show
(1) that apparently the oxidation state of Cu is not responsible for the exceptional metallic behavior of Cu salts,
(2) that the charge carrier motion in the Cu salts is much less one-dimensional than in non-copper salts and
(3) that internal perturbations within the crystals induced by alloying, change the conductive behavior appreciably.

1. Introduction

The salts of DCNQI with metals (Li, Na, K, Rb, Ag, Tl, Cu) as counterions have aroused considerable interest after the first reports [1]. Many of them have a metal-like conductivity at room temperature. Whereas most of the salts get semiconducting at low temperatures, the Cu-salts are unique: many of them are metallic down to very low temperatures. Applied pressure induces a metal-insulator transition at low temperature [2,3]. It is not clear why only the copper salts show this behavior. It was argued that the mixed valence state of copper ($Cu^{+1.3}$) is responsible [4,5]. In order to understand the exceptional behavior of Cu-salts better, we performed three types of experiments:
 - we tried to find out if the oxygen content of the crystals is important
 - we tried to disturb the crystal structure by alloying
 - we tried to compare the dimensionality of copper salts with that of non-copper salts.

2. Crystals

Crystals of $(DMe-DCNQI)_2Cu$ and other Cu salts and alloys were grown in the usual way by electrocrystallisation. In order to exclude oxygen, the growth was performed in a glove box under a N_2 atmosphere. Crystals grown under these conditions are less uniform and apparently of poorer quality.

3. Influence of oxygen

Valence band spectroscopy (UPS) [6] and ESR [5] show that crystals of $(DMe-DCNQI)_2Cu$ are affected by air. The normal oxidation state of copper should be

Cu^{+1}. In reality, air-grown crystals have the oxidation state $Cu^{+1.3}$. Crystals grown anaerobically contain less than 5% Cu^{+2}, in other words an oxidation state close to Cu^{+1}.

From room temperature down to 100K the conductivity of these crystals is very much the same as that of crystals grown under air. Also below 100K they remain metallic, with a less pronounced increase of σ towards very low temperatures, as that of air grown crystals. This might be due to the poor crystal quality which may be responsible for more defects and as a consequence for a poorer low temperature conductivity according to Matthiessen's rule. But there is no indication of a major difference in the conductivity of crystals grown with and without oxygen.

4. Alloys

We investigated different alloys of Cu salts of the composition 2,5 - X,Y - DCNQI with X,Y = CH_3 (≡ Me), Cl, Br, I, OCH_3 [7]. Using alloys with different substituents, we introduced a distribution of different environments of the Cu - DCNQI bonds, which is responsible for an interstack bridging inside the crystals via the CN-group. Fig. 1 gives some examples. The striking result is that there are alloys which remain metallic down to very low temperatures, whereas other alloys become semiconducting somewhere between 100K and 150K. Most interesting is the fact that the phase transition of the salt $(Br,Me-DCNQI)_2Cu$ at 150K can be removed by alloying with $(I,Me-DCNQI)_2Cu$, see fig. 1, at a mixing concentration of > 2:3. With decreasing (I,Me-DCNQI) content, the alloy exhibits the phase transition. The measurements with alloys show clearly that the metallic state

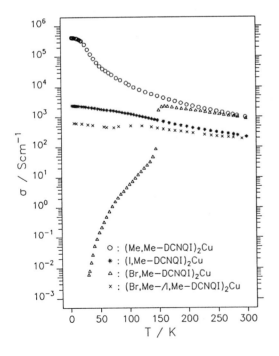

Fig. 1: The temperature dependence of the conductivity of Cu-DCNQI salts with different substituents. For comparison the curve of an alloy is given too.

within the crystals is very sensitive to small variations in environment and structure.

5. Dimensionality

It is well known that the dimensionality of charge- and spin-motion can be measured by frequency-dependent nuclear relaxation, i.e. of the protons. The relaxation rates of the protons are determined by several contributions, being individually dominant in different temperature ranges: For example, in systems undergoing a metal-insulator transition, the proton spin lattice relaxation is composed of (coming from low temperatures) a) the interaction with localized paramagnetic states, b) the reorientation of methyl-groups in the DCNQI and c) the interaction with delocalized spins in the conducting region. In the following we concentrate on the high temperature range (T > 100K), in which T_1 is due to interaction with the conducting electrons. The proton relaxation rates of $(DMe-DCNQI)_2Li$ and $(DMe-DCNQI)_2Cu$ are compared in fig. 2. In $(DCNQI)_2Li$ the three T_{1H}^{-1} - contributions mentioned above are clearly seen in a powder sample (poor crystal quality). Using an assembly of many good single crystals, the low temperature peak (T = 30K) due to the paramagnetic impurities is absent.

The CH_3-group peak at about 70K is most pronounced. The interaction with the electrons (above 100K) is frequency dependent and Korringa-like, which means $T_1 \cdot T \cdot \chi^2$ = const. In comparison, the relaxation rates of the copper salts i.e. $(DMe-DCNQI)_2Cu$ are Korringa-like, an order of magnitude lower and - even up to ν = 300MHz - not frequency dependent in the high conductivity range. All other copper salts, with a phase transition in the conductivity, exhibit at T_c an abrupt increase of the relaxation rates due to the localisation of the charge carriers.

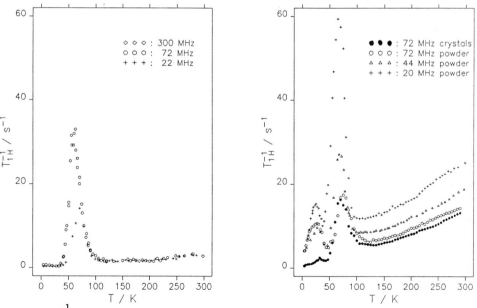

Fig. 2: T_{1H}^{-1} as a function of T and ν_H: $(DMe-DCNQI)_2Cu$ left
$(DMe-DCNQI)_2Li$ right.

So, via the frequency dependence of the NMR, it is possible for the first time to compare the dimensionality of copper salts with that of other salts in the metallic state. A strong frequency dependence with $T^{-1} \sim \omega^{-1/2}$ means a onedimensional charge carrier motion and the lack of a frequency dependence points to a 3-dimensional one [8].

With common conductivity values of $\sigma \simeq 100$ Scm^{-1} we obtain $\tau_{\parallel} < 10^{-14}$ sec (via the Drude approximation), therefore $\omega_e \tau_{\parallel} \ll 1$ even at the highest magnetic fields available. Therefore just $\tau_{\perp}$, which is usually much larger than $\tau_{\parallel}$, can cause a dispersion. In the case of (DCNQI)$_2$Cu, there is no dispersion at all, which means $\omega_e \tau_{\perp} < 1$ even at $B_0 = 7T$ ($\nu = 300$MHz) and consequently $\tau_{\perp} < 8 \cdot 10^{-13}$ sec. This high interstack scattering rate causes the high conductivity of $\sigma \simeq 1000$ Scm^{-1} at room temperature, passing defects on the DCNQI stack via interstack motion.

For (DCNQI)$_2$Li, $T_{1H}^{-1} \sim \omega^{-1/2}$ is given even at the lowest frequency measured. That means, that $\omega_e \tau_{\perp} > 1$ at 0.5T ($\nu = 20$ MHz) and $\tau_{\perp} > 10^{-11}$ sec. If only these values enter the conductivity, the anisotropy $(\tau_{\parallel}/\tau_{\perp})^{-1}$ is in the case of (DCNQI)$_2$Li a factor of 100 larger than for the copper complex, confirming the assumed higher dimensionality of the latter.

6. Summary

In conclusion, the exceptional metallic behavior of the copper salts of DCNQI is apparently not due to the mixed valence state Cu$^{+1.3}$ or due to the influence of oxygen. It is very sensitive to internal perturbations like alloying. It is probably due to the fact that Cu salts are less one-dimensional than those with other metal ions.

7. Acknowledgement

This work was supported by the Deutsche Forschungsgemeinschaft (SFB 329) and by the Stiftung Volkswagenwerk (SH).

References

[1] A. Aumüller, P. Erk, G. Klebe, S. Hünig, J.U. von Schütz and H.-P. Werner; Angew. Chemie 98 (1986) 159; Angew. Chemie Int. Ed. Engl. 25 (1986) 740.
[2] S. Tomic, D. Jérome, A. Aumüller, P. Erk, S. Hünig and J.U. von Schütz; Europhys. Lett. 5 (1988) 553.
[3] S. Tomic, D. Jérome, A. Aumüller, P. Erk, S. Hünig and J.U. von Schütz; J. Phys. C 21 (1988) 203.
[4] A. Kobayashi, R. Kato, H. Kobayashi, T. Mori and H. Inokuchi; Solid State Comm. 64 (1987) 45.
[5] H.-P. Werner, J.U. von Schütz, H.C. Wolf, R. Kremer, M. Gehrke, A. Aumüller, P. Erk and S. Hünig; Solid State Comm. 65 (1988) 809.
[6] D. Schmeisser, K. Graf, W. Göpel, J.U. von Schütz, P. Erk and S. Hünig; Chem. Phys. Lett. 148 (1988) 423.
[7] H.J. Gross, U. Langohr, J.U. von Schütz, H.-P. Werner, H.C. Wolf, S. Tomic, D. Jérome, P. Erk, H. Meixner and S. Hünig; J. Phys. France 50 (1989) 2347.
[8] G. Soda, D. Jérome, M. Weger, J. Alizon, J. Gallice, H. Robert, J.M. Fabre and L. Giral; J. Physique 38 (1977) 931.

Anomalous Temperature Dependence of the Resistivity of (DMeO-DCNQI)$_2$Cu at High Pressure

H. Kobayashi[1], *A. Miyamoto*[1], *R. Kato*[1], *Y. Nishio*[1], *K. Kajita*[1], *W. Sasaki*[1], *and A. Kobayashi*[2]

[1]Departments of Chemistry and Physics, Faculty of Science, Toho University, Funabashi, Chiba 274, Japan

[2]Department of Chemistry, Faculty of Science, The University of Tokyo, Hongo, Bunkyo-ku, Tokyo 113, Japan

Abstract. The electrical resistivity of (DMeO-DCNQI)$_2$Cu was measured up to 12.5 kbar and down to 1.5 K. Similar to (DMe-DCNQI)$_2$Cu, (DMeO-DCNQI)$_2$Cu exhibits a pressure-induced metal instability followed by the reappearance of the metallic state at low temperature. It is suggested that (DMeO-DCNQI)$_2$Cu is a molecular metal system where the itinerant electrons and the localized magnetic moments (Cu^{+2}) coexist at low temperature and high pressure.

1. Introduction

The crystalline molecular metal systems obtained so far can be classified into two groups. One is the group of organic metals and the other is that of the conductors based on the transition metal complexes. In the 1970s, the band origins of these two groups were considered to be different to each other. But Underhill's work on the metallic compounds of M(mnt)$_2$ complexes (M=Ni, Pt; mnt= 1,2-dicyano-1,2-ethylenedithiolato) with π-metal bands have removed the gap between these two groups [1]. From the view-point of the electronic band structure, almost all the molecular metals and superconductors currently studied can be regarded as members of the same family of π-metal systems.
Recent observation of the mixed-valency of Cu in (R_1,R_2-DCNQI)$_2$Cu (R_1,R_2-DCNQI= 2,5-substituted N,N'-dicyanoquinonediimine) has revealed an existence of another type of molecular conducting system with pπ-d mixing bands [2]. In this report, we will present an anomalous resistivity behavior of (DMeO-DCNQI)$_2$Cu at high pressure, suggesting the possibility of an appearance of the localized magnetic moments (Cu^{+2}) in the itinerant electrons.
The prototype of the DCNQI-Cu. systems has been reported by Aumüller et al. in 1986 [3]. There are two types of (R_1,R_2-DCNQI)$_2$Cu complexes : (I) the systems with stable metallic states down to low temperatures (R_1,R_2= CH_3, CH_3 ((DMe-DCNQI)$_2$Cu); CH_3O, CH_3O ((DMeO-DCNQI)$_2$Cu); CH_3, I ((Me,I-DCNQI)$_2$Cu)) and (II) the systems exhibiting a sharp metal-insulator (MI) transitions (R_1 and/or R_2 =Br, Cl) [4,5]. The group-I system is converted into the group-II system by applying pressure. The pressure-induced metal instability of (DMe-DCNQI)$_2$Cu has been found independently by two groups [6,7]. At first sight, this instability is not normal in view of the stacking structure of the planar DCNQI molecules (i.e. one-dimensional (1D) nature of the electronic structure), since pressure usually increases the interchain interaction and suppresses the 1D metal instability. In order to make clear the origin of this anomalous situation, we have proposed the pπ-d mixing band model [2] and pointed out the good correlation between T_{MI} and the magnitude of the distortion of the coordination tetrahedron around Cu [4]. The mixed-valency of Cu (Cu$^+$, Cu^{+2} [8]) suggests that the highest 3d orbital of Cu is located near the Fermi level of 1D 2pπ metal band of

DCNQI to form a $p\pi$-d mixing band [2,4]. This is an important new aspect of the molecular metal systems. At high temperatures, all the DCNQI-Cu systems are in Pauli paramagnetic states. But in the group-II system, the magnetic state changes abruptly at T_{MI}. The localized magnetic moments (Cu^{+2}) appear and the susceptibility increases with lowering temperature [5,9]. That is, the spin of $3d_{xy}$ orbital electron of Cu cannot make its appearance due to the $p\pi$-d interaction in the metallic state but in the insulating state without $p\pi$-d mixing, the valence of Cu cation is fixed in either Cu^+ or Cu^{+2}. Therefore, it may be of special interest to examine what is the valence state of Cu at low temperatures in the "reentrant pressure region".

2. Experimental

Long needle crystals of $(DMeO-DCNQI)_2Cu$ used in this experiments were prepared electrochemically. Resistivities were measured up to 12.5 kbar using the conventional four-probe method and clamp type high pressure cell. The mixture of kerosene and silicone oil was used as a pressure medium.

3. Results and Discussions

As shown in Fig. 1, the system dose not show metal instability up to 8 kbar. The residual resistivity increases suddenly at 8 kbar. Between 8 and 10 kbar, "reentrant resistivity behavior" (metal→ semiconductor→ metal) was observed. In this pressure region, the residual resistivity increased with pressure and was sample dependent. Extremely sharp recovery of the metallic conductivity discovered in $(DMe-DCNQI)_2Cu$ has not been observed [7]. Above 10 kbar, the reentrant resistivity anomaly changes into a sharp MI transition closely similar to that of group II compounds. Besides the large difference in the critical pressure (P_c) above which the metal instability occurs, the general feature of the pressure dependence of the electrical resistivity of $(DMeO-DCNQI)_2Cu$ resembles that of $(DMe-DCNQI)_2Cu$ [7] (P_c= 8 kbar $(DMeO-DCNQI)_2Cu$ and about 50 bar $(DMe-DCNQI)_2Cu$). Since the pressure medium freezes at low temperature, there will be some inhomogeneity of the pressure. However, the effect of such an inhomogeneity of the pressure will be very small, because the separation between electrical leads is extremely narrow ($\sim$0.1 mm).

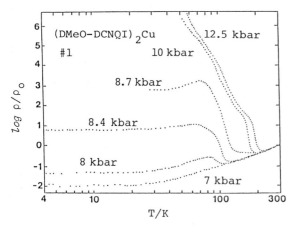

Fig. 1. Temperature-pressure dependence of the resistivity (ρ) of $(DMeO-DCNQI)_2Cu$.

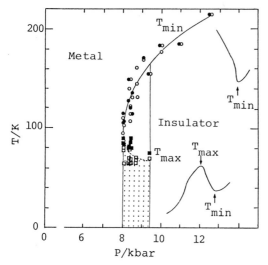

Fig. 2. Phase diagram of $(DMeO-DCNQI)_2Cu$. T_{min} and T_{max} are the temperatures, where ρ takes its minimum and maximum, respectively. The dotted area indicates the region where the metallic state reappears.

The pressure-temperature phase diagram is given in Fig. 2. Since the sharp MI transition of the group-II compound can be regarded as an extreme case of the "reentrant resistivity anomaly", it will be useful to recall what happens at T_{MI} of group-II compounds. Above T_{MI}, all the Cu sites are crystallographically equivalent, suggesting the oxidation state of every Cu atom to be +1.3. Below T_{MI}, the threefold lattice distortion (axbx3c) develops, which makes all the Fermi surfaces vanish. It is natural to consider that Cu^+ and Cu^{+2} produced below T_{MI} form a 3D lattice. At low temperature ($\sim$ 10 K), the magnetic moments on Cu^{+2} sites are antiferromagnetically ordered [5]. Broadly speaking, the valence of the Cu appears to change according to the following equilibrium equation: $Cu^{+1.3} = 2Cu^+ + Cu^{+2}$. The content of Cu^{+2} ($n(Cu^{+2})$) is considered to increase rapidly above 8 kbar with increasing pressure and to reach its saturated value around 10 kbar. The Cu atoms are considered to be in the intermediate state between the "homogeneous valence state" ($Cu^{+1.3}$) and "heterogeneous valence state" (Cu^+, Cu^{+2}) in the "reentrant pressure region",[10]. In an earlier paper, we presented one plausible picture of the valence state of Cu. The existence of Cu^{+2} will be gradually disclosed with increasing pressure. Since the system retains metallic state down to low temperatures, the $p\pi$-d interaction must persist also in the "reentrant pressure region". Then the metal electrons and magnetic moments may coexist and the system can be regarded as a kind of Kondo system. According to this picture, we have tried to explain the linear dependence of ρ on log T between T_{min} and T_{max} followed by the reappearance of the metallic phase (see Fig. 3) [10]. Of course, this is an oversimplified picture because the MI transition of the group-II compound is considered to be induced by the cooperative effect of the distortion of the coordination tetrahedron around Cu and CDW lattice distortion. It is not clear whether these two types of the distortions take place simultaneously at the "reentrant pressure region". In this connection, it may be interesting that around 9 kbar, ρ seems to increase sharply after the linear increase with decreasing log T between T_{max} and T_{min}. Although definite evidence of the existence of the localized magnetic moments in the metallic DCNQI-Cu system has not been obtained yet, there are some anomalous magnetic behaviors possibly related to the existence of Cu^{+2} in $(DMe-DCNQI)_2Cu$ [5,11].

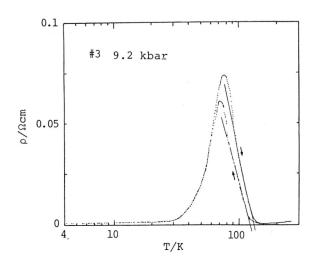

Fig. 3. An example of the ρ vs. log T curve.

4. Acknowledgement

We are indebted to Profs. T. Takahashi, S. Kagoshima and H. Fukuyama for illuminating discussions. Thanks are also due to Dr. T. Mori for the collaboration and fruitful discussions.

References

[1] A. E. Underhill and M. M. Ahmad, J. Chem. Soc., Chem. Commun., 1981, 67; A. Kobayashi, T. Mori, Y. Sasaki, H. Kobayashi, M. M. Ahmad, and Underhill, Bull. Chem. Soc. Jpn., **57**, 3262 (1984).
[2] A. Kobayashi, R. Kato, H. Kobayashi, T. Mori, and H. Inokuchi, Solid State Commun., **64**, 45 (1987).
[3] A. Aumüller, P. Erk, G. Klebe, S. Hünig, J. U. von Schütz, and H. -P. Werner, Angew. Chemie. Int. Ed. Engl. **25**, 740 (1986).
[4] R. Kato, H. Kobayashi, and A. Kobayashi, J. Am. Chem. Soc., **111**, 5224 (1989).
[5] T. Mori, H. Inokuchi, A. Kobayashi, R. Kato, and H. Kobayashi, Phys. Rev. B **38**, 5913 (1988).
[6] T. Mori, K. Imaeda, R. Kato, A. Kobayashi, H. Kobayashi, and H. Inokuchi, J. Phys. Soc. Jpn., **56**, 3429 (1987).
[7] S. Tomic, D. Jérome, A. Aumüller, P. Erk, S. Hünig, and J. U. von Schütz, J. Phys. C: Solid State Phys., **21**, 1203 (1988).
[8] D. Schmeisser et al. have claimed that the valence of Cu in the crystal prepared under the inert atmosphere is not mixed-valent(Chem. Phys. Lett., **148**, 423 (1988)). But this does not conflict with our description because the oxidation state of Cu in air-grown crystal is +1.3.
[9] A. Kobayashi, R. Kato, and H. Kobayashi, Synthetic Metals, **27**, B275 (1988).
[10] H. Kobayashi, A. Miyamoto, R. Kato, Y. Nishio, K, Kajita, and W. Sasaki, Solid State Commun., to be published.
[11] H. Kobayashi et al., to be published; S. Kagoshima, private communications.

Reflectance Spectra of DCNQI Salts

H. Tajima[1], G. Ojima[1], T. Ida[1], H. Kuroda[1], A. Kobayashi[1], R. Kato[2], H. Kobayashi[2], A. Ugawa[3], and K. Yakushi[3]

[1]Department of Chemistry, Faculty of Science, University of Tokyo, Hongo, Bunkyo-ku, Tokyo 113, Japan
[2]Department of Chemistry, Faculty of Science, Toho University, Funabashi, Chiba 274, Japan
[3]Institute of Molecular Science, Okazaki 444, Japan

Abstract. The reflectance spectrum of four DCNQI salts, (i.e. $(Me_2DCNQI)_2Ag$, $(Me_2DCNQI)_2Na$, $(Me_2DCNQI)_2Cu$, and $(MeBrDCNQI)_2Cu$), were measured in the region between 450–25000 cm^{-1}. The first and the second salts were found to be one-dimensional metals. The third salt was found to be a three-dimensional metal over all the temperatures measured. In the fourth salt the drastic change of reflectance spectrum, suggesting the phase transition from a three-dimensional metal to a one-dimensional insulator, was observed around 150K.

1. Introduction

The charge-transfer salts of DCNQI derivatives, $(R_1R_2DCNQI)_2M$ (R_1,R_2:Me,MeO,Br,Cl,I), are new type organic conductors[1]. They can be divided into three groups according to the electrical behaviors[2]. The first group is the one-dimensional metals which exhibit a broad metal-insulator transition. The salts having the cation other than Cu fall into this group. The second group is the three-dimensional metals without metal-insulator transition. $(Me_2DCNQI)_2Cu$ belong to this group. The third group behaves as a three-dimensional metal at room temperature, but undergoes a metal-insulator transition at a low temperature. $(MeBrDCNQI)_2Cu$ belongs to this group. In this study, we present the spectroscopic evidence to show the difference in the nature of the electronic structure among these three groups.

2. Results and discussion

$(Me_2DCNQI)_2Ag$, $(Me_2DCNQI)_2Na$: Figure 1 shows the optical conductivity spectrum of $(Me_2DCNQI)_2Ag$. The broad infrared dispersion appears only for the polarization parallel to the stacking axis, i.e. c-axis. A similar spectrum was observed for $(Me_2DCNQI)_2Na$[3]. Although these materials have metallic properties at least at room temperature, an optical gap appears below 1500 cm^{-1}. This feature is common for the

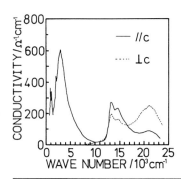

Fig.1 The conductivity spectrum of $(Me_2DCNQI)_2Ag$ at room temperature

one-dimensional metal having a large electron-electron interaction [4]. Since the broad peak at 3000cm^{-1} (0.4eV) in the c// conductivity spectrum can be attributed to the nearest neighbor Coulomb repulsion V, its value of V is roughly estimated to be 0.4eV.

(Me$_2$DCNQI)$_2$Cu: Figure 2 shows the reflectance spectrum of (Me$_2$DCNQI)$_2$Cu, and Fig. 3 shows the conductivity spectrum obtained from the Kramers-Kronig transformation of the reflectance spectra shown in Fig. 2. The broad absorption band above ~ 4000cm^{-1} is clearly observed in the conductivity spectrum together with the three-dimensional intra-band transition. On lowering the temperature, the band about 4000 cm^{-1} grows up in the conductivity spectrum, and at the same time a dispersion due to Drude-like intra-band transition appears more clearly. The new band at 4000 cm^{-1} may be attributed to the inter-band transition between the separated bands made from LUMO

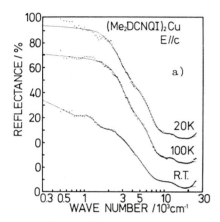 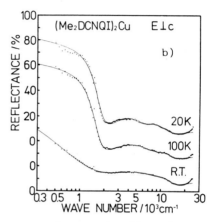

Fig. 2 Reflectance spectra of (Me$_2$DCNQI)$_2$Cu: a) E//c, b) E⊥c. The solid line represents the Drude fit to the reflectance spectra.

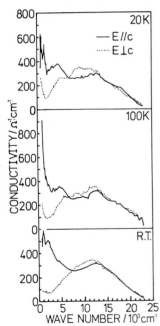

Fig. 3 Temperature dependence of the conductivity spectrum of (Me$_2$DCNQI)$_2$Cu.

Table 1 Results of Drude fit((Me$_2$DCNQI)$_2$Cu)

		ε_c	γ	ω_p (/10^3cm^{-1})	m*/m$_e$	ω_j	Γ_j (/10^3cm^{-1})	Ω_{pj}
//c	20K	3.07	0.42	10.7	1.7	4.3	7.1	11.9
						12.9	7.5	9.3
	100K	3.04	0.71	11.2	1.6	4.8	6.3	10.4
						13.0	8.6	10.8
	R.T.	2.87	3.24	9.4	2.3	1.0	0.4	1.9
						2.1	0.9	2.2
						4.4	9.8	10.3
						12.8	8.2	9.9
//a	20K	3.46	0.67	5.0	7.9	4.4	3.3	5.6
						10.2	9.7	13.6
	100K	3.47	0.60	4.7	9.2	6.1	7.7	10.2
						12.6	7.9	10.4
	R.T.	3.25	2.38	3.6	15.3	7.1	7.8	8.9
						12.0	7.4	10.2

of DCNQI and d$_{xy}$ of Cu. In order to find out the optical mass we analyzed the reflectance spectra by use of the Drude-Lorentz model, assuming the complex dielectric function $\varepsilon(\omega)$ expressed by the following equation:

$$\varepsilon(\omega) = \varepsilon_c - \omega_p^2/(\omega^2 + i\gamma\omega) - \sum_j \Omega_{pj}^2/(\omega^2 - \omega_j^2 + i\Gamma_j\omega)$$

where ε_c, ω_p, γ, and [ω_j^2, Ω_{pj}, Γ_j] denote the background dielectric constant, the plasma frequency of free carriers, the relaxation rate of free carriers, and the parameters of the Lorentz oscillator for the j-th excitation, respectively. The optical mass tensor was calculated from the plasma frequency. The thus obtained optical mass is the average of the effective mass of all the electrons in the conduction band. The obtained parameters are listed in Table 1. There are some ambiguities in the estimation of the optical mass at room temperature because of the large relaxation rate, γ, at room temperature. However the parameters obtained from the low temperature spectrum are relatively reliable. The anisotropic ratio of optical mass (m$\perp_c$/m$_{//c}$) is roughly estimated to be 0.2. This value is somewhat smaller than the corresponding value of β-(BEDT-TTF)$_2$I$_3$ ($\sim$ 0.3)[5].

(MeBrDCNQI)$_2$Cu: Figures 4 and 5 show the reflectance and conductivity spectra of (MeBrDCNQI)$_2$Cu, respectively. The line shape of the reflectance spectrum of (MeBrDCNQI)$_2$Cu is almost same as that of (Me$_2$DCNQI)$_2$Cu at room temperature. We also estimated the Drude parameters for the metallic phase by using Drude formula. The results are summarized in the Table 2. A drastic change from the spectral shape characteristic of a three-dimensional band system to that of a one-dimensional band system was observed at 160K. As shown in the conductivity spectrum the Drude-like intra-band transition disappears and a new band strongly polarized along c-axis appears at 3000 cm^{-1}. It should be noted that this change is in contrast to the spectral change which is frequently observed in the metal-insulator transition of the BEDT-TTF salts, where a two-dimensional inter-band transition takes over the two-dimensional Drude like intra-band transition below the phase transition temperature[6]. This unusual behaviour of

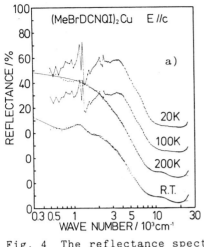
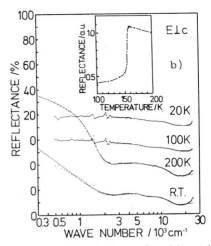

Fig. 4 The reflectance spectrum of $(MeBrDCNQI)_2Cu$: a) E//c, b)E⊥c. The solid line represents Drude fit to the reflectance spectra (200K, R.T.) or guide for the eye (100K, 20K). The inset in Fig. 4a shows change of reflectivity accompanied with phase transition for the light polarization perpendicular to the c-axis at 850cm^{-1}

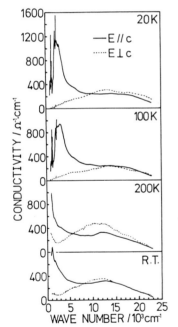

Table 2 Results of Drude Fit($(MeBrDCNQI)_2Cu$)

		ε_c	γ	ω_p	m^*/m_e	ω_j	Γ_j	Ω_{pj}
			($/10^3$ cm^{-1})			($/10^3$ cm^{-1})		
//c	200K	3.15	0.88	10.3	1.9	13.0	9.4	10.8
						3.7	8.5	12.1
	R.T.	3.15	4.23	10.6	1.8	1.0	0.5	3.7
						1.9	1.0	3.2
						4.2	8.3	8.3
						12.7	8.2	10.5
//a	200K	4.62	1.17	5.6	6.4	6.3	9.1	10.8
						11.7	9.4	13.9
	R.T.	3.64	3.15	4.7	9.2	5.5	3.4	4.6
						11.4	9.5	13.2

Fig. 5 Temperature dependence of the conductivity spectrum of $(MeBrDCNQI)_2Cu$

$(MeBrDCNQI)_2Cu$ suggests that the electron transfer mediated by the $Cu^{1.3+}$ is inhibited below the phase transition temperature. In addition to the change of the intra-band transition, the broad absorption above 4000cm^{-1} is also weakened in the insulating phase, but does not completely disappear. This suggests that the CT interaction between Cu and DCNQI is weakened but still exists in the insulating phase. This conclusion is somewhat inconsistent with the one derived from the change of the intra-band transition at least within the simple one-electron band picture.

References.

[1] A. Aumüller, P. Erk, G. Klebe, S. Hünig, J. U. von Schutz, and H.-P. Werner, Angew. Chem. Int. Ed. Engl., 25, 740(1986).
[2] T. Mori, H. Inokuchi, A. Kobayashi, R. Kato, and H. Kobayashi, Phys. Rev. B 38, 5913(1988).
[3] K. Yakushi, G. Ojima, A. Ugawa, and H. Kuroda, Chem. Lett. 1988, 95.
[4] K. Yakushi, S. Aratani, K. Kikuchi, H. Tajima, and H. Kuroda, Bull. Chem. Soc. Jpn., 59, 363(1986).
[5] H. Tajima, H. Kanbara, K. Yakushi, H. Kuroda, and G. Saito, Solid State Commun. 57, 911(1986).
[6] For example, see H. Tajima, H. Kanbara, K. Yakushi, H. Kuroda, G. Saito, and T. Mori, Synthetic Metals, 25, 323(1988).

Optical Spectra of Highly Conducting Phthalocyanine Salts

K. Yakushi[1], H. Yamakado[2], T. Ida[1], A. Ugawa[1], H. Masuda[1], and H. Kuroda[3]

[1]Institute for Molecular Science, Myodaiji-cho, Okazaki, Aichi 444, Japan
[2]Department of Structural Molecular Science,
  Graduate University for Advanced Studies
[3]Department of Chemistry, Faculty of Science, University of Tokyo,
  Hongo, Bunkyo-ku, Tokyo 113, Japan

Abstract. The polarized reflectance spectra are measured on the single crystals of unoxidized PbPc; partially oxidized phthalocyanine salts, $NiPc(AsF_6)_{0.5}$, $CoPc(AsF_6)_{<0.5}$, $H_2Pc(AsF_6)_{0.67}$; and neutral radical LiPc. The absorption bands in the visible region are assigned by comparing these spectra. Based on this assignment, the oxidation part of $CoPc(AsF_6)_{<0.5}$ and the pressure dependence of $NiPc(AsF_6)_{0.5}$ are discussed.

1. Introduction

Some phthalocyanine molecules involve d-electrons in the conjugated π-electron system. Most importantly the highest occupied molecular orbital, which is responsible for electrical conduction in solid state, is not hybridized with the d_{z2} orbital, although they are energetically close to each other.[1] Owing to this special nature, the conductive phthalocyanine salts in solid state have distinct natures from conventional organic conductors.[2-4] The solid state properties of iodine salts of metallo-phthalocyanine have been extensively studied by the research groups of Northwestern University.[2-5] However, the systematic study of the optical spectra has not been conducted so far. In this paper, we present the optical spectra of a series of phthalocyanine single crystals having different degree of oxidation.

2. Sample preparations and crystal structures

The single crystals of monoclinic-form lead phthalocyanine (PbPc) were grown by a slow sublimation on glass substrate heated at 250 °C. $NiPc(AsF_6)_{0.5}$, $CoPc(AsF_6)_{<0.5}$, $H_2Pc(AsF_6)_{0.67}$, and LiPc were obtained by the use of electrochemical technique. The crystal of $CoPc(AsF_6)_{<0.5}$ belongs to a tetragonal system.[6] The crystals of PbPc[7], $NiPc(AsF_6)_{0.5}$[8], and LiPc[9] approximately have a tetragonal arrangement of phthalocyanine molecules. The crystal structure of $H_2Pc(AsF_6)_{0.67}$[10] is isostructural to $NiTBP(AsF_6)_{0.67}$,[11] involving a trimer of H_2Pc in a unit cell. In all these crystals, phthalocyanine molecules are stacked along the c-axis with their molecular planes perpendicular to the c-axis. We can therefore directly compare the solid-state spectra of these materials, since the crystal spectra do not suffer the factor group splitting.

3. Polarized reflectance spectra

Figure 1 shows the optical conductivity spectra polarized perpendicular to the stacking direction of unoxidized PbPc, two-thirds oxidized $H_2Pc(AsF_6)_{0.67}$, and fully oxidized LiPc. The absorption band of PbPc at 14200 cm^{-1} (A) corresponds to the so-called Q-band (a transition from a_{1u} to e_g). The small band at 19300 cm^{-1} in Fig. 1a is inherent in PbPc, corresponding to the weak absorption band at about 22000 cm^{-1} in the solution spectrum, which is not observed in other metallo-phthalocyanine. The absorption band at 14900 cm^{-1} (A) of LiPc shown in Fig 1b was assigned by Homborg and Teske to the Q-band, and the other band around 21000 cm^{-1} (B) to the transition from lower e_g to half-occupied a_{1u}.[12] We therefore assigned the absorption bands, A and B of $H_2Pc(AsF_6)_{0.67}$ in Fig. 1c to the electronic transitions from a_{1u} to upper e_g and from lower e_g to a_{1u}, respectively. This assignment is reasonable, because the intensity of the band A decreases with the increase of the degree of oxidation while the intensity of the band B increases. Obviously, the oxidation of H_2Pc and LiPc occurs in the ligand part, so that the systematic change of the spectra corresponds to the change in a ligand π-conjugated system accompanying the oxidation. Conversely, the appearance of the transition B means the ligand-oxidation.

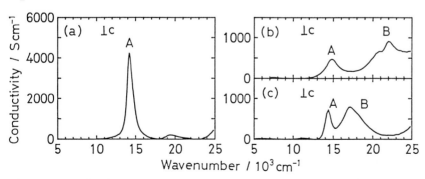

Fig. 1 Optical conductivity spectra polarized perpendicular to the stacking axis of (a)PbPc, (b) LiPc and (c) $H_2Pc(AsF6)_{0.67}$.

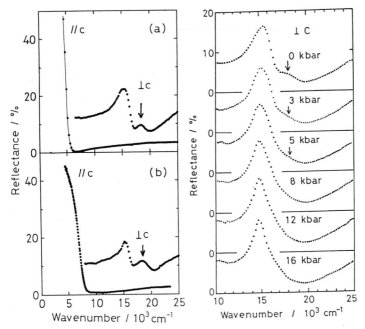

Fig. 2 left: Polarized reflectance spectra of (a)NiPc(AsF$_6$)$_{0.5}$ and (b)CoPc(AsF$_6$)$_{<0.5}$.

Fig. 3 right: Pressure dependence of the reflectance spectrum of NiPc(AsF$_6$)$_{0.5}$.

Figure 2 shows the reflectance spectra of NiPc(AsF$_6$)$_{0.5}$ and CoPc(AsF$_6$)$_{<0.5}$, having the transition B shown by arrows. Application of the knowledge obtained above leads to the view that the ligand oxidation occurs more or less in both salts. This interpretation is consistent with other experimental results of NiPc(AsF$_6$)$_{0.5}$[8], but contradictory to CoPcI[3] which is the analogous salt to CoPc(AsF$_6$)$_{<0.5}$. In this connection, the reflectance spectra of CoPc(AsF$_6$)$_{<0.5}$ parallel to the stacking axis shown in Fig. 2b is quite different from the Drude-like line shape of NiPc(AsF$_6$)$_{0.5}$. The Kramers-Kronig transformation of CoPc(AsF$_6$)$_{<0.5}$ manifests another weak electronic transition around 10000 cm^{-1}. Since Co^{2+} have a d^7 configuration, the 3d$_{z^2}$ band is probably split by the on-site Coulomb energy. We tentatively assigned this weak transition to the charge-transfer transition from the lower 3d$_{z^2}$ Hubbard band to the upper one. The complete assignment of CoPc(AsF$_6$)$_{<0.5}$ including the low-energy transition should be conducted based on the spectrum in the infrared region.

Figure 3 shows the pressure dependence of the polarized reflectance spectrum of NiPc(AsF$_6$)$_{0.5}$ measured by the use of diamond anvil cell. Upon applying pressure, the transition B indicated by arrows progressively disappears. The assignment of the transition B leads to the interpretation that the oxidized part in $NiPc(AsF_6)_{0.5}$ changes gradually from the ligand to the nickel ion. This interpretation is rationalized, if the $3d_{z^2}$-orbital raises its energy level against the top of the conduction band made up of ligand molecular orbitals due to the contraction of the Ni-Ni distance of the nearest neighbor phthalocyanine.

References

[1] F. W. Kutzler and D. E. Ellis, J. Chem. Phys., **84**, 1033 (1986).
[2] J. Martinsen, L. J. Pace, T. E. Phillips, B. M. Hoffman, and J. A. Ibers, J. Am. Chem. Soc., **104**, 83 (1982).
[3] J. Martinsen, J. L. Stanton, R. L. Greene, J. Tanaka, B. M. Hoffman, and J. A. Ibers, J. Am. Chem. Soc., **107**, 6915 (1985).
[4] M. Y. Ogawa, J. Martinsen, S. M. Palmer, J. L. Stanton, J. Tanaka, R. L. Greene, B. M. Hoffman, and J. A. Ibers, J. Am. Chem. Soc., **109**, 1115 (1987).
[5] J. Martinsen, S. M. Palmer, J. Tanaka, R. C. Greene, and B. M. Hoffman, Phys. Rev. B, 30, 6269 (1984).
[6] H. Yamakado and K. Yakushi, unpublished data.
[7] K. Ukei, Acta Cryst., B**29**, 2290 (1973).
[8] K. Yakushi, H. Yamakado, M. Yoshitake, N. Kosugi, H. Kuroda, T. Sugano, M. Kinoshita, A. Kawamoto, and J. Tanaka, Bull. Chem. Soc. Jpn. **62**, 687 (1989).
[9] H. Sugimoto, M. Mori, H. Masuda, and T. Taga, Chem. Commun., 962 (1986).
[10] T. Ida and K. Yakushi, unpublished data.
[11] K. Yakushi, M. Yoshitake, H. Kuroda, A. Kawamoto, J. Tanaka, T. Sugano, and M. Kinoshita, Bull. Chem. Soc. Jpn., **61**, 1571 (1988).
[12] H. Homborg and C. L. Teske, Z. anorg. allg. Chem., **527**, 45 (1985).

Preparation of Electroconducting Materials Containing Copper Compounds: A Copper Oxide with 1,3,4,6-Tetrathiapentalene-2,5-dione

M. Inoue[1]*, C. Cruz-Vázquez*[1]*, M.B. Inoue*[1,2]*, K.W. Nebesny*[2]*, and Q. Fernando*[2]

[1]Centro de Investigación en Polímeros y Materiales, Universidad de Sonora, Apdo. Postal 130, Hermosillo, Sonora, México
[2]Department of Chemistry, University of Arizona, Tucson, AZ 85721, USA

<u>Abstract</u>. The reaction of 1,3,4,6-tetrathiapentalene-2,5-dione ($C_4S_4O_2$) with copper(I) ions generated in situ yielded a copper oxide, $(C_4S_4O_2)Cu_3O_3Cl_{0.3} \cdot 1.2H_2O$, that complexed with the organic molecules. The X-ray photoelectron spectrum showed that the copper atoms were in the Cu(I) state. The powder conductivity was 2 S cm^{-1} at 300 K, and the activation energy for the charge transport was 0.04 eV at high temperatures. The use of copper compounds is a versatile method for the preparation of conducting materials.

1. Introduction

One of the features of highly conducting materials is that the constituent molecules are in a mixed-valence state. The use of an oxidizing reagent that has an appropriate redox potential is, therefore, important for the preparation of highly conducting materials. Some copper(II) compounds meet this requirement, because their oxidation capacity can be controlled over a wide range by changing coordinated ligands and also by selecting solvents in which the synthesis is performed [1]. In our previous papers, we reported that a variety of conducting charge-transfer complexes were readily obtained by the use of copper(II) compounds as oxidizing reagents: tetracyanoquinodimethan (TCNQ) complexes were obtained by reactions between LiTCNQ and copper(II) chelates [1,2-4], and copper halides with tetrathiafulvalene (TTF) or tetramethyltetraselenafulvalene (TMTSF) were formed by direct oxidations of appropriate donors with copper(II) halides [1,5-7]. Some copper(II) compounds can be used also as oxidative coupling reagents for the preparation of conducting polymers: polythiophene, polypyrrole and polyaniline obtained by this method were highly conductive with a small number of chain defects and cross linkings [8-10].

Some copper(I) compounds are expected to be useful as reducing reagents for the preparation of conducting materials. For example, Cu(I) ions generated in situ are potentially useful for the purpose. In the present work, we have studied a reaction of 1,3,4,6-tetrathiapentalene-2,5-dione (or thiapenedione) with copper(I) ions, and obtained a new copper oxide compound in which the organic molecules are complexed.

2. Results and Discussion

When a mixture of thiapenedione and $CuCl_2 \cdot 2H_2O$ was refluxed with copper metal in tetrahydrofuran (THF), a black powder was obtained. The elemental analysis of the product showed that the following reaction occurred:

$C_4S_4O_2$ + $CuCl_2 \cdot 2H_2O$ + Cu → $(C_4S_4O_2)Cu_3O_3Cl_{0.3} \cdot 1.2H_2O$.

$C_4S_4O_2$: 1,3,4,6-tetrathiapentalene-2,
5-dione (or thiapenedione)

When copper metal was absent, no reaction proceeded between thiapenedione and copper(II) chloride. Thiapenedione has been reported to undergo a ring-opening reaction or a deoxidation dimerization depending on reaction conditions [11,12]. In the present reaction, however, thiapenedione molecules are maintained intact in the product. This was evidenced by the following facts: 1) the elemental analysis showed that the ratio of carbon and sulfur involved was unity, 2) the product was soluble in organic solvents such as acetonitrile and acetone, whereas the corresponding dimeric compound, bis(carboxydithio)tetrathiafulvalene, is insoluble in common organic solvents [12], and 3) an unresolved 2p peak of sulfur was observed for the X-ray photoelectron spectrum (XPS) as shown in Fig. 1, suggesting that all sulfur atoms are equivalent. When the solution of the product was exposed to air, a green powder precipitated. The powder in solid state, however, was stable enough to handle in air.

The electrical conductivity of the compressed pellets was 2 S cm^{-1} at 300 K. The temperature dependence is shown in Fig. 2. At high temperatures, it can be explained by the thermally activated temperature dependence, $\sigma = \sigma_\infty \exp(-E/kT)$, with an activation energy of 0.04 eV. The charge transport is due to positive holes, because the thermoelectric power is positive: 10 µV K^{-1} at 300 K.

In the XPS, a copper $2p_{3/2}$ core electron peak was observed at a binding energy of 933.9 eV and the Cu $2p_{1/2}$ peak at 952.9 eV. Each peak was not accompanied by shake-up satellites (Fig. 3), which are characteristic of a paramagnetic copper(II) species. The copper atoms are, therefore, in the Cu(I) state. The infrared spectrum showed an intense electronic absorption band extending from 4000 to 1000 cm^{-1}; vibrational bands were masked by the intense band.

The product contains a large amount of oxygen. Its origin is not clear. There is no doubt, however, that a copper oxide lattice is formed in the material. This is a new example of a copper oxide complex that is formed with organic molecules. The present synthetic method is expected to be applied to the preparation of a variety of copper oxides that may be highly conductive.

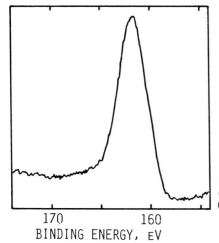

Fig. 1. X-ray photoelectron spectrum of sulfur 2p core electrons of $(C_4S_4O_2)Cu_3O_3Cl_{0.3} \cdot 1.2H_2O$.

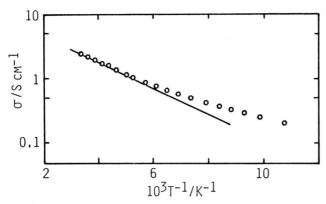

Fig. 2. Temperature dependence of electrical conductivity. The straight line was calculated by $\sigma = \sigma_\infty \exp(-E/kT)$ with $E = 0.04$ eV.

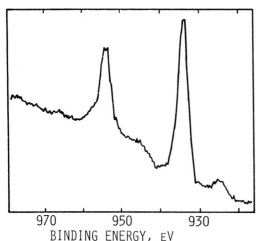

Fig. 3. Copper 2p X-ray photoelectron spectrum.

3. Experimental

The material was obtained by refluxing a tetrahydrofuran (THF) solution (50 ml) that dissolved thiapenedione (200 mg) and $CuCl_2 \cdot 2H_2O$ (246 mg) in the presence of copper wire for four days. The black precipitate was collected on a fritted glass filter, washed with THF and dried in vacuum. The analytical data slightly fluctuated from one sample to another. The C:S ratio, however, was unity for all samples studied. Typical analytical data were: C, 10.02; H, 0.51; S, 26.67; Cl, 2.08; Cu, 40.1%. The corresponding formula is $(C_4S_4O_2)$-$Cu_3O_3Cl_{0.3} \cdot 1.2H_2O$ (Calcd: C, 10.03; H, 0.51; S, 26.76; Cl, 2.22; Cu, 39.8%).

The electrical conductivity was determined on the compressed pellets by van der Pauw's four-probe method. The XPS data were obtained with the aid of a Vacuum Generators ESCALAB MK II with Mg K_α X-rays (1254.6 eV). The binding energy was calibrated by using the carbon 1s peak as an internal standard: the binding energy was assumed to be 284.6 eV.

4. Acknowledgements

The work at the Universidad de Sonora was supported by the Dirección General de Investigación Científica y Superación Académica, SEP, Mexico (Grant No. C-88-01-0231-3), and the work at the University of Arizona was funded by the Center for Advanced Studies in Copper Recovery and Utilization under Defence National Stockpile Center, U.S.A. (Grant No. DN-004).

5. References

[1] M. Inoue and M.B. Inoue, Rev. Inorg. Chem., 9, 219 (1988).
[2] M. Inoue and M.B. Inoue, Mol. Cryst. Liq. Cryst., 86, 139 (1982).
[3] M. Inoue and M.B. Inoue, Inorg. Chem., 25, 37 (1986).
[4] M.B. Inoue, M. Inoue, Q. Fernando and K.W. Nebesney, J. Phys. Chem., 91, 527 (1987).
[5] M.B. Inoue, M. Inoue, Q. Fernando and K.W. Nebesny, Inorg. Chem., 25, 3976 (1986).
[6] M. Inoue, C. Cruz-Vázquez, M.B. Inoue, S. Roberts and Q. Fernando, Synth. Met., 19, 641 (1987).
[7] M.B. Inoue, C. Cruz-Vázquez, M. Inoue, G.J. Pyrka, K.W. Nebesny and Q. Fernando, Synth. Met., 22, 231 (1988).
[8] M.B. Inoue, E.F. Velázquez and M. Inoue, Synth. Met., 24, 223 (1988).
[9] M.M. Castillo-Ortega, M.B. Inoue and M. Inoue, Synth. Met., 28, C65 (1989).
[10] M. Inoue, R.E. Navarro and M.B. Inoue, Synth. Met., 30, 199 (1989).
[11] C. Faulmann, P. Cassoux, R. Vicente, J. Ribas, C.A. Jolly and J.R. Reynolds, Synth. Met., 29, E557 (1989).
[12] R.R. Schumaker and E.M. Engler, J. Am. Chem. Soc., 99, 5521 (1977).

Part III

**TMTSF Family:
Superconductivity and
Spin Density Waves**

A Hidden Low-Temperature Phase in the Organic Conductor (TMTSF)$_2$ReO$_4$

S. Tomić[1,2] *and D. Jérome*[1]

[1]Laboratoire de Physique des Solides, Université de Paris-Sud,
 F-91405 Orsay, France
[2]Institute of Physics of the University, P.O. Box 304, YU-41001 Zagreb, Yugoslavia

Abstract. We show, using an appropriate pressure-temperature cycling, that it is possible to maintain the non-centrosymmetric anion ReO$_4$ in an ordered configuration that does not give rise to an induced insulating phase at low temperatures. If such a cooling procedure is followed, the low-temperature phase diagram of (TMTSF)$_2$ReO$_4$ is very reminiscent of the diagram observed with centrosymmetric anions such as PF$_6$, AsF$_6$ and so on. We discuss the phase diagram of (TMTSF)$_2$ReO$_4$ in the framework of a unique model for (TMTSF)$_2$X organic conductors.

1. Introduction

(TMTSF)$_2$X organic conductors are single-chain materials in which a nominally quarter-filled band is created by a charge delocalization on the organic chain /1/. At low temperatures, the ground state can be either insulating, metallic or superconducting (SC). The origin of the insulating ground state appears to be closely related to the choice of anion X. In the case of centrosymmetric anions such as PF$_6$, AsF$_6$ and SbF$_6$, the ground state is due to the formation of a spin-density wave (SDW) phase (T$_c$ ≤ 12K), while in the compounds with non-centrosymmetric anions, such as ReO$_4$, FSO$_3$ and BF$_4$, the metal-to-insulator phase transition is driven by an ordering of the anions (AO) according to the wavevector $q_2 = (1/2, 1/2, 1/2)$ (40K ≤ T$_c$ ≤ 180K). A finite external pressure suppresses the insulating phase in both cases and leads to a metallic behaviour in the whole temperature region and eventually to a SC ground state. Moreover, at approximately the same critical pressure, a new anion-ordered superstructure is induced with a different wave vector $q_3 = (0, 1/2, 1/2)$ in (TMTSF)$_2$ReO$_4$ /2/.

However, the specific role of anions with respect to the ground state is far from being understood. The purpose of this paper is to review and discuss relevant experimental data and to argue that the electronic interactions responsible for the stabilization of the ground state (SDW or SC) in the family of (TMTSF)$_2$X organic conductors depend essentially on the organic stack and are only very weakly affected by the anion sublattice.

2. Experimental results

First, we recall two examples of compounds with non-centrosymmetric anions in which the ground state is a SDW phase. These are the quenched state of (TMTSF)$_2$ClO$_4$, and (TMTSF)$_2$NO$_3$ /3/, /4/. In the former, the anions are frozen

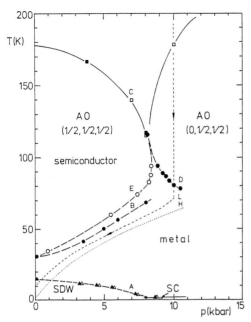

Fig.1. The pressure-temperature phase diagram of $(TMTSF)_2ReO_4$. Open and full squares: cooling and warming, respectively. After p-T cycling path L: open triangles for cooling; full triangles, circles and squares for warming. Open and full hexagons for decreasing and increasing pressure, respectively.

in in a disorder at low temperatures and in the latter are ordered below 40K with a wavevector (1/2,0,0).

In what follows, we summarize the complete temperature-pressure phase diagram of another member with the non-centrosymmetric anion $(TMTSF)_2ReO_4$ obtained recently (Fig.1.) /5/, /6/, /7/, /8/. The metal-to-insulator phase transition due to q_2 AO that exists at ambient pressure is suppressed to pressures below 8 kbar in favour of a metallic phase in which anions are ordered with a wavevector q_3. The respective ground states are a non-magnetic anion-driven insulator and a superconductor. However, if the q_3 AO (i.e. metallic phase) is maintained at pressures lower than 8 kbar, a new semiconducting phase is stabilized as the ground state (Fig.2.). The pressure dependence of the transition temperature strongly resembles that of the SDW ground state found in the compounds with centrosymmetric anions and also that of the NO_3 compound (Fig.3.). Furthermore, the border line with a SC ground state occurs at about same critical pressure ($p_c \simeq 8$ kbar). Moreover, in the narrow pressure region around the critical pressure, the re-entrance of the SC phase is also observed. Therefore, it rather appears justified to identify the low-temperature semiconducting state in $(TMTSF)_2ReO_4$ with a spin-density wave phase, although no direct proof has yet been given with magnetic measurements.

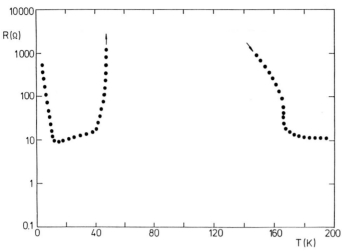

Fig.2. Overall temperature behaviour of resistance at p=3.5 kbar after p-T cycling L on warming.

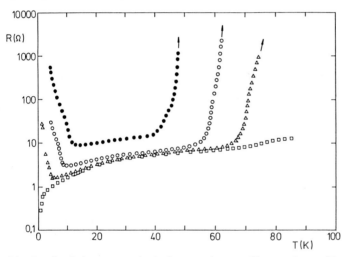

Fig.3. Resistance against temperature after p-T cycling L on warming up to 100K at four different pressures: full and open circles, triangles and squares for 3.5, 6, 8 and 9 kbar, respectively.

3. Discussion

Since similar low-temperature (SDW-SC) phase diagrams are found for the centrosymmetric anions as well as for several non-centrosymmetric ones, the anion-ordering phenomenon has probably a much weaker impact on the physics of the organic stacks than what was believed from a theoretical viewpoint /9/ (except for the anion ordering giving rise to a $2k_F$ potential along the stacking axis with the concomitant opening of a gap at the Fermi level). We may also recall that the q_3 AO in the ReO_4 compound under pressure removes the inversion centre between molecules in the same stack and consequently the electron-electron Umklapp coupling should be enhanced. The latter is expected to enhance the stability of the SDW phase against SC /10/. Instead, we find that q_3 AO is compatible with both ground states, the external pressure being the only relevant variable.

Two distinct theoretical models have been proposed to describe zero-field SDW phase and its evolution under pressure. The theory by C. Bourbonnais et al. /11/ assumes the antiferromagnetic interchain exchange (IEX) mechanism to be responsible for the stabilization of the SDW phase at low pressures. Moreover, this coupling generates interchain pairing fluctuations over a wide range of temperatures in the one-dimensional region. The applied pressure acts to reduce the IEX coupling constant and amplitude with concomitant decrease of the SDW transition temperature. Above a critical pressure the SDW phase is suppressed and SC is established.

In the standard nesting model, the metallic phase of quasi-one-dimensional systems depends on the geometry of the Fermi surface. The refined theory, first proposed by Yamaji /12/, introduces a second harmonic component of the $\vec{b}\cdot\vec{k}$ term in the energy dispersion relation denoted later as t'_b /13/. At ambient pressure the nesting is perfect i.e. $t'_b=0$ and the SDW ground state is stabilized. The deviation from perfect nesting (t'_b finite) increases with increasing pressure and at some critical value $t'^{*}_b \simeq T_c$ (1 bar) the SDW state is suppressed. Note again that the SDW transition temperature in all measured materials (except the Q-ClO$_4$) is around 10K at 1 bar and the critical pressure about 8 kbar, in agreement with the prediction of the nesting model.

In conclusion, a unique theoretical model, based on the common low-temperature phase diagram, may be used to explain the electronic structure and the stabilization of a particular ground state (SDW or SC) in the organic conductors (TMTSF)$_2$X.

References

1. "Low-Dimensional Conductors and Superconductors" (edited by D.Jérome and L.G.Caron, NATO ASI Series B, Plenum, New York, 1987).
2. R.Moret, S.Ravy, J.P.Pouget, R.Comes and K.Bechgaard, Phys.Rev.Lett.57, 1915 (1986).
3. W.M.Walsh, F.Wudl, E.Aharon-Shalom, L.W.Rupp, J.M.Vandenberg, K.Andres and J.B.Torrance, Phys.Rev.Lett.49, 885 (1982).
4. S.Tomić, D.Jérome, J.R.Cooper and K.Bechgaard, Synth.Met.27, B645 (1988).
5. S.S.P.Parkin, D.Jérome and K.Bechgaard, Mol.Cryst.Liq.Cryst.79, 21 (1981).
6. S.Tomić, D.Jérome and K.Bechgaard, J.Phys.C: Solid State Phys.17, L11, (1984).
7. S.Tomić, D.Jérome and K.Bechgaard, in ref.1., p.335.
8. S.Tomić and D.Jérome, J.Phys.:Condens.Matter 1, 4451 (1989).
9. V.J.Emery, J.de Physique Coll.44, C3-977 (1983).
10. V.J.Emery, R.Bruinsma and S.Barišić, Phys.Rev.Lett.48, 1039 (1982).
11. C.Bourbonnais, in ref.1., p.155.
12. K.Yamaji, J.Phys.Soc.Jpn.55, 860,(1986).
13. G.Montambaux, to be published in Phys.Rev.B (1989).

NMR Evidence for the Existence of 1D Paramagnons in Organic Conductors

P. Wzietek[1],*, F. Creuzet[1], C. Bourbonnais[1],(*), D. Jérome[1], P. Batail[1], and K. Bechgaard[2]

[1]Laboratoire de Physique des Solides, Université de Paris-Sud,
F-91405 Orsay, France
[2]H.C. Ørsted Institute, Universitetsparken 5, DK-2100 Copenhagen, Denmark
*Also at (*)permanent address: C.R.P.S., Dept. de Physique,
Université de Sherbrooke, Sherbrooke, Québec, Canada, J1K-2R1

Abstract. In this short review paper we show the existence of a scaling relation between the nuclear relaxation and the magnetic susceptibilty in the (TMTTF)2X and (TMTSF)2X series as a function of temperature. Dynamic scaling arguments show that this relation results from 1D paramagnon effects over quite a large high temperature domain. The analysis of the T_1^{-1} data as a function of pressure has allowed extraction of the value of the short range coupling constant $g_1/\pi v_F$ which is found to be *smaller* in sulphur than for the selenide compounds.

1. Introduction

Nuclear relaxation rate experiments obtained by NMR are well known to be useful for the study of spin correlations in metallic or insulating materials. Recently a lot of interest has been devoted to using this technique for the characterization of spin fluctuations in organic compounds [1,2]. This has been proved to be particularly useful in the context of the (TMTTF)2X and (TMTSF)2X series. Looking at their combined phase diagram, these compounds present a variety of phase transitions. The (TMTTF)2X series at low pressue for example, can present either a spin-Peierls (X= PF_6, AsF_6...) or an antiferromagnetic (AF) phase at low temperature (X=Br...). Both are preceded by a Mott-Hubbard type of insulating behaviour at much higher temperature. Under pressure, the spin-Peierls state is observed to be suppressed and at moderate pressure an antiferromagnetic state with the characteristics of the bromine salt is restored [1b]. At higher pressure, the insulating behaviour of the paramagnetic state is in turn suppressed and then the AF phase transition becomes similar to the one taking place in the selenide (TMTSF)2X series [1a,3,4]. In order to fully understand the origin of such a sequence of phase transitions, a complete description of correlations in the paramagnetic phase is clearly needed. Here we shall briefly summarize the results of a recent nuclear relaxation rate data analysis made on typical compounds of both series [5].

2. Results and Analysis

Electronic spin fluctuations are well known to influence the nuclear relaxation rate in correlated metals which according to the Moriya expression reads [6]

$$T_1^{-1} = 2\gamma_N^2 |A|^2 T \int d^dq\, \chi_\perp''(\vec{q},\omega_N)/\omega_N, \qquad (1)$$

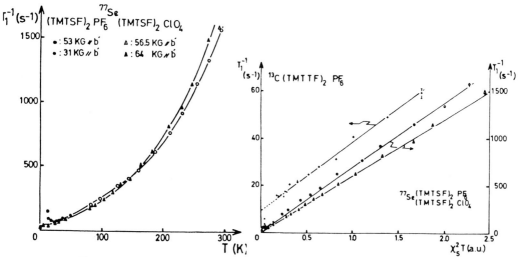

Fig. 1. T_1^{-1} vs T for $(TMTSF)_2PF_6$ and $(TMTSF)_2ClO_4$ at P=1bar (left). T_1^{-1} vs $T\chi_s^2(T)$ of refs [1b,7] for $(TMTTF)_2PF_6$, $(TMTSF)_2PF_6$ and $(TMTSF)_2ClO_4$ at P=1bar (right).

where χ'' is the imaginary part of the retarded magnetic susceptibility. From (1), T_1^{-1} is sensitive to both uniform ($q \sim 0$) and AF ($q \sim Q_0$) correlations. The dynamics which are quite different for each type of correlations will impose a different temperature profile for $T_1^{-1}[q \sim 0]$ and $T_1^{-1}[q \sim Q_0]$ [5]. Most importantly from (1), the nuclear relaxation will also be sensitive to the *dimensionality* d of spin correlations.

If we now turn our attention to the ^{77}Se T_1^{-1} vs T data [1a,5] of figure 1 obtained for the $(TMTSF)_2PF_6$ at P=1bar above the AF critical point (300K > T > $T_N \approx$ 12K) and for the normal phase (300K > T ≥ 6K) of the superconductor $(TMTSF)_2ClO_4$ at ambient pressure, we observe that for a quite large temperature domain (T ≥ 30K), T_1^{-1} profiles present an upward curvature . Such a growing enhancement with T is reminiscent of what has already been observed by Miljak and Cooper [7] for the static magnetic susceptibility χ_s for the same compounds in similar pressure conditions. Using the data of Figure 1 and those of ref.[7], one can establish a clear-cut scaling relation between T_1^{-1} and $\chi_s(T)$ in this temperature domain. As shown on the right scale of figure 2, a relation of the form

$$T_1^{-1} = C_0 T \chi_s^2(T) + C_1 \qquad (2)$$

with $C_0 > 0$ and $C_1 \approx 0$ remarkably holds above 30K or so. One first notes that the $\chi_s^2(T)$ dependence here should not be confused with a Korringa law for non-interacting electrons in three dimensions for which $(T_1T)^{-1} \alpha N(0)^2$. Here N(0) is the density of states at the Fermi level, a temperature independent quantity at low temperature . Taking into account the full thermal smearing of the Fermi distribution at very high temperature *does not* introduce a $\chi_s^2(T)$ dependence for T_1^{-1}. Applying the dynamic scaling hypothesis to the uniform contribution $T_1^{-1}[q \sim 0]$, it can be easily shown that the Coulomb interaction and the low dimensional character are at the origin of this behaviour [5]. Indeed, let us assume that the wave vector q scales as ξ_F^{-1} with ξ_F as the uniform correlation length (F stands for ferromagnetic) and ω as ξ_F^{-z} where z is the dynamical exponent [8]. For the dynamic

69

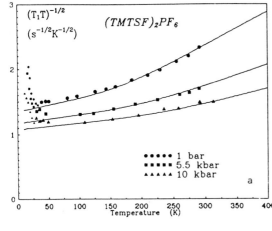

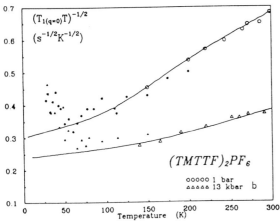

Fig. 2. $(T_1T)^{-1/2}$ vs T data for $(TMTSF)_2PF_6$ (left) and $(TMTTF)_2PF_6$ (right) under pressure. The low temperature points are from refs [1a] (a: T < 50K) and [1b] (b: full circles and triangles).

susceptibility, the scaling form $\chi''(\vec{q},\omega) \approx \chi_s(T) \, D(\vec{q}\circ\vec{\xi}_F, \omega\xi_F^{\bar{z}})$ can be used where D is a scaling function. The quantum theory of *paramagnons* justifies the use of the gaussian type of exponents at $d \geq 1$ [9] namely, $\bar{z} = 3$ and $\chi_s(T) \propto \xi_F^2$. One thus obtains for the uniform part of the relaxation [5]

$$T_1^{-1}[q \sim 0] \approx T[\chi_s(T)]^{(5-d)/2}, \qquad (3)$$

which depends on the dimensionality of uniform spin correlations. Therefore the data of figure 2 give strong evidence for the existence *one-dimensional* harmonic (gaussian) paramagnon excitations for both T_1^{-1} and χ_s. Actually for a 1D electron gas, paramagnons are precisely the spin boson excitations for which the harmonic character is well known to be dominant in one dimension [10].

Deviations from the $T_1^{-1} \propto T\chi_s^2(T)$ law at low temperature (T ≤ 30K) come from the AF part which gives a quite huge enhancement to $(T_1T)^{-1}$ below 30K for $(TMTSF)_2ClO_4$ and which becomes singular for $(TMTSF)_2PF_6$ near T_N [1a]. This will give a non-trivial temperature dependence to T_1^{-1} through C_1 in (2). At this point, a comparison with the more correlated spin-Peierls salt $(TMTTF)_2PF_6$ is worthwhile. The data of figure 2 show the ^{13}C T_1^{-1} vs EPR $T\chi_s^2(T)$ data of Creuzet et al.[1b] above the lattice softening domain and below the Mott-Hubbard charge localization (40K ≤ T ≤ 200K). It is again clear that the relation given in (2) is very well satisfied. In contrast to the selenide compounds however, T_1^{-1} does not extrapolate to zero as $T \to 0$ which means that C_1 is constant. This temperature-independent contribution turns out to be the one expected for a 1D quantum antiferromagnet. In such a case, the scaling hypothesis leads to a power law behaviour in d=1:

$$T_1^{-1}[q \approx 2k_F] \propto T\; T^{-\gamma} \; . \qquad (4)$$

For a 1d Mott insulator, one has the well-known result $\gamma=1$ [10] so that (4) becomes temperature independent, in remarkable agreement with the data of figure 2. Under pressure, metallic properties of $(TMTTF)_2PF_6$ are gradually restored so that the value of C_1 at high temperature should evolve, as expected, toward that of the $(TMTSF)_2X$ series [3,4,5].

It turns out however, that this is not strictly true for *uniform* correlations. Indeed, using the RPA theoretical expression [5b]

$$\chi_s(T) = \chi_s^o(T)\;[1 - \tfrac{1}{2}g_1(T)\chi_s^o(T)]^{-1} \qquad (5)$$

for the uniform susceptibility in T_1^{-1}, where χ_s^o is the tight binding bare suceptibility that takes into account the thermal smearing of the Fermi surface, and $g_1(T)$ is the backward scattering term between electrons which is temperature dependent in 1d due to many-body effects [$g_1(T)= g_1/(1+g_1/\pi v_F^* \ln E_F^*/T)$] [5b,9], one can fit the $(T_1T)^{-1}$ vs T data at high temperature in order to obtain an estimation of the ratio $g_1/\pi v_F$ for both series. In figure 2a and b, we give the ^{77}Se and ^{13}C $(T_1T)^{-1}$ vs T data of $(TMTSF)_2PF_6$ and $(TMTTF)_2PF_6$ respectively under pressure. Using the respective band calculation values for the Fermi energy [11] namely, $E_F=3100K$ and $1600K$ and taking into account the lattice dilatation [11] which leads to the Fermi velocity change $dv_F/v_FdT \approx .07/300K$ and $dv_F/v_FdT \approx .15/300K$ for Se-PF$_6$ and S-PF$_6$ respectively, very good fits were obtained using (5) and (2) with the values of $g_1/\pi v_F \approx 1.16$ (Se-PF$_6$) and $g_1/\pi v_F \approx 1$ (S-PF$_6$) at P=1bar. Under pressure, the fits allow determination of the pressure coefficients $dv_F/v_FdP \approx 2,21\%/kbar$ (Se-PF$_6$) and $dv_F/v_FdP \approx 2,75\%/kbar$. The data used for the fits were taken above 50K. There, the AF contribution to T_1 is either negligible ($C_1 \approx 0$) as for the selenides or temperature independent ($C_1 \approx const$) as for the sulphur compounds so that it can be easily subtracted. From these results, we arrive at the surprising result that, as far as the uniform spin correlations are concerned, the selenide compounds are *more* correlated than the sulphur series with a sizeably higher value of $g_1/\pi v_F$. According to the microscopic 1D theory, a smaller g_1 is consistent with more pronounced $4k_F$ charge and $2k_F$ (AF) spin fluctuations. This is well known to be the case for $(TMTTF)_2X$ compounds [3,4].

In conclusion, the present analysis of the data demonstrated that 1D paramagnon effects are stronger in the selenide than for the sulphur series. This indicates that the short range coupling constant $g_1/\pi v_F$ is more likely to reflect an intrinsic molecular characteristic and therefore, for this coupling constant, there are apparently no scaling properties between the two series under pressure.

Acknowledgments

One of the authors (P. W) would like to thank the *Centre de Recherche en Physique des Solides* (C.R.P.S) of the Université de Sherbrooke and the *Canadian Institute of Advanced Research* (C.I.A.R) for financial support during his stay at C.R.P.S.

References

[1] (a) F. Creuzet, C. Bourbonnais, L. G. Caron, D. Jérome, Synth. Met. 19, 277 (1987); (b) F. Creuzet, C. Bourbonnais, L. G. Caron, D. Jérome, and K. Bechgaard, Synth. Met. 19, 289 (1987).

[2] T. Takahashi, H. Kawamura, T. Ohyama, Y. Maniwa, K. Murata, and G. Saito, J. Phys. Soc. Jpn, 58, 703 (1989).

[3] V. J. Emery, R. Bruisma, and S. Barisic, Phys. Rev. Lett. 48, 1039 (1982).

[4] C. Bourbonnais, in *Low-Dimensional Conductors and Superconductors*, edited by D. Jérome and L. G. Caron, NATO Advanced Study Institute, Ser. B, Vol. 155 (Plenum, New York, 1987), p. 155.

[5] (a) C. Bourbonnais, P. Wzietek, F. Creuzet, D. Jérome, P. Batail and K. Bechgaard, Phys. Rev. Lett. 62, 1532 (1989); (b) F. Creuzet, D. Jérome, L. Valade, P. Cassoux, Europhys. Lett. 6, 177 (1988)

[6] T. Moriya, J. Phys. Soc. Jpn. 18, 516 (1963).

[7] M. Miljak and J. Cooper, Mol. Cryst. Liq. Cryst. 119, 141 (1985); Ibid., J. Phys. (Paris), Colloq. 44, C3-893 (1983).

[8] P. Hohenberg and B. I. Halperin, Rev. Mod. Phys. 49, 435 (1977).

[9] J. Hertz, Phys. Rev. B 14, 1165 (1976).

[10] J. Solyom, Adv. Phys. 28, 209 (1979); V. J. Emery, in *Highly Conducting One-dimensional Solids*, edited by J. T. Devreese et al. (Plenum, New York 1979), p. 327.; K. Efetov, Sov. Phys. JETP 43, 1221 (1976).

[11] B. Gallois and A. Abderrabba, Thesis, Bordeaux, (1988), unpublished.

Long-Range Spin-Fluctuations and Superconductivity in the Quasi-One-Dimensional Hubbard Model

H. Shimahara

Institute of Physics, University of Tsukuba, Ibaraki 305, Japan

Abstract. The superconductivity mediated by antiferromagnetic spin-fluctuations (AFSF) is studied in the quasi-one-dimensional (Q1D) Hubbard model. The superconducting transition temperatures and the momentum dependence of the order parameter are calculated numerically based on an RPA for the spin-fluctuations. It is found that the long-range nature of the AFSF plays an essential role in the superconductivity in the Q1D case. The obtained phase diagram is compared with those of the organic superconductors $(TMTSF)_2X$ and $(DMET)_2X$, and some qualitative agreements between the theory and experiments are found.

1. Introduction.

The quasi-one-dimensional organic superconductors in the TMTSF and DMET families, discovered recently, are of current interest because of the various experimental facts and theoretical problems. Among such problems, we study an interplay between antiferromagnetism and superconductivity in this paper. Some compounds in their families, $(TMTSF)_2X$ ($X=PF_6$, AsF_6, $\cdots$) and $(DMET)_2X$ ($X=Au(CN)_2$, $\cdots$), exhibit an SDW transition at ambient pressure and superconductivity under pressure (>8kbar for $(TMTSF)_2PF_6$, for example) [1]. The well-known phase diagrams in the pressure-temperature plane show the superconducting phase on the border of the SDW phase and a sensitive decrease of the superconducting transition temperature T_c with increasing pressure. The appearance of the SDW transition itself is also of interest, because most other Q1D organic compounds exhibit a Peierls transition, which suggests strong electron-phonon coupling in such organic compounds. The SDW transition seems to suggest that SDW fluctuations would be dominant rather than CDW fluctuations in these systems, and that superconductivity would occur in such a situation. Thus the phase diagrams give rise to a theoretical interest of a possibility of superconductivity induced by AFSF exchange interactions [2-8].

Furthermore, the NMR relaxation rate of $(TMTSF)_2ClO_4$ exhibits no peak just below the T_c, indicating lines of zeros of the superconducting gap function on the Fermi-surface [9-10]. Such an anisotropic superconductivity can be naturally explained by the AFSF exchange mechanism, but it seems to be difficult to explain it with mechanisms based only on electron-phonon interactions.

On the other hand, the SDW transitions seem to be well-explained in a nesting model with nesting vectors $Q \sim (\pm\pi/2, \pm\pi, 0)$ [11]. Then in the AFSF exchange mechanism of superconductivity, two electrons with wave vectors $(k,-k)$ near the Fermi-surface exchange the AFSF, which have sharp peaks at Q, and are scattered to $(k',-k')$ with $k' \sim k \pm Q$ near opposite sides of the Fermi-surface. In this process the pair wave function changes its sign, because the AFSF function is positive in the momentum space, and thus the gap function has lines of zeros at $k \sim Q/2$ for a singlet pairing.

The AFSF exchange interactions have characteristics as follows: (1) Their momentum dependence has sharp peaks at Fermi-surface nesting vectors **Q**. (2) They have no definite small cutoff-energy like a Debye frequency in the case of electron-phonon interactions. (3) Their properties sensitively depend on the band structures of the electrons, since the AFSF are enhanced by Fermi-surface nesting. (4) The interactions become strong and of long-range as one approaches the SDW transition points. The pairing interactions themselves become effective accordingly, but simultaneously pseudo-gap and mass enhancement due to the SDW fluctuations suppress the superconductivity. Thus it is difficult to obtain even rough behavior of the T_c on phase diagrams without a concrete calculation which fully takes into account the above characteristics. For example if the suppression of the superconductivity due to the pseudo-gap and the mass enhancement were too serious, the superconducting phase would disappear near the SDW phase boundary in contrast to the experimental phase diagrams. In this paper, we calculate superconducting transition temperatures and construct a theoretical phase diagram of SDW and superconductivity in the Q1D case.

2. Model and Approximations

We start with the Q1D Hubbard model:

$$H = \sum_{i,j,\sigma} t_{ij} c_{i\sigma}^{\dagger} c_{j\sigma} + U \sum_i n_{i\alpha} n_{i\beta} , \tag{1}$$

with $n_{i\sigma} \equiv c_{i\sigma}^{\dagger} c_{i\sigma}$ and an electron dispersion

$$\varepsilon_p = \sum_i t_{ij} e^{-i p \cdot R_{ij}} = -2t\cos(p_x) - 2t'\cos(p_y) - 2t''\cos(p_z) . \tag{2}$$

Here we take t=0.25eV bearing the TMTSF family in mind. The hopping constant t' is an effective parameter which expresses distortion of the Fermi-surface under pressure and is assumed to increase with increasing pressure. It would be appropriate to take t' as about 0.1t at ambient pressure, and t" is assumed to be much smaller than t' but much larger than the temperatures of concern in this paper.

Based on a perturbation theory, we adopt an RPA for spin- and charge-fluctuations following Yamaji [11] and Scalapino et al. [5]. We regard the AFSF exchange vertex as a boson propagator and calculate electron self-energies and two electron vertices in the lowest order self-consistent scheme in this electron-boson system. The superconducting transition is signalled by the first divergence of the two-electron vertex as temperature decreases. We neglect the frequency dependence of the interaction in the calculation of the two-electron vertex and partly in that of the self-energy. However, we take into account the long-range nature of the interaction through the momentum dependence. We take into account mass renormalization effects through the frequency dependence of the self-energy as well as those of pseudo-gaps through the momentum dependence. Details of our approximation are explained in ref.8.

We take U=1.48253t so that the SDW transition temperature in the RPA (T_{SDW}) is equal to 20K at t'=0. Then we obtain T_{SDW}=12K at about t'=0.136t.

3. Numerical Results

First, we calculate the self-energy part and obtain normal state properties [8]. It is found from the result that the density of states around the Fermi-surface is reduced by the SDW fluctuations

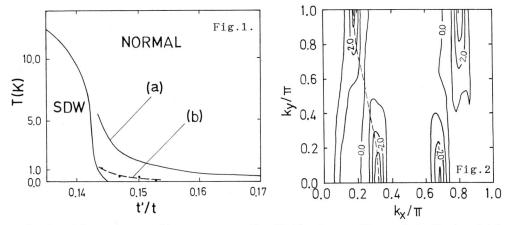

Fig.1. The phase diagram on the T-t' plane. The unlabelled solid line shows the SDW transition temperatures. The solid line (a) and the closed circles show the T_c without and with self-energy effects respectively. The broken line (b) is to guide the eyes.

Fig.2. Contour lines of the gap function $\Delta(k)/\Delta$ in momentum space for $t'=0.15t$, $T=0.0002t$ $(\sim T_c)$, and $U=1.48253t$. Self-energy effects for the electron Green's functions are included. Here Δ is a normalization factor such that $N^{-1}\sum_k \Delta(k)^2/\Delta^2 = 1$. The broken line shows the unperturbed Fermi-surface.

near SDW transition points as is expected. Moreover a band narrowing is also found.
Next, we calculate the superconducting transition temperature T_c. The result is shown in Fig.1. It is found that the T_c does not tend to be enhanced noticeably in the vicinity of the SDW phase boundary. We also find sensitivity of T_c to t' and noticeable suppression due to the pseudo-gap and the mass enhancement, suppression which is shown by the difference of the lines (a) and (b) in Fig.2. Our phase diagram coincides with experimental ones qualitatively and semiquantitatively. In the calculation of T_c shown in Fig.1, the long-range nature of the spin-fluctuations is fully taken into account. However if we neglect it, then the T_c will be estimated to be vanishingly small. For example, that is verified by a numerical calculation in which long-range fluctuations along the conductive chains beyond the 8×(lattice constant) are neglected [8].
Furthermore we calculate the momentum dependence of the gap function. It is found from Fig.2 that the gap function has sharp peaks around the Fermi-surface and lines of zeros on the Fermi-surface as expected [8].

4. Conclusion and Discussion.

In the above calculation, it was found that the phase diagram coincides with the experimental ones qualitatively. Such a phase diagram can be obtained only in treatments in which one fully takes into account the long-range nature of the AFSF along each conductive chain and the self-energy effects near SDW transition points. We found a noticeable suppression of the superconductivity due to the pseudo-gap and the mass renormalization through the electron self-energy. We think that the present mechanism is still possible in the TMTSF and DMET families considering the agreement of the theory and the experiments, although for the BEDT·TTF family the present mechanism would not be applicable at least straightforwardly.

We think that our result has clarified the qualitative behavior of the superconducting transition temperature as one approaches the SDW phase boundary on a phase diagram, at least semiphenomenologically in the system dominated by the AFSF. However, for quantitative verification of this mechanism in the Q1D organic superconductors, the approximations mentioned in the previous section should be improved. For example, contributions of electron-phonon interactions are most important. This problem remains for future studies.

Acknowledgments. The author would like to thank Profs. S.Takada, K.Maki, K.Yamaji, C.Bourbonnais, Y.Suzumura and K.Bechgaard for valuable discussions. He also wishes to thank Iwanami Fuju-kai for financial support.

References

1. R.Brusetti, M.Ribault, D.Jérome and D.Bechgaard: J.Phys.(Paris) 43,801 (1982).
2. V.J.Emery: Synth.Met.13,21 (1986).
3. K.Miyake, S.Schmitt-Rink and C.M.Varma: Phys.Rev.B34,6554 (1986).
4. M.T.Beal-Monod, C.Bourbonnais, and V.J.Emery: Phys.Rev.B34,7716 (1986).
5. D.J.Scalapino, E.Loh,Jr. and J.E.Hirsch: Phys.Rev.B34,8190 (1986); 35,6694 (1987).
6. C.Bourbonnais and L.G.Caron: Europhys.Lett.5,209 (1988).
7. H.Shimahara and S.Takada: J.Phys.Soc.Jpn.57,1044 (1988).
8. H.Shimahara: J.Phys.Soc.Jpn.58,1735 (1989).
9. M.Takigawa, H.Yasuoka and G.Saito: J.Phys.Soc.Jpn.56,873 (1987).
10. Y.Hasegawa and H.Fukuyama: J.Phys.Soc.Jpn.56,877 (1987).
11. K.Yamaji: J.Phys.Soc.Jpn.51,2787 (1982).

Transport Properties of Impure Anisotropic Quasi-One-Dimensional Superconductors

Y. Suzumura

Department of Physics, Tohoku University, Sendai 980, Japan

Abstract. Transport properties for quasi-one-dimensional superconductors having a line of zeros of the gap parameter have been studied in order to examine the effect of the nonmagnetic impurity. It is shown that the impurity takes a significant role for the NMR relaxation rate. The temperature dependence of coefficients for the thermal conductivity and the longitudinal ultrasonic attenuation are also calculated.

In organic conductors of $(TMTSF)_2X$ family indicating quasi-one-dimensional properties, it has been shown that the superconducting (SC) state exists next to the spin density wave state (SDW) as a function of the pressure or the magnetic field[1]. Since the repulsive interaction gives rise to the SDW state, it may be reasonable to consider that the SC state originates from an attractive interaction between chains and then has an anisotropic order parameter with a line of zeros on the Fermi surface [2]. Actually some remarkable evidence for such a SC state was found in the temperature dependence of the NMR relaxation rate [3,4] for which the importance of the effect of the nonmagnetic impurity has been maintained. In the present paper, in order to examine the effect of nonmagnetic impurity, we calculate the transport properties of such a SC state based on the previous calculation [5,6].

We consider a Hamiltonian which consists of an array of one-dimensional chains,

$$\mathcal{H} = \sum_{\mathbf{k},\sigma} \epsilon_{\mathbf{k}} C^+_{\mathbf{k},\sigma} C_{\mathbf{k},\sigma} - \sum_{\mathbf{k}} \Delta_{\mathbf{k}} (C^+_{\mathbf{k}\uparrow} C^+_{-\mathbf{k}\downarrow} + h.c.) + N\Delta^2/g \\ + V_0/N \sum_{\mathbf{k},\mathbf{k}',\sigma} \sum_j C^+_{\mathbf{k},\sigma} C_{\mathbf{k}',\sigma} \exp[i(\mathbf{k}-\mathbf{k}')\cdot \mathbf{R}_j] \; , \quad (1)$$

where $\epsilon_{\mathbf{k}} = v_F(|k_x| - k_F) - 2t_b \cos k_y - 2t_c \cos k_z$, $v_F = 2t_a \sin k_F$, and k_F is the Fermi momentum of the one-dimensional chain. The condition that $t_a \gg t_b \gg t_c$ is assumed for the hopping energy. The quantity $\Delta_{\mathbf{k}}$ denotes an order parameter for the singlet d-wave pairing given by $\Delta_{\mathbf{k}} = \Delta \cos k_y$ and $\Delta = (g/N)\Sigma_{\mathbf{k}} \cos k_y < C_{-\mathbf{k}\downarrow} C_{\mathbf{k}\uparrow} >_H$, where $g(>0)$ is the coupling constant of the attractive interaction. The last term of eq. (1) which denotes the impurity scattering is treated in the Born approximation. The self-consistency equation for Δ is given by

$$\frac{1}{\lambda} = \int_0^\infty dz \left\langle \int_{-\omega_D}^{\omega_D} d\epsilon_{\mathbf{k}} \frac{\cos^2 k_y}{\pi} \mathrm{Im} \frac{\tanh(z/2T)}{-\tilde{z}^2 + \epsilon_{\mathbf{k}}^2 + \Delta_{\mathbf{k}}^2} \right\rangle , \quad (2)$$

where $\lambda = g/\pi v_F$ and T is the temperature. The quantity ω_D is the cutoff energy and $\langle \cdots \rangle = \pi^{-2} \int_0^\pi dk_y \int_0^\pi dk_z \cdots$. We use k in place of k_y or $\mathbf{k}$. In eq. (2), $\tilde{z}$ is given by

$$\tilde{z} = z + < i\tilde{z}/\{2\tau(\tilde{z}^2 - \Delta_k^2)^{1/2}\} > , \tag{3}$$

where $\mathrm{Im}\tilde{z} > 0$ and $1/\tau = 2n_i V_0^2/v_F$. The quantity n_i is the impurity concentration.

In the case where $\omega_D \gg t_b \gg 1/\tau$, static properties have been examined previously based on eqs.(2) and (3) [5]. For the clean case, the transition temperature ,T_c, is given by $T_{c0} = (2e^\gamma/\pi)\,\omega_D\exp[-2/\lambda]$ and Δ is equal to $2e^{-1/2}\Delta_0$ at $T = 0$ where $\Delta_0 = 2\omega_D\exp[-2/\lambda]$. As $1/\tau$ increases, T_c decreases monotonically and becomes zero at $\tau = \tau_0 = 1/\Delta_0$. Thermodynamic properties are related essentially to the density of states, where the normalized quantity , $D(\omega)$, is given by $2\tau\mathrm{Im}\tilde{z}$. The normalized quantity is defined as the the quantity which is normalized by that of the normal state. The quantity ,$D(\omega)$, with the small ω is obtained as follows. When $1/\tau = 0$, $D(\omega) \simeq |\omega|/\Delta$ which is characteristic of Δ_k having a line of zeros on the Fermi surface. When $1/\tau \neq 0$, one obtains $D(\omega) \simeq D(0)\{1 + const \cdot (\tau^2\omega\Delta/D(0))^2\}$ where $D(0) \simeq 8\tau\Delta\exp[-\pi\tau\Delta]$ for $\tau\Delta \gg 1$ and $D(0) \simeq 1 - (12/5)|\ln\tau_0/\tau|$ for $\tau\Delta \ll 1$ [5].

First we examine the NMR relaxation rate given by [7]

$$T_1^{-1} = const \cdot T \lim_{\omega \to +i0} \frac{1}{\omega} \sum_{\mathbf{q}} \mathrm{Im}\chi(\mathbf{q},\omega) . \tag{4}$$

In eq.(4), $\chi(\mathbf{q},\omega)$ is the dynamical spin susceptibility where the normal state is proportional to T. The normalized quantity , R , is calculated as [4,6]

$$R = \int_{-\infty}^\infty dz\, (-\frac{\partial f(z)}{\partial z})D(z)^2 + \frac{F}{4\pi\tau^2 t_b t_c} , \tag{5}$$

where $f(z) = 1/(\exp[z/T]+1)$. The second term is due to the vertex correction which can be disregarded by noting that $F \sim o(1)$ and that $t_b \gg t_c \gg 1/\tau$. When T decreases, R takes a maximun just below T_c for $\tau_0/\tau < 0.15$ while R decreases monotonically for the larger τ_0/τ [6]. At $T = 0$, one obtains $R = D(0)^2$. In Fig.1, the numerical result of $T_1(T_c)/T_1 \equiv RT/T_c$ is shown with $\tau_0/\tau = 0.2(1)$ and $0(2)$ and is compared with the experimental data for $(TMTSF)_2ClO_4$ [3]. It should be noted that the good agreement near $T = T_c$ is obtained by taking account of the effect of nonmagnetic impurity.

Next we study the thermal conductivity which is given by [8]

$$K = const \cdot \frac{1}{T} \lim_{\omega \to +i0} \frac{1}{\omega} \mathrm{Im}\langle[j_h,j_h]\rangle(0,\omega) , \tag{6}$$

where j_h is the heat current operator and the normal state is proportional to T. The normalized quantity , $\overline{K}$, is calculated as

$$\overline{K} = \frac{6}{\pi^2 T^2} \int_{-\infty}^\infty dz\, (-\frac{\partial f(z)}{\partial z})z^2 \left\langle \frac{1}{2\tau\mathrm{Im}(\tilde{z}^2 - \Delta_k^2)^{1/2}}(\frac{|\tilde{z}|^2 - \Delta_k^2}{|\tilde{z}^2 - \Delta_k^2|}+1)\frac{1}{2}\right\rangle . \tag{7}$$

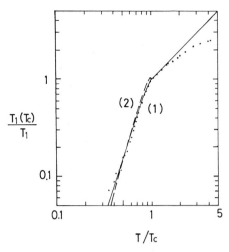

Fig.1 NMR relaxation rate $1/T_1$ as a function of T/T_c in the case of $\tau_0/\tau = 0.2(1)$ and $0(2)$ (dashed curve), where the closed circles denote the experimental data by Takigawa et al. [3].

Since eq.(7) cannot be expressed simply in terms of the density of states, there is the qualitative difference between R and $\overline{K}$ especially at low temperatures. In the case of $T = 0$, eq.(7) is given by $\overline{K} \simeq (e^{1/2}/2\pi)(\tau_0/\tau)$ for $0 < \tau_0/\tau \ll 1$ and $1 - (24/5)|\ln\tau_0/\tau|$ for $\tau_0/\tau \sim 1$ where there is a discontinuity of $\overline{K}$ for $1/\tau \to 0$ i.e., $\overline{K} = 1/2$ at $\tau_0/\tau = 0$. In Fig.2, $\overline{K}$ as a function of T is shown with some choices of τ_0/τ. The large value of $\overline{K}$ at low temperatures even in the case of small τ_0/τ may be characteristic of the weak impurity potential i.e., the small V_0 [9].

Finally, we study the attenuation of the ultrasonic wave which propagates along the one-dimensional chain with the wave vector $\mathbf{q}$ and the frequency ω. The coefficient is given by [10].

$$\alpha_L = const \cdot \frac{q^2}{\omega} \mathrm{Im} \left\{ <[\tau_x, \tau_x]> (\mathbf{q}, \omega) - \frac{(<[\tau_x, n]> (\mathbf{q}, \omega))^2}{<[n, n]> (\mathbf{q}, \omega)} \right\}, \tag{8}$$

where $< [\,,\,] > (\mathbf{q}, \omega)$ denotes the correlation function for the density operator, n, and/or $\tau_x = \Sigma_{\mathbf{k}\sigma} (k_x + q_x/2)^2 C^+_{\mathbf{k}+\mathbf{q}\sigma} C_{\mathbf{k}\sigma}$. By substituting the value at the Fermi surface for k_x in eq.(8), we calculate the dominant contribution in which the vertex correction cancels. Since the energy conservation in the process from the phonon system into the electron system is not valid for the clean limit in the present case, eq.(8) becomes of the order of $(t_b/t_a)^2$ compared with the usual three dimensional case. The normalized quantity, $\overline{\alpha}_L$, is calculated as

$$\overline{\alpha}_L = (1+(ql)^2) \int_{-\infty}^{\infty} dz \, (-\frac{\partial f(z)}{\partial z}) \left\langle a(k) \frac{\tau \mathrm{Im}(\tilde{z}^2 - \Delta_k^2)^{1/2}}{(2\tau \mathrm{Im}(\tilde{z}^2 - \Delta_k^2)^{1/2})^2 + (ql)^2} (\frac{|\tilde{z}|^2 - \Delta_k^2}{|\tilde{z}^2 - \Delta_k^2|} + 1) \right\rangle, \tag{9}$$

where $a(k) = 2(\cos k)^2$ and $l = v_F \tau$. Equation(9) can be understood somehow in terms of the damping of the phonon because eq.(9) becomes equal to $(1 + (ql)^2)\omega^{-1}\mathrm{Im} < [n, n] > (\mathbf{q}, \omega)$ for $a(k) \to 1$. However the role of $a(k)$ is crucial at low temperatures. The limiting value of $\overline{\alpha}_L$ in the case of $1/\tau \to 0$ is obtained as follows. In the case of $ql \ll 1 (\gg 1)$, $\overline{\alpha}_L \propto T^2 (\propto T^4)$ for $T \ll T_c$ and $\overline{\alpha}_L$ is monotonic (takes a maximum) near T_c. In Fig.3, $\overline{\alpha}_L$ ($ql = 1$) is shown with some choices of τ_0/τ where the monotonic variation is obtained as the function of both T/T_c and τ_0/τ.

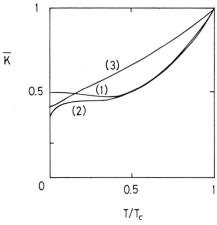 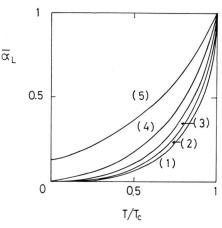

Fig.2 Normalized thermal conductivity $\overline{K}$ as a function of T/T_c with $\tau_0/\tau = 0(1), 0.2(2),$ and $0.8(3)$.

Fig.3 Normalized longitudinal ultrasonic attenuation as a function of T/T_c with $ql = 1$ where $\tau_0/\tau = 0(1), 0.2(2), 0.4(3), 0.6(4)$ and $0.8(5)$.

In summary, by paying attention to the role of the nonmagnetic impurity, we examined transport properties which show the characteristic of the quasi-one-dimensional SC state with a line of zeros of the gap parameter.

The author thanks N. Toyota, K. Miyake and T. Tsuzuki for useful comments.

References

[1] D. Jérome and H.J. Schulz, Adv. Phys. **31** 299 (1982).

[2] V.J. Emery, J. Physique (Colloque) **44** C3-997 (1983); Y. Hasegawa and H. Fukuyama, J. Phys. Soc. Jpn. **55** 3978 (1986); C. Bourbonnais and L.G. Caron, Europhys. Lett. **5** 209 (1988).

[3] M. Takigawa, H. Yasuoka, and G. Saito, J. Phys. Soc. Jpn. **56** 873 (1987).

[4] Y. Hasegawa and H. Fukuyama, J. Phys. Soc. Jpn. **56** 877 (1987).

[5] Y. Suzumura and H.J. Schulz, Phys. Rev. B **39** 11398 (1989).

[6] Y. Suzumura, J. Phys. Soc. Jpn **58** 2642 (1989).

[7] T. Moriya, J. Phys. Soc. Jpn. **18** 516 (1963).

[8] V. Ambegaokar and A. Griffin, Phys. Rev. **137** A1151 (1965).

[9] S. Schmitt-Rink, K. Miyake and C.M. Varma, Phys. Rev. Lett. **57** 2575 (1986).

[10] T. Tsuneto, Phys. Rev. **121** 402 (1961); L. Kadanoff and I. Falko, Phys. Rev. **136** A1170 (1964).

Some Recent Experiments on the Field Induced Spin Density Wave States in the Bechgaard Salts

P.M. Chaikin[1,2], *J.S. Brooks*[3], *S.T. Hannahs*[3], *W. Kang*[1], *G. Montambaux*[4], *and L.Y. Chiang*[2]

[1]Department of Physics, Princeton University, Princeton, NJ 08544, USA
[2]Exxon Research and Engineering Co., Route 22E, Annandale, NJ 08801, USA
[3]Department of Physics, Boston University, Boston, MA 02215, USA
[4]Université de Paris-Sud, F-91405 Orsay, France

Abstract The Field Induced Spin Density Wave (FISDW) transitions which are observed in the Bechgaard salts have provided a number of exciting surprises since their discovery. The theoretical model which has developed over the past several years has painted a fascinating picture: a cascade of transitions between different SDW semi-metal states with Quantum Hall Effect like transport properties leading finally to a high field SDW semiconducting phase. Extensive experiments on the ClO_4 salt show qualitative agreement but also substantial departures from the predictions of the standard model. New experiments on the PF_6 salt at a particular pressure show beautiful confirmation of virtually all of the standard model predictions, including the first unambiguous observation of the QHE in a bulk crystal and the presence of the long anticipated semiconducting state. The mysteries which remain include "fast" magneto-oscillations and why the standard model is not more universally obeyed.

1 Introduction

The Bechgaard salts have supplied condensed matter physics with some of the most interesting discoveries in the past decade[1,2,3]. The PF_6 salt was the first organic superconductor under pressure and at ambient pressure was the first material to show a clear metal-Spin Density Wave (SDW) insulator phase transition. However the most surprising and most interesting of the phenomena associated with these salts is their behavior in moderate to strong magnetic fields at low temperature. In this regime we find the first evidence of the Quantum Hall Effects (QHE) in a bulk material and the discovery of completely new effects, the magnetic Field Induced Spin Density Wave transitions, phenomena which test our basic understanding of some of the most subtle questions in condensed matter.

Because they were the first family of organic superconductors the Bechgaard salts were heavily studied in terms of their electronic properties. The band structure is well known both experimentally and theoretically (bandwidths are $t_a : t_b : t_c \approx $ 1eV: 0.1eV: .003eV). From these studies we know that the quasi-one dimensional Fermi surface consists of warped non-intersecting sheets and has no closed orbits in the highly conducting a-b plane. With only extended states resulting from the open orbits, the energy spectrum must be continuous and thus lacks the wealth of magneto-oscillatory phenomena associated with the discrete spectrum of localized closed orbit Landau levels.

The initial discovery of magneto-oscillations in the PF_6[4] salt under pressure were therefore tremendously puzzling, especially since they only occurred after a temperature dependent threshold field had been surpassed. The mystery grew deeper after Hall measurements[5] on the ClO_4 salt indicated that the oscillations actually corresponded to a set of phase transitions and that the Hall resistance appeared step like much as in the Quantum Hall Effect (QHE)[6]. Based on these early experiments theories were developed and have evolved to paint the beautiful and fascinating picture which we now know as the "standard model"[7].

2 The Standard Model

Consider the application of a magnetic field to the open orbit Fermi surface. The magnetic field has two effects. First it causes the electrons to move in trajectories which are extended in the open orbit direction but oscillate over a finite width in the perpendicular direction both in k space and in real space. The result is electronic motion which is increasingly more one dimensional as the field is increased. We know that a one dimensional metal is unstable at low temperature against a density wave distortion which opens a gap at the Fermi surface. Now the second effect of the magnetic field is to produce a magnetic length, $\lambda = hc/eHb$, the period of the oscillation of the electron along the extended orbit direction. The density wave of wavevector Q wants to mix states that differ by $2k_f$, the Fermi momentum, but the presence of the magnetic wavevector $G = 2\pi/\lambda$ causes mixing of states by $Q \pm nG$. The result is a series of gaps at $q \pm nG$ where $q = Q/2 \sim k_f$. The system gains energy by putting k_f and E_f in the largest of the gaps. The gaps oscillate and change relative size with magnetic field and the wavevector q jumps discontinuously to keep E_f in the largest gap. Thus intrinsic in the instability of the system is a series of transitions. Each subphase is characterized by the number of subbands separating q from $2k_f$. The predicted phase boundaries and gap structures associated with each phase are shown schematically in Fig. 1. The final high field state in this cascade of phase transtions is to the n=0 state where $2k_f = Q$, a single gap remains and the system is insulating. All preceding states are semimetals with no extended states at E_f and must therefore show the QHE[8]. To understand how we go from an open orbit metal to something which exhibits the QHE by applying a magnetic field it is instructive to look at the orbits in the spin density wave state.

We turn the problem around. Suppose we turn on the periodic potential $Q = 2k_f$ for the spin density wave before we turn on the magnetic field. (The lack of a gap across the entire Fermi surface means that the undistorted metal has lower energy than the SDW without the magnetic field.) This would give

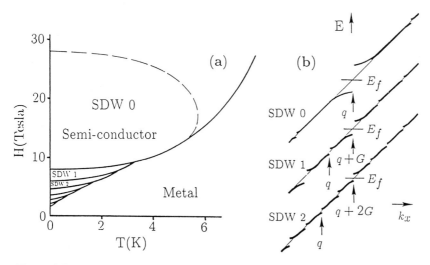

Fig. 1 a) Standard Model phase diagram (dashed line indicates experimentally observed reentrance in ClO_4). b) Dispersion relation corresponding to the three highest field phases.

a semi-metal. On the other hand the nesting of the Fermi surface does lead to closed orbits. When we turn on a magnetic field the semimetal splits up into discrete Landau levels. Now it is possible to have the Fermi level sit inside a gap. Note that neither the spin density wave potential alone, nor the magnetic field alone will give a gap at the Fermi surface for this particular bandstructure. However, the combination of the two conspire to give a series of gaps any one of which can be at the Fermi level. In order to complete this picture, it is necessary to realise that the magnetic energy, $\hbar\omega_c$, is comparable to and often much larger than the gap induced by the periodic spin density wave potential Δ_{SDW} and therefore we are in the limit of very strong magnetic breakdown. In that case electrons can tunnel from one Landau level to another in k space. This gives rise to Landau bands, which are just the bands shown in figure 1b. Since the Fermi level sits in the gap between the Landau bands, we are forced to the situation of having only filled Landau bands - the condition required for the QHE.

3 Comparison with Previous Experiments

After the initial experiments on PF_6 most of the work was done on ClO_4 because the effects can be observed at ambient pressure. A comparison between the experimental data on the ClO_4 salt and the "standard model" shows substantial agreement: – A cascade of field induced SDW transitions with a 2nd order transition from metal to SDW and 1st order transitions between SDW subphases. – A low temperature threshold field of magnitude $\sim t_c^2/E_f$. Plateaux in the Hall resistance ρ_{xy} identified with ratios 1:2:3. – General agreement of the calculated and measured phase boundary and many of the transport and thermodynamic properties.

However, there were also substantial disagreements: – The "Ribault" anomalies – intervening phases characterized by sign changes in ρ_{xy}. These phases have different temperature dependent transport, than the "standard model" phases and are considerably more sensitive to impurities, the degree of anion order and electric fields. – The "last transition" from the n=1 semimetal to the n=0 semiconducting phase was never seen. Rather, above the expected boundary in the phase diagram a different semi-metal phase was found. This phase is also highly nonlinear and yields a very stable hall plateau from 8 to 27 Tesla with a value that seems to indicate a 1/3 Fractional Quantum Hall Effect. – Possibly least expected was the reentrance of the "normal metal" phase at low temperature as the field was increased above 27 Tesla.

There were interesting observations that were seemingly outside the purview of the standard model: – "Fast" oscillations ubiquitous in transport and thermodynamic measurements. Although there is no general consensus on the mechanism for the fast oscillations[9] all models have oscillations appearing periodically with $t_b/eHv_f b$. Some models require anion ordering as in the ClO_4 salt. – A magnetoresistance in the "normal metal" phase that varied $\sim e^{-H/T}$ indicating localization in the magnetic field.

It was time to reinvestigate the PF_6 salt and take it to higher magnetic field. Although less well studied than $(TMTSF)_2ClO_4$, $(TMTSF)_2PF_6$ had a similar set of positives for the standard model. The negatives included Ribault-like anomalies with many more sign changes (at particular pressures), but the n=1 state had never been exceeded in field, there was no indication of the fast oscillations and PF_6 has no anion ordering transition.

4 New Results on $(TMTSF)_2PF_6$

The most striking result of the present study of $(TMTSF)_2PF_6$ is shown in figure 2. Here we show the Hall resistance versus field. The resemblance of this data to the quantum Hall effect is unmistakable. There are 8 transitions observable with 5 well defined plateaus in the ratio 5:4:3:2:1. Finally, above 18 Tesla there is a transition to a state where the Hall coefficient is not particularly well defined since the resistance is very rapidly increasing with field. The temperature dependence of the transport coefficients in this high field state indicate that it is the n=0 state long predicted but previously unseen. Considerably more data on this and other PF_6 samples are described in the literature[10], but the important results are an almost perfect confirmation of the standard model predictions as far as the transport and phase boundary are concerned. The longitudinal resistance ρ_{xx} shows peaks between the Hall steps but remains large where the steps are flat. ρ_{xx} should tend to zero on the steps, as T → 0, but we do not know what it should do during the first order transition between SDW

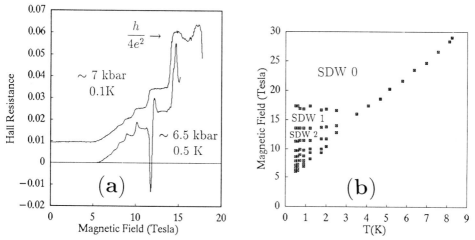

Fig. 2 a) Hall resistance for two samples of $(TMTSF)_2PF_6$. b) The phase diagram is essentially that predicted in Fig.1a.

subphases. (Unlike the usual QHE there is no need for the Fermi level to pass through extended states in the middle of a Landau band.)

Note that when the pressure is slightly changed we begin to see some of the more complex behavior associated with ClO_4 and previous PF_6 studies. With equal interest we note that fast oscillations were also observed in these samples (but are not shown here). This is a clear indication that the explanations involving the effect of the anion ordering gap are at best inappropriate for the present case.

5 Conclusions

The renewed studies of the PF_6 salt under pressure open some interesting avenues for understanding these marvelous phenomena. The remaining mystery at this pressure, the fast oscillations, is truly fundamental, it contradicts virtually our entire experience with magneto-oscillations. An interesting comparison will be the pressure dependence of the fast oscillation frequency (supposedly $\propto t_b$) and the FISDW frequency (supposedly $\propto$ the unnested part of t_b often taken as t_b^2/E_f). Also unresolved is the value of ρ_{xy} at the n=1 plateau. We find $\sim h/4e^2$ whereas theory suggests $h/2e^2$[9].

But the agreement we have found with the standard model in this case makes the disagreement with previous results more intriguing. It is clearly not just the anion ordering which makes the ClO_4 different, small pressure changes make for worse agreement for PF_6. Only further experiments will answer whether the fractional QHE and the reentrance are related to the anion order or will be seen in PF_6 at higher fields or different pressures. One of the interesting possibilities is that pressure changes the electronic interactions as well as the bandstructure. This could make the FQHE more or less stable. Traditionally, the FQHE is a property of partially filled Landau levels. One might expect the delocalization represented by the widths of the Landau bands in our case to compete with the Coulomb effects and destroy the FQHE for certain values of the parameters. Thus there remain important theoretical questions to be answered.

We would like to acknowledge support from NSF DMR 88-18510 and 88-22532, NATO travel grants 03351/88 and 0191/89 and useful discussions with R. Guertin and F. M. D. Haldane.

References

[1] D. Jérome, and H. J. Schultz, Adv. in Phys. **31**, 299 (1982).

[2] See review articles by M. Ribault, P. M. Chaikin, G. Montambaux, M. Héritier in NATO ASI Series, Low Dimensional Conductors and Superconductors, ed. by D. Jérome and L. G. Caron, (Plenum Press, New York, 1987).

[3] See articles by K. Yamaji, P. Lederer, F. Pesty, X. Yan, G. Montambaux, P. M. Chaikin, P. Garoche, J. Brooks, M. J. Naughton in proceedings ICSM .....Mol. Cryst. Liq. Cryst. **27**, (1989).

[4] J. F. Kwak et al., Phys. Rev. Lett. **46**, 1296 (1981) and Mol. Cryst. Liq. Cryst **79**, 121 (1981).

[5] M. Ribault et al., J. Phys. Lett. **44**, L-953 (1983), P. M. Chaikin et al., Phys. Rev. Lett. **51**, 2333 (1983).

[6] K. von Klitzing, G. Dorde and M. Pepper, Phys. Rev. Lett. **45**, 494, (1980).

[7] L. P. Gor'kov, A. G. Ledel, J. de Phys. Lett. **45**, L433 (1984), P. M. Chaikin, Phys. Rev. **B31**, 4770 (1985), G. Montambaux et al., J. Phys. Lett. **45**, L-533 (1984), M. Héritier, et al., J. Physique Lett. 45, (1984) L-943, K. Yamaji, J. Phys. Soc. Japan 54, 1034 (1985), M. Ya Azbel, et al., Phys. Lett. **A117**, 92 (1986), K. Maki, Phys. Rev. **B33**, 4826, (1986).

[8] D. Poilblanc et al., Phys. Rev. Lett. **58**, 270 (1987)

[9] K. Yamaji, Physica, **143B+C**, 439 (1986);J. Phys. Soc. Jpn., **55**, 1424 (1986), S. A. Brazovskii, and V. M. Yakovenko, Pis'ma Z.E.T.F. **43**, 102 (JETP Lett. **43**, 134) (1986), M. Ya. Azbel, P. M. Chaikin, Phys. Rev. Lett. **59**, 582 (1987), T. Osada, and N. Muira, Solid State Commun. **69**, 1169 (1989)

[10] S.T. Hannahs et al., Bull. Am. Phys. Soc. **34**, 3 (1989) and Phys. Rev. Lett. **63**, (1989) in press and J. R. Cooper et al. Phys. Rev. Lett. **63**, (1989) in press.

Phase Diagram of the Spin Density Waves Induced by the Magnetic Field in Organic Metals

F. Pesty, P. Garoche, and M. Héritier

Laboratoire de Physique des Solides, Associé au CNRS, U.P.S., Bât. 510, F-91405 Orsay Cedex, France

Abstract. We have studied the novel properties of the phase diagram of the field-induced spin density waves (FISDW). Most of our considerations will concern the $(TMTSF)_2ClO_4$ compound. The phase diagram has been built up from the measurement of specific heat and the magnetocaloric effect, thermodynamic quantities that are directly sensitive to second order phase transitions. A splitting of the lines is observed at low temperatures when the sample is prepared in a very well ordered state. A single transition line, which separates two adjacent parent SDW, gives rise to a pair of lines separating a new SDW from parent ones. The process iterates, leading to a treelike phase diagram. The lack of even harmonics supports a description in terms of fractional quantized nesting.

1. Introduction

The SDW induced by the magnetic field in the Bechgaard family of molecular compounds has been extensively studied both experimentally and theoretically [1], but the precise description of the phase diagram, and especially the investigation of the transition lines which separate SDW sub-phases is not straightforward. The first problem deals with the nature of the phase transitions: if the limits between sub-phases are temperature independent, a large change in the magnetization can be observed without any change in the entropy, and the sudden modification of the quantization condition will not be actually related to a real phase transition. We are thus left with a mere Landau diagram, as for conventional quantum oscillations such as the Shubnikov-de Haas effect. On the other hand, if we can follow the temperature evolution of the transition lines' slopes, we will be able to describe the coupling of the quantization condition together with the SDW order parameter. In this respect, the investigation of the electronic entropy is of first importance, but high resolution measurements are necessary because the opening of a Peierls-like gap at the metal-SDW transition exponentially reduces the electronic entropy.

For the $(TMTSF)_2ClO_4$ compound, a difficulty arises from the anion-ordering transition. Owing to a supercooling effect a partial disorder can be frozen in at low temperatures by increasing the cooling rate in the 30 to 10 kelvins range [2]. We are hence allowed to arbitrarily reduce the low-temperature electronic mean free path. This effect yields two drastic changes in the FISDW phase diagram. Firstly, the T_C value for the appearance of the FISDW is shifted toward low temperature as the cooling rate is increased [3], corresponding to a change in the coupling strength. Secondly, a qualitative modification of the diagram occurs when the mean free path reaches a value large enough for the electrons to explore the complex quantization condition. In that case, for a cooling rate of about one kelvin per hour, the Hall effect exhibits negative plateaus in between the integer transitions [4,5], new structures appear in the magnetization data [6], and magnetocaloric measurements reveal the splitting of the transition lines leading to a treelike phase diagram [7]. We will study here this splitting process.

2. Simultaneous measurements of C_B and $(\partial M/\partial T)_B$

In order to construct a coherent phase diagram, the experiments on the $(TMTSF)_2ClO_4$ salt must be performed at the very same cooling rate. For this purpose a new experimental setup has been designed so as to measure simultaneously as functions of the magnetic field the specific heat, C_B, and the isothermal coefficient of the magnetization, $\alpha_T = (\partial M/\partial T)_B$.

The two-milligram single crystal sample is first cooled down at a constant rate of 1.3 kelvins per hour between 30 and 10 kelvins. It is connected to a very well regulated thermal bath (a few microkelvins at $T_0 = 0.4$ kelvin), through a thermal conductance K_0. The small periodic power supplied

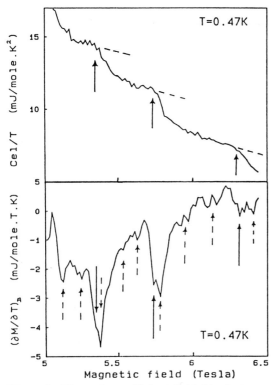

Figure 1: Electronic specific heat divided by the temperature (upper curve) and magnetocalo- rimetric effect (lower one), in molar units, for the (TMTSF)$_2$ClO$_4$ at 0.47K, and for a 1.3 K/h cooling rate.

to the sample produces a $\delta T=1$ mK thermal oscillation around the mean temperature <T>. The temperature is then recorded as a function of the magnetic field, swept at a constant rate. The temperature can be expressed according to this expression:

$$T(t) = (T_0 + P_0/K_0) + \Delta T(t) + \delta T. \cos (\omega t + \varphi) ,$$

where the first term represents the constant part of <T>, fixed by the mean power P_0. The second term, $\Delta T(t)$, arises from the magnetocaloric effect. It is related for a reversible process, as is the case here, to the field evolution of the entropy $(\partial S/\partial B)_T$. Finally the cosine term is inversely proportional to the specific heat [8], $C_B=T(\partial S/\partial T)_B$. As the field is varied, the temperature is sampled in phase with the thermal excitation, and a digital treatment is used to extract the zero and the ω Fourier components.

Figure 1 displays the result at T=0.47K, for a cooling rate of 1.3 K/h and for a field increased at a 3 mT/sec rate. The upper curve represents the electronic specific heat divided by the temperature,Cel/T, whereas the lower one shows the corresponding magnetocaloric effect. These two thermodynamic quantities clearly give complementary information on the new structure of the phase diagram. Since Cel/T is the first derivative of the electronic entropy with respect to the temperature, it describes the thermal filling of energy levels. In particular, its exponential decay is to be related with the gap opening at the metal-FISDW transition. For instance, the dashed lines on the upper curve mark the discontinuous changes in the gap value as the system goes through the integer SDW transitions (solid arrows).

On the other hand, the magnetocaloric effect can be expressed, through the use of Maxwell's relation, to the α_T coefficient. This quantity is thus directly sensitive to the magnetic degrees of freedom

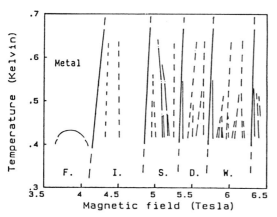

Figure 2: Treelike phase diagram built up from simultaneous measurements of C_B and $(\partial M/\partial T)_B$. Solid lines represent the integer transitions, split at low temperature. Dashed lines feature some of the fractional transition lines.

of the electronic system. It is well adapted to investigating quasi vertical transition lines in a fixed temperature experiment, whereas the specific heat is best suited to studying the horizontal parts of the metal-FISDW line. This is why the splitting process is much better described in the lower curve (dashed arrows): it gives rise to large split anomalies in α_T, but only to small bumps in the specific heat, located at some "noisy" field regions. Moreover it must be noted that the fractional transitions, at variance with the integer ones, do not lead to a significant change in the gap value, the slope of the specific heat in between an integer sub-phase remaining approximately constant.

The resulting phase diagram, as built up from the simultaneous measurement of C_B and α_T, is presented in Figure 2. The solid lines correspond to the phase transitions between integer SDWs. They are in good agreement at high temperatures with the main plateaus of the Hall effect [9] and with the paramagnetic jumps of the magnetization [10]. As the temperature is decreased, they separate into sub-lines, in an iterative splitting process. The dashed lines in Fig. 2 mark the fractional transitions. Most of the lines are too dense to be properly followed in temperature, but a three-stage process can be identified. A tentative indexation of these lines is difficult, firstly because of their finite slope, and secondly because temperature lowering, unlike conventional transitions, does not reduce the transition width but rather gives rise to a new set of lines. Nevertheless, it must be noted that the main sub-phases cannot be indexed as low even harmonics, but rather to third and maybe fifth ones. This emergence of odd harmonics can be explained in the framework of the fractional quantized nesting model [11].

According to the standard model of the FISDW [12-16], the integer cascade of phase transitions is believed to result from an interference effect between two competing periodicities of the electronic system in the longitudinal direction a. The first periodicity is $x_0=1/eBb$, where b is the interatomic distance in the transverse direction and B the magnetic induction. It arises from the orbital effect of the field on the phase of the electronic wave function. The second one is produced by the period of the self-consistent potential generated by the SDW. Because of nesting properties, the relevant periodicity is not $Q_{//}$, the longitudinal component of the SDW wave vector, but rather the deviation from perfect nesting: $q_{//}=Q_{//}-2k_F$, where k_F is the Fermi wave vector. These two periodicities open complex families of gaps in the quasi-particle spectrum, as independent Bloch electrons do in a magnetic field [17]. The SDW can adjust its periodicity in order to maintain its Fermi level in between one of these gaps and the nesting vector is quantized: $q_{//}=n/x_0$, where n is an integer number.

In fact, because the real space length $L=2\pi/q_{//}$ is not commensurate with the lattice period a, we cannot limit our analysis to the first Brillouin zone, and a multi-zone model of the nesting quantization has been developed, taking into account umklapp processes [11,18]. As is clearly indicated by our specific heat measurements along the metal-FISDW transition line [19], the weak coupling limit $\lambda\rho(E_F)\ll 1$ cannot be used for the description of this phase transition. As a consequence, the SDW gap width is of the same order as the distance between gaps: $\Delta_{SDW}\approx\hbar\omega_c$, and the periodic potential of the SDW can open a gap near the Fermi level, not only to first order as in the single zone model, but to any

89

odd order of perturbation, the even harmonics not being permitted since they are not related to electron-hole pairing. As a result, the SDW nesting vector has rational quantization conditions: $q_{//}=n/mx_0$ to the $2m^{th}$ order of perturbation, where n is an integer and m an odd number, in good agreement with present data.

References

1. For a review of experimental and theoretical results on the Bechgaard salts, see: Low Dimensional Conductors and Superconductors, Eds D. Jérome and L.G. Caron (NATO ASI Plenum Press) 155 (1986); see also: Proc. of ICSM'88, Santa Fe, Syn. Metals 27-29 (1988-89) and references therein
2. P. Garoche, R. Brusetti and K. Bechgaard, Phys. Rev. Lett. 49 (1982) 1346
3. F. Pesty and P. Garoche, to be published in Lower-Dimensional Systems and Molecular Devices, Eds R.M. Metzger, P. Day and G. Papavassiliou (NATO-ASI Plenum Press) (1989)
4. M. Ribault, Mol. Cryst. Liq. Cryst 119 (1985) 91
5. B. Piveteau, L. Brossard, F. Creuzet, D. Jérome, R.C. Lacoe, A. Moradpour and M. Ribault, J. Phys. C19 (1986) 4483
6. P.M Chaikin, J.S. Brooks, R.V. Chamberlin, L.Y. Chiang, D.P. Goshorn, D.C. Johnston, M.J. Naughton and X. Yan, Physica 143B (1986) 383
7. G. Faini, F. Pesty and P. Garoche, J. Phys. (Paris) 49 (1988) C8-807
8. P.F. Sullivan and G. Seidel, Phys. Rev. 173 (1968) 679; F. Pesty and P. Garoche, to be published
9. M. Ribault, D. Jérome, J. Tuchendler, C. Weyl and K. Bechgaard, J. Phys. (Paris) Lett. 44 (1983) L953
10. M.J. Naughton, J.S. Brooks, L.Y. Chiang, R.V. Chamberlin and P.M. Chaikin, Phys. Rev. Lett. 55 (1985) 969
11. M. Héritier, in Low Dimensional Conductors and Superconductors, Eds D. Jérome and L.G. Caron (NATO ASI Plenum Press) 155 (1986) 243
12. L. P. Gor'kov and Lebed', J. Phys. (Paris) Lett. 45 (1984) L433
13. M. Héritier, G. Montambaux and P. Lederer, J. Phys. (Paris) Lett. 45 (1984) L943, and ibid 46 (1985) L831
14. K. Yamaji, Syn. Metals 13 (1986) 29
15. M.Ya. Azbel, P. Bak and P.M. Chaikin, Phys. Rev. A34 (1986) 1392
16. A. Virosztek, L. Chen and K. Maki, Phys. Rev. B34 (1986) 3371
17. D.R. Hofstadter, Phys. Rev. B14 (1976) 2239
18. M. Héritier, F. Pesty and P. Garoche, to be published in New Trends in Magnetism, Eds S. Rezende and M. Cotinhio Filho (World Scientific, Singapore, 1989)
19. F. Pesty, P. Garoche and M. Héritier, Proc. of the MRS 1989 Fall Meeting, Boston, USA (Nov. 1989)

Spin Density Wave and Field Induced Spin Density Wave Transport

K. Maki

Department of Physics, University of Southern California,
Los Angeles, CA 90089, USA

Abstract. The spin density wave (SDW) and the field induced spin density (FISDW) in (TMTSF)$_2$ salts and possibly in (DMET)$_2$ salts will provide a new class of systems which exhibit the Fröhlich conduction. We summarize some theoretical results on electric conductivity and magnetotransport in SDWs and FISDWs.

1. Introduction

It is now well-established that both the pressure dependence of the SDW transition temperature of a Bechgaard salt like (TMTSF)$_2$PF$_6$ and the appearance of a series of FISDWs in strong magnetic fields for another salt (TMTSF)$_2$ClO$_4$ are described in terms of anisotropic Hubbard model first introduced by Yamaji [1]. Although the high field behavior of (TMTSF)$_2$ClO$_4$ appears to deviate significantly from the prediction of the simple model, more recent experiments on (TMTSF)$_2$PF$_6$ under high pressure appear to confirm the theoretical prediction.

Let us consider a Hamiltonian given by

$$H = \sum_{p\sigma} \varepsilon(p) C_{p\sigma}^+ C_{p\sigma} + U \sum_q n_{\uparrow q} n_{\downarrow -q} \tag{1}$$

with

$$\varepsilon(p) = -2t_a \cos ap_1 - 2t_b \cos bp_2 - 2t_c \cos cp_3 - \mu$$

$$\approx v(p_1 - p_F) - 2t_b \cos bp_2 - \varepsilon_0 \cos 2bp_2 \tag{2}$$

and

$$\varepsilon_0 = -1/4 \, t_b^2 \cos ap_F / t_a \sin^2 ap_F \tag{3}$$

and μ is the chemical potential and we take

$$t_a / t_b / t_c \sim 10 / 1 / 0.03 \tag{4}$$

as assumed usually for Bechgaard salts.

Here we neglect the t_c term. Further we have traded the second order term in $(p_1 - p_F)$ with the ε_0 term [2,3]. We note that the quasi-two dimensionality of the system appears only through the ε_0 term. Further the ε_0 term breaks the perfect nesting associated with the nesting vector $Q = (2p_F, \pi/b, \pi/c)$. Then the pressure dependence of the SDW transition temperature is understood as due to the increase in ϵ_0 under pressure. Perhaps the effect of the two dimensionality is most readily seen in the electron tunneling. Unlike the one dimensional model, the peaks in the density of states appear at $E = \pm (\Delta + \epsilon_0)$ where Δ is the order parameter [2,4] and it is symmetric at $E = 0$; $N(-E) = N(E)$. In a FISDW, on the other hand, the density of states develops a series of energy gaps [5,6].

2. Phason Dynamics

The order parameter in a SDW or in a FISDW is given by

$$<\vec{S}(x)> = 4 U^{-1} \Delta(T) \hat{n} \cos(\vec{Q}.\vec{X} + \phi(x)) \tag{5}$$

where $\hat{n}$ is a unit vector and $\phi(x)$ is the phase of the order parameter; the rotation of $\hat{n}$ gives rise to spin wave while time dependent $\phi(x)$ describes the sliding SDW. The spatio-temporal variation of $\phi(x)$ is described by the following Hamiltonian density [7]

$$H_\phi = 1/4 \ N_0 \ f \ [(\frac{\partial \phi}{\partial t})^2 + v^2(\frac{\partial \phi}{\partial x})^2 + v_2^2(\frac{\partial \phi}{\partial y})^2 + v_3^2(\frac{\partial \phi}{\partial z})^2]$$

$$- enf \ Q^{-1} \ \phi \ E + V_{pin}(\phi) \tag{6}$$

with

$$V_{pin}(\phi) = - (\pi/2 \ N_0 V_2)^2 \ \Delta(T) \ \tanh(\Delta(T)/2T) \times$$

$$\sum_i \cos[2(\vec{Q}.\vec{x} + \phi(x))] \ \delta(\vec{x} - \vec{x}_i) \tag{7}$$

where N_0 is the electron density of states, v, v_1, and v_3 are the Fermi velocities in three directions, V_2 is the Fourier transform of the impurity potential with $q = Q$ and f is a complicated function of ω and q the frequency and the wave vector associated with $\phi(x)$. Also slow spatio-temporal variation of $\phi(x)$ gives rise to the electric charge and current

$$\rho_{SDW} = - en \ f \ Q^{-1} \ \partial \phi/\partial x \tag{8}$$

and

$$J_{SDW} = en \ f \ Q^{-1} \ \partial \phi/\partial t \ , \tag{9}$$

which satisfy the charge conservation

$$\partial/\partial t \ \rho_{SDW} + \partial/\partial x \ J_{SDW} = 0 \ . \tag{10}$$

When $\omega, vq \ll \Delta_0$, the condensate density f takes two particular values:

$$f = \begin{cases} f_0 & \text{for } \omega > vq \\ f_1 & \text{for } \omega < vq \end{cases} \tag{11}$$

We note that conventionally f_O and f_1 are used in Eqs. (9) and (8) respectively. However, the assignment is not valid in the general context to be discussed here. In general, one of f_O or f_1 should use all of Eqs. (6), (8) and (9) depending on whether $\omega/vq \gtrless 1$. In this consideration, the spatial variation in $\phi(x)$ is most crucial. In a pinned SDW $\phi(x)$ is distorted with length scale L the Fukuyama-Lee-Rice length [8]. Then in the circumstance with $\omega < \omega_C = vL^{-1}$, $f = f_1$ has to be used. If we take $L = 10^{-2}$cm for example, we will have $\omega_C = 1$ GHz. Therefore in the analysis of the dc experiment $f = f_1$ has to be used as long as the narrow band noise frequency in the nonohmic regime does not exceed ω_C. On the other hand in the microwave experiment $f = f_O$ has to be used describing the phason resonance, which appears to take place around $\omega \sim 10$GHz [9].

3. Electric Conductivity

The crucial parameters in the electric conductivity are the threshold electric field E_T in the dc experiment and the pinning frequency in the microwave experiment [10]. We give here only the threshold electric field in a SDW [11]. It is important to distinguish the strong pinning limit and the weak pinning limit.

In the strong pinning limit we obtain

$$E_T^S(0) = (Q/e)(n_i/n)(\pi N_0 V_2)^2 \Delta(o) \tag{8}$$

and

$$E_T^S(T)/E_T^S(o) = (\Delta(T)/\Delta(o))\tanh(\Delta(T)/2T) f_1^{-1} \tag{9}$$

while in the weak pinning limit

$$E_T^W(o) = (4-D)/4D \, (Q/e)(n_i/n)^{2/4-D} (\bar{v}/\pi v)^{D/4-D} \times$$

$$[D(\pi/2 \, N_0 V_2)^2 \Delta(o)]^{4/D-4} \tag{10}$$

and

$$E_T^W(T)/E_T^W(o) = (E_T^S(T)/E_T^S(o))^{4/4-D} \tag{11}$$

where n_i and n are the impurity and the electron concentration, D is the dimensionality of the SDW and $\bar{v}$ is the geometric mean of the Fermi velocities involved. The temperature dependence of $E_T(T)/E_T(0)$ in Eqs (9) and (11) are shown in Fig. 1.

Recently, the Fröhlich conduction is observed in the SDW of $(TMTSF)_2NO_3$ by Tomić et al.[12]. The magnitude and the temperature dependence of $E_T(T)$ are consistent with the present prediction.

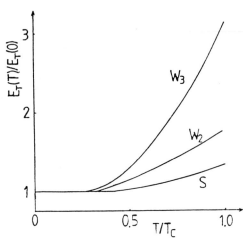

Fig. 1 The temperature dependence of the threshold field in the strong pinning limit and the weak pinning limit are shown.

4. Magnetotransport

We shall limit ourselves to the FISDW and to the pinned SDW. More general cases are analyzed elsewhere [13]. At low temperatures where the quasi-particle density is small, we have the resistivity tensor [13]

$$\rho_{xx} = (e^2 n/m \; K \; \cot^2(ap_F))^{-1} \; \Gamma_n'(1-f_c)^{-2} \tag{12}$$

$$\rho_{xy} = (e^2 n \; b/ma \; K \; \cot(ap_F))^{-1} \; \Omega \; (1-f_c)^{-1} \tag{13}$$

where $K = 1/2 \; [t_b/t_a \; \sin(ap_F)]^2$, $\Omega = vbeH$,

$$\Gamma_n' = 4\Gamma_2 \; f(\Delta) \; \{ 1 + (\pi\Delta^2/2\Gamma_2 T) \; f(-\Delta)\} \tag{14}$$

and

$$f_c = (2\Delta)^2 \; \Omega^{-1} \; [(2\Delta)^2 + \Omega^2]^{-1/2} \; \sinh^{-1}(\Omega/2\Delta) \tag{15}$$

and $f(E) = (1 + e^{BE})^{-1}$ is the Fermi distribution function.

ρ_{XY} develops a series of steps at each FISDW transition with $\rho_{XY} = h/4e^2 n$ per an a-b plane with n integer. This behavior may be qualitatively understood, since Eq(13) is rewritten

$$\rho_{xy} = (\pi c/8e^2) \; (H/H_0) \; (1-f_c)^{-1} \tag{16}$$

where the last FISDW (the N = 0 state) is entered around $H = H_0$ [6]. On the other hand, Eq (16) cannot describe rather extended plateaus in ρ_{XY} observed experimentally [14, 15].

5. Sound Propagation

The electromechanical effect (i.e. the softening of the crystal due to the depinning of the SDW by an external electric field) will give another signature of the Fröhlich conduction. We limit here to the simplest case $\omega \ll Dq^2$ where ω and q are the frequency and the wave vector of the sound wave and $D = v^2/2\Gamma_2$ is the diffusion constant of the electron system. In this limit, the sound velocity in a pinned SDW is given by [16]

$$C/C_o = 1 - \lambda(1 - f_1) \qquad (17)$$

where C_o is the bare sound velocity and λ is the dimensionless electron phason coupling constant. The sound velocity increases in a pinned SDW, since the phason cannot screen the ionic potential.

Such a behavior has been already observed in a SDW of $(TMTSF)_2PF_6$ by Chaikin et al.[17]. When the SDW is depinned by an electric field, the phason participates in the screening and we obtain

$$C/C_o = 1 - \lambda \ .$$

This result agrees with an earlier calculation by Nakane and Takada [18].

6. Concluding Remarks

We have shown that mean field treatment of anisotropic Hubbard model describes not only the phase diagram of Bechgaard salts under pressure and/or magnetic field but also a variety of transport properties. Though experimental studies of the Fröhlich conduction are just started, the results are already very encouraging. However, further work both experimental and theoretical is required to understand the magnetotransport in SDWs and FISDWs.

Acknowledgements. Work reported here is done with close collaboration with Atilla Virosztek, Liang Chen and Xiaozhou Huang. This work is supported by the National Science Foundation under grant No. DMR86-11829.

References:

1. K. Yamaji, J. Phys. Soc. Jpn 51 2787 (1982); 52 1361 (1983)
2. K. Yamaji, J. Phys. Soc. Jpn 54 1034 (1985); Synth. Met. 13 29 (1986)
3. H. Hasegawa and H. Fukuyama, J. Phys. Soc. Jpn 55 3978 (1986)
4. X.-Z. Huang and K. Maki, Phys. Rev. B 40 2575 (1989)
5. D. Poilblanc, M. Héritier, G. Montambaux, and P. Lederer, J. Phys. C 19 L 321 (1986)
6. A. Virosztek, L. Chen, and K. Maki, Phys. Rev. B 34 3371 (1986)
7. A. Virosztek and K. Maki, Phys. Rev. B 27 2028 (1988)

8. H. Fukuyama and P. A. Lee, Phys. Rev. B <u>17</u> 535 (1978); P. A. Lee and T. M. Rice, Phys. Rev. B <u>19</u> 3970 (1979)
9. T. W. Kim, J. P. Carini, G. Gruner, K. Maki and F. Wudl, Phys. Rev. B (submitted)
10. K. Maki and A. Virosztek, Phys. Rev. B <u>39</u> 2511 (1989)
11. K. Maki and A. Virosztek, Phys. Rev. B <u>39</u> 9640 (1989)
12. S. Tomic, J. R. Cooper, D. Jérome and K. Bechgaard, Phys. Rev. Lett. <u>62</u> 2446 (1989)
13. A. Virosztek and K. Maki, Phys. Rev. B <u>39</u> 616 (1989)
14. M. Ribault, Mol. Cryst. Liq. Cryst. <u>119</u> 91 (1985)
15. R. V. Chamberlin, M. J. Naughton, X. Yan, L. Y. Chiang, S. Y. Hsu and P. M. Chaikin, Phys. Rev. Lett. <u>60</u> 1189 (1988)
16. K. Maki and A. Virosztek, Phys. Rev. B <u>36</u> 2910 (1987)
17. P. M. Chaikin, T. Tiedje, and A. N. Bloch, Sol. State Commun. <u>41</u> 739 (1982)
18. Y. Nakane and S. Takada, J. Phys. Soc. Jpn. <u>54</u> 977 (1985)

Magnetothermodynamics and Magnetotransport in (TMTSF)$_2$ClO$_4$

G. Montambaux

Laboratoire de Physique des Solides, Université Paris XI, F-91405 Orsay, France

Abstract. Some particular aspects of the thermodynamics of the Field Induced Spin Density Wave phases of (TMTSF)$_2$ClO$_4$ are reviewed. A critical comparison between experiments and theory is given. New magnetotransport data in a tilted field open new questions.

1 Magnetothermodynamics

The Magnetic Field Induced Spin Density Wave (FISDW) phases of Bechgaard salts are well described within a weak coupling nesting model [1,2,3,4,5,6]. In a magnetic field H, the electronic motion acquires a new periodicity so that the electronic spectrum of the ordered phase is different from usual spin (or charge) density wave phases. Instead of having one gap at the Fermi level, it has a series of gaps which open at quantized value of the wave vector $\pm(k_F + nG/2)$, G being the wave vector which characterizes the new periodicity: $G = eHb/\hbar$. b is the distance between conducting chains. These gaps are labelled by a quantum number. They vary with the field and the largest one sits at the Fermi level. Successive subphases occur, characterized by different values of the quantum number. Using a simple form of the Fermi surface, the thermodynamics has been derived[3] and describes fairly well the essential features of the experimental data in low field. However some results are still a mystery such as the reentrances of the metallic phase in low field[7], the destruction of the SDW ordering in high field[8], the diamagnetic excursions of the magnetization[9], the specific heat results [10], phases with change in sign in the Hall effect[11] etc...

To describe these results in a coherent way, a simple thermodynamic description has been given recently[12]. It has been shown that, in a system with many gaps, the thermodynamics is strongly altered by the presence of the secondary gaps[3,12]. But a simple relation holds between the ground state energy E and the critical temperature T_c. It still has the BCS form $E = -0.236\gamma_e T_c^2$, even if E and T_c deviate from their BCS values[12]. γ_e is the linear coefficient of the specific heat in the normal phase. This result is powerful since it connects the structure of the transition line $T_c(H)$ to the variation with the field of the ground state properties such as the magnetization $M(H) = 0.236\gamma_e dT_c^2/dH$. It is not based on a specific model. As a result, the magnetothermodynamics of the FISDW's can be derived with only the $T_c(H)$ line as input and vice versa. For example, the reentrances of the metallic phase in low field have been shown to be related to the negative excursions of the magnetization and the destruction of the SDW phase in high field to be connected with the large diamagnetic variation of the magnetization. This has been used to support the evidence of the reentrance of the metallic phase in high field. It provides also interesting indications on the

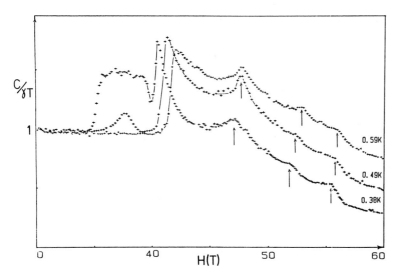

Fig.1. Specific heat $C_p(H)$ for various temperatures[7].

role of the cooling rate on the phase diagram. From the analysis of the magnetization data, the reentrance in high field has been predicted to depend on the cooling rate[12]. Preliminary data seem to confirm this dependence[13]. The critical role of the cooling rate and of the anion ordering is still a mystery. A proper theory of the reentrance should account for this dependence[14,15].

Here, I want to show that the specific heat results can also be described simply from the knowledge of the $T_c(H)$ line. In the past, the specific heat data have been used with success to get the precise structure of the phase diagram[7]. Typical curves $C_p(H)$ at constant temperatures are recalled in fig.1. The bumps signal the first order transitions. The dip around 4T on the $T = .38K$ curve signals the reentrance of the metallic phase between subphases. From these curves, the phase diagram of fig.2 has been derived. Here, I want to describe quantitatively the variation $C_p(H)$. This can be done using a simple BCS picture : the specific heat $C_p(T)$ for a fixed field H is assumed to have the BCS variation i.e. it is a unique function $C_p(T/T_c)$ where T_c is the critical temperature. The variations of C_p at fixed T simply result from the variation of T_c with the magnetic field. In the condensed phase, at fixed T, C_p is a decreasing function of T_c. As a result, the variations $C_p(H)$ at fixed T directly reflect the variations $T_c(H)$. Fig.3a shows the variation $C_p(H)$ obtained from this analysis. The agreement with fig.1 is satisfactory given the roughness of our approximation and could be easily improved. For example the jump at the metal-SDW transition is larger than found experimentally. But, it is known that taking into account the complete gap structure, this jump becomes smaller[16] and its evolution with the field agrees fairly well with the experiments at least in low field[10]. The overall variation $C_p(H)$ is well understood. The peaks which signal the transitions between subphases are not related to the first order character of the transition. They simply reflect the reentrances of the metallic phase. This is proved by the increase of these peaks when approaching $T_c(H)$.

It is also worth looking at the specific heat in higher field, directly from $T_c(H)$. The variation shown on fig.3b is predicted. Recent investigation of $C_p(H)$ in high field

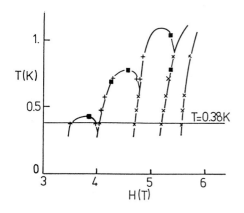

Fig.2. Phase diagram deduced from specific heat data[7].

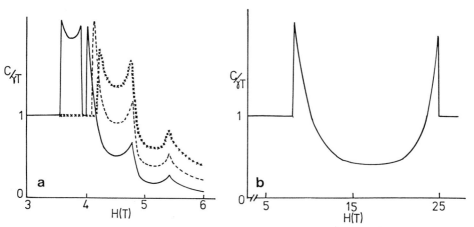

Fig.3. a: Specific heat $C_p(H)$ deduced from the $T_c(H)$ line for various temperatures.
Fig.3. b: Predicted specific heat in high field, as deduced from the $T_c(H)$ line. $T = 2.6K$.

agrees fairly well with this double peak structure, at least qualitatively[17]. This proves definitely that the transition in high field is a second order transition towards a metallic phase.

2 Magnetotransport

If the extraordinary rich magnetothermodynamics of $(TMTSF)_2ClO_4$ still exhibits unexplained and fascinating features, the transport in the FISDW phases is even less understood. The study of its angular dependence [18,19,20] was expected to provide clues for a better understanding. We have studied systematically the variation with the field of the magnetoresistance $\Delta\rho(H)$ for several inclinations of the field. Our results are twofold[21]. In the metallic phase, the magnetoresistance is always found to vary as $\Delta\rho \propto H^\alpha$ where the exponent α depends on the inclination of the field and varies between 1.8 and less than 2. In the FISDW region, the positions of the anomalies which signal the transition between subphases is found to vary following a $\cos\theta$ law, in agreement with previous data[20] and with the nesting model which gives a 2D description of the SDW ordering. But the amplitude of the anomalies depends on the angle θ. This shows that a complete description of the FISDW ordering should include 3D effects.

3 Conclusion

Although there is not yet a complete understanding of the magnetothermodynamics of $(TMTSF)_2ClO_4$, a simple analysis can describe coherently many of the experimental results and give predictions which have been verified experimentally such as the dependence of the high field phase diagram on the cooling rate and the behavior of the specific heat in high field. Many other results are yet unexplained. The magnetotransport is still a puzzle. Experiments in tilted field show that the situation may be even more complex than previously suspected.

The author acknowledges discussions with G. Boebinger, J. Brooks, P. Chaikin, N. Fortune, P. Garoche, M. Naughton, F. Pesty and X. Yan. Part of these discussions have been possible with NATO grant 19189. The magnetotransport experiments quoted in section 2 have been performed by G. Boebinger.

References

[1] L. P. Gor'kov and A. G. Lebed, J. Physique Lett. **45**, L 433 (1984)

[2] M. Héritier et al., J. Physique Lett. **45**, L 943 (1984); G. Montambaux et al., Phys. Rev. Lett. **55**, 2078 (1985)

[3] G. Montambaux and D. Poilblanc, Phys. Rev. **B37**, 1913 (1988)

[4] Y. Yamaji, Synth. Met. **13**, 29 (1986)

[5] M. Azbel, P. Bak and P. Chaikin, Phys. Rev. **A39**, 1392 (1986)

[6] A. Viroztek, L. Chen and K. Maki, Phys. Rev. **B34**, 3371 (1986)

[7] F. Pesty, P. Garoche and K. Bechgaard, Phys. Rev. Lett. **55**, 2495 (1985)

[8] M. J. Naughton et al., Phys. Rev. Lett.. **61**, 621 (1988)

[9] M. J. Naughton, Ph. D. Thesis, Boston University (1986) ; R. V. Chamberlin et al., Jpn J. Appl. **26**, 575 (1987)

[10] F. Pesty et al., J. Appl. Phys. **63**, 3061(1988)

[11] M. Ribault, Mol. Cryst. Liq. Cryst.,**119**, 91 (1985).

[12] G. Montambaux et al., Phys. Rev. **B39**, 885 (1989)

[13] P. M. Chaikin and M. J. Naughton, private communication

[14] V. Yakovenko, Phys. Rev. Lett. **61**,2276 (1988)

[15] M. Héritier et al., in New Trends in Magnetism, Recife (World Scientific, 1989)

[16] G. Montambaux, J. Phys. **C20**, L 327 (1987)

[17] N. Fortune et al., to be published
[18] R. Brusetti et al. J. Physique, **44**, C3-1055 (1983)
[19] K. Murata et al., Mol. Cryst. Liq. Cryst. **119**, 131 (1985)
[20] X. Yan et al., Solid State Comm. **66**,905 (1988)
[21] G. S. Boebinger et al., to be published

Quantized Hall Effect in Spin-Wave Phases of Two-Dimensional Conductors

M. Kohmoto

Institute for Solid State Physics, University of Tokyo,
7-22-1 Roppongi, Minato-ku, Tokyo 106, Japan

Abstract. A tight-binding Hamiltonian is used to study integer quantization of Hall conductivity σ_{xy} in spin-density-wave (SDW) ordered phases of two-dimensional conductors. A fractional number of magnetic flux quanta per unit cell creates energy gaps within tight-binding bands: σ_{xy} values of both signs arise naturally, even in the presence of only one type of charge carrier.

Recent Hall conductivity measurements for the Bechgaard salts $(TMTSF)_2X$, $X = ClO_4$ and PF_6, have revealed complex behavior reminiscent of the quantized Hall effect[1-9]. These salts order in triclinic crystals with the donor TMTSF molecules and acceptor X anions stacked in separate columns parallel to the **a**-axis. The resulting electron transfer integrals are in the ratio $t_a:t_b:t_c$ 300:30:1, so that the salts approximate anisotropic two-dimensional conductors. Identification of the experimental data as an instance of the integer quantized Hall effect is, however, complicated by several features. The salts display ordinary hole-like metallic conduction below a threshold magnetic field H_t of a few Tesla; for $H > H_t$, the transverse resistivity ρ_{xy} changes sign and increases rapidly in magnitude, exhibiting a series of plateaus which depend sensitively both on temperature and rate of cooling. For $X = ClO_4$ under ambient pressure, for example, initial experiments[2-4] revealed only positive (electron-like) plateaus in ρ_{xy}, but in a later experiment[5] where 70 hours were taken to cool the sample from 30 K to 4.2 K, *negative* plateaus in ρ_{xy} near $H = 5T$ and 6.7T appeared as well, accompanied by dips toward zero in the longitudinal resistivity ρ_{xy}. For $X = PF_6$ under a pressure of several kbar, many positive and negative plateaus in ρ_{xy} appear[7-9] at $H > H_t$, even for fairly rapid cooling rates[6], but this structure disappears quickly as the temperature is raised[8,9], being well-developed at $T = 0.15$ K but completely absent at $T \gtrsim 1.20$ K. Theses resistivity plateaus exhibit hysteretic behavior[6,8], and have been identified with a series of first-order transitions into spin-density-wave (SDW) ordered phases[1,3,10]: the fact that these SDW's involve anion ordering for $X = ClO_4$ but not for $X = PF_6$ may explain why such slow cooling rates are needed to equilibrate the ClO_4 salt.

Theoretical study of these systems to date[10-13] has focused on tight-binding models. For a fully quantum-mechanical treatment, which leads to the same secular equation as a Peierls-Onsager approach[14], consider a two-dimensional tight-binding Hamiltonian with only nearest-neighbor hopping. With **H** ∥ **c** and hence **A** = (0, Hx, 0), it may be written in momentum space as $\mathcal{H} = \int (\mathcal{H}_0 + \mathcal{H}_\Delta) d^2k/(2\pi)^2$. In the kinetic term

$$\mathcal{H}_0(\mathbf{k}) = \sum_{\sigma=\uparrow\downarrow} \{-2t_a(\cos k_x) c_\sigma^+(\mathbf{k}) c_\sigma(\mathbf{k})$$

$$-t_b [e^{-ik_y} c_\sigma^+(k_x + 2\pi\phi, k_y) c_\sigma(\mathbf{k}) + \text{h.c.}]\}, \qquad (1)$$

the phase change $2\pi\phi = eH/c\hbar$ accompanies y-translations in the presence of **A**: ϕ is the number of flux quanta per unit cell. (Here the electron charge is $-e < 0$, and lattice constants are set to $a = b = 1$ without qualitatively changing the physics.) The other term, $\mathcal{H}_\Delta$, embodies the Hubbard repulsion U. For SDW ordering treated in the mean-field approximation, $U\langle c^+_{r\downarrow} c_{r\uparrow}\rangle \equiv \Delta \cos \mathbf{Q} \cdot \mathbf{r}$, and thus

$$\mathcal{H}_\Delta(\mathbf{k}) = -\frac{1}{2}\Delta [c^+_\uparrow(\mathbf{k}+\mathbf{Q}) c_\downarrow(\mathbf{k}) + c^+_\downarrow(\mathbf{k}+\mathbf{Q}) c_\uparrow(\mathbf{k}) + \text{h.c.}]. \qquad (2)$$

(The Zeeman splitting term in $\mathcal{H}$ is ignored, since experiments show the appearance of the SDW phases to depend only on the effect of **H** on the *orbital* motion of the electrons[1].)

To explain why both positive and negative plateaus in σ_{xy} are seen, it has been argued that, under small changes in **Q**, the carriers in the pocket left over by the imperfect Fermi surface nesting may alternate between electrons and holes[8,11,13]. But a less structure-specific explanation is desirable for these sign changes, since they are common to both the ClO4 salt, which exhibits anion ordering, and the PF6 salt, which does not. For the case $\Delta \equiv 0$ (no SDW ordering), it is well-known that a *fractional* number of flux quanta per unit cell, ϕ, leads to gaps *within* a single tight-binding band[15]. Thouless *et al.* [16] proved that σ_{xy} must be quantized whenever the Fermi level lies within one of these gaps; as the Fermi level moves from one gap to another, values of σ_{xy} of both signs arise naturally.

The present work extends these results to systems with SDW ordering. As noted above, **Q** may be approximated by $(\pi/2, \pi)$. As long as **Q** is *commensurate* with respect to the reciprocal lattice, so that the system retains overall periodicity, the quantization of σ_{xy} must follow[17]. In the present case, the SDW amplitude Δ may be allowed to vary independently of magnetic field H (in response to e.g., changes in applied pressure). The value of σ_{xy} in any particular magnetic subband is a topological invariant of this system determined by the overall phase change in the electron wave function around an appropriate closed contour in the magnetic Brillouin zone[17].

With $\phi = p/q$ (p,q relatively prime integers), the computationally simplest cases to consider are q equal to a multiple of four. (Small multiples will be considered here even though, as noted above, $\phi \simeq 10^{-4}$ in experimental situations.) One can then work within a magnetic reduced Brillouin zone (RBZ) $0 \leq k_x < 2\pi/q$, $0 \leq k_y < \pi$, and express the eigenstates $|u(\mathbf{k})\rangle$ of $\mathcal{H}$ in terms of 4q components ψ_{nm}, φ_{nm} via

$$|u(\mathbf{k})\rangle = \sum_{n=0}^{q-1} \sum_{m=0}^{1} [\psi_{nm} c^+_\uparrow(k_x^{(n)}, k_y^{(m)}) + \varphi_{nm} c^+_\downarrow(k_x^{(n-q/4)}, k_y^{(m-1)})] |0\rangle, \quad (3)$$

where $k_x^{(n)} = k_x + 2\pi n/q$ and $k_y^{(m)} = k_y + m\pi$, with $\mathbf{k} \in$ RBZ. The Hall conductivity may be computed from the Kubo formula as in Section III of Ref. 17,

$$\sigma_{xy} = \frac{ie^2}{2\pi h} \sum_{\text{bands}} \int_{\text{RBZ}} d^2k \{ [\frac{\partial}{\partial k_x}\langle u|][\frac{\partial}{\partial k_y}|u\rangle] - [\frac{\partial}{\partial k_y}\langle u|][\frac{\partial}{\partial k_x}|u\rangle]\}, \quad (4)$$

where the sum is over all occupied subbands.

Expecting the behavior to be qualitatively independent of the value of t_b/t_a, as is true when $\Delta = 0$[16], let $t_a = t_b = 1$. Symmetries then reduce the problem of diagonalizing $\mathcal{H}$ to that of solving the secular equation

$$-[2 \cos k_x^{(n)} + \frac{\Delta}{2}] \chi_n - e^{-ik_y}\chi_{n-p} - e^{ik_y}\chi_{n+p} - \frac{\Delta}{2}\chi_{n+q/2} = E\chi_n, \quad (5)$$

for $n = 0, 1, \ldots, q-1$. In particular, when p is odd, $\mathcal{H}$ has four eigenvectors corresponding to each of the q solutions (E, $\{\chi_n\}$) of (5). Some algebra shows that Stokes theorem can then be used to evaluate (4) in terms of the change in the phase θ_{q-1} of χ_{q-1} along an edge of the RBZ, *except* that each zero of χ_0 in the RBZ is a singularity according to our global phase convention and so must be isolated[14,17]. This leaves the result

$$\sigma_{xy} = \frac{e^2}{\pi h} \sum_{\text{bands}} [\Delta \theta_{q-1} [k_x = \frac{2\pi}{q}, 0 \leq k_y < \pi] - \sum_i \Delta\theta_{q-1}(\partial A_i)], \quad (6)$$

where the first term denotes the change in θ_{q-1} as k_y is increased from 0 to π

along the right-hand edge of the RBZ, and each of the other terms is the phase change around a small contour A_i enclosing the ith zero of χ_0 in the RBZ.

Equation (6) has been used to compute the contribution to σ_{xy} from each of the q subbands of (5). Consider first the simplest case, $p/q = 1/4$. Starting from $\Delta = 0$, where $\sigma_{xy} = Ne^2/h$ with $N = 1$ (in agreement with the Diophantine equation of Thouless *et al.* [16]), σ_{xy} remains fixed at this value as Δ is increased, *except* at the special point $\Delta=2$, where zeros of χ_0 and χ_{q-1} are seen to coalesce, making $N = -1$. As Δ is increased beyond 2, these zeros again separate, and N returns to 1.

In conclusion, it is apparent that a fractional number of flux quanta per unit cell ϕ, and a SDW vector **Q** commensurate with respect to the underlying lattice, constitute general conditions under which both positive and negative quantized Hall conductivities can arise in systems with SDW ordering, without need of a special Fermi surface shape. Moreover, these conductivity values are sensitive to the SDW order parameter Δ even at fixed magnetic field, and so may be affected by other external variables like pressure or temperature, as indeed has been seen experimentally. Detailed analysis of experiments is however, complicated by the small values of ϕ attainable, which imply large q and hence the presence of many subbands. Additionally, a small degree of three-dimensional character may not only account for imperfect quantization of σ_{xy} and the failure of ρ_{xx} to vanish completely, but it may also either increase or decrease the number of subbands themselves[18].

This work is based on a collaboration with A.M. Szpilka.

References

1. For a review, see P.M. Chaikin, E.J. Mele, L.Y. Chiang, R.V. Chamberlin, M.J. Naughton, and J.S. Brooks, Synth. Met. **13**, 45 (1986).
2. P.M. Chaikin, M.-Y. Choi, J.F. Kwak, J.S. Brooks, K.P. Martin, M.J. Naughton, E.M. Engler, and R.L. Greence, Phys. Rev. Lett. **51**, 2333 (1983).
3. M. Ribault, D. Jérome, J. Tuchendler, C. Weyl and K. Bechgaard, J. Phys. (Paris) **44**, L953 (1983).
4. M. Ribault, J. Cooper, D. Jérome, D. Mailly, A. Moradpour, and K. Bechgaard, J. Phys. (Paris) **45**, L935 (1984).
5. M. Ribault, Mol. Cryst. Liq. Cryst. **119**, 91 (1985).
6. J.F. Kwak, J.E. Schirber, P.M. Chaikin, J.M. Williams, and H.-H. Wang, Mol. Cryst. Liq. Cryst. **125,** 375 (1985).
7. M. Ribault, F. Pesty, L. Brossard, B. Piveteau, P. Garoche, J. Cooper, S. Tomic, A. Moradpour, and K. Bechgaard, Physica (Utrecht) **143** B, 393 (1986).

8. L. Brossard, B. Piveteau, D. Jérome, A. Moradpour, and M. Ribault, Physica (Utrecht) **143** B, 406 (1986).
9. B. Piveteau, L. Brossard, F. Creuzet, D. Jérome, R.C. Lacoe, A. Moradpour, and M. Ribault, J. Phys. C**19**, 4483 (1986).
10. L.P. Gor'kov and A.G. Lebed', J. Phys. (Paris) **45**, L433 (1984).
11. M. Héritier, G. Montambaux, and P. Lederer, J. Phys. (Paris) **45**, L943 (1984).
12. K. Yamaji, J. Phys. Soc. Jpn. **54**, 1034 (1985); Synth. Met. **13**, 29 (1986).
13. D. Poilblanc, G. Montambaux, M. Héritier, and P. Lederer, Phys. Rev. Lett. **58**, 270 (1987).
14. M. Kohmoto, Phys. Rev. B**39**, 11943 (1989).
15. D.R. Hofstandter, Phys. Rev. B**14**, 2239 (1976)
16. D.J. Thouless, M. Kohmoto, M.P. Nightingale, and M. denNijis, Phys. Rev. Lett. **49**, 405 (1982).
17. M. Kohmoto, Ann. Phys. (NY) **160**, 343 (1985).
18. G. Montambaux and M. Kohmoto, "Quantized Hall Effect in Three Dimensions," ISSP Report Ser. A, No.2074 (1988).

^{1}H Spin-Lattice Relaxation in the SDW State of $(TMTSF)_2PF_6$ Under Pressure

T. Takahashi[1], T. Ohyama[1], T. Harada[1], K. Kanoda[1], K. Murata[2], and G. Saito[3]

[1]Department of Physics, Gakushuin University, Mejiro, Toshima-ku, Tokyo 171, Japan
[2]Electrotechnical Laboratory, Umezono 1-1-4, Tsukuba, Ibaraki 305, Japan
[3]Institute for Solid State Physics, University of Tokyo, Roppongi, Minato-ku, Tokyo 106, Japan

Abstract. The SDW state of $(TMTSF)_2PF_6$ has been studied by the measurements of ^{1}H nuclear spin-lattice relaxation rate T_1^{-1} and resistivity. Successive anomalies in T_1^{-1}, which may suggest the existence of transitions of magnetic nature, were observed well below the SDW transition temperature. In the resistivity measurement, slight breaks in the activation plot were found at the temperatures where the T_1^{-1} anomalies were observed. We suggest that the low-field SDW state is divided into three or more sub-phases.

$(TMTSF)_2PF_6$ has been considered as a typical material of the first generation of organic superconductors, the TMTSF family [1,2]. The ground state of this material at ambient pressure has been revealed to be an SDW state, which is driven by a Fermi surface instability, characteristic of low dimensional conductors [1]. While the essential features of the SDW transition are quite well understood by the nesting theory [3], there are several serious problems waiting for more sophisticated explanation. Several years ago, we found a clear anomaly in the temperature dependence of T_1^{-1} well below the SDW transition [4-6]. Under pressure, the T_1^{-1} anomaly becomes much pronounced and occurs at lower temperature. However, there seems no change in the static characteristics of the SDW around that temperature. The origin of this anomalous behavior is thus far unclear.

The purpose of the present work is to get further information to investigate the origin of the inner structure of the SDW state in this material. We have extended the measurements of T_1^{-1} below 1 K and carried out a precise measurement of resistivity below the SDW transition. We found anomalies in resistivity and relaxation rate, which may suggest that the low-field SDW state is divided into three or more sub-phases, as is the high-field SDW state [7].

The experimental results of T_1^{-1} presented in the previous papers [4-6] are summarized as follows:

i) In addition to the T_1^{-1}-enhancement (the first anomaly) at the SDW transition, another anomaly (the second anomaly) in ^{1}H-NMR T_1^{-1} was observed at lower temperatures (at 3.5 K at ambient pressure [4]): As temperature decreases, T_1^{-1} tends to saturate and then suddenly starts to decrease exponentially. The apparent activation energy was about 11 K. The temperature at which the anomaly was observed and the activation energy seem to be isotropic; at all measured field directions, no appreciable change in these parameters was observed within experimental errors.

ii) The T_1^{-1}-anomaly becomes much more pronounced under pressure [5]. The break-type anomaly at ambient pressure gradually changes to a peak under pressure. The 'transition temperature', T_a, was found to shift towards the lower temperature side, as the pressure increases. The coefficient was estimated as $dT_a/dP = -0.2$ K/kbar. The new phase boundary has been drawn on the well-known T-P phase diagram of this

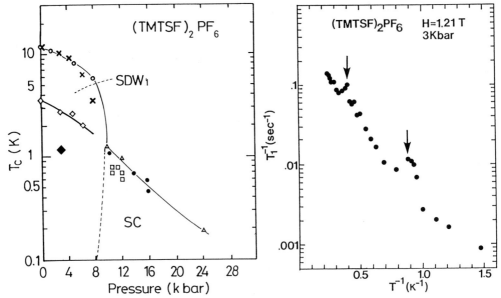

Fig. 1 The proposed phase diagram of $(TMTSF)_2PF_6$ by using the $^1H-T_1^{-1}$ data. Symbols, ×, ◇ and ◆, indicate the first, the second and the third anomalies, respectively (see text). The original T-P phase diagram was taken from ref.[1].

Fig. 2 Logarithmic plot of $^1H-T_1^{-1}$ vs the inverse of temperature, measured at 3 kbar. Arrows indicate the relaxation anomalies below the SDW transition.

material (Fig. 1). The activation energy below the transition was found to be insensitive to pressure.

iii) There was no appreciable change in the absorption line shape between above and below T_a [4,8]. Since the ^1H-NMR lineshape is very sensitive to the SDW parameters, the SDW wave number and the amplitude [8,9], this clearly indicates that the low temperature phase is also an SDW state with the same properties as the previous one.

The earlier relaxation measurements were restricted to the temperatures above 1.4 K. We have extended the same measurements down to 0.5 K. The results at 3 kbar are shown in Fig. 2. One can see a clear peak anomaly (the third anomaly !) in T_1^{-1} at 1.2 K, well below T_a observed before. This suggests that there is the third phase boundary; the SDW state may be divided into three or more sub-phases. The new point is plotted in Fig. 1. Below 1 K, the temperature dependence of T_1^{-1} becomes slower; the apparent activation energy is 2.6 K.

At this stage, we recognized that no precise measurement of resistivity well below the SDW transition had been published. In order to know whether there exists some anomaly in the resistivity at the temperatures where the NMR relaxation has the peaks, we have measured resistivity in that temperature region.

The results are shown in Fig. 3. Below the SDW transition, the resistivity at 1.5 kbar, e.g., increases exponentially, with an activation energy of 24.6 K, which becomes 17.6 K below about 3 K. A break in the logarithmic plots of the resistivity against 1/T can be

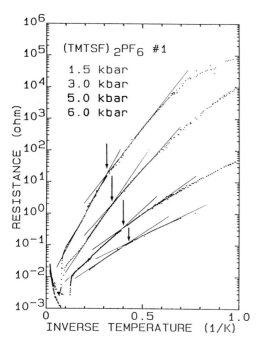

Fig. 3 Logarithmic plots of the resistivity against the inverse of temperature. Straight lines are guides to the eye. Arrows indicate the break points, which agree with the relaxation anomaly.

easily seen. The break around 3 K may be associated with the 'second transition' observed in the T_1^{-1} measurements. The change in the slope under pressure also seems to take place at the temperature of the relaxation anomaly at each pressure. Moreover, the resistivity tends to be saturated below 1 K. It is possible that the third anomaly at 1.2 K is related to this saturation of resistivity.

The question is whether the second and the third anomalies of ^{1}H-NMR relaxation correspond to real thermodynamic transitions or not. As in a liquid sample, a peak-like anomaly of T_1^{-1} appears when there exist thermal fluctuations of the local field seen by the nuclei and their correlation time agrees with the inverse of the Larmor frequency [10]. The SDW dynamics, e.g., sliding motions or spin dynamics, may be the origin of local field fluctuations, since the lattice dynamics has been completely frozen. Considering the sharpness of the peaks of T_1^{-1}, however, we are inclined to believe the existence of some kind of real transition of magnetic nature. Moreover, the double peak structure cannot be explained with simple local field fluctuations. The successive anomalies revealed in the present work remind us of a series of sub-phases of the field-induced SDW state, observed at high fields [7]. The situation is, of course, completely different from the present case.

The experimental fact which may be related to the present anomaly is the one reported by Ulmet et al. [11]; they found an oscillation of magnetoresistance in this material below 10 K, which suddenly vanished below about 4 K. While the external field was much higher than in the present work, we suppose that both have the same origin. They have suggested a transition accompanied by deformation of the Fermi surface. If it is the case, however, the nesting vector must shift drastically at the transition and we should have detected some change in the ^{1}H-NMR lineshape [8].

The 'second phase boundary' seems to tend continuously to the metal-superconducting one. It may suggest some relation between the superconducting instability and the new transition. However, the new anomalies were always observed in the SDW state and the field dependence seems to be very small, as just mentioned. Thus a direct relation between the present phase and the superconducting phase is not likely. Recently, it has been revealed that the superconductivity in $(TMTSF)_2X$ family has a gapless nature [12]. The anisotropic gap may be a consequence of non-local attractive force [13], and the possibility of the interaction mediated by spin fluctuations are discussed, as in heavy fermion system [14]. We wonder whether the non-local attractive force might modify the nature of the SDW ordering at low temperatures.

The origin of the new anomalies observed well below the SDW transition is still an open question. The sharp enhancement of the nuclear relaxation clearly indicates that there should remain some degrees of freedom responsible for the local-field dynamics even below the SDW ordering. This is an important problem for full understanding of the low temperature behaviors of the organic conductors. The SDW state in this material is not as simple as always believed.

References.

1. D. Jérome and H.J. Schulz: Adv. Phys. **31**, 299 (1982).
2. Low Dimensional Conductors and Superconductors, ed. D. Jérome and L.G. Caron (NATO ASI series, Plenum Press, 1987)
3. K. Yamaji: J. Phys. Soc. Jpn. **51**, 2787 (1982); Superconductivity in Magnetic and Excitonic Materials, eds. T. Matsubara and K. Kotani (Springer-Verlag, 1984) p. 149.
4. T. Takahashi, Y. Maniwa, H. Kawamura and G. Saito: Physica **143B**, 417 (1986); Synth. Met. **19**, 225 (1987); T. Takahashi: ref.[2], p.195.
5. H. Kawamura, T.Ohyama, Y. Maniwa, T. Takahashi, K. Murata, G, Saito, Jpn. J. Appl. Phys. **26**, Suppl. 26-3, 583 (1987).
6. T. Takahashi, Synth. Met. **27**, B397 (1988).
7. P.M. Chaikin, E.J. Mele, L.Y. Chiang, R.V. Chamberlin, M.J. Naughton and J.S. Brooks: Synth. Met. **13**, 45 (1986); earlier references therein.
8. T. Takahashi, H. Kawamura, T. Ohyama, Y. Maniwa, K. Murata and G. Saito, J. Phys. Soc. Jpn. **58**, 703 (1989).
9. T. Takahashi, Y. Maniwa, H. Kawamura and G. Saito: J. Phys. Soc. Jpn. **55**, 1364 (1986).
10. C.P. Slichter, Principles of Magnetic Resonance (Springer-Verlag, 1978) p.166.
11. J.P. Ulmet, P. Auban, A. Khmou, S. Askenazy and A. Moradpour: J. Physique Lett. **46**, L-535 (1985).
12. M. Takigawa, H. Yasuoka and G. Saito: J. Phys. Soc. Jpn. **56**, 873 (1987).
13. Y. Hasegawa and H. Fukuyama: J. Phys. Soc. Jpn. **56**, 877 (1987).
14. F.J. Ohkawa and H. Fukuyama: J. Phys. Soc. Jpn. **53**, 4344 (1984); J.E. Hirsch: Phys. Rev. **B31**, 4403 (1985); Phys. Rev. Lett. **54**, 1317 (1985).

Non-ohmic Electrical Transport in the Spin-Density Wave State of Organic Conductors

S. Tomić[1,2], *J.R. Cooper*[1,2], *W. Kang*[2], *and D. Jérome*[2]

[1]Institute of Physics of the University, P.O. Box 304, YU-41001 Zagreb, Yugoslavia
[2]Laboratoire de Physique des Solides, Université de Paris-Sud, F-91405 Orsay, France

Abstract. We have searched for electric-field-dependent conductivity in the spin-density wave (SDW) ground state of the organic conductors $(TMTSF)_2X$, $X=NO_3$ and PF_6. We have found that the non-ohmic conductivity appears above a finite threshold field (E_T) whose minimum values measured at 4.2K are 5-40 mV/cm. E_T is temperature independent below $T_c/2$ (where T_c is the transition temperature) and increases close to T_c. The excess conductivity is smaller in samples with a lower resistivity ratio. A sliding SDW mode, depinned under high enough electric fields, might be responsible for the observed electric-field-dependent response. We discuss our results in the framework of recent theories for a sliding SDW mode pinned to nonmagnetic impurities and show that they agree rather well with theoretically predicted behaviour. Finally, we compare electric-field-dependent transport in the SDW ground state with that observed in the charge-density wave (CDW) state, where CDW sliding is a well established phenomenon.

1. Introduction

Various highly anisotropic conductors, both inorganic and organic, are ideal systems for studying collective transport phenomena /1/. Depending on the material and applied pressure, there is usually a phase transition to a superconducting (SC), a charge-density wave (CDW), or a spin-density wave (SDW) ground state at low temperatures. A translational mode of the CDW ground state couples to an applied electric field and gives collective transport. The essential properties of the CDW current-carrying state are as follows: the dc electrical conductivity increases sharply above a finite threshold field (E_T), the conductivity is frequency dependent and the non-linear current-voltage characteristics are accompanied by narrow and broad band noise.

Theoretically, similar behaviour might be expected for a SDW state, because collective transport does not depend on the nature of the underlying interaction mechanism /2/. The quasi one-dimensional SDW model systems are some members of the $(TMTSF)_2X$ family in which the SDW nature of the ground state with a critical temperature of about 10K has been firmly established by various magnetic measurements /3/, /4/, /5/.

The purpose of this paper is to review and discuss recent experiments performed to look for one of the properties of a possible SDW current-carrying state: namely a dc electrical conductivity which increases above a finite threshold field /6/, /7/. We have investigated two materials: the NO_3 and the PF_6 compounds with SDW transition temperatures of 11 and 11.5K and SDW single-particle gaps of approximately 16 and 32K, respectively. As far as the frequency-dependent conductivity is concerned, the results obtained by

G.Grüner et al. /8/, clearly show the existence of a collective mode in the SDW state of the PF_6 compound with a pinning frequency of about 30GHz and with a relaxation time and effective mass similar to those of the metallic state. In addition, K.Nomura et al. /9/ very recently reported the first observation of narrow band noise in the SDW state of quenched ClO_4 crystals.

2. Experimental results

The electric-field-dependent conductivity observed in the NO_3 and PF_6 compound is shown in Fig.1. In the metallic state the conductivity stays constant in the whole field range measured (up to about 0.7V/cm). However, in the SDW state, the conductivity is constant until a threshold field is reached, above which the conductivity increases. Values of the threshold field measured at 4.2K are 40 and 7.5mV/cm for the NO_3 and PF_6 compound, respectively. The sharpness of the threshold field was checked by continuous current measurements (see insert of Fig.1.b.) and by dynamic resistance measurements (Fig.2.). The excess conductivity is smaller in samples with a lower resistivity ratio $\rho(RT)/\rho(min)$. In addition, a certain amount of impurities

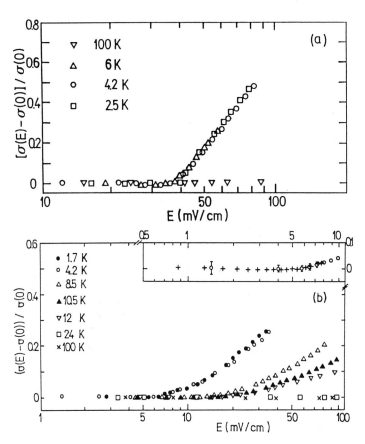

Fig.1. Non-ohmic conductivity $\sigma(E)-\sigma(E\to 0)/\sigma(E\to 0)$ versus logarithm of electric field (E) at various temperatures for (a) $(TMTSF)_2NO_3$ and (b) $(TMTSF)_2PF_6$.

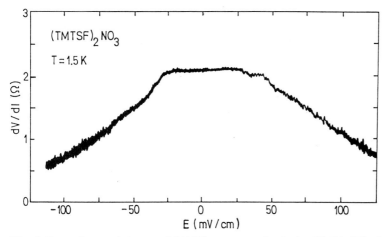

Fig.2. Dynamic resistance (dV/dI) versus electric field (E) for a $(TMTSF)_2NO_3$ crystal immersed in superfluid helium at 1.5K.

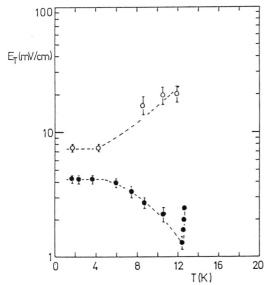

Fig.3. Threshold field (E_T) versus temperature (T) for $(TMTSF)_2PF_6$. Open and full circles for samples with painted and clamp contacts, respectively.

which is large enough to broaden the SDW transition, but does not affect T_c, strongly increases E_T giving a value as high as $E_T \simeq 140$ mV/cm at 1.7K for the PF_6 compound. Furthermore, for both NO_3 and PF_6 the value of the threshold field is temperature independent below $T_c/2$. For the latter we also established the overall temperature dependence of E_T as presented in Fig.3. Its value is constant in temperature until about 5K, but then it changes on further approaching T_c. The changes depend on the type of contacts used, for clamp contacts E_T increases only very close to T_c.

3. Discussion

The onset of non-ohmic conductivity at the three-dimensional SDW ordering temperature strongly suggests that the nonlinearity is associated with the establishment of a SDW indeed. It is difficult to explain the observed effects using models based on a single-particle picture like Zener breakdown and hot-electron effects /6/. Our results are reminiscent of those in the CDW systems where the nonlinearities have been attributed to the sliding CDW becoming depinned in high enough electric fields.

However, the threshold field does not seem to diverge at T_c, as for most CDW materials and in addition, the increase of E_T at low temperatures, which has been observed for most CDW materials is clearly absent for the SDW. Above $T_c/2$, for samples with painted contacts E_T shows a steady increase towards T_c: $E_T(T_c)/E_T(1.7K) \simeq 2.5$. Such a behaviour has been predicted by Maki and Virosztek /10/ in the framework of the mean-field model for a sliding SDW mode pinned to nonmagnetic impurities. The theoretically expected values for E_T rise are 1.33 and between 1.77 and 3.13 for the strong and weak pinning limits, respectively. In addition, the observed values of threshold fields are close to the ones theoretically expected. However, for samples with strain-free clamp contacts, E_T displays a minimum above $T_c/2$ before increasing very close to T_c. It is worth noting that for these samples the behaviour of the low field resistivity close to T_c is far from that expected in mean-field theory. A similar, extremely sharp SDW transition was also observed by NMR measurements /11/. This point remains to be clarified. Finally, the observation that the excess conduction is smaller in samples with a lower resistivity ratio and the threshold field is larger in samples with higher impurity concentrations indicate the important influence of the nonmagnetic pinning centers on the SDW sliding mechanism.

References

1. "Low-Dimensional Conductors and Superconductors" (edited by D.Jérome and L.G.Caron, NATO ASI Series B, Plenum, New York, 1987)
2. P.A.Lee, T.M.Rice and P.W.Anderson, Solid State Commun.14, 703 (1974).
3. K.Mortensen, Y.Tomkievicz, T.D.Schultz and E.M.Engler, Phys.Rev.Lett.46, 1234 (1981).
4. W.M.Walsh, F.Wudl, E.Aharon-Shalom, L.W.Rupp, J.M.Vandenberg, K.Anders and J.B.Torrance, Phys.Rev.Lett.49, 885 (1982).
5. P.Baillargeon, C.Bourbonnais, S.Tomić, P.Vaca and C.Coulon, Synth.Met.27, B83 (1988).
6. S.Tomić, J.R.Cooper, D.Jérome and K.Bechgaard, Phys.Rev.Lett.62, 462 (1989).
7. S.Tomić, W.Kang, J.R.Cooper and D.Jérome, submitted to Phys.Rev.B.
8. G.Grüner, Synth.Met.29, F453 (1989).
9. K.Nomura, T.Shimizu, K.Ichimura, T.Sambongi, M.Tokumoto, H.Anzai and N.Kinoshita, to be published in Solid State Commun. (1989).
10. K.Maki and A.Virosztek, Phys.Rev.B39, 9640 (1989).
11. T.Takahashi, H.Kawamura, T.Ohyama, Y.Maniwa, K.Murata and G.Saito, J.Phys. Soc.Jpn.58, 703 (1989).

Part IV

**BEDT-TTF Family:
Superconductivity**

Superconductivity in BEDT-TTF Based Organic Metals: An Overview

M. Tokumoto

Electrotechnical Laboratory, Tsukuba, Ibaraki 305, Japan

Abstract. Superconducting and normal state properties characteristic of the BEDT-TTF based organic metals, including β-(BEDT-TTF)$_2$X and κ-(BEDT-TTF)$_2$Cu(NCS)$_2$, are reviewed. In contrast to the quasi one-dimensional TMTSF based superconductors, the two-dimensional nature of the electronic structure of BEDT-TTF salts has produced a wealth of ambient-pressure superconductors. Out of a systematic study of the β-(BEDT-TTF)$_2$X salts with the superconducting transition temperature (Tc) varying from 8 K to below 1 K, empirical rules for various factors governing Tc in this class of organic metals have been compiled. The two-dimensionality has also brought about recent progress in the field of the Fermi surface study (Fermiology) of organic synthetic metals through observations of the Shubnikov-de Haas effect, the de Haas-van Alphen effect and the oscillatory angular dependence of magnetoresistance as a new phenomenon. Construction of a reliable band structure consistent with the Fermi surface is expected to provide a solid foundation for chemical design of organic metals and superconductors.

1. Introduction

In the history of organic metals and superconductors, "*dimensionality*" or "*anisotropy*" has played a key role. The first organic metal TTF-TCNQ and its derivatives turned out to be a prototype to exhibit physics involved in one-dimensional conductors, such as the Peierls transition, charge density wave, Kohn anomaly, etc. Although these organic metals provided interesting physics, the metal-insulator transition inherent to one-dimensional electronic systems was a hindrance to the realization of superconductivity. In order to realize organic metals which do not undergo metal-insulator transitions and stay metallic down to low temperatures, a search for higher dimensionality was made and led to a new family of organic metals based on TMTSF. (TMTSF)$_2$PF$_6$, which shows a metal-insulator transition at about 12 K under ambient pressure, was found to stay metallic down to low temperatures with application of pressure and to exhibit superconductivity at about 1 K — the first organic substance to do so. As a quasi-one-dimensional conductor, (TMTSF)$_2$X was found to exhibit a variety of physical phenomena, including superconductivity, spin density waves (SDWs), anion ordering and field induced SDWs, and the related phase diagram including a reentrant metallic state.

A further step into higher dimensionality led us to a whole variety of organic conductors based on BEDT-TTF. Organic metals based on BEDT-TTF were found to show a wealth of structural modifications or morphology. In contrast to the quasi-one-dimensional TMTSF based superconductors, the two-dimensional nature of the electronic structure of BEDT-TTF salts has produced a wealth of ambient-pressure superconductors. The two-dimensionality has also brought about recent progress in the field of Fermi surface study (Fermiology) of these organic synthetic metals through observations of the Shubnikov-de Haas effect, the de Haas-van Alphen effect and the oscillatory angular dependence of magnetoresistance as a new phenomenon. The present status of BEDT-TTF based organic superconductors is summarized in Table I.

Table I. BEDT-TTF based organic superconductors.

Formula	Phase	Operation	P_c(kbar)	T_c(K)
(BEDT-TTF)$_2$ReO$_4$			4	2
(BEDT-TTF)$_2$I$_3$	α	I$_2$ Doped	0	3.3
(BEDT-TTF)$_2$I$_3$	β$_L$		0	1.5
(BEDT-TTF)$_2$I$_3$	β$_L$	Annealed	0	2
(BEDT-TTF)$_2$I$_3$	β$_H$		0.5	8.1
(BEDT-TTF)$_2$I$_3$	α$_t$	Tempered	0	8.1
(BEDT-TTF)$_2$IBr$_2$	β		0	2.2-3
(BEDT-TTF)$_2$AuI$_2$	β		0	3.4-5
(BEDT-TTF)$_2$(I$_3$)$_{2.5}$	γ		0	2.5
(BEDT-TTF)$_2$I$_3$	θ		0	3.6
(BEDT-TTF)$_2$I$_3$	κ		0	3.6
(BEDT-TTF)$_4$Hg$_{2.89}$Br$_8$			0	4
(BEDT-TTF)$_4$Hg$_{2.89}$Cl$_8$			>12	1.8
(BEDT-TTF)$_3$Cl$_2$(H$_2$O)$_2$			16	2
(BEDT-TTF)$_2$Cu(NCS)$_2$	κ		0	10.4

This paper is intended to give an overview on the structural aspects and superconducting properties characteristic of the BEDT-TTF based organic metals, with particular emphasis on the two most extensively studied systems, i.e. β-(BEDT-TTF)$_2$X and κ-(BEDT-TTF)$_2$Cu(NCS)$_2$.

2. Structural Aspects

It is to be noted that both crystal and electronic structures of charge transfer salts of BEDT-TTF are closely related to the unique structure of the BEDT-TTF molecule itself.
  The structure of the BEDT-TTF molecule is nearly planar for the central fulvalene portion of the molecule, with large deviations from planarity for the terminal ethylene groups. It is this steric effect of the ethylene groups that prevents infinite face-to-face stacking of the molecules. Consequently the side-by-side interaction is very important and comparable to the face-to-face interaction in BEDT-TTF salts. The intrastack S-S distances, which are often larger than their van der Waals radii sum (3.60Å), suggests that the BEDT-TTF molecules may be considered to be loosely connected along the columns. On the other hand, short interstack S-S contacts which are often significantly less than 3.60Å are observed between adjacent columns. Thus the crystal packing of the BEDT-TTF molecules in BEDT-TTF based superconductors is in sharp contrast to the crystal packing of the quasi-one-dimensional (TMTSF)$_2$X, where the intrastack interactions are very significant. This difference of the packing between the two radical cation salts is reflected in the electronic structures of these salts.
  This loose intrastack coupling gives a variety of morphology in BEDT-TTF salts. We can see some typical examples of packing patterns of donor molecules within the conducting plane, i.e. β, θ and κ, from a variety of (BEDT-TTF)$_2$I$_3$ salts. As we go from β to θ and κ, the concept of stacking, which is clear in β-type, becomes less important in θ-type and eventually loses its meaning in the κ-type molecular arrangement. An extraordinary example of a structural property related to molecular packing is the transformation of the α-phase of (BEDT-TTF)$_2$I$_3$ to the β-phase with Tc=6-7 K by heating above 70°C [1]. Just imagine how a rearrangement of molecular packing from α-type, which is similar to θ-type, to β-type occurs! The resulting phase caused by thermal tempering, called α$_t$, was found to show a bulk superconductivity at 8.1 K [2].

A similar but *reversible* transformation was found in (BEDT-TTF)$_2$IBr$_2$, where α' to β transformation occurs by heating up to T=153-155 K, whereas the β to α' transformation occurs by application of pressure above 18 kbar [3].

As a consequence of the highly two-dimensional molecular packing within the conducting donor plane, the electronic structure of BEDT-TTF salts is highly two-dimensional and shows a small anisotropy within the conducting plane. This feature gives us a number of organic metals having a 2-D electronic structure with a *closed orbit*. Another feature inherent to the crystal structure is that it has a layered structure consisting of alternating conducting plane of donors and semi-insulating anion sheets. This out-of-plane anisotropy results in a quasi-two-dimensional electronic structure consisting of a Fermi surface which is a *"slightly warped cylinder"*. Many interesting properties closely related to this specific electronic structure, such as a giant quantum oscillations [4] and a new oscillatory angular dependence of magnetoresistance [5-7] are being discovered.

3. Superconducting Properties

Two outstanding features characteristic of the BEDT-TTF based organic superconductors, i.e. ambient-pressure superconductivity and relatively high-Tc, have made it possible to study their superconducting properties extensively and quantitatively.

3.1 Effect of Defects, Disorder and Impurities

Superconductivity of BEDT-TTF based organic metals was found to be sensitive to defects, disorder and impurities. In quasi-one-dimensional metals, such as (TMTSF)$_2$X, the conduction path of electrons along the stacking donors can be seriously disturbed by the presence of defects or disorder. However, in two-dimensional systems like BEDT-TTF salts, the electronic conduction path forms a two-dimensional network so that the presence of point defects cannot have a serious effect on the electrical conduction. Therefore it is not obvious why the superconductivity of BEDT-TTF based organic metals is so sensitive to the presence of *non-magnetic* impurities and defects.

In the following, we show some typical examples which exhibit the effect of defects, disorder and impurities on the superconducting properties in BEDT-TTF based organic metals. They can be classified into different grades, from unknown factors such as in θ-(BEDT-TTF)$_2$I$_3$ to well-characterized ones as in the case of β-(BEDT-TTF)$_2$ trihalide mixed crystals.

Superconductivity in the θ-(BEDT-TTF)$_2$I$_3$ salt is reported to show a partial superconducting transition by resistance measurements. Some of the crystals show a complete resistive transition at 3.6 K, while others show a partial transition or no transition at all [8]. Recently a bulk superconductivity in θ-(BEDT-TTF)$_2$I$_3$ was observed by magnetization and Meissner effect measurements, while superconducting volume fractions of some samples were significantly suppressed [9]. The origin of the difference is unknown. Another example of unknown factors which suppress the superconductivity is a disappearance of superconductivity in polycrystalline compressed samples [10].

Some of the moderately known factors are seen in the β-(BEDT-TTF)$_2$I$_3$ salt. Incommensurate lattice modulation [11], which appears below 175 K at ambient pressure, is considered to lower the Tc from 8 K to 1-1.5 K. Another important feature in β-(BEDT-TTF)$_2$I$_3$ is the disordered ethylene group, designated as A-type or B-type[12] (staggered or eclipsed [13]) at one end of the molecule and the ordered group at the opposite end.

Recently, annealing at about 110 K was found to result in a change of the incommensurate superstructure [14]. It was also found that a new superconducting state with Tc=2 K appears as a result of annealing [15], although its origin or the structural difference responsible for the change of Tc has not been identified yet.

An example of well-characterized factors would be the substitution of anions in β-(BEDT-TTF)$_2$X. Complete substitution of IBr$_2$ and AuI$_2$ for I$_3$ suppressed the incommensurate superstructure and realized the high-Tc state in β-(BEDT-TTF)$_2$X. However, replacement of I$_3$ with asymmetric anion I$_2$Br did not give superconductivity although the "lattice pressure" model[16] predicted higher Tc and the incommensurate superstructure was missing. Partial substitution of anions corresponds to alloying in metals. It was found that we can prepare β-(BEDT-TTF)$_2$ trihalide mixed crystals for a wide composition range[17].

3.2 Correlation Between Tc and Conductivity

We can show two examples where the correlation between Tc and conductivity seems to be close. The first example is β-(BEDT-TTF)$_2$I$_3$ in which two superconducting states are known to exist at ambient pressure, i.e. the low-Tc state (Tc=1-1.5K) and the high-Tc state (Tc=7-8K). The difference between the two states is considered to be the incommensurate lattice modulation[11] which appears below 175K, where the resistance derivative changes [18,19]. A more direct difference in resistivity at low temperature between the two states was observed [20].

Concerning the new 2 K state in β-(BEDT-TTF)$_2$I$_3$, annealing at about 110K is also accompanied by a decrease in resistance in addition to the change of incommensurate superstructure [15]. However, the relation between the change of Tc and resistance at low temperature has not been clarified yet.

The second example is seen in β-(BEDT-TTF)$_2$ trihalide mixed crystals, where we see a clear correlation between residual conductivity, i.e. conductivity at low temperature, and Tc for a wide range of composition of β-(BEDT-TTF)$_2$(I$_3$)$_{1-x}$(IBr$_2$)$_x$, β-(BEDT-TTF)$_2$(IBr$_2$)$_{1-x}$(I$_2$Br)$_x$ and β-(BEDT-TTF)$_2$(I$_2$Br)$_{1-x}$(I$_3$)$_x$ [19]. Another interesting result is the presence of a boundary representing the *minimum conductivity* (6000 S/cm) required for realization of superconductivity in this system [19].

Figure 1 shows the Ginzburg Landau coherence length $\xi_{GL}(0)$ along the conducting plane in various β-(BEDT-TTF)$_2$X superconductors as a function of Tc [21]. The coherence length was estimated from the temperature dependence of the upper critical field $H_{c2}(T)$ perpendicular to the conducting plane, using the relation

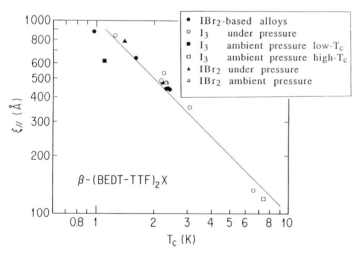

Fig. 1. Coherence length along the conducting plane($\xi_{//}$) in various β-(BEDT-TTF)$_2$X.

$$-[dH_{c2\perp}/dT]_{T=T_c} = \phi_0/2\pi\xi_{//}(0)^2 T_c.$$

It is noteworthy that almost all points are on a universal relation $\xi_{//}(\text{Å})=1000/T_c(K)$ with some exceptions such as the low-Tc state of β-(BEDT-TTF)$_2$I$_3$ and some high concentration alloys. This overall agreement demonstrates that the entire β-(BEDT-TTF)$_2$X family is characterized as pure (or clean) superconductors, with some exceptions including the low Tc state of the I$_3$ salt.

3.3 Effect of Pressure

It is now well known that the superconductivity in organic metals (BEDT-TTF)$_2$X is very sensitive to pressure. In other words, Tc decreases rapidly with application of hydrostatic pressure. The large pressure dependence is very important since it should be closely related to the mechanism of superconductivity in these organic materials. The origin of the pressure dependence of Tc is, however, not clarified yet, although some models to explain it have been proposed [22,23]. Here we classify them into three categories.

1. Density of states $N(E_F)$ <*Intraplane Interaction*>
2. Interaction between the conducting planes <*Interplane Interaction*>
3. Yamaji Model[23] (Pairing through intramolecular vibration)

The first one relates it to the change of the density of states of the 2-D electron system. Since the pressure generally reduces the intermolecular distance, it is expected to increase the transfer integrals and consequently the bandwidth, which in turn reduces the density of states at the Fermi level in a simple tight-binding picture of 2-D electron system. And a simple formalism based on the BCS theory tells us that the Tc decreases with decreasing density of states. This explanation is qualitatively consistent with the observed decrease in Tc, but has a difficulty in quantitative agreement. For example, the effect of pressure on the density of states estimated from spin susceptibility (~-3%/kbar [24]) is too small to explain the large change in Tc.

The possibility of the 2nd origin is interesting. Since a pure 2-D electronic system is not expected to show superconductivity, we need to introduce the interaction between conducting layers, which possibly governs the Tc in organic superconductors with a layered structure. In order to understand the effect of pressure, we must know the change of lattice parameters in each case. But, only a few compressibility studies have been made. In β-(BEDT-TTF)$_2$I$_3$ hydrostatic pressure results in rather isotropic compression in each direction [25]. Therefore, the effect of hydrostatic pressure involves a decrease of both in-plane and out-of-plane lattice constants. On the other hand, uniaxial pressure perpendicular to the plane is expected to cause a decrease in the distance between the conducting planes accompanied by some increase in in-plane lattice constants. Thus combination of the two experiments is expected to identify whether *intraplane* or *interplane* interactions dominate.

Some preliminary measurements on the effect of uniaxial pressure, perpendicular to the plane, on Tc in κ-(BEDT-TTF)$_2$Cu(NCS)$_2$ revealed that Tc decreases by uniaxial compression as ΔT_c=-0.01~0.02 K/bar, suggesting that interplane interaction is important in this case.

In the third model, Tc is expected to change via a change in the frequency of intramolecular vibration as well as electron-molecular vibration coupling constants. We need more experimental studies to settle down this important problem.

4. Summary

We have reviewed the structural and electrical properties to extract the following empirical rules characteristic of the superconductivity in BEDT-TTF based organic metals.

1. Correlation between Tc and conductivity in β-(BEDT-TTF)$_2$X.
2. Minimum conductivity of 6000 S/cm at low temperature, necessary for superconductivity in β-(BEDT-TTF)$_2$ trihalides.
3. The whole β-(BEDT-TTF)$_2$X family is a pure (or clean) superconductor, except for a few cases, including the low Tc state of the I$_3$ salt.
4. Tc decreases by application of hydrostatic pressure. Uniaxial pressure experiments are expected to enlighten its origin.

References

1. G.O.Baram, L.I.Buravov, L.S.Degtyarev, M.E.Kozlov, V.N.Laukhin, E.E.Laukhina, V.G. Onishchenko, K.I.Pokhodnya, M.K.Sheinkman, R.P.Shibaeva and E.B.Yagubskii, JETP Lett., **44**, 376 (1986).
2. D. Schweitzer, P. Bele, H. Brunner, E. Goku, U. Haeberlen, I. Hennig, I. Klutz, R. Swietlik and H. J. Keller, Z. Phys., **B67**, 489 (1987).
3. N. V. Avramenko, A.V.Zvarykina, V.N.Laukhin, E.E.Laukhina, R.B.Lyubovskii and R.P.Shibaeva, JETP Lett., **48**, 472 (1988).
4. W. Kang, G.Montambaux, J.R.Cooper, D.Jérome, P.Batail and C.Lenoir, Phys. Rev. Lett., **62**, 2559 (1989).
5. M. V. Kartsovnik, P.A.Kononovich, V.N.Laukhin and I.F.Shchegolev, JETP Lett., **48**, 541 (1988).
6. K. Kajita, Y.Nishio, T.Takahashi, W.Sasaki, R.Kato, H.Kobayashi, A.Kobayashi and Y.Iye, Sold State Commun., **70**, 1189 (1989).
7. K. Yamaji, J. Phys. Soc. Jpn., **58**, 1520 (1989).
8. H. Kobayashi, R. Kato, A. Kobayashi, Y. Nishio, K. Kajita and W. Sasaki, Chem. Lett. 833 (1986); K. Kajita, Y. Nishio, S. Moriyama, W. Sasaki, R. Kato, H. Kobayashi, and A. Kobayashi, Solid State Commun. **64**,1279(1987)
9. M. Tamura, H. Tajima, H. Kuroda and M. Tokumoto, in preparation.
10. D. Schweitzer, S. Gärtner, H. Grimm, E. Goku and H. J. Keller,Solid State Commun., **69**, 843 (1989).
11. T.J.Emge, P.C.W.Leung, M.A.Beno, A.J.Schultz, H.H.Wang, L.M.Soma and J.M.Williams, Phys. Rev. **B30**, 6780 (1984).
12. P. C. W. Leung, T.J.Emge, M.A.Beno, H.H.Wang,J.M.Williams,V.Petricek and P.Coppens, J. Am. Chem. Soc., **107**, 6184 (1985)
13. J. M. Williams, H.H.Wang, T.J.Emge, U.Geiser, M. A. Beno, P. C. W. Leung, K. D. Carlson, R. J. Thorn, A. J. Schultz and M. -H. Wangbo, Prog. Inorg. Chem., **35**, 51 (1987).
14. S. Kagoshima, Y. Nogami, M. Hasumi, H. Anzai, M. Tokumoto, G. Saito and N. Mori, Solid State Commun., **69**, 1177 (1989).
15. S. Kagoshima, M. Hasumi, Y. Nogami, N. Kinoshita, H. Anzai, M. Tokumoto and G. Saito, Solid State Commun., **71**, 843 (1989).
16. M. Tokumoto, H. Bando, K. Murata, H. Anzai, N. Kinoshita, K. Kajimura, T. Ishiguro and G. Saito, Synthetic Metals, **13**, 9 (1986).
17. H. Anzai, M. Tokumoto, K.Takahashi and T.Ishiguro, J. Cryst. Growth, **91**, 225(1988).
18. B. Hamzic, G. Creuzet and C. Lenoir, Europhys. Lett., **3**, 373 (1987).
19. M. Tokumoto, H. Anzai, K. Murata, K. Kajimura and T. Ishiguro, Jap. J. Appl. Phys., **26-S3**, 1977 (1987); Synthetic Metals, **27**, A251 (1988).
20. V. B. Ginodman, A. V. Gudenko, L. N. Zherikhina, V. N. Laukhin, E. B. Yagubskii, P. A. Kononovich and I. F. Shegolev, Acta Polymerica **39**, 533 (1988).
21. Based on data from papers including K. Murata, M. Tokumoto, H.Bando, H.Tanino, H. Anzai, N. Kinoshita, K. Kajimura, G. Saito and T. Ishiguro, Physica, **135B**, 515 (1985)
22. A. Nowak, U.Poppe, M.Weger, D.Schweitzer and H.Schwenk, Z. Phys., **B68**, 41 (1987).
23. K. Yamaji, Solid State Commun., **61**, 413 (1987).
24. B. Rothaemel, H.Brunner, D. Schweitzer and H. J. Keller, Phys. Rev., **B34**, 704 (1986); Y. Maniwa, T. Takahashi, K. Murata and G. Saito, Physica, **143B**, 506 (1986)
25. H.Tanino, K.Kato, M.Tokumoto, H.Anzai and G.Saito,J. Phys. Soc. Jpn., **54**, 2390(1985).

T-P Phase Diagram of β-$(ET)_2I_3$

V.N. Laukhin[1], V.B. Ginodman[2], A.V. Gudenko[2], P.A. Kononovich[3], and I.F. Schegolev[3]

[1]Institute of Chemical Physics, USSR Academy of Sciences,
 SU-142 432 Chernogolovka MD, USSR
[2]P.N. Lebedev Physics Institute, USSR Academy of Sciences,
 SU-117 334 Moscow, USSR
[3]Institute of Solid State Physics, USSR Academy of Sciences,
 SU-142 432 Chernogolovka MD, USSR

Abstract. The T-P phase diagram of β-$(ET)_2I_3$ is investigated. The origin of the T_c difference between β_L ($T_c \simeq 1.5$ K) and β_H ($T_c = 8$ K) phases is discussed.

One of the most interesting properties of the organic metal β-$(ET)_2I_3$ is the existence of two superconducting modifications, β_L and β_H, with transition temperatures, T_c, of 1.5 and 8 K, respectively. It was earlier found that the high T_c state may be obtained either by applying a small hydrostatic pressure [1] or by using special conditions of synthesis [2] and treatment [3]. The question arises of the origin of the difference in T_c. The investigation of the T-P phase diagram of β-$(ET)_2I_3$ undertaken in [4] throws some light on this question.

In Fig. 1 the T-P phase diagram of β-$(ET)_2I_3$ is shown. β_D designates the β-phase in which the end ethylene groups of the ET molecule randomly occupy one of two possible positions on either side of the molecule plane. β_L designates the β-phase after the superstructure transition, when the end groups are deflected on one side in one half-period of the superstructure, and on the other side in the other half-period ($T_c \simeq 1.5$ K). β designates the β-phase in which the superstructure is absent and the end ethylende groups are completely ordered ($T_c \simeq 8$ K at ambient pressure). Curve (a) is the line of the reversible superstructure second-order phase transition $\beta_D \rightleftharpoons \beta_L$. Curves (b) and ($c_1$-$c_2$) are the hysteresis branches of the phase transition $\beta_L \rightleftharpoons \beta_H$. The presumable position of the equilibrium line of this phase transition is shown by the dashed line. The β_H-phase is stable in the region to the right of the curve (c_1-c_2). The curve (b) encloses the region of the metastable state of this phase. Point O with coordinates $T_0 = 160 \pm 2$ K and $P_0 = 420 \pm 5$ bar is the critical point of the second-order phase transition.

The phase transitions in specially selected very perfect crystals of β-$(ET)_2I_3$ were registered either by a jump of the resistance along the a-axis, R_a, and the c^*-axis, R_c, or by jumps of the temperature derivative, dR_a/dT, at various temperatures and the pressures produced in a helium gas bomb.

The temperature dependences of R_a (Fig. 2) and R_c (Fig. 3) were measured at ambient pressure for both β_L and β_H. The low-temperature parts of these curves are shown in insets. It should be noted that the curves for β_H have been obtained only with increasing temperature, the β_H-phase being formed by slow cooling under pressure

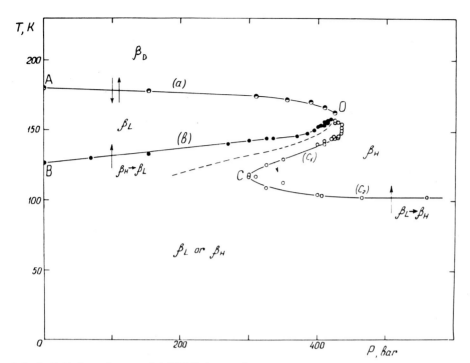

Fig. 1. T-P phase diagram of β-(ET)$_2$I$_3$ (see text)

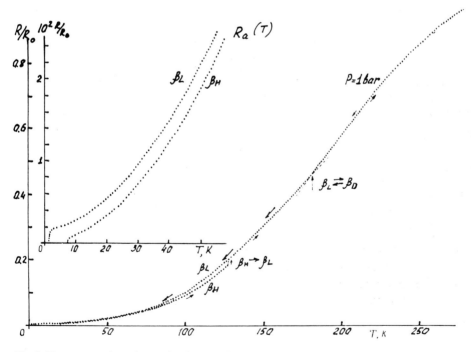

Fig. 2. Temperature dependence of resistance along the a-axis for β_L- and β_H-phases at ambient pressure. Low temperature parts of the curves are shown in the inset

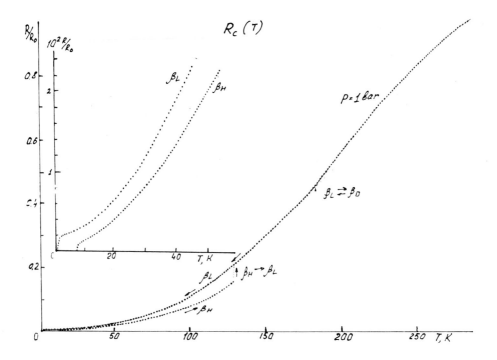

Fig. 3. The same dependence as in Fig. 2 for resistance along the c^*-axis

($P \geq 400$ bar) down to 4.2 K and then release of the pressure. It is clearly seen that the resistivity of the β_H-phase is lower than that of the β_L-phase, and the first-order transition $\beta_H \to \beta_L$ is accompanied by a jump of both R_a and R_c. It is evident that the formation of the superstructure in the β_L-phase gives rise to a random potential (due to the incommensurability of its period with the period of the basic lattice), switching on an additional mechanism of carrier scattering. In the β_H-phase, as was mentioned above, the superstructure is absent, the additional mechanism of carrier scattering is switching off and the resistivity decreasing. So, the resistance jumps accompanying the $\beta_L \rightleftharpoons \beta_H$ transformation are associated only with the formation and disappearance of the superstructure in the first-order phase transition.

On the other hand, it was shown in [5] that the superstructure vector in real space has the direction [1, 1, 0.3]. This means that the carrier scattering due to the superstructure random potential must be stronger in the c direction than in the a direction. Indeed, the resistance jump along the c^*-axis is $\Delta R_c/R_c$ = 25%–30% (Fig. 3), while that along the a-axis is $\Delta R_a/R_a$ =5%–10% (Fig. 2). Consequently, the resistivity anisotropy in the β_H-phase is lower than in the β_L-phase.

Acknowledgements. We thank E.B. Yagubskii and E.E. Laukhina for interest in the work and for high quality samples. L.N. Bulaevskii and L.N. Zherikhina are thanked for a helpful discussion.

References

1. V.N. Laukhin, E.E. Kostyuchenko, Yu.V. Sushko, I.F. Schegolev, E.B. Yagubskii: Pis'ma Zh. Eksp. Teor. Fiz. **41**, 68 (1985) [JETP Lett. **41**, 81 (1985)]
2. V.N. Merzhanov, E.E. Kostyuchenko, V.N. Laukhin, R.M. Lobkovskaya, M.K. Makova, R.P. Shibaeva, I.F. Schegolev, E.B. Yagubskii: Pis'ma Zh. Eksp. Teor. Fiz. **41**, 146 (1985)
3. A.V. Zvarykina, M.V. Kartsovnik, V.N. Laukhin, E.E. Laukhina, R.B. Lyubovskii, S.I. Pesotskii, R.P. Shibaeva, I.F. Schegolev: Zh. Eksp. Teor. Fiz. **94**, 277 (1988)
4. V.B. Ginodman, A.V. Gudenko, P.A. Kononovich, V.N. Laukhin, I.F. Schegolev: Zh. Eksp. Teor. Fiz. **94**, 333 (1988) [Sov. Phys. – JETP **67**, 1055 (1988)]
5. P.C.W. Leung, T.J. Emge, M.A. Beno, H.H. Wang, I.M. Williams, V. Petricek, P. Coppens: J. Am. Chem. Soc. **107**, 6184 (1985)

"2K-Superconducting State" in the Organic Superconductor β-(BEDT-TTF)$_2$I$_3$

S. Kagoshima[1], M. Hasumi[1], Y. Nogami[1,*], N. Kinoshita[2], H. Anzai[2], M. Tokumoto[2], and G. Saito[3]

[1]Department of Pure and Applied Sciences, University of Tokyo, Komaba 3-8-1, Meguro, Tokyo 153, Japan
[2]Electrotechnical Laboratory, Umezono, Tsukuba, Ibaraki 305, Japan
[3]Institute for Solid State Physics, University of Tokyo, Roppongi 7-22-1, Minato, Tokyo 106, Japan
*Present address: Department of Physics, Kyoto University, Sakyo-ku, Kyoto 606, Japan

Abstract. Annealing of the title compound at about 110K for 20-40 hr causes a change in the superstructure, and gives rise to two superconducting states with the critical temperature of 2K and $\sim$ 7.5K. The former state is found to occupy most of the sample volume. This state having $T_c \sim$ 2K, found in the present study for the first time, has a lower resistance and a narrower EPR linewidth in its normal state than the compound without annealing has. Furthermore the critical field anisotropy decreases in this 2K superconducting state. These results suggest that the randomness of this system decreases or the interlayer coupling of molecules increases during annealing. The other state with $T_c \sim$ 7.5K is interpreted to be nothing but the "high-T_c" state having no superstructure.

1. Introduction

In the field of organic metals, β-(BEDT-TTF)$_2$I$_3$ is one of the best studied materials because of the presence of two superconducting states with different critical temperatures and an incommensurate superstructure. The superconducting critical temperature of this organic salt is about 1.5K in the pressure range lower than 0.4 - 0.5 kbar. This state is called "low-T_c" state. However, the critical temperature goes up to about 8K discontinuously when the pressure is higher than this range (so called "high-T_c" state). At ambient pressure an incommensurate superlattice has been found to appear below 175K but it seems to be absent in the higher pressure range. We have been expecting that the superlattice is present in the whole low-pressure range, and it plays a key role in understanding the difference of T_c between two superconducting states. [1]

We made X-ray diffraction studies under high pressure in order to investigate properties of the superlattice. We found that the superlattice was actually present in the pressure range lower than 0.4 kbar but was absent in the higher pressure range. Thus we verify experimentally that the difference between "low-T_c" and "high-T_c" states is specified by the superlattice.

In addition we found another phenomenon: The wave vector of this superlattice was found to become short below about 110K. This structural change needs a very long time to complete when the pressure approaches ambient pressure. The characteristic time for the change appears to be one day or two. Details of the X-ray work will be published separately. [2]

Samples annealed at about 110K have a superlattice which is different from the original. Then we made electrical and some other measurements in order to find possible effects of annealing on the superconductivity. [3]

2. Experiments

Sample crystals were prepared by a conventional electro-chemical method. Dc resistance of samples was measured by a conventional four probe method with gold wires of 25μm dia. glued with gold paint onto the two-dimensional ab-plane of samples. Magnetizaion measurements were made under the magnetic field of 10Oe in a SQUID magnetometer. Electron paramagnetic resonance (EPR) measurements of carriers were made in the X-band on a single crystal, whose c^* axis was parallel to the magnetic field. The temperature dependence of H_{c2} was determined by extrapolation of field-induced transition curves to zero resistance. These measurements were made at ambient pressure.

3. Results and Discussions

First, we made dc resistance measurements of a sample with and without annealing. As shown in **Fig.1**, the sample annealed at about 106K for about 100hr shows a resistance drop at about 7.5K and 2.0K. The former suggests a partial superconductivity in a small volume of sample, while the latter seems to indicate a superconducting transition of the whole volume of the sample at about 2.0K. In contrast, the same sample without annealing, cooled down with a conventional cooling rate through the temperature range 120-100K, just starts to become superconducting at about 1.5K (it is the so- called "low-T_c" state.) as shown in the same figure. Thus two superconducting states with T_c of about 7.5K and 2.0K appear as a result of annealing. [4]

In order to evaluate the superconducting volume fraction of the sample, we made magnetization measurements of a sample annealed at about 110K for 135hr. We found a few percent of sample volume becomes superconducting below about 7.5K and most of has the volume T_c of 2.0K which

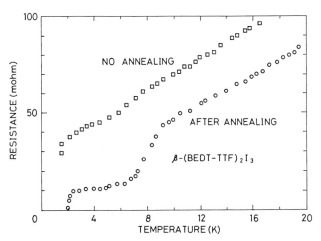

Fig.1 Temperature dependence of the dc resistance of a sample of β-(BEDT-TTF)$_2$I$_3$ with and without annealing. The annealing was made at about 106K for 100hr.

is higher than the T_c of the so-called "low-T_c" state, 1.0 - 1.5K. Thus we conclude that the annealing at about 110K predominantly brings about an increase of the superconducting critical temperature from $\sim$ 1.5K to 2.0K. The superconductivity with Tc $\sim$ 7.5K is possibly a by-product of the annealing. This problem will be discussed later.

Next we measured the temperature dependence of upper critical field H_{c2} in order to investigate properties and possible origins of these superconducting states. In the higher T_c state, our data points of an annealed sample show a good agreement with the results obtained by Orsay group in samples which are pressurized and cooled down to low temperature followed by pressure release; that is the so-called Orsay process. From this result we conclude that the 7.5K superconducting state of the annealed samples will be nothing but the so-called "high-T_c" state which has no superlattice. The superlattice is considered to disappear in a small volume of sample during annealing.

The result of H_{c2} measurements of the lower T_c state of annealed samples having T_c of 2.0K, which is different from the so-called "low-T_c" state, suggests the reduction of the anisotropy in the a-c* plane. Detailed results will be reported separately by Sasaki et al. [5]

Then what is happening during the annealing at about 110K ? We measured, as shown in **Fig.2**, the wave vector of the superlattice, the sample resistance and the EPR linewidth of independent samples as functions of time during the annealing at about 110K. They show similar behaviors. The results suggest that these changes in structure, resistance and spin-relaxation have a common origin. At least two of them, the X-ray and EPR studies probe bulk properties of samples. Therefore we consider these results are ascribed to the 2K- superconducting state which occupies most of the sample volume. The decrease of the resistance and the EPR linewidth suggests the reduction of carrier scattering.

Fig.3 shows the temperature dependence of resistance of a sample during annealing. Each set of results was obtained by suspending the annealing followed by cooling down to low temperature. The "high-T_c" state is easily found even in the early stage of annealing. We find the 2K-superconducting state has the T_c of 2K from the beginning but the T_c does not come from

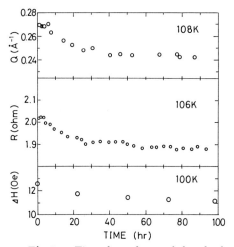

Fig.2 Time dependence of the absolute value of the wave vector Q specifying the superstructure, the sample resistance R and the EPR linewidth H of independent samples of β-(BEDT-TTF)$_2$I$_3$ during the annealing at about 110K.

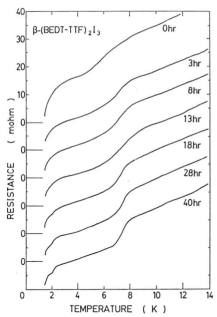

Fig.3 Temperature dependence of the resistance of a sample of β-(BEDT-TTF)$_2$I$_3$ during annealing. The figures 3hr, 8hr, 13hr, ... 40hr shown in this figure denote the total time of annealing at about 106K.

1.5K up to 2.0K continuously. So we consider the 2K-superconducting state is presumably another phase of superconductivity.

What is brought about by the annealing? The reduction of carrier scattering suggests the decrease of randomness in this system. On the other hand, the reduction of anisotropy suggests the increase of interlayer coupling. We consider that these effects are strongly related with a possible configurational ordering of the terminal ethylene groups in the BEDT-TTF molecules. If this idea is correct, we will have three superconducting states, which are specified by the superlattice and the conformation of the ethylene group.

Acknowledgments — We are grateful to T.Takahashi of Gakushuin University and K.Murata of Electrotechnical Laboratory for useful discussions on the normal state properties of this material. We are grateful to N.Toyota and T.Sasaki of Institute for Materials Research, Tohoku University for discussions on the anisotropy of critical field.

References

1. As a review see for example, M.Tokumoto, this Symposium.
2. Y.Nogami, S.Kagoshima, N.Mori, H.Anzai and G.Saito, to be published.
3. S.Kagoshima, Y.Nogami, M.Hasumi, H.Anzai, M.Tokumoto, G.Saito and N.Mori, Solid State Commun. **69**, 1177 (1989).
4. S.Kagoshima, M.Hasumi, Y.Nogami, N.Kinoshita, H.Anzai, M.Tokumoto and G.Saito, Solid State Commun. **71**, 843 (1989)
5. T.Sasaki, N.Toyota, M.Hasumi, T.Osada, S.Kagoshima, H.Anzai, M.Tokumoto and N.Kinoshita, to be published in J. Phys. Soc. Jpn.

A Change of the Incommensurate Superstructure in the Organic Superconductor β-(BEDT-TTF)$_2$I$_3$

Y. Nogami[1,*], S. Kagoshima[1], H. Anzai[2], M. Tokumoto[2], G. Saito[3], and N. Mori[3]

[1]Department of Pure and Applied Sciences, University of Tokyo, Komaba 3-8-1, Meguro, Tokyo 153, Japan
[2]Electrotechnical Laboratory, Umezono, Tsukuba, Ibaraki 305, Japan
[3]Institute for Solid State Physics, University of Tokyo, Roppongi, 7-22-1, Minato, Tokyo 106, Japan
*Present address: Department of Physics, Faculty of Science, Kyoto University, Kitashirakawa-oiwakecho, Sakyo, Kyoto 606, Japan

Abstract. A change of the wavevector of the incommensurate superstructure from $Q_0 = (0.075, 0.275, 0.205)$ to $Q' = (0.067, 0.248, 0.187)$ is observed in the organic superconductor β-(BEDT-TTF)$_2$I$_3$ by X-ray study under pressure. The wavevector change is found mainly in the range between 100K and ~120K, and it slows down at low pressures. This wavevector change is accompanied by occurrence of a new "2K superconducting" state in this compound.

1. Introduction

The organic superconductor β-(BEDT-TTF)$_2$I$_3$ is characterized by both an incommensurate superstructure and two or more superconducting states. Below 175K the incommensurate superstructure ($Q_0 = (0.075, 0.275, 0.205)$) is formed at ambient pressure.[1-3] Under pressure higher than about 400bar, the so-called "high-T_c" superconducting state ($T_c = $ ~7K) is realized[4,5] without the superstructure.[6] Near ambient pressure the so-called "low-T_c" superconducting state is obtained ($T_c = $ ~1.5K) with the presence of the superstructure.

The most interesting feature in the crystal structure of this compound is the randomness in conformation of terminal ethylenes.[1] One of two terminal ethylenes in a BEDT-TTF molecule can take either A or B conformation. At ambient pressure the coexistence of two conformations is observed even when the superstructure is formed below 175K.[7]

Except at ambient pressure, systematic structural study is not performed on the incommensurate superstructure. One purpose of the present study is to investigate properties of the incommensurate superstructure closely in all pressure ranges up to about several kbar. Another purpose is to investigate possible correlations among the superstructure and two or more superconducting states.

2. Experimental

Sample crystals of β-(BEDT-TTF)$_2$I$_3$ were obtained by conventional electrochemical method. X-ray diffraction measurements were made on the single crystals in a new-type diamond anvil pressure-cell. This anvil cell is characterized by a large volume of sample space (about 0.2mm^3) and a large allowance of the diffraction angle (40 degrees). The maximum pressure reached is about 10kbar. The copper pressure-cell holder was attached to a cold head of an Oxford CF-1108 cryostat.

The lowest temperature reached is about 10K, and the accuracy of temperature control was about 0.1K. We made pressure calibration of sample crystals through the onset temperature of the incommensurate superstructure which has been hypothesized as a function of pressure by resistivity[8] and thermal measurements.[9]

3. Results and Discussions

Figure 1 shows temperature dependence of the X-ray satellite reflection intensity from the superstructure and the absolute value of the wave vector Q specifying the superstructure. Circles, triangles and squares denote experimental results of samples under pressures of ~400bar, ~300bar and 1bar, respectively. The onset temperature of the superstructure decreases from 175K to 150K with increasing pressure. No superstructure is found when the pressure is estimated to be higher than ~400bar. Figure 1 also shows clearly that the absolute value of the wave vector changes from $Q_0=0.275\text{Å}^{-1}$ to $Q'=0.24\text{Å}^{-1}$ below 110K under pressure higher than 300bar.

We found that this wavevector change occurred even at ambient pressure when temperature of samples was kept constant for a long enough time. This annealing effect developed mainly between 100K and 110K. Figure 2 shows wavevector components of the incommensurate superstructure as functions of time during the annealing at about 109.5K and ambient pressure. The new wavevector after annealing,

$$Q'=(0.067\pm0.003, 0.248\pm0.0025, 0.187\pm0.0025),\qquad(1)$$

is also incommensurate at least in the a and c components. A commensurability locking may occur along the b axis at about 30hr. Each component of the wavevector stays constant in the initial 4hr. After this stage, the wavevector change proceeds continuously. No split of the satellite reflection peak profile was found in the annealing process. This suggests that the superstructure during the annealing is specified not by superposition of Q_0 and Q', but by a well-defined intermediate wavevector. Figure 2 also shows that the absolute value of reciprocal lattice vector $(3 0 \bar{1})$ is lengthened during the annealing indicating the shrinkage of the real lattice vector.

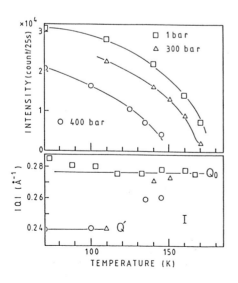

Fig. 1. Temperature dependences of the X-ray intensity of satellite reflection $(3\bar{1}\bar{1})+Q$ from the superstructure, the absolute value of the wave vector Q specifying the superstructure. The onset temperature of the superstructure decreases from 175K to 150K with increasing pressure up to 400bar. The absolute value of the wave vector Q is independent of temperature ($\sim 0.275\text{Å}^{-1}$) at ambient pressure, but under pressures higher than 300bar, it changes to about 0.24 Å^{-1} below 110K. The vertical bar denotes an estimation of probable error.

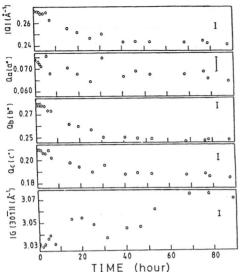

Fig. 2. Wave vector components of the incommensurate superstructure as functions of time during the annealing at about 109.5 K and ambient pressure. Also shown is the absolute value of the reciprocal lattice vector $(30\bar{1})$. The vertical bars denote estimations of probable error.

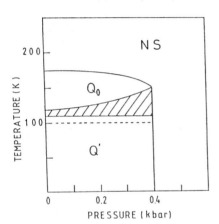

Fig. 3. A proposed phase diagram of the incommensurate wave vector specifying the superstructure. The wave vector Q_0 is stable just below the onset temperature of the superstructure, but Q' becomes stable below about 110K with a long enough annealing. The shaded area denotes the regime where the intermediate wave vector between Q_0 and Q' can be stabilized. NS denotes the regime where no superstructure is found. The dotted line denotes the lowest temperature above which the annealing effect on the wavevector change develops.

We found this wavevector change to Q' proceeded principally in the range 100-110K and 1-400bar. Under high pressures the change is completed in a short time, while it takes more than 30hr at ambient pressure. A new wave vector intermediate between Q_0 and Q' was found to be stabilized when the annealing temperature was between ~110K and ~120K near ambient pressure. As shown in Fig. 1, at ~145K and 400bar, the intermediate wavevector is also stable. A proposed phase diagram of the incommensurate structure is obtained as shown in Fig. 3.[10]

The wavevector of the incommensurate superstructure stays constant below 100K, where the motion of terminal ethylenes is frozen according to NMR.[11] Judging from a presence of the key temperature common to these phenomena, there must be close correlation between the incommensurate superstructure and the terminal ethylenes. The long relaxation time(~30hr) of the wavevector change suggests the presence of a high potential barrier. This potential barrier appears to correspond to the fairly high potential barrier (~1000K) of thermal motion of ethylenes between A and B conformations.[11]

The peak-to-peak width of EPR signal and the electrical resistivity decrease similarly with the wavevector change during the annealing.[12] This is interpreted as some disorder being removed by the annealing. The commensurability locking along b axis is one of the possible mechanisms causing the decrease of the disorder. The number of discommensurations, which can be an origin of the disorder should decrease when the b component of the wavevector approaches 0.25 during the annealing. Another possible mechanism is the rearrangement of all terminal ethylenes into the A conformation.[12]

We found a new "2K superconducting" state in the annealed sample with the presence of the superstructure Q', while the same sample has T_c of about 1.5K without the annealing.[10] It can be said that the wavevector change is accompanied by an increase of the superconducting critical temperature. The reduction of the disorder observed by EPR and resistivity measurements can lead to this increase of superconducting critical temperature.

We express our sincere thanks to N. Kinoshita, T. Yagi, H. Takahashi, W. Utsumi, M. Tanaka, M. Hasumi and Y. Kuramoto. This work is supported by Grants-in-Aid for Scientific Research, No. 62460027 and No. 63790175, of the Ministry of Education, Science and Culture.

References

[1] P. C. W. Leung, T. J. Emge, M. A. Beno, H. H. Wang and J. M. Williams: J. Am. Chem. Soc. **106**, 7644(1984).
[2] Y. Nogami, S. Kagoshima, T. Sugano and G. Saito: Synth. Metal, **16**, 367(1986).
[3] S. Ravy, J. P. Pouget, R. Moret and C. Lenoir: Phys. Rev. **B37**, 5113(1988).
[4] K. Murata, M. Tokumoto, H. Anzai, H. Bando, G. Saito, K. Kajimura and T. Ishiguro: J. Phys. Soc. Jpn. **54**, 2084(1985).
[5] V. N. Laukhin, E. E. Kostychenko, Yu. V. Suchenko, I. F. Schegolev and E. B. Yagubski: JETP Lett. **41**, 81(1985).
[6] A. J. Schultz, M. A. Beno, H. H. Wang and J. M. Williams: Phys. Rev. **B33**, 7823(1986).
[7] T. Takahashi and Y. Maniwa: private communication.
[8] I. F. Schegolev: Proc. LT-18, Kyoto, 1987, Jpn. J. Appl. Phys. Suppl. **26-3**, 1972(1987).
[9] W. Kang, G. Creuzet, D. Jérome and C. Lenoir: J. Phys. (Paris), **48**, 1035(1987).
[10] Y. Nogami, S. Kagoshima, H. Anzai, M. Tokumoto, N. Mori, N. Knoshita and G. Saito: to be published in J. Phys. Soc. Jpn.
[11] Y. Maniwa, T. Takahashi and G. Saito: J. Phys. Soc. Jpn. **55**, 47(1986).
[12] S. Kagoshima, M. Hasumi, Y. Nogami, N. Kinoshita, H. Anzai, M. Tokumoto and G. Saito: to be published in Solid State Commun.

Effect of Annealing on the Superconductivity of β-(BEDT-TTF)$_2$I$_3$

K. Kanoda[1], K. Akiba[1], K. Suzuki[1], T. Takahashi[1], and G. Saito[2]

[1] Department of Physics, Gakushuin University, Mejiro, Toshima-ku, Tokyo 171, Japan
[2] Institute for Solid State Physics, University of Tokyo, Roppongi, Minato-ku, Tokyo 106, Japan

Abstract. Kagoshima et al. have recently found that the title compound, when annealed at 100-110 K, exhibits two superconducting transitions at 2 and 7.5 K. We have pursued the time evolution of the two phases depending on the annealing time at 104 K by means of complex susceptibility measurements. Drastic growth of the 7.5 K component was found. The present results suggest that the 7.5 K phase is the most stable state at low temperatures even at ambient pressure.

1. Introduction

One of the puzzles in the organic superconductor, β-(BEDT-TTF)$_2$I$_3$, is the existence of two different superconducting states [1,2]; the one with a transition temperature of 1-1.5 K (the low-T_c state) and the other with 7-8 K (the high-T_c state). Although the origin of the difference in T_c has not been understood, it has been known that there exists a structural difference between the two superconducting phases of this compound: An incommensurate lattice modulation, which was observed below 175 K in the low-T_c state [3], was found to be suppressed in the high-T_c state under pressure [4]. In addition to slight molecular displacements, the lattice modulation is accompanied by a change in the conformation of the ethylene groups in the BEDT-TTF molecules. In the former phase (low-T_c), the so-called A and B conformations coexist while the latter phase is believed to contain only the A conformation [5].

Kagoshima et al. have recently found metastability of the low-T_c state with x-ray, resistivity and susceptibility measurements [6,7]. When the sample was held at a temperature in the range of 100-120 K, the wave number of the superstructure gradually decreased by about 10% for all components. After this annealing process, two superconducting transitions appeared at 2 and 7.5 K. They claimed that the 2-K phase was a well-defined phase different from the low-T_c state and occupied most of the sample volume while the 7.5-K phase occupied only a small portion. Characteristics of the superconductivity generated by annealing should be useful information to understand the puzzling superconductivity in this compound.

We have studied the development of superconductivity in β-(BEDT-TTF)$_2$I$_3$ as a function of annealing time, where the evolution of the superconductivity was characterized by complex susceptibility measurements. Drastic growth of the high T_c phase is reported. Details will be published elsewhere.

2. Experimental and Results

Experiments were performed with single crystals which were grown by electrochemical method. The complex susceptibility was measured with a Hartshorn-type mutual inductance bridge with the ac field of 195 Hz applied perpendicular to the conductive ab plane of the crystals. The

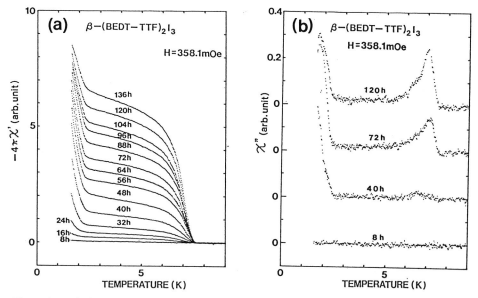

Fig. 1 (a) Real part and (b) imaginary part of complex susceptibility of β-(BEDT-TTF)$_2$I$_3$ annealed at 104.4 K.

measurements were made down to 1.6 K. First, the single crystal sample was cooled down to the liquid He temperature with the usual cooling rate. After the measurement the sample temperature was increased up to 104.5 ± 0.3 K and held there for 8 hours. Then the sample was cooled again and the susceptibility was measured. This sequence of annealing and measurement was repeated.

The results are given in Fig. 1, where the temperature dependence of (a) the real part χ' and (b) the imaginary part χ'' are shown with the accumulated annealing time as a parameter. The amplitude of the ac field was 358 mOe. Before annealing, there is no observed superconductive diamagnetism in the measured temperature range, since the T_c of the low-T_c state is lower than 1.6 K. It is clear in the figure that the annealing causes two superconducting transitions, the onsets of which are found at 7.5 and 2.3 K. The imaginary part, χ'', clearly has non-vanishing values in the regions below 2.3 K and between 5 and 7.5 K, corresponding to the two superconducting transitions. A rough estimate of the absolute value of the complete diamagnetism corresponds to the full scale of Fig. 1 (a).

The overall view of our results is as follows: The initial appearance of both the transitions are followed by rather rapid growth of the 2.3 K transition. Then 7.5 K transition finally becomes so predominant that the superconducting diamagnetism approaches a complete value.

3. Discussion

The onset temperature (7.5 K) remains unchanged all through the development of this phase. This means that a well-defined superconducting phase nucleates from the initial stage of annealing.

Most remarkable is a drastic growth of this phase. Time evolution of the 7.5-K phase is shown in Fig. 2, where the value of χ' at 5 K is plotted as a function of annealing time. One can see that the time evolution is not monotonous. The initial slope of the curve is rather

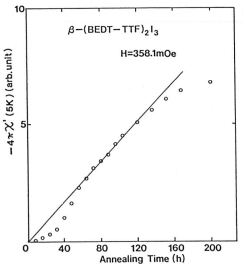

Fig. 2 Time evolution of the 7.5-K phase in β-(BEDT-TTF)$_2$I$_3$ annealed at 104.4 K. $-4\pi\chi'$ at 5 K is plotted as a function of annealing time.

small but χ' then starts to increase more rapidly up to 50 hours. In the region between 50 and 120 hours, χ' fits well a straight line passing through the origin. Finally χ' starts to show a tendency to saturate for further annealing. The saturation value of χ' appears to reach a considerable fraction of the complete diamagnetism.

It is difficult to make a reliable estimate of the superconducting volume fraction from diamagnetic susceptibility. One of the reasons is that no one knows the actual shape of the superconducting grains, which is important to estimate demagnetization factor. If the morphology of the grains does not change so much in the annealing process, however, the relative variation of χ' should correspond to that of the superconducting volume fraction.

Another difficulty is the possibility of a shielding effect due to superconducting couplings between the grains. If this effect is important, χ' should increase more rapidly than the linear dependence on annealing time, since the shielded area increases much faster than the actual superconducting volume. The observed linearity in time rather leads us to consider that the inter-grain shielding is ineffective in the present case. The deviation from the linear dependence of χ' at the initial stage (0-50 hours) can be explained by considering magnetic field penetration into the grains whose sizes are smaller or comparable to the penetration depth. This is consistent with the profile of χ'', but the details will be published separately.

From the above considerations, we conclude that the 7.5 K phase develops to a considerable portion of the volume. This is in contrast to the previous report [7]. We believe that the difference comes from the fact that the annealing temperature $\sim$ 104 K in the present work was lower than 108-110 K in the previous measurements.

It is an open question what is going on during the annealing process, but one hundred Kelvin is a characteristic temperature for freezing of thermal motions of the ethylene groups [8]. Since the low-T_c state is metastable, transitions to the real ground state should be brought about by thermal activation through a finite energy barrier. Therefore, the relative occupations and the thermal relaxation rate between the metastable and the ground states are expected to be strongly temperature dependent. This may explain the sensitivity to the annealing temperature and the discrepancy between the present and the previous results. It seems reasonable to us that the conformational degree of freedom of the ethylene groups is responsible for the structural change and the generation of the two superconducting phases.

Discussion is now directed to the 2.3 K transition. The annealed sample should be considered as an inhomogeneous superconductor, as seen in Fig. 1. In such a system care must be taken in confirming whether the lower transition indicates another superconducting phase or not; a granular network, in principle, can undergo an intergranular (phase-coherence) transition below the individual intragranular transition. Indeed the observed profile in Fig. 1 has a strong resemblance to that of the proximity-coupled NbTi fibers, where the transition at a lower temperature occurs due to the phase coherence [9,10]. However, we can conclude that the 2.3 K transition is not of this type for the following reason: The phase coherence transition occurs when the thermal disturbance kT becomes less than the coupling energy. Therefore, the transition temperature should be quite sensitive to the strength of the coupling. In our case, annealing should make the coupling stronger through the growth of the 7.5 K grains. However, the onset of the lower transition stays constant (2.3 K) at any stage of annealing process, as seen in χ' and χ'' of Fig. 1. This fact evidences that 2.3 K transition is another well-defined superconducting phase. χ'' at a lower field (116 mOe) forms a clear peak at 2.0 K which is reasonably identified as a midpoint of the transition. This transition temperature is clearly different from the accepted values (1.0-1.5 K) of the low-T_c state and agrees with that reported by Kagoshima et al. [7].

In conclusion, the present work revealed the time evolution of the two superconducting phases in the β-(BEDT-TTF)$_2$I$_3$ annealed at 104 K. Drastic growth of the 7.5 K phase implies that this is the most stable state at ambient pressure. To get further information on the structure of these phases, NMR study, which probes the conformation of the ethylene group, is under way.

Acknowledgments. The authors would like to thank Professor Kagoshima for useful interactions. This work was supported in part by a Grant-in-aid for Scientific Research from the Ministry of Education, Science, and Culture of Japan.

References

1. K. Murata, M. Tokumoto, H. Anzai, H. Bando, G. Saito, K. Kajimura, and T. Ishiguro, J. Phys. Soc. Jpn. **54**, 2084 (1985).
2. V.N. Laukhin, E.E. Kostyuchenko, Y.V. Sushko, I.F. Schegorev, and E.B. Yagubskii, JETP Lett. **41**, 81 (1985).
3. P.C.W. Leung, T.J. Emge, M.A. Beno, H.H. Wang, and J.M. Williams, J. Am. Chem. Soc. **106**, 7644 (1984).
4. A.J. Schultz, M.A. Beno, H.H. Wang, and J.M. Williams, Phys. Rev. B **33**, 7823 (1986).
5. A.J. Schultz, H.H. Wang, and J.M. Williams, J. Am. Chem. Soc. **108**, 7853 (1986).
6. S. Kagoshima, Y. Nogami, M. Hasumi, H. Anzai, M. Tokumoto, G. Saito, and N. Mori, Solid State Commun. **69**, 1177 (1989).
7. S. Kagoshima, M. Hasumi, Y. Nogami, N. Kinoshita, H. Anzai, M. Tokumoto, and G. Saito, Solid State Commun. **71**, 843 (1989).
8. Y. Maniwa, T. Takahashi, and G. Saito, J. Phys. Soc. Jpn. **55**, 47 (1986).
9. Y. Oda, G. Fujii, and H. Nagano, Jpn. J. Appl. Phys. **21**, L37 (1982).
10. T. Ishida and H. Mazaki, in Proceedings of the International Cryogenic Materials Conference, Kobe, 1982, eds. K. Tachikawa and A. Clark, (Butterworth, Lodon, 1982) p.430.

Evolution of the "High-T_c" States at Ambient Pressure in β-(BEDT-TTF)$_2$I$_3$

M. Tokumoto, Y. Yamaguchi, N. Kinoshita, and H. Anzai

Electrotechnical Laboratory, Tsukuba, Ibaraki 305, Japan

Abstract. Annealing at about 110 K causes a change in the incommensurate superstructure in β-(BEDT-TTF)$_2$I$_3$. The effect of annealing on the ambient-pressure superconductivity in β-(BEDT-TTF)$_2$I$_3$ was studied by magnetic susceptibility measurement. Both shielding and Meissner measurements revealed that two new superconducting states with Tc at 7.5 K and 2 K appear as a result of annealing at ambient pressure. We assign the former to correspond to the "high-Tc" state. On the other hand, the latter superconducting state with Tc=2 K should be regarded as a new superconducting state different from the original "low-Tc" state with Tc=1.0-1.5 K. Thus β-(BEDT-TTF)$_2$I$_3$ was found to exhibit three superconducting states with different Tc. Both time- and temperature-evolution of these superconducting states were studied by *in situ* magnetization measurements.

1. Introduction

The superconducting critical temperature(Tc) vs. pressure phase diagram of β-(BEDT-TTF)$_2$I$_3$ shows two superconducting states with quite different Tc, namely the "low-Tc" state with Tc below 1.5 K and the "high-Tc" state with Tc up to 8 K.[1,2] At ambient pressure, an incommensurate lattice modulation appears below 175 K[3-5], and a superconducting transition with Tc=1.0-1.5 K is commonly observed. However, application of pressure above about 0.5 kbar suppresses the incommensurate superstructure and a new superconducting state with Tc=7-8 K appears. The correlation between the drastic rise in Tc and the disappearance of superstructure lead one to postulate that the intrinsic Tc of β-(BEDT-TTF)$_2$I$_3$ is 8 K, and the low Tc at ambient pressure is caused by suppression of Tc by the appearance of the incommensurate lattice modulation.

Recently, Kagoshima et al.[6] reported a change of the superstructure and an associated rise of Tc in β-(BEDT-TTF)$_2$I$_3$. They made an X-ray diffraction measurement of the superstructure in β-(BEDT-TTF)$_2$I$_3$ under pressure, and found that the wave vector of the superstructure become short below about 100 K. The corresponding change was found to occur even at ambient pressure when the sample is kept in the range 100-120 K for a long time, say, 24 hrs. Actually, the sample annealed for 65 hrs at ambient pressure showed a small Meissner effect (~6x10^{-5} emu/g·Oe) due to superconducting transition below about 7 K. For the annealing time of 20, 40 and 65 hrs, they found Tc of ~4.5, ~6 and ~7 K, respectively[6]. This experimental result is very interesting since it shows a sort of coexistence of the superstructure and the high-Tc state, which implies that the presence of superstructure itself may not be fatal to the realization of the "high-Tc" state in this material. Rather, it indicates that the wave vector of the superstructure might play an important role in the determination of Tc in this material. Thus the above exciting report stimulated us and led us to raise the following questions. (1) Is it a true bulk superconductivity? The reported superconducting volume fraction seems to be too small to assure the coexistence of the "high-Tc" state with the superstructure. (2) Can we realize a superconducting state with continuously controllable Tc? If so, what is the recipe? etc. In order to clarify these points, we performed measurements of dc magnetization and studied the effect of annealing on the superconductivity in β-(BEDT-TTF)$_2$I$_3$ at ambient pressure.

In this paper, we report evidence for evolution of bulk "high-Tc" states by extended annealing at ambient pressure. Both shielding and Meissner signal in dc magnetization measurement revealed (1) a coexistence of two superconducting states, i.e. a new superconducting state with Tc=2 K and the "high-Tc" state with Tc=7.5 K, and (2) extinction of the 2 K state as the 7.5 K state grows by prolonged annealing.

2. Experimental

A single crystal of β-(BEDT-TTF)$_2$I$_3$ of distorted hexagon shape with rough dimensions of 2x3x0.4 mm^3 and weight of 0.48 mg, grown by electrochemical oxidation was used for the magnetization measurement. Annealing conditions, i.e. annealing temperature(Ta) and annealing time, for each step of annealing are shown in Fig.1. After each step of annealing, dc magnetization was measured at temperatures between 1.8 K and 9 K, using a SQUID susceptometer SHE model 905. Continuous *in situ* annealing was realized and the sample was cooled down and kept at low temperatures when annealing was interrupted once a week in order to replenish liquid helium. The sample was kept in the susceptometer for more than 50 days. Magnetic field up to 10 Oe was applied along the c* axis, i.e. perpendicular to the wide surface(a-b plane) of the crystal.

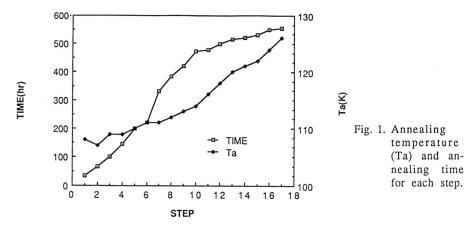

Fig. 1. Annealing temperature (Ta) and annealing time for each step.

3. Results and Discussions

Since the annealing was performed successively without heating the sample during the interval between any two consecutive annealing processes, the effect of annealing is expected to accumulate additively with successive annealing. We observed that the effect of annealing performed in 17 steps in total can be classified into the following three stages. *In the first stage*, i.e. steps 1 through 7, corresponding to the first 331 hrs of annealing at temperatures from 107 K up to 111 K, we observed coexistence of two superconducting states, i.e. one with Tc=2 K and the other "high-Tc" state with Tc=7.5 K. Figure 2 shows the temperature dependence of dc magnetization after the annealing of step 5. Note that before annealing, i. e. when the sample is cooled down quickly, no diamagnetic signal due to superconductivity can be observed in this temperature range, since Tc in the normal "low-Tc" state is too low to be measured in the present experimental system. Thus the superconducting state with Tc=2 K is a new superconducting state which is distinctly different from the original "low-Tc" state.

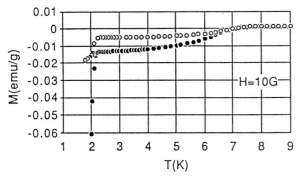

Fig. 2. Diamagnetic shielding(●) and Meissner effect(○) after the annealing of step 5.

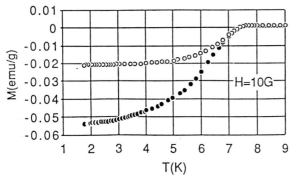

Fig. 3. Diamagnetic shielding(●) and Meissner effect(○) after the annealing of step 8.

However, since the diamagnetic shielding of the 2 K state is quite large ($\sim 1.3 \times 10^{-1}$ emu/g·Oe), comparable with the whole sample volume, we speculate that the Tc of the "low-Tc" state was raised from 1.0-1.5 K to 2 K by annealing, and the major part of the sample has become the 2 K state at least at the beginning of the first stage.

It is highly conceivable that the observed rise in Tc is closely related to the change in the wave vector of the superstructure observed by Kagoshima et al.[6] Since the observation by X-ray scattering must be a bulk effect, we can hardly correlate the change of the wave vector to the "high-Tc" state with Tc =7.5 K. *In the second stage*, i. e. steps 8 through 12, corresponding to the 168 hrs of annealing at temperatures from 112 K up to 118 K, we observed only the 8 K "high-Tc" state. The volume fraction of the 8 K state seems to become saturated and stays quite stable upon further annealing. Figure 3 shows the temperature dependence of dc magnetization after the annealing of step 8. The high-Tc state did not show any appreciable decay until the annealing temperature exceeded 118 K. *In the third stage*, i. e. steps 13 through 17, corresponding to the 57 hrs of annealing at temperatures from 120 K up to 126 K, further annealing at higher temperatures started to ruin the "high-Tc" state and decrease the superconducting volume fraction without appreciable decrease in Tc. Throughout the second and the third stages, only the 8 K superconducting state was observed. However, since the volume fraction of the 8 K state is smaller than that for the 2 K state, it is quite reasonable to assume that the rest of sample is in the "low Tc" state with Tc=1.0-1.5 K. Some of the physical properties of the 2 K state are reported separately.[7] Our new finding is that the effect of extended annealing in the first stage results in a decrease in the volume fraction of the 2 K superconducting state. On the other hand, the "high-Tc" state with Tc=7.5 K shows a very small diamagnetic signal corresponding to the superconducting volume fraction of less than 1% at the beginning of the first stage. The successive annealing in the first stage causes an increase in the volume fraction of the 8 K superconducting state in contrast to the decrease in the volume fraction of the 2 K superconducting state. Eventually, the 2 K state was found to disappear leaving the "high-Tc" state. The resistance measurement of β-$(BEDT-TTF)_2I_3$ at ambient pressure quite often exhibits a superconducting transition accompanied by a partial "high-Tc" transition around 8 K as reported earlier[8,9]. The latter has not been regarded as a bulk superconductivity since no evidence of Meissner signal has been detected so far. On the other hand, a bulk "high-Tc" state can be realized under pressure, and also at ambient pressure by application of pressure followed by a subsequent release of pressure at low temperature. This "high-Tc" state without the incommensurate superstructure is metastable and goes back to the "low-Tc" state upon heating above about 120 K. Therefore the possibility that the "high-Tc" state in the second and third stage of annealing correspond to the state without the superstructure is still an open question and should be examined by structural study using either X-ray or neutrons. Regarding the questions we raised, we found (1)Yes. The new state is a bulk superconducting state corresponding to the observed change of wave vector. On the other hand, the "high-Tc" state with Tc=7.5 K seems to have a different origin although the amount of the superconducting volume is appreciably large. (2)No. We cannot control the Tc by annealing. Tc is either 1-1.5 K, 2 K or 7.5 K. The difference in the annealing temperature or time seems to result in different volume fractions of each superconducting states. It is not clear how the intermediate value of the wavevector can be related to the change of volume fraction of superconducting states.

4. Summary

The effect of annealing at about 110 K on the ambient-pressure superconductivity was studied by magnetization measurement. In addition to the "high-Tc" state with Tc=7.5 K, a new superconducting state with Tc=2 K was found to exist as a bulk state corresponding to the change of the wave vector of the incommensurate superstructure. Both time- and temperature-evolution of these superconducting states were revealed by *in situ* magnetization measurements.

References.

1. K. Murata, M. Tokumoto, H. Anzai, H. Bando, G. Saito, K. Kajimura and T. Ishiguro: J. Phys. Soc. Jpn., **54**, 1236, 2084 (1985).
2. V. N. Laukhin, E. E. Kostyuchenko, Yu. V. Sushko, I. F. Schegolev and E. B. Yagubskii: JETP Lett. **41**, 81 (1985).
3. T. J. Emge, C. W. Leung, M. A. Beno, A. J. Schultz, H. H. Wang, L. M. Sowa and J. M. Williams, Phys. Rev. **B30**, 6780 (1984).
4. Y. Nogami, S. Kagoshima, T. Sugano and G. Saito: Synth. Met. **16**, 367 (1986).
5. S. Ravy, R. Moret, J. P. Pouget and R. Comes: Synth. Met. **19**, 237 (1987).
6. S. Kagoshima, Y. Nogami, M. Hasumi, H. Anzai, M. Tokumoto and G. Saito, Solid State Commun. **69**, 1177 (1989).
7. S. Kagoshima, M. Hasumi, Y. Nogami, N. Kinoshita, H. Anzai, M. Tokumoto, G. Saito and N. Mori, Solid State Commun. **71**, 843 (1989).
8. M. Tokumoto, K. Murata, H. Bando, H. Anzai, G. Saito, K. Kajimura and T. Ishiguro, Solid State Commun. **54**, 1031 (1985).
9. M. Tokumoto, H. Bando, K. Murata, H. Anzai, N. Kinoshita, K. Kajimura, T. Ishiguro and G. Saito, Synthetic Metals **13**, 9 (1986).

Electrical Resistance and Upper Critical Field in the "2K-Superconducting State" of β-(BEDT-TTF)$_2$I$_3$

T. Sasaki[1], N. Toyota[1], M. Hasumi[2], T. Osada[2], S. Kagoshima[2], M. Tokumoto[3], N. Kinoshita[3], and H. Anzai[3]

[1]Institute for Materials Research, Tohoku University, Katahira 2-1-1, Aoba, Sendai 980, Japan
[2]Department of Pure and Applied Science, University of Tokyo, Komaba 3-8-1, Meguro, Tokyo 153, Japan
[3]Electrotechnical Laboratory, Umezono 1-1-4, Tsukuba, Ibaraki 305, Japan

Abstract. The electrical resistance and the upper critical field H_{c2} have been measured at ambient pressure on a layered organic superconductor β-(BEDT-TTF)$_2$I$_3$ in the states before and after annealing at 109K for 120 hours. The resistance decreases and exhibits saturation during annealing. This can be attributed to the shrinking of the lattice, which induces the enhancement of the intermolecular transfer integrals, rather than the reduction of the impurity scattering. The critical temperature T_c increases from 1.4 to 2.08K, while the $H_{c2\perp\text{plane}}$ increases and the anisotropy, $H_{c2//\text{plane}}/H_{c2\perp\text{plane}}$, decreases on annealing. Empirically, the increase in T_c, including pressurized states, is associated with the decrease of $H_{c2//\text{plane}}/H_{c2\perp\text{plane}}$.

β-(BEDT-TTF)$_2$I$_3$, β-di[bis(ethylenedithio)tetrathiafulvalene]triiodide, undergoes a superconducting transition at 1.1~1.5K at ambient pressure.[1] The superconducting critical temperature T_c rises from 1.1~1.5K to 7~8K under relatively small pressure ~1kbar.[2][3] It is reported that an incommensurate superstructure develops below 175K at ambient pressure.[4][5] With the development of the superstructure, the terminal ethylene groups of the BEDT-TTF molecules become disordered. However, this superstructure and disordering of the terminal ethylene groups does not appear under pressure. Therefore this structural modulation is expected to play an important role in determining T_c.

Recently, Kagoshima et al. found that the incommensurate superstructure became unstable below about 110K at ambient pressure and a wave vector **Q** characterizing the superstructure changed from 0.27 to 0.24 Å^{-1} on keeping the temperature at about 110K for 20~40 hours.[6][7][8] Simultaneously the resistivity decreases. This annealing state exhibits two superconducting transitions at 2 and 7.5K. The purpose of the present paper is to report the effect of annealing on the resistivity and the superconducting upper critical field H_{c2} in the "2K-Superconducting State" of β-(BEDT-TTF)$_2$I$_3$.[9]

The single crystals were prepared by the usual electrochemical oxidation method. The shape of a crystal is a hexagonal plate with size of about 2.0×1.2×0.4 mm^3. The crystallographic a-axis runs along the long direction. We define the c^*-axis as the direction perpendicular to the crystal plane. The

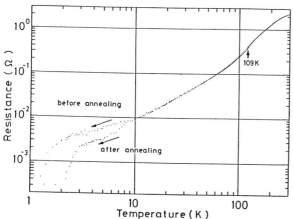

Figure 1 Temperature dependence of dc resistance of β-(BEDT-TTF)$_2$I$_3$ before and after annealing at 109K for 120 hours.

resistance measurements were carried out by a standard four probe method with dc or ac (31.3Hz) current flowing along the a-axis.

Figure 1 shows a logarithmic plot of the temperature dependence of the resistance. First, the sample was cooled from room temperature to 0.5K at a conventional cooling rate, taking about 10 hours. The T_c is found to be 1.4K, defined at the zero resistance temperature. The residual resistance ratio ($RRR \equiv \rho(300K)/\rho(2K)$) is about 650. Then the sample was heated up to about 110K and the temperature was kept at 109K, illustrated by an arrow in Fig.1. The resistance decreases and exhibits saturation during an annealing at 109K for 120 hours. The total amount of the resistance decrease is about 8% at the end of annealing (for details, see Ref. [9]). The sample which is again cooled exhibits two superconducting transitions at about 8 and 2K. The former transition, which is associated with a slight resistance drop, shows that the so-called "high-T_c" phase exists partially in the sample. By contrast, the latter transition with $T_c = 2K$ induced by annealing is a bulk superconducting transition, confirmed by the magnetization measurements.[7] One of the explanations of the resistance decrease at 109K could be based on an idea that an increased magnitude of ordering, for example, of the terminal ethylene groups of the BEDT-TTF molecules occurs, and then, the contribution of the impurity scatterings to the resistance might be reduced. This idea, however, is unlikely to apply to this resistance decrease. Because the effect of the reduction of the impurity scatterings should be enhanced at lower temperature, the low temperature resistance after annealing is expected to be smaller than that before annealing. However, as seen from Fig. 1, the resistance in the two states (T>8K; normal states) is little different. As an alternative, we propose that the transfer energy between the BEDT-TTF molecules might be enhanced by annealing-induced shrinking of the lattice. From the X-ray diffraction analysis [8], the lattice shrinks with decreasing temperature and, in addition, it becomes smaller during annealing at ambient pressure. Then, the distances between the BEDT-TTF molecules become short and the overlap integrals become large.

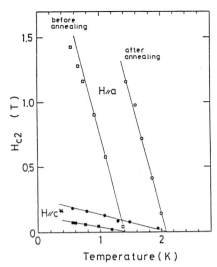

Figure 2 Temperature dependence of the upper critical field H_{c2} along the c^*- and the a- axis of β-(BEDT-TTF)$_2$I$_3$ before and after annealing.

Table I Fundamental parameters of β-(BEDT-TTF)$_2$I$_3$ before and after annealing (present study) and pressurized states (after ref. [10]).

	T_c (K)	P(kbar)	$\xi_{GL//}(0)$ (Å)	$\xi_{GL\perp}(0)$ (Å)	$H_{c2//}/H_{c2\perp}$
Before annealing	1.40	0	508.5	24.3	20.9
After annealing	2.08	0	354.6	24.5	14.5
Pressurized state	6.55	1.6	131	10	13.1
	3.06	3.5	355	22.7	15.7
	2.18	5	488	31	15.6

This change favors increased conductivity after annealing. It is crucial to confirm the structural differences at low temperatures between states before and after annealing in order to make certain of being able to apply this idea to the small difference of the temperature dependence of the resistance in the two states at low temperatures.

Figure 2 shows the temperature dependence of H_{c2}. In this figure, the squares and circles indicate the data before and after annealing, respectively. Table I lists the G-L coherence length and other parameters for both states and pressurized states.[10] The anisotropy of H_{c2}, $H_{c2//\text{plane}}$ / $H_{c2\perp\text{plane}}$, shows a significant decrease from 20.9 to 14.5, caused by annealing.

This change in the anisotropy is mainly affected by the increase of $H_{c2\perp}$plane after annealing. As found in Table I , T_c, the G-L coherence length and the anisotropy of the 2K-phase are comparable to those of the pressurized states at 3.5~5kbar.[10] It is interesting that the 2K-superconducting phase might be understood by the "lattice pressure" model.[11] The shrinking of the lattice during annealing, which might cause the resistance decrease at 109K, is not inconsistent with this model. Besides the above idea, taking into account the fact that $\xi_{//}$ is proportional to $m_{//}^{-1/2}$, the observed 30%-decrease in $\xi_{//}$ is expected to significantly enhance the density of states at the Fermi level, dominantly determined by the in-plane effective mass $m_{//}$. This change is favorable to an increase in T_c. Table I leads us to the empirical rule that an increase in T_c is associated with a decrease of $H_{c2//\text{plane}} / H_{c2\perp\text{plane}}$.

Finally, it might be difficult to explain the increase in T_c in terms of the weak localization effect [12], because even the superconductivity of the sample before annealing with the mean free path $\ell \simeq \xi_{//}$ [9][13] is unlikely to be understood in the dirty limit.

The authors would like to thank Prof. K. Takanaka, Drs. Y. Suzumura and Y. Nogami for useful discussions and Profs. T. Fukase and Y. Muto for their encouragement. This work was supported in part by a Grant-in-Aid for Special Project Research from the Ministry of Education, Science and Culture of Japan.

References.

[1] E. B. Yagubskii, I. F. Shchegolev, V. N. Laukhin, P. A. Kononovich, M. V. Kartsovnik, A. V. Zvarykina and L. I. Buravov: Pis'ma Zh. Eksp. Teor. Fiz. 39 (1984) 12 [JETP Lett. 39 (1984) 12].
[2] K. Murata, M. Tokumoto, H. Anzai, H. Bando, G. Saito, K. Kajimura and T. Ishiguro: J. Phys. Soc. Jpn. 54 (1985) 1236.
[3] V. N. Laukhin, E. E. Kostyuchenko, Yu. V. Sushko, I. F. Shchegolev and E. B. Yagubskii: Pis'ma Zh. Eksp. Teor. Fiz. 41 (1985) 68 [JETP Lett. 41 (1985) 81].
[4] T. J. Emge, P. C. W. Leung, M. A. Beno, A. J. Schultz, H. H. Wang, L. M. Sowa and J. M. Williams: Phys. Rev. B30 (1984) 6780.
[5] S. Ravy, J. P. Pouget, R. Moret and C. Lenoir: Phys. Rev. B37 (1988) 5113.
[6] S. Kagoshima, Y. Nogami, M. Hasumi, H. Anzai, M. Tokumoto, G. Saito and N. Mori: Solid State Commun. 69 (1989) 1177.
[7] S, Kagoshima, M. Hasumi, Y. Nogami, N. Kinoshita, H. Anzai, M. Tokumoto and G. Saito: Solid State Commun. 71 (1989) 843.
[8] Y. Nogami, S. Kagoshima, H. Anzai, M. Tokumoto, N. Mori, N. Kinoshita and G. Saito: J. Phys. Soc. Jpn. 59 (1990) 259.
[9] T. Sasaki, N. Toyota, M. Hasumi, T. Osada , S. Kagoshima, H. Anzai, M. Tokumoto and N. Kinoshita: J. Phys. Soc. Jpn. 58 (1989) 3477.
[10] K. Murata, M. Tokumoto, H. Anzai, H. Bando, K. Kajimura, T. Ishiguro and G. Saito: Synth. Metals 13 (1986) 3 and Synth. Metals 19 (1987) 151.
[11] M. Tokumoto, H. Bando, K. Murata, H. Anzai, N. Kinoshita, K. Kajimura, T. Ishiguro and G. Saito: Synth. Metals 13 (1986) 9.
[12] Y. Hasegawa and H. Fukuyama: J. Phys. Soc. Jpn. 55 (1986) 3717.
[13] N. Toyota, E. W. Fenton, T. Sasaki and M. Tachiki: Solid State Commun. 72 (1989) 859.

Bulk Superconductivity at Ambient Pressure in Polycrystalline Pressed Samples of Organic Metals

D. Schweitzer[1], S. Kahlich[1], S. Gärtner[2], E. Gogu[2], H. Grimm[2], R. Zamboni[2,*], and H.J. Keller[3]

[1] 3. Physikalisches Institut der Universität Stuttgart, Pfaffenwaldring 57, D-7000 Stuttgart 80, Fed. Rep. of Germany
[2] MPI für Medizinische Forschung, AG: Molekülkristalle, Jahnstraße 29, D-6900 Heidelberg, Fed. Rep. of Germany
[3] Anorganisch Chemisches Institut der Universität Heidelberg, Im Neuenheimer Feld 270, D-6900 Heidelberg, Fed. Rep. of Germany
* On leave from Istit. di Spectr. Molecolare del C.N.R., Bologna, Italy

Abstract. Bulk superconductivity in polycrystalline pressed samples of α_t-(BEDT-TTF)$_2$I$_3$ and β_p-(BEDT-TTF)$_2$I$_3$ is reported. This finding shows that organic superconductors can be used in principle for the preparation of electronic devices and superconducting cables.

1. Introduction

Organic metals and superconductors usually grow as single crystals at an electrode in an electrochemical cell. Therefore physical investigations of organic metals and superconductors are performed on single crystals. This is certainly an important fact for the understanding of the electronic properties of such materials. On the other hand, in the case of possible applications of such organic metals certainly very rarely single crystals could be used. Since up to now superconducting organic polymers are also not available, the only possible way to use organic superconductors at least in principle for applications would be as polycrystalline powders, which might be pressed to form larger samples.

From a physical point of view superconductivity in such polycrystalline pressed samples should be observable because it has been shown that the coherence lengths in such quasi-two-dimensional organic superconductors are typically of the order of 10−100 Å, that means of the order of the dimensions of the unit cell. A problem for the observation of superconductivity in polycrystalline pressed samples might arise from the fact that organic metals are usually relatively soft compared to inorganic superconductors. Therefore the organic materials might undergo phase transitions when a pressure is applied to the powder in order to obtain mechanically stable samples, and the development of annealing processes might be necessary for observing bulk superconductivity.

Here we report the preparation of such polycrystalline pressed samples of organic superconductors which show bulk superconductivity at ambient pressure.

2. Experimental and Results

Mechanically stable samples of the size of 4x1x0.5 mm³ were prepared from carefully pulverized single crystals of organic metals such as α-(BEDT-TTF)$_2$I$_3$, α_t-(BEDT-TTF)$_2$I$_3$, β-(BEDT-TTF)$_2$I$_3$ and (BEDT-TTF)$_2$Cu(NCS)$_2$ (the crystallites resulting from the pulverisation process had typical diameters of 0.5-10µm) by applying a pressure of about 1 kbar to the powder. The resistivity of the samples was measured by the usual four point method.

In the case of the polycrystalline pressed samples of (BEDT-TTF)$_2$·Cu(NS)$_2$ we were not able to observe bulk superconductivity [1], not even after annealing the samples at 80°C for several days. This fact might be due to a phase transition during the preparation of the sample.

In contrast to the samples of (BEDT-TTF)$_2$Cu(NCS)$_2$ in the case of the polycrystalline pressed samples of α_t-(BEDT-TTF)$_2$I$_3$, which were prepared from a powder of α-(BEDT-TTF)$_2$I$_3$ and annealed at 75°C for at least 3 days directly after the preparation, bulk superconductivity was observed (at 2 K about 50 % volume superconductivity with respect to an ideal superconductor [1]). The superconducting transition is relatively broad. While the onset for superconductivity in the resistivity curve is found near 9 K, zero resistivity is observed at 2.2 K. In the case of crystals of α_t-(BEDT-TTF)$_2$I$_3$ zero resistivity appears already at 6 K [2].

A question which arose was whether it is possible to obtain bulk superconductivity in samples of α_t-(BEDT-TTF)$_2$I$_3$ when the preparation of the samples starts from α_t-(BEDT-TTF)$_2$I$_3$ powder. Curve a in fig. 1 shows the resistivity versus temperature for such a polycrystalline pressed sample of α_t-(BEDT-TTF)$_2$I$_3$. A metal-like behaviour was found over the whole temperature range between 300 and 1.3 K but no bulk superconductivity could be found. In contrast after annealing the sample at 75°C for 3 days again a broad superconducting transition could be observed (curve b in fig.1) and ac susceptibility measurements indicate at 2 K a 50 % volume superconductivity with respect to an ideal superconductor. The behaviour of pressed α_t-(BEDT-TTF)$_2$I$_3$ samples prepared in this way was more or less identical with that of samples prepared from α-(BEDT-TTF)$_2$I$_3$ powder [1].

In order to obtain some more information about such phase transitions which occur under pressure during the preparation of the polycrystalline pressed samples, resonance Raman investigations, in particular on the most intensive vibrational symmetric stretching mode of the I$_3^-$ anions, were carried out. Earlier measurements on single crystals of α-, α_t- and β-(BEDT-TTF)$_2$I$_3$ [3] had shown that the resonance Raman-spectra are very sensitive to the symmetry of the I$_3^-$ anions. The symmetric stretching mode of the linear symmetric I$_3^-$ anions is usually found at about 10 cm^{-1} higher energy than that of the asymmetric and non-linear I$_3^-$ anions [3]. In the resonance Raman

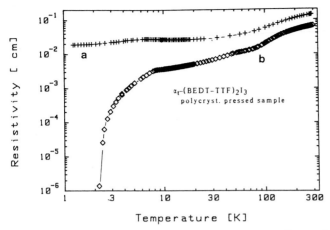

<u>Figure 1:</u> Resistivity versus temperature of a polycrystalline pressed sample of α_t-(BEDT-TTF)$_2$I$_3$ (curve a) and of a sample of α_t-(BEDT-TTF)$_2$I$_3$ which was annealed after the preparation (curve b).

spectra of the polycrystalline pressed sample of α-(BEDT-TTF)$_2$I$_3$ symmetric, linear and asymmetric I$_3^-$ anions are observed [4]. This indicates that the pressure during the preparation of the samples deforms the I$_3^-$ anions partially. A similar result can be observed for the polycrystalline pressed samples which were preprared directly from powdered α_t-crystals and not annealed after the preparation. In contrast to this finding the resonance Raman spectra of the annealed polycrystalline pressed samples of α_t-(BEDT-TTF)$_2$I$_3$ (which become superconducting) show only the stretching mode of the linear and symmetric I$_3^-$ anions, indicating again the higher symmetry and higher order of the structure.

The most surprising observation was made by measuring the temperature dependence of the resistivity of polycrystalline pressed samples of β-(BEDT-TTF)$_2$I$_3$ (in the following called β_p-(BEDT-TTF)$_2$I$_3$) [5]. Without annealing the samples showed an onset of superconductivity at 9 K, zero resistivity at 3.2 K and the middle of the resistive transition at 7.5 K (see fig. 2). This observation is surprising because single crystals of β-(BEDT-TTF)$_2$I$_3$ show a rather sharp superconducting transition at 1.2 K [6] and a metastable superconducting state at 8 K [7]. Here in the polycrystalline samples of β_p-(BEDT-TTF)$_2$I$_3$ the superconducting state at 7.5 K is stable and a bulk effect of the sample, as can be seen from the change of the ac susceptibility (see fig. 2) which corresponds at 2 K to about 50 % of that expected for a perfect superconductor.

In the samples of β_p-(BEDT-TTF)$_2$I$_3$ a structural phase transition occurring under pressure again plays a role. As a consequence of the phase transition here the transition temperature into the superconducting state is increased. This behaviour reemphasizes that organic superconductors might also be of interest for industrial applications.

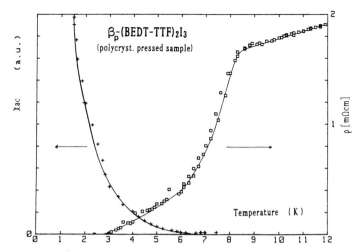

Figure 2: Resistivity and change in ac-susceptibility versus temperature of a polycrystalline sample of β_p-(BEDT-TTF)$_2$I$_3$ (below 12 K).

References.

[1] D. Schweitzer et al. Solid State Comm. 69, 843 (1989).
[2] D. Schweitzer et al. Z. Phys. B - Cond. Matt. 67, 489 (1987).
[3] R. Swietlik et al., Phys. Rev. B. 36, 6881 (1987).
[4] R. Zamboni et al., in "Lower Dimensional Systems and Molecular Electronics", Plenum Press 1989, in print.
[5] D. Schweitzer et al., Angew. Chem. Adv. Mater. 101, 977 (1989).
[6] E. B. Yagubskii et al., Sov. Phys. JETP Lett. 39, 12 (1984).
[7] F. Creuzet et al., J. Phys. (Paris) Lett. 46, L-1079 (1985).

An Ambient Pressure Organic Superconductor κ-(BEDT-TTF-h$_8$ and -d$_8$)$_2$Cu(NCS)$_2$ with T_c Higher than 10 K

H. Mori[1], S. Tanaka[1], H. Yamochi[2], G. Saito[2], and K. Oshima[3]

[1] International Superconductivity Technology Center, 2-4-1 Mutsuno, Atsuta-ku, Nagoya 456, Japan
[2] Institute for Solid State Physics, 7-22-1, Roppongi, Minato-ku, Tokyo 106, Japan
[3] Okayama University, 1-1-1 Tsushimanaka, Okayama 777, Japan

Abstract. The structure, chemical and physical properties of κ-(BEDT-TTF)$_2$Cu(NCS)$_2$ [BEDT-TTF: bis(ethylenedithio)tetrathiafulvalene] with T_c higher than 10 K are discussed. A higher and sharper superconducting transition was clearly observed when a higher-purity starting anion was used.

1. Introduction

So far, extensive studies on κ-(BEDT-TTF)$_2$Cu(NCS)$_2$ with T_c higher than 10 K have been carried out [1–11]. In the preparation, BEDT-TTF was oxidized electrochemically with KSCN, CuSCN, and 18-crown-6 ether in 1,1,2-trichloroethane under 1–5 μA. With this procedure, three phases were produced: α-(BEDT-TTF)$_2$Cu(NCS)$_2$, which exhibits a metal-insulator transition around 200 K [12]; (BEDT-TTF)Cu$_2$(SCN)$_3$, which shows semiconducting behavior with E_a = 0.017 eV [13]; and κ-(BEDT-TTF)$_2$Cu(NCS)$_2$, with $T_c \sim$ 10 K (κ-phase). In this paper we present the structure, and chemical and physical properties of the κ-phase and finally show how T_c increases in a purified sample.

2. Crystal Structure [3, 8]

As in usual organic superconductors, the κ-phase has a layered structure: the anion sheet and the donor sheet are stacked alternately along the a-axis (Fig. 1). In the donor layer, two crystallographically independent donors form a pair and the pair is arranged almost perpendicularly to neighboring pairs to construct a two-dimensional network in the bc plane. In the anion sheet, a Cu$^+$ is coordinated by two nitrogen atoms in a unit (SCN-Cu-NCS) and one sulfur in a neighboring unit to form a one-dimensional zig-zag polymer along the b-axis. The trigonal coordination of Cu$^+$ is unique.

Due to the absence of the inversion center (the space group P2$_1$), the absolute structure can be determined. From Bijvoet pair measurements, two possible structures were obtained; samples 1 and 2 correspond to Fig. 1a and Fig. 1b, respectively. Also, cut-out portions of these samples showed optical isomerism: sample 1 was levo-rotatory and 2 was dextro-rotatory. $(\alpha)_{632.8\,nm}^{25\,°C}$ was determined to be $\sim \pm 230°$.

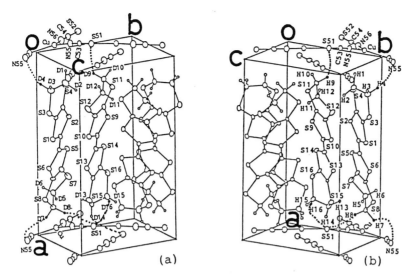

Fig. 1. Crystal structure of κ-(BEDT-TTF)$_2$Cu(NCS)$_2$

3. ESCA, ESR, and Thermopower [5]

Figure 2a shows an ESCA spectrum at room temperature. These binding energies are reasonable for Cu$^+$ and neither shoulder peak nor shake-up satellites of Cu^{2+} were observed at room temperature.

ESR signals from 298 K to 4 K were also obtained. At 298 K, one broad Lorentzian signal attributed to the BEDT-TTF radical cation was observed. No Cu^{2+} signal which usually appears from $g = 2.05$ to 2.50 was observed, so that we determine the state of copper as the +1 anion over the entire temperature range.

Since the anion sheet is an insulator, there are conducting carriers only in a donor plane. Therefore the thermopower should be positive assuming the simple tight-binding approximation. However, Fig. 2b shows that the thermopower along the c-axis is positive while that along the b-axis is negative from 300 K to 10 K. The

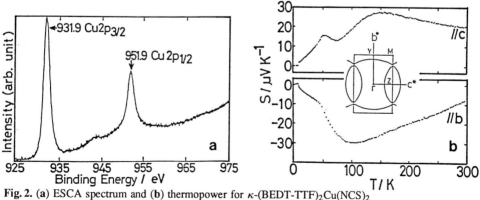

Fig. 2. (a) ESCA spectrum and (b) thermopower for κ-(BEDT-TTF)$_2$Cu(NCS)$_2$

band calculation based upon an extended Huckel method explained that the open Fermi surface around ΓY is electron-like, which is consistent with the negative thermopower along the b-axis, while the closed Fermi surface around Z is hole-like, which is in good agreement with the positive thermopower along the c-axis. Therefore the anisotropy of the thermopower is a result of the anisotropic band structure of this κ-phase salt [14].

4. Electrical Resistivity, H_{c2}, and Shubnikov–de Haas Signal [2, 4]

Figure 3 shows the temperature dependence of the electrical resistivity. Around room temperature metallic behavior appears, and the resistivity increases with decreasing temperature. After the resistivity-peak around 90–100 K, the metallic state was found again, accompanied by a superconducting transition at 10.4 K in the non-deuterated

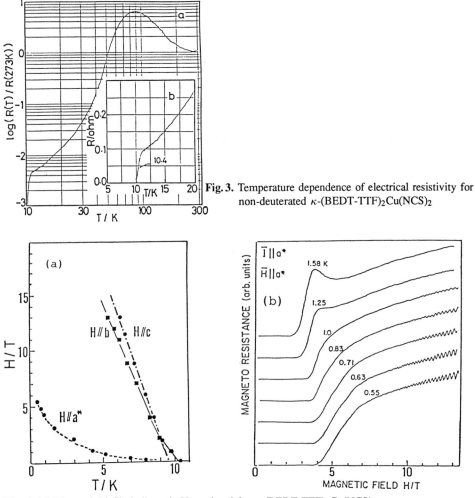

Fig. 3. Temperature dependence of electrical resistivity for non-deuterated κ-(BEDT-TTF)$_2$Cu(NCS)$_2$

Fig. 4. (a) H_{c2} and (b) Shubnikov–de Haas signal for κ-(BEDT-TTF)$_2$Cu(NCS)$_2$

salt. The T_c of the deuterated salt is a little higher, $T_c = 11.0$ K, in spite of the higher isotope mass of the donor, suggesting an inverse isotope effect in this salt.

In Fig. 4a, the temperature dependence of H_{c2} is shown. The behaviors along the b- and c-axes make clear the two-dimensionality in the bc plane. The calculated coherence length is $\xi_c(0) : \xi_{a*}(0) = 182$ Å: 9.6 Å = 19 : 1.

The first clearly observed Shubnikov–de Haas signal was obtained below 1 K above 8 T in the κ-phase (Fig. 4b). The period of oscillation ($0.0015\,\text{T}^{-1}$), which corresponds to 18% of the first Brillouin zone, agrees well with 18% of the calculated closed Fermi surface.

5. How does T_c Increase with a Purified Sample?

In order to pursue a higher T_c and sharper superconducting transition, we measured the T_c with purified samples. The purification was carried out in an anion; KSCN was recrystallized in EtOH twice. CuSCN was purified with recrystallized KSCN three times. 18-crown-6 ether was recrystallized in CH_3CN. Transparent crystals of crown salt, $Cu(SCN)_2K$(18-crown-6 ether), were obtained with their purified starting materials and were used as a supporting electrolyte in electrocrystallization. As shown in Fig. 5, the most purified sample (a) shows the highest ($T_c = 11.1$ K) and the sharpest transition. Therefore, it is necessary to purify the starting materials in order to get the correct intrinsic property of the κ-phase. Observations of H_{c2} and the Shubnikov–de Haas signal with the most purified sample are in progress.

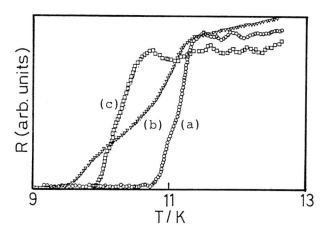

Fig. 5. Superconducting transition behavior for (a) most purified, (b) purified, and (c) not purified κ-$(BEDT\text{-}TTF\text{-}d_8)_2Cu(NCS)_2$ samples

References

1. H. Urayama, H. Yamochi, G. Saito, K. Nozawa, T. Sugano, M. Kinoshita, S. Sato, K. Oshima, A. Kawamoto, J. Tanaka: Chem. Lett. **1988**, 55

2   K. Oshima, H. Urayama, H. Yamochi, G. Saito: J. Phys. Soc. Jpn. **57**, 730 (1988)
3   H. Urayama, H. Yamochi, G. Saito, S. Sato, A. Kawamoto, J. Tanaka, T. Mori, Y. Maruyama, H. Inokuchi: Chem. Lett. **1988**, 463
4   K. Oshima, T. Mori, H. Inokuchi, H. Urayama, H. Yamochi, G. Saito: Phys. Rev. B **37**, 938 (1988)
5   H. Urayama, H. Yamochi, G. Saito, T. Sugano, M. Kinoshita, T. Inabe, T. Mori, Y. Maruyama, H. Inokuchi: Chem. Lett. **1988** 1057
6   K. Oshima, H. Urayama, H. Yamochi, G. Saito: Physica C **154**, 1148 (1988)
7   G. Saito, H. Urayama, H. Yamochi, K. Oshima: Synth. Met. **27**, 331 (1988)
8   H. Urayama, H. Yamochi, G. Saito, S. Sato, T. Sugano, M. Kinoshita, A. Kawamoto, J. Tanaka, T. Inabe, T. Mori, Y. Maruyama, H. Inokuchi: Synth. Meth. **27**, 393 (1988)
9   K. Oshima, T. Mori, H. Inokuchi, H. Urayama, H. Yamochi, G. Saito: Synth. Met. **27**, A413 (1988)
10 K. Oshima, H. Urayama, H. Yamochi, G. Saito: Synth. Met. **27**, A419 (1988)
11 K. Oshima, H. Urayama, H. Yamochi, G. Saito: Synth. Meth. **27**, A473 (1988)
12 N. Kinoshita, K. Takahashi, K. Murata, M. Tokumoto, H. Anzai: Solid State Commun. **67**, 465 (1988)
13 U. Geiser, M.A. Beno, A.M. Kini, H.H. Wang, A.J. Shultz, B.D. Gates, C.S. Cariss, K.D. Carlson, J.M. Williams: Synth. Met. **27**, A375 (1988)
14 T. Mori, H. Inokuchi: J. Phys. Soc. Jpn. **57**, 3674 (1988)

Nuclear Spin-Lattice Relaxation in the Organic Superconductor (BEDT-TTF)$_2$Cu(NCS)$_2$: Measurements by the Field Cycling Technique

T. Takahashi[1], K. Kanoda[1], K. Sakao[1], M. Watabe[1], H. Mori[2], and G. Saito[2]

[1]Department of Physics, Gakushuin University, Mejiro, Toshima-ku, Tokyo 171, Japan
[2]Institute for Solid State Physics, University of Tokyo, Roppongi, Minato-ku, Tokyo 106, Japan

Abstract. ^{1}H nuclear spin-lattice relaxation in the organic superconductor, (BEDT-TTF)$_2$Cu(NCS)$_2$, was investigated at zero field by field cycling technique as well as at the fields of 3.28 and 11.7 kOe. Below the superconducting transition, anomalous behaviors of the relaxation were observed; a large enhancement of T_1^{-1} in the field and a strongly non-single-exponential relaxation with extremely fast component at nominal zero field. These results are discussed in light of vortex dynamics including liquid-like behavior relevant to two dimensional superconductivity.

The nuclear spin-lattice relaxation in metallic systems generally takes place through hyperfine couplings between the nuclear and the conduction-electron spins and therefore probes the pairing state of electrons when applied to the study of superconductivity. There have been observed two different types of temperature dependence of the nuclear spin-lattice relaxation rate, T_1^{-1}, in the superconducting states, depending on the type of electron pairing: For s-wave pairing with isotropic nature as in the BCS model, T_1^{-1} increases just below T_c, reaching a maximum around, say, 0.9 T_c, and then decreases exponentially due to the presence of finite gap [1]. In anisotropic superconductivity with a gapless nature, the enhancement below T_c does not appear and the low temperature variation follows a power law [2]. The latter behavior of the unconventional superconductivity has been reported in a typical organic superconductor, (TMTSF)$_2$ClO$_4$ [3]. The organic conductors are now intriguing materials from the viewpoint of the physics of superconductivity.

In recent years, we have been doing NMR studies on (BEDT-TTF)$_2$-Cu(NCS)$_2$ [4], which has the highest T_c among the organic superconductors to date. This paper reports the measurements of ^{1}H nuclear spin-lattice relaxation for a polycrystalline sample at nominal zero field by using field cycling technique as well as the results at the finite fields.

First, the temperature dependences of T_1^{-1} at the fields of 3.28 and 11.7 kOe are given in Fig. 1. The recovery of the nuclear magnetization fitted well to a single exponential curve for all the measurements. At the field of 3.28 kOe, T_1^{-1} is greatly enhanced below T_c, forming a peak around 4 K, and then decreases rapidly at lower temperatures toward zero. The peak value is about 30 times as large as that at 10 K. At a higher field, 11.7 kOe, the enhancement of T_1^{-1} is reduced and the peak position shifts to a lower temperature, 3 K. With reference to the upper critical field data [5], T_c at 3.28 kOe is distributed between 8 and 10 K; the anomaly occurs well below the superconducting transition. The large field dependence indicates a close relation to the superconductivity.

The enhancement is exceptionally large compared with the usual superconductors and cannot be explained by any theory either for the isotropic or anisotropic superconductivity, mentioned above. At first

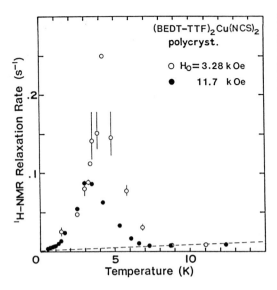

Fig. 1 ^{1}H-NMR relaxation rate at the fields of 3.28 and 11.7 kOe. The broken line indicates the Korringa behavior, $T_1T = 1100$ secK, determined at high temperatures.

sight, the observed peak is reminiscent of SDW or some other magnetic transitions [6]. According to the band calculation, this compound has two parts of Fermi surfaces; one is a closed cylinder and the other is a warped open sheet [7]. Thus it may not be unreasonable to speculate that the latter has some kind of Fermi surface instability. In this case, however, the interplay with the superconductivity should be a serious problem. Since the superconductivity does not show any anomaly around that temperature, according to our recent measurement of weak-field penetration depth [8], this possibility is unlikely.

At present we consider that the enhancement is caused by the field fluctuations generated by the vortex dynamics. This type of motion, in general, causes the so-called BPP-type relaxation, which is given by [9]; $T_1^{-1} \sim \gamma_n^2 h_0^2 \tau_c / (1+\omega_n^2 \tau_c^2)$, where γ_n, h_0, τ_c and ω_n are the gyromagnetic ratio of the nucleus, the amplitude and the correlation time of the field fluctuation, and the Larmor frequency. The maximum of $T_1^{-1} \sim \gamma_n^2 h_0^2 / 2\omega_n$ appears when the temperature dependent τ_c satisfies $\tau_c \omega_n = 1$. In order to explain the observed peak value, 0.25 s^{-1}, at 3.28 kOe, h_0 should be 0.4 Oe, which is not unrealistic. Considering the sharpness of the peak, vortex lattice melting is favorable as the origin of the required dynamics. Note that this compound has strong two dimensional electronic states. Indeed, the estimated melting temperature using the transport data yields 2.7 K which is close to the temperature where the anomaly occurs [4].

Several experimental results available support this conjecture. The temperature dependence of NMR line width was found to increase abruptly below 2-3 K, while it did not change through T_c. The increase of the line width means the appearance of the inhomogeneity of the field. In the vortex state, the field undulates spatially and the distribution of the field is determined by the penetration depth. The abrupt increase at low temperatures, however, cannot be explained by the temperature dependence of the penetration depth. This puzzle is easily solved if we assume that at higher temperatures the vortices are mobile or in a liquid phase and average out the field distribution at the nuclear sites, while at low temperatures the formation of vortex lattice generates undulation of the field. Moreover, the dc susceptibility measurements by Sugano et al. [10], in field-cooled and zero-field cooled conditions, seem to us to imply the existence of a well defined temperature, above which vortices are free from pinning but below which the pinning is effective.

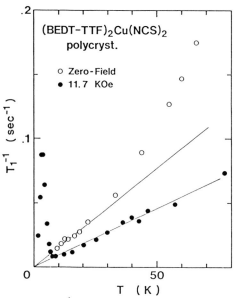

Fig. 2 ^{1}H-NMR relaxation rate at zero-field and at 11.7kOe. The two straight lines show the Korringa behaviors, $T_1T = 650$ secK and $T_1T = 1100$ secK, at zero and at 11.7kOe, respectively.

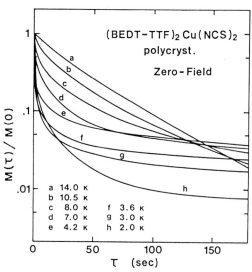

Fig. 3 Relaxation profile of the nuclear magnetization at nominal zero field. Non-single exponential decays are observed below T_c.

Next we give the results of the zero-field relaxation rate. Above T_c, the recovery of the nuclear magnetization in time is single exponential as was in the field. Figure 2 shows the temperature dependence of T_1^{-1} at zero magnetic field together with that at 11.7 kOe. In the temperature range between 10 and 30 K, T_1^{-1} follows the Korringa relation, $(T_1T)^{-1} = (650 \text{ secK})^{-1}$. The value of $(T_1T)^{-1}$ at zero field is enhanced by the factor of 1.7 compared with that at 11.7 kOe. This agrees well the theoretical value of 2 for uncorrelated local fields [11]. Above 30 K, T_1^{-1} gets to deviate from the linearity while the deviation at 11.7 kOe occurs above 100 K. This is reasonably attributed to an additional relaxation due to ethylene thermal motions.

Below T_c, the decay of the nuclear magnetization becomes non-single exponential. The profile of the decay curve is shown in Fig. 3. As the temperature decreases, the non-single exponential nature becomes more pronounced. In particular, the initial slope of the decay is surprisingly steep and is even larger by two orders of magnitude than the relaxation rate at 11.7 kOe. Furthermore, the long component is much longer than the measured time range; it may rather seem that a finite value of nuclear magnetization remains in the long-time limit. If this were the case, however, a considerable part of the external field (of the order of several per cent of the initial value!) should have been left in our nominal zero field. The field dependence of the relaxation profile was precisely investigated at 4.2 K by the field cycling technique; on increasing the field, the non-single exponential decay at zero field was found to turn gradually into a single exponential curve; the relaxation rate becomes smaller for further increase of the field, and tends continuously to the results at 3.28 and 11.7 kOe.

These results are not inconsistent with the vortex dynamics discussed above. The non-single exponential decay means inhomogeneous relaxation of nuclei. Since the relaxation in the normal state is homogeneous, the vortex structure is the most probable candidate for the origin of the inhomogeneity. Even at zero (applied) field in the process of the field cycling, we must suppose the existence of a considerable number of vortices. There should be vortex motions, such as flux creep or, although speculative, the motions of bound vortex-antivortex pairs. The amplitude of the field fluctuation, h_0, should be proportional to that of instantaneous spatial variation of the field and therefore is determined by the vortex density; the smaller the applied field, the larger the amplitude of the field distribution. (Strictly speaking, only the transverse components of the field fluctuation contribute to the relaxation.) This explains the observed field dependence that the smaller field gives the larger relaxation rate.

In order to confirm the picture, the relaxation measurements for ^{13}C-enriched sample are under way. We also hope the ^{13}C relaxation will reveal the hyperfine contribution which has been masked in the present ^1H relaxation.

Acknowledgments. This work was supported by Grant-in-aid for scientific research from the Ministry of Education, Science and Culture of Japan.

References

1. L.C. Hebel and C.P. Slichter, Phys. Rev. **113**, 1504 (1959).
2. Y. Hasegawa and H. Fukuyama, J. Phys. Soc. Jpn. **56**, 877 (1987); ibid, **56**, 2619 (1987).
3. M. Takigawa, H. Yasuoka, and G. Saito, J. Phys. Soc. Jpn. **56**, 873 (1987).
4. T. Takahashi, T. Tokiwa, K. Kanoda, H. Urayama, H. Yamochi, and G. Saito, Physica C**153-155**, 487 (1988); Synth. Met. **27**, A319 (1988).
5. K. Oshima, H. Urayama, H. Yamochi, and G. Saito, J. Phys. Soc. Jpn. **57**, 730 (1988).
6. T. Takahashi, Y. Maniwa, H. Kawamura, and G. Saito, Physica **143B**, 417 (1986).
7. K. Oshima, T. Mori, H. Inokuchi, H. Urayama, H. Yamochi, and G. Saito, Phys. Rev. B**38**, 938 (1988).
8. K. Kanoda, K. Akiba, T. Takahashi, and G. Saito, in this volume.
9. C.P. Slichter, _Principles of Magnetic Resonance_ (Springer-Verlag, 1980) p.167.
10. T. Sugano, K. Nozawa, H. Hayashi, K. Nishikida, K. Terui, T. Fukasawa, H. Takenouchi, S. Mino, H. Urayama, H. Yamochi, G. Saito, and M. Kinoshita, Synth. Met. **27**, A325 (1988).
11. A. Abragam, _Principles of Nuclear Magnetism_ (Oxford University Press, 1961) p.363.

Magnetic-Field Penetration Depth of (BEDT-TTF)$_2$Cu(NCS)$_2$ Determined by Complex Susceptibility

K. Kanoda[1], K. Akiba[1], T. Takahashi[1], and G. Saito[2]

[1]Department of Physics, Gakushuin University, Mejiro, Toshima-ku, Tokyo 171, Japan
[2]Institute for Solid State Physics, University of Tokyo, Roppongi, Minato-ku, Tokyo 106, Japan

Abstract. The magnetic-field penetration depth, λ, of the organic superconductor, (BEDT-TTF)$_2$Cu(NCS)$_2$, has been investigated by means of the complex susceptibility measurements on single crystals. The temperature dependence of λ showed a power-law behavior at low temperatures, instead of the BCS behavior. The present result strongly suggests anisotropic superconductivity with the nodes of gap parameter on the Fermi surface.

1. Introduction

Extensive research on material science has provided several exotic metals such as heavy electron, organic and oxide superconductors. It has become a practical problem whether the superconductivity in these materials is of conventional BCS-type or not. Among them, the heavy electron systems have increasing evidence that these are not conventional superconductors. This finding came in part from the measurements of temperature dependences of specific heat [1], nuclear spin-lattice relaxation rate [2], ultrasonic attenuation [3] and magnetic field penetration depth [4].

The title compound, (BEDT-TTF)$_2$Cu(NCS)$_2$, has the highest transition temperature, $T_c \sim 10.4$ K, among the organic superconductors to date [5]. This compound consists of alternating sheets of BEDT-TTF molecules and Cu(NCS)$_2$ ions, and is characterized by a quasi-two-dimensional electronic state. Several experiments aiming at the clarification of the type of superconductivity have been made. Specific heat was measured by Katsumoto et al. [6], who found that it follows the T^3 law at low temperatures. However, a possible large contribution of phonons did not allow them to deduce the temperature dependence of electronic part. The ^{1}H nuclear spin-lattice relaxation rate, T_1^{-1}, was measured by our group [7]. The observed temperature dependence of T_1^{-1}, however, was far from any behaviors already predicted. This strongly suggests the existence of some other relaxation mechanism, never considered; the question still remains open.

Magnetic-field penetration depth, λ, probes the type of superconductivity. In the present work, we made precise measurements of complex susceptibility, χ, for single crystals of (BEDT-TTF)$_2$Cu(NCS)$_2$ and derived the penetration depth in two directions of external field. Evidence of anisotropic superconductivity of a gapless nature is reported.

2. Results

The complex susceptibility χ ($\chi=\chi'-i\chi''$) was measured down to 1.5 K, with a Hartshorn-type mutual inductance bridge, resolution of which is ~ 5 nH. The ac field was applied either perpendicular or parallel

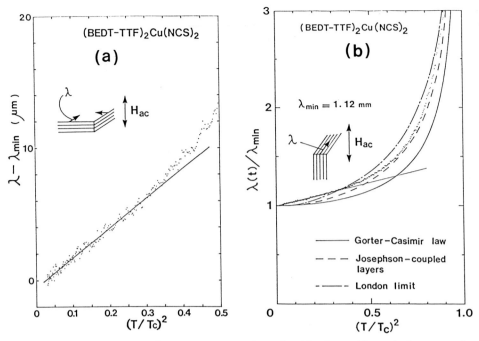

Fig.1 (a) Deviation of the penetration depth from the minimum value, $\lambda-\lambda_{min}$, plotted as a function of $(T/T_c)^2$. Inset shows the directions of the applied ac field and the penetration with respect to the crystal layer. (b) Normalized penetration depth, λ/λ_{min}, plotted as a function of $(T/T_c)^2$. Note that the directions of the applied ac field and the penetration are different from those in (a).

to the crystal layer. In the parallel case, where the demagnetizing effect can be neglected, the absolute value of the susceptibility was determined. For calibration, we used the complete diamagnetism of Sn films (7-500 μm thick).

At perpendicular fields, χ' exhibits a sharp transition at 9.2 K. On increasing the field, χ' around T_c becomes depressed gradually but the low-temperature values lie on a "universal curve" in the zero field limit when the applied ac field is less than 2 Oe. The imaginary part, χ'', forms a sharp peak only around the transition region. Vanishing χ'' except in the vicinity of T_c implies that no vortices enter into the samples. In the perpendicular direction of the field, χ' shows almost a perfect diamagnetism, the deviation from which is quite small. Moreover, the demagnetization effect is large and the correction is not easy for crystals of irregular shape. Thus one cannot determine the absolute value of λ from the susceptibility data. It is, however, possible to obtain the deviation from the minimum value of λ in the case of thin crystal with a diameter of $2R$ and a thickness of $2d$ ($R \gg d$), by taking into account the χ'-dependence of the demagnetization factor. The calculation is straightforward [8] and the final formula for $R \gg \lambda_{min}$ is given as $\lambda(T)-\lambda_{min}=R[1-[\chi'(T)/\chi'(T_{min})]^{1/3}]$. The results thus obtained from χ' at 116 mOe are plotted as a function of $(T/T_c)^2$ in Fig. 1(a). (χ'' at 116 mOe is considered as those in the zero-field limit in the temperature range, $T/T_c < 0.95$.) The temperature dependence of λ is well approximated by the T^2-law at low temperatures of $(T/T_c)^2 < 0.35$, i.e. $T/T_c < 0.6$. Experimentally, the exponent, α, of the power-law, T^α, is in the range of 1.7-2.2. Such a power law is in remarkable contrast to the BCS behavior, where λ varies exponentially in temperature, as considered later.

Next, the results in the ac field parallel to the crystal layers are given. In this geometry of the field, we can determine the absolute value of the penetration depth since the absolute values of χ' are measured. The results were independent of the amplitude of the ac field at least up to 1.2 Oe. The imaginary part, χ'', is zero throughout the entire temperature range at the field below 1.2 Oe. The deviation from the perfect diamagnetism ($-4\pi\chi' < 1$) in the low field region is caused by the field penetration, as in the perpendicular geometry. It should be noted that when the field is applied parallel to the crystal surface the penetration can take place both from the surface in a direction perpendicular to the layers and from the edges in the parallel direction. In layered materials, the latter is often dominant. Detailed measurements and analysis were made for several crystals with different dimensions and revealed that the parallel penetration shown in the inset of Fig.1(b) determines χ' in the present system. (Details were described in Ref. 8.) λ is calculated by the equation $-4\pi\chi' = 1 - (2\lambda/D)\tanh(D/2\lambda)$, where D is the dimension of the sample in the direction of field penetration. The obtained penetration depth, which is normalized to the minimum value, is shown as a function of $(T/T_c)^2$ in Fig.1(b), as well as the empirical Gorter-Casimir law, $\lambda \sim [1-(T/T_c)^4]^{-1/2}$, and the model calculation of the Josephson-coupled layers based on the BCS framework [9]. The latter gives the same temperature dependence as in the dirty local limit, namely, $\lambda \sim \Delta(T)\tanh(\Delta(T)/2kT)$, which yields exponential dependence in T at low temperatures. As seen in the figure, our results exhibit T^2-dependence, $\lambda(T)/\lambda(0) = 1 + 0.45(T/T_c)^2$, at low temperatures. Now the temperature dependence of λ is again of a power-law, instead of an exponential BCS behavior. The results were fairly reproducible for four measurements on different crystals.

3. Discussion

The temperature dependence of the penetration depth at low temperatures generally probes the gap structure in the density of states of the quasi-particle excitation. The BCS theory predicts the exponential temperature dependence due to the finite gap in the excitation. For example, in the local limit which is relevant to our compound, $\lambda(T)-\lambda(0) = \lambda(0)(2\pi\Delta/kT)^{1/2}e^{-\Delta/kT}$ at $T \ll \Delta/k$. On the other hand, a power-law dependence means the existence of non-vanishing density of states above the ground state and is expected for anisotropic superconductors with nodes of the gap parameter on the Fermi surface. The present results give evidence that the superconductivity in the organic conductor, $(BEDT-TTF)_2Cu(NCS)_2$, has a gapless nature of this kind similar to the heavy electron systems.

The possibility of anisotropic superconductivity and its characteristics, if any, in the quasi-two-dimensional organic systems were discussed in the context of nonlocal attractive and on-site Coulomb interactions by Hasegawa and Fukuyama [10]. Exciton-mediated inter-layer interaction was proposed by Nakajima [11].

In conclusion, the organic superconductor $(BEDT-TTF)_2Cu(NCS)_2$ exhibits power-law temperature dependence of the penetration depth. Anisotropic pairing with nodes of the gap parameter on the Fermi surface was suggested for this material.

Acknowledgments

The authors would like to thank K. Suzuki for sample preparation and M. Mori for technical assistance. This work was supported by a Grant-in-aid for Scientific Research from the Ministry of Education, Science, and Culture of Japan.

References

[1] H. R. Ott, H. Rudigier, T. M. Rice, K. Ueda, Z. Fisk, and J. L. Smith, Phys. Rev. Lett. **52**, 1915 (1984).
[2] D. E. McLaughlin, Cheng Tien, W. G. Clark, M. D. Lan, Z. Fisk, J. L. Smith, and H. R. Ott, Phys. Rev. Lett. **53**, 1833 (1984).
[3] D. J. Bishop, C. M. Varma, B. Batlogg, and E. Bucher, Phys. Rev. Lett. **53**, 1009 (1984).
[4] D. Einzel, P. J. Hirschfeld, F. Gross, B. S. Chandrasekhar, K. Andres, H. R. Ott, J. Beuers, Z. Fisk, and J. L. Smith, Phys. Rev. Lett. **56**, 2513 (1986).
[5] H. Urayama, H. Yamochi, G. Saito, K. Nozawa, T. Sugano, M. Kinoshita, S. Sato, K. Oshima, A. Kawamoto, and J. Tanaka, Chem Lett. **1988**, 55 (1988).
[6] S. Katsumoto, S. Kobayashi, H. Urayama, H. Yamochi, and G. Saito, J. Phys. Soc. Jpn. **57**, 3672 (1988).
[7] T. Takahashi, T. Tokiwa, K. Kanoda, H. Urayama, H. Yamochi, and G. Saito, Physica C **153-155**, 487 (1988); Synth. Met. **27**, A319 (1988).
[8] K. Kanoda, T. Takahashi, G. Saito, in Proceedings of the International Conference on M^2S-HTSC, Stanford, 1989 (to be published in Physica C); K. Kanoda, K. Akiba, S. Suzuki, T. Takahashi, G. Saito, (to be published)
[9] G. Deutscher and O Entin-Wohlman, J. Phys. C **10**, L433 (1977).
[10] Y. Hasegawa and H. Fukuyama, J. Phys. Soc. Jpn. **56**, 2619 (1987).
[11] S. Nakajima, J. Phys. Soc. Jpn. **56**, 871 (1987).

Tunneling Spectroscopic Study of the Superconducting Gap of $(BEDT-TTF)_2Cu(NCS)_2$ Crystals

Y. Maruyama[1], T. Inabe[1], H. Mori[2], H. Yamochi[3], and G. Saito[3]

[1] Institute for Molecular Science, Myodaiji, Okazaki 444, Japan
[2] International Superconductivity Technology Center, Mutsuno, Atsuta-ku, Nagoya 456, Japan
[3] Institute for Solid State Physics, University of Tokyo, Roppongi, Tokyo 106, Japan

Abstract. The superconducting gaps of κ-(BEDT-TTF-h_8 and -d_8)$_2$Cu(NCS)$_2$ crystals have been measured by a tunneling spectroscopic method. One of the observed gap data, 4 meV, for an -h_8 salt gives 4.5 for $2\Delta/k_BT_c$ (2Δ = 4 meV and Tc = 10.4 K), which is a little larger than that of the BCS ratio, 3.52. The other spectra observed in different crystals, however, exhibit much smaller gap structure. A similar measurement for a -d_8 crystal has revealed that there seem to be three different superconducting gaps, 0.8, 2.1 and 4.3 meV in the tunneling spectrum. These observations are suggestive of the anisotropic nature of the superconductivity in these crystals.

1. Introduction

The organic superconductors, κ-(BEDT-TTF-h_8 and -d_8)$_2$Cu(NCS)$_2$ (BEDT: bis(ethylenedithiolo)-tetrathiafulvalene), have the highest transition temperatures T_c, 10.4 and 11 K respectively, among the organic superconductors reported so far [1]. The crystal and electronic structures [2] and the magnetoresistance oscillations [3] have clearly revealed their 2-dimensional metallic character. Anisotropy of the thermoelectric powers is also suggestive of strong 2-dimensionality [4].
  Tunneling spectroscopic studies on organic superconductors have been carried out for TMTSF salts (TMTSF: tetramethyl-tetraselenafulvalene) [5, 6, 7], β-(BEDT-TTF)$_2$AuI$_2$ [8], β-(BEDT-TTF)$_2$I$_3$ [9], and κ-(BEDT-TTF)$_2$Cu(NCS)$_2$ [10]. Some of the observed data lead to an extraordinarily large value of $2\Delta/k_BT_c$ compared with the BCS weak coupling limit. In TMTSF salts, such an anomaly has been ascribed to a superconducting fluctuation in the one-dimensional columns and/or an SDW transition on quick cooling, and in β-(BEDT-TTF)$_2$AuI$_2$ and κ-(BEDT-TTF-h_8)$_2$Cu(NCS)$_2$ to the anisotropic electronic structure [8, 10]. In β-(BEDT-TTF)$_2$I$_3$, a rather normal value of the ratio was reported [9]. Very recently, Bando et al. have observed tunnel spectra for κ-(BEDT-TTF)$_2$Cu(NCS)$_2$ by a scanning tunneling microscope technique [11].
  We have measured tunneling spectra for the normal direction to the bc plane of κ-(BEDT-TTF-d_8)$_2$Cu(NCS)$_2$ single crystals and compared them with the result of the -h_8 salt.

2. Experimental

Single crystals were prepared by the method reported before [1, 2]. The typical size of crystals used is 3×1.5×0.1 mm^3. A crystal was fixed on a quartz plate by an epoxy resin. Very thin (less than 50 Å) insulator film was deposited on the surface (bc plane) of a crystal by the evaporation of alumina (Al_2O_3) with electron-beam heating in a high vacuum vessel. A gold film was evaporated subsequently on an alumina coated surface of the crystal as a counter electrode to form an $Au/Al_2O_3/(BEDT-TTF)_2Cu(NCS)_2$ tunneling cell [6]. dI/dV vs V curves were obtained by a conventional voltage modulation method [6].

3. Results and Discussion

Our tunnel junctions are not a point-contact type but a face-to-face contact. The surface of the crystal used was not completely flat and it has a few very thin step-wise terrace structures which may occasionally influence the observed spectra. The tunneling current may flow through the large flat surface in some situation, but in another situation it may flow through the micro-contact of edges or side-faces of the terraces.

A typical spectrum for κ-$(BEDT-TTF-h_8)_2Cu(NCS)_2$ at 1.5 K is shown in Fig. 1. No clear change in the tunneling spectra was observed around the transition temperature, 10.4 K, and also some reliable spectra above 4.2 K were obtained. Only below 4.2 K could we observe significant structures in the spectra, which may be assigned to a superconducting gap structure. In Fig. 1, we can estimate the gap energy by drawing a line so as to balance the areas of the one valley and two peaks. Thus, we obtain 4 meV as the gap, 2Δ. In accordance with the BCS relation, $2\Delta/k_BT_c = 3.52$, in the weak coupling limit, 4 meV for 2 gives 4.5 for T_c = 10.4 K. Tunnel spectra of a different crystal are shown in Fig. 2, whose general appearance seems to be quite different from that in Fig. 1. We can recognize two or three gap structures in these spectra; $\lesssim 1$ meV, ~ 2 meV and ~ 4 meV. In this case, the tunneling current may flow through three types of micro-contacts.

Similar results were obtained for a -d_8 salt crystal. The temperature dependence of tunnel spectra for a κ-$(BEDT-TTF-d_8)_2Cu(NCS)_2$ crystal is shown in Fig. 3. No significant structure appeared above 4.2 K, and some structures can be recognized only in the 1.5 K spectrum. In order to make them clear, we subtracted the smooth spectrum line of 4.2 K from the curve of 1.5 K as shown in Fig. 4. The resultant curve which is shown in the bottom part of

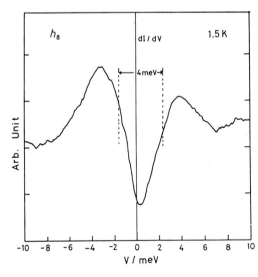

Fig. 1. Tunneling spectrum of a κ-$(BEDT-TTF-h_8)_2Cu(NCS)_2$ single crystal at 1.5 K.

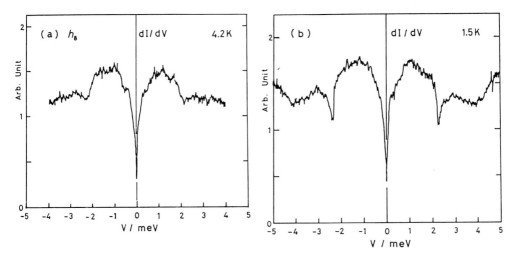

Fig. 2. Tunneling spectra of a different κ-(BEDT-TTF-h_8)$_2$Cu(NCS)$_2$ crystal. (a) At 4.2 K, and (b) at 1.5 K.

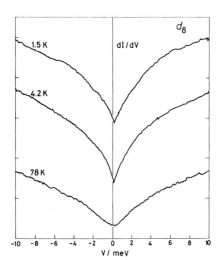

Fig. 3. Temperature dependence of a κ-(BEDT-TTF-d_8)$_2$Cu(NCS)$_2$ tunneling spectra.

the figure may correspond to the gap structure, which is almost similar to the structure in Fig. 2. The largest gap for the -d_8 salt, 4.3 meV, is a little larger than the corresponding value, 4 meV, for -h_8 salt, and the difference between them is just in the range of that expected from their critical temperatures. However, this range of difference is within the experimental error bar, ±0.2 meV, so at present we can't say that the gap of the -d_8 salt is definitely larger than that of -h_8 salt. Anyhow, we can tentatively conclude that κ-(BEDT-TTF)$_2$Cu(NCS)$_2$ has anisotropic superconducting gaps. The largest gap is about 4 meV and the smallest one is less than 1 meV.

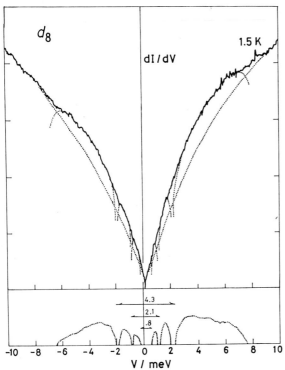

Fig. 4. Subtraction deconvolution of the $-d_8$ tunneling spectrum at 1.5 K.

References

[1] H. Urayama, H. Yamochi, G. Saito, K. Nozawa, T. Sugano, M. Kinoshita, S. Sato, K. Oshima, A. Kawamoto, and J. Tanaka, Chem. Lett., **1988**, 55.
[2] H. Urayama, H. Yamochi, G. Saito, S. Sato, A. Kawamoto, J. Tanaka, T. Mori, Y. Maruyama, and H. Inokuchi, Chem. Lett., **1988**, 463.
[3] K. Oshima, T. Mori, H. Inokuchi, H. Urayama, H. Yamochi, and G. Saito, Phys. Rev. B**38**, 938 (1988).
[4] H. Urayama, H. Yamochi, G. Saito, T. Sugano, M. Kinoshita, T. Inabe, T. Mori, Y. Maruyama, and H. Inokuchi, Chem. Lett., **1988**, 1057.
[5] C. More, G. Roger, J. P. Sorbier, D. Jérome, M. Ribault, and K. Bechgaard, J. Phys. (Paris) Lett., **42**, 313 (1981).
[6] Y. Maruyama, R. Hirose, G. Saito, and H. Inokuchi, Solid State Commun., **47**, 273 (1983).
[7] H. Bando, K. Kajimura, H. Anzai, T. Ishiguro, and G. Saito, in Proceedings of the Seventeenth International Conference on Low-Temperature Physics, ed. by U. Ecken, A. Schmid, W. Weber, and H. Wudl (North-Holland, Amsterdam, 1984), p. 713.
[8] M. E. Hawley, K. E. Gray, B. D. Terris, H. H. Wang, K. D. Carlson, and J. M. Williams, Phys. Rev. Lett., **57**, 629 (1986).
[9] A. Nowack, U. Poppe, M. Weger, D. Schweitzer, and H. Schwenk, Z. Phys. B**68**, 41 (1987).
[10] Y. Maruyama, T. Inabe, H. Urayama, and G. Saito, Solid State Commun. **67**, 35 (1988).
[11] H. Bando, private communication.

STM Measurements of Superconducting Properties in κ-(BEDT-TTF)$_2$Cu(NCS)$_2$

H. Bando, S. Kashiwaya, T. Tokumoto, H. Anzai, N. Kinoshita, M. Tokumoto, K. Murata, and K. Kajimura

Electrotechnical Laboratory, Tsukuba, Ibaraki 305, Japan

Abstract. Tunneling spectroscopy on κ-(BEDT-TTF)$_2$Cu(NCS)$_2$ single crystals by a low-temperature scanning tunneling microscope (STM) is reported. The superconducting energy gap at 1.9 K was $2\Delta = 4.8 \pm 1.1$ meV. Both its magnitude and temperature dependence were consistent with the BCS theory. The line shape of dI/dV - V characteristics deviated from the BCS theory, indicating some distribution in the energy gap magnitude.

1. Introduction

Using an STM we can image the shape of electron clouds on sample surfaces with a spatial resolution in the atomic scale. Employing its capability of tunneling spectroscopy, we can also measure the local density of states of electrons within a microscopic surface area, as was indicated by Binnig and Rohrer[1]. Tunneling spectroscopy at low temperature through the well-defined vacuum barrier of STM gives us information of density of states as well as the superconducting energy gap 2Δ from the dI/dV - V characteristics.

We previously observed the two-dimensional surface of κ-(BEDT-TTF)$_2$Cu(NCS)$_2$ crystals by STM in air at room temperature. The topograph reflected the periodicity of the crystal structure known by X-ray measurements. Then it was confirmed that the surfaces of crystals were conducting and allow tunneling spectroscopy measurement by the STM at least at room temperature.

Recently we have constructed a low-temperature STM and performed tunneling spectroscopy on the two-dimensional surface of the crystals at temperatures down to 1.8 K[2]. Experimental results of dI/dV - V characteristics and discussions on its line shape as well as the superconducting energy gap are presented in the following.

2. Experimental

Detailed descriptions of the low-temperature STM apparatus are presented in a separate article[2]. The microscope was suspended in an indium-sealed can of a cryostat insert. After the interior of the insert was evacuated and a little amount of helium gas was introduced as a heat exchange gas, the cryostat insert was dipped into the 1 K pot of a cryostat.

Single crystals of κ-(BEDT-TTF)$_2$Cu(NCS)$_2$ were grown electrochemically. Their nominal size was 2 mm × 1 mm × 0.2 mm. Before STM measurements crystals were first washed with organic solvents to get rid of the monomers, rinsed with acetone and ethanol, then Au ohmic contacts were evaporated on the corners. The surface to be measured was again etched with organic solvents and rinsed with acetone and ethanol.

Since the organic superconductors are sensitive to mechanical frictions, the crystal was tightly attached to the sample holder at one corner well separated from the position where the tunneling tip accessed. The methods to hold the crystal and bond the leads were verified by detection of the superconducting resistance transition. Tunneling tips were mechanically ground platinum wires, 0.3 mm in diameter, coated with Au of thickness 15 nm.

3. Results and Discussion

When the tip accessed to the sample surface and magnitude of tunneling current first exceeded the noise level of the preamplifier (about 10 pA$_{pp}$), the I-V characteristics were linear though the shape of the tip was sharp to allow topographic measurement. Nonlinear I-V characteristics as shown in Fig. 1 were observed after the tip had proceeded a little more, by about 5 nm, so that the derivative dlog$I$/ds was reduced. Here, s is the vertical position of the tip above the sample surface.

In an STM measurement the tunneling current increases exponentially with s. The derivative dlog$I$/ds is approximately proportional to $(\phi)^{1/2}$, with ϕ the work function. If the system is clean, often dlog$I$/ds as high as 200 dB/nm is observed. On the other hand, when the tip is in touch with the sample, with some insulating layer in between, apparent value of dlog$I$/ds is much reduced. The present result means that the nonlinear dI/dV characteristics were first observed after the tip touched the sample surface which was presumably covered with a thin insulating barrier layer. In that case the I-V characteristics are affected by the quality of barrier. Good results are obtained if the barrier contains no internal excitation mode to assist an excessive tunneling process.

Since the leakage portion of dI/dV in Fig. 1 is small, we supposed that we could analyze the data as typical characteristics of an intrinsic superconductor-insulator-normal tunnel junction. Simply drawing a line between two minima at high bias voltages and measuring the distance between the two points of intersection, we obtained the value $2\Delta = 4.8 \pm 1.1$ meV and $2\Delta/k_B T_c = 5.1 \pm 1.2$. Here, we adopted the value $T_c = 11$ K.

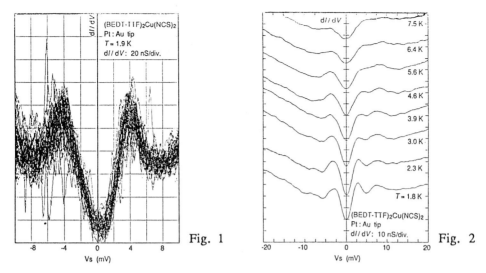

Fig. 1. Differential conductance measured at 1.9 K. Traces of dI/dV - V characteristics of 60 contiguous sweeps were plotted.

Fig. 2. Temperature dependence of differential conductance.

Results of a continuous measurement of the dI/dV - V characteristics as the temperature was raised up to 7.5 K are shown in Fig. 2. In this case a barrier condition comparable with that in Fig. 1 was not achieved. The features such as finite leakage current, asymmetrical nonlinear background, or secondary peaks at high bias voltages are possibly ascribed to the properties of the tunneling barrier but not the superconductor. Still the characteristics of the energy gap were followed. Throughout the temperature range no appreciable change in 2Δ was recognized except the thermal smearing in proportion to $k_B T$. Above T_c the structures related to 2Δ disappeared, leaving the nonlinear background. As a whole, both the magnitude and the temperature dependence of the gap are comparable with the BCS theory.

Finally, the line shape of dI/dV - V characteristics in Fig. 1 is considered. It was difficult to reproduce the line shape using the density of states by Dynes et al.[3],

$$D(\varepsilon) \propto \mathrm{Re}\{|\varepsilon - i\Gamma|/\sqrt{(\varepsilon - i\Gamma)^2 - \Delta^2}\}, \tag{1}$$

where Γ represents the life-time broadening. To follow the long tails inside the gap large Γ was required, which rounded off the sharp peaks. On the other hand if a simple distribution in Δ such as

$$P(\Delta) = \begin{cases} 1/(\Delta_{max} - \Delta_{min}) & [\Delta_{min} \leq \Delta \leq \Delta_{max}] \\ 0 & [\text{otherwise}] \end{cases} \tag{2}$$

is assumed, density of states function takes the form

$$D(\varepsilon) \propto \int d\Delta P(\Delta)\{|\varepsilon|/\sqrt{\varepsilon^2-\Delta^2}\}$$

$$= \begin{cases} \varepsilon\{\mathrm{Sin}^{-1}(\Delta_{max}/\varepsilon)-\mathrm{Sin}^{-1}(\Delta_{min}/\varepsilon)\}/(\Delta_{max}-\Delta_{min}) & [\Delta_{max} \leq |\varepsilon|] \\ \varepsilon\{\mathrm{Cos}^{-1}(\Delta_{min}/\varepsilon)\}/(\Delta_{max}-\Delta_{min})\} & [\Delta_{min} \leq |\varepsilon| \leq \Delta_{max}] \\ 0 & [|\varepsilon| \leq \Delta_{min}] \end{cases} \quad (3)$$

The function has onsets at $|\varepsilon| = \Delta_{min}$, increases linearly, shows cuspidate maxima at $|\varepsilon| = \Delta_{max}$, and then converges to unity. The feature is common to various forms of $P(\Delta)$ having clear onset and offset with respect to Δ. Equation (3) suggests that we probed a superconducting system where the energy gap value has finite distribution. The contribution to the tunneling current from each electronic state at the Fermi energy depends on the junction configuration since it is determined by the tunneling matrix elements. If the configuration is controlled by an STM, one can selectively measure the energy gap of the electrons having velocity nearly parallel to a specified direction. We suppose it is also possible to interpret the previously reported results of tunneling spectroscopy by Hawley et al. [4] or Maruyama et al.[5] if they sensed subsets of the distribution according to the configurations.

In conclusion, the superconducting energy gap of κ-(BEDT-TTF)$_2$Cu(NCS)$_2$ measured at 1.9 K by a low-temperature STM was 2Δ = 4.8 ± 1.1 meV. Its magnitude and temperature dependence were consistent with the BCS theory. The line shape of the spectra indicated some distribution in the energy gap magnitude. To perform spectroscopic measurements with atomic resolution, we must improve surface treatment and tunneling current detection.

References

1. G. Binnig and H. Rohrer: IBM J. Res. Develop. **30**, 355 (1986).
2. H. Bando, S. Kashiwaya, H. Tokumoto, H. Anzai, N. Kinoshita, and K. Kajimura: *to be published in* J. Vac. Sci. Technol. A**8**, no.1 (1990).
3. R.C. Dynes, V. Narayanamurti, and J.P. Garno: Phys. Rev. Lett. **41**,1509 (1978).
4. M.E.Hawley, K.E. Grey, B.D. Terris, H.H. Wang, K.D. Carlson, and J.M. Williams: Phys. Rev. Lett. **57**, 629 (1986).
5. Y. Maruyama, T. Inabe, H. Urayama, H. Yamochi, and G. Saito: Solid State Commun. **67**, 35 (1988).

Effect of Tensile Stress on the Superconducting Transition Temperature in (BEDT-TTF)$_2$Cu(NCS)$_2$

H. Kusuhara[1], *Y. Sakata*[1], *Y. Ueba*[1], *K. Tada*[1], *M. Kaji*[1], *and T. Ishiguro*[2]

[1] R & D Group, Sumitomo Electric Industries, 1-1-3 Shimaya, Konohanaku, Osaka 554, Japan
[2] Department of Physics, Kyoto University, Sakyoku, Kyoto 606, Japan

Abstract. The effect of tensile stress on the superconducting transition temperature T_c of κ-(BEDT-TTF)$_2$Cu(NCS)$_2$ is studied by suppressing the thermal contraction. The T_c increased by 0.5–1.2 K for b-axis elongation. By comparing the calculated T_c values with those observed, the significance of the interlayer interaction is elucidated. The resistivity peak around 90 K was extraordinarily enhanced by the elongation.

It has been empirically revealed that the superconducting transition temperature T_c of BEDT-TTF salts decreases with decreasing anion size or the effective volume per BEDT-TTF molecule [1,2]. This is consistent with the pressure dependence of T_c for these salts, where T_c decreases with increasing pressure. Based on this fact it is expected that if we could expand the volume of the BEDT-TTF salts by some means the T_c would be raised.

To check this idea we applied a tensile uniaxial stress to κ-(BEDT-TTF)$_2$Cu(NCS)$_2$, exhibiting the highest T_c, in the following way. According to the measurements of the temperature dependence of the lattice constant down to 104 K by *Saito* et al. [3], the salt possesses a large and anisotropic thermal contraction, e.g. $\Delta b/b = 0.45\%$, $\Delta c/c = 2.2\%$, while $\Delta a/a = -0.82\%$, at 104 K, where the contraction is given as a ratio of the decrement of the lattice constant on cooling from room temperature to 104 K to the lattice constant at room temperature: a positive sign denotes a contraction. By extrapolating these values smoothly to 0 K, we get the following values: $\Delta b/b = 0.95\%$, $\Delta c/c = 2.7\%$ and $\Delta a/a = -0.90\%$. When the crystal is stuck to a substrate with a small thermal contraction, such as Cu or fused quartz, we can induce tensile strain in the sample automatically. The crystal was fixed at two ends with epoxy, crossing a groove in the substrate plane. Since the sample has typical dimensions of a thin plate, 2 mm long (along b-axis), 1 mm wide (c-axis) and 0.05 mm thick (a-axis), the tensile stress is applied along the b-axis by fixing the two ends of the longest direction of the crystal. By sticking the crystal plane to the substrate plane, we consider that the crystals are elongated in the b-c plane. In any case the samples are left free along the a-axis. Since the thickness of the sample is very thin compared with other directions along which the strain is induced, it can be assumed that the stress is applied rather uniformly.

The crystals were prepared by electrochemical crystallization [4]. Measurements of the resistivity and the dc susceptibility have been carried out.

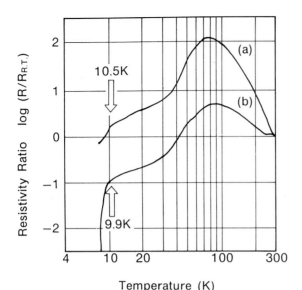

Fig. 1. Temperature dependence of resistivity in (BEDT-TTF)$_2$Cu(NCS)$_2$ (a) with tensile stress along the b-axis and (b) without stress

Figure 1 illustrates the results of typical measurements. By applying the tensile strain along the b-axis, T_c was raised about 0.6 K. Here we evaluated T_c with a clear onset of the resistance decrease, since the tensile stress is not necessarily applied uniformly and the associated T_c may be distributed, and further the resistance decrease should not appear without the occurrence of superconductivity. It should also be noted that the resistance maximum appearing in the normal conductivity region near 90 K is enhanced extraordinarily and shifts to lower temperature a little.

In Table 1 we summarize the results of the resistivity measurements for several samples. Note that the T_c's of the samples vary from sample to sample, especially when they come from different preparation batches.

The increase in T_c is 0.5–1.2 K when the tensile stress is applied along the b-axis but the increase is suppressed for samples with planar elongation, although a much larger tensile effect is expected, since the contraction along the c-axis is three times larger than along the b-axis for the free crystal. To confirm that the increase in T_c is due to the intrinsic bulk effect, the dc magnetization measurements were carried out, and a typical result is shown in Fig. 2. The strained sample shows a broader transition but its onset to superconductivity is raised compared to the stress-free sample. The result for the dc susceptibility measurements are summarized in Table 2.

We also evaluated the change in T_c theoretically, based on the band calculation and the T_c formula for the superconducting mechanism through the interaction between the HOMO and the a_g mode of the BEDT-TTF molecule [5]. The band calculations were carried out using the program developed by *Mori* [6] on the basis of the extended Hückel method by taking account of the tensile strain. Here we assumed that the lattice constant along the a direction is not modified by the elongation in the b-c plane. The results of the calculation are shown in Fig. 3. To calculate T_c, we adopted

Table 1. Results of resistivity measurements of $(BEDT-TTF)_2Cu(NCS)_2$

Sample No.	Direction of Elongation	holder	Transition Temperature (K)	$\triangle T_c$ (K)
8-1	b	Cu	11.8	0.7
8-2	b,c	Cu	9.9	-1.2
8-3	-	-	11.1	
15-1	b	Cu	10.5	0.6
15-2	b	Cu	10.4	0.5
15-3	b	Quartz Glass	11.9	2.0
15-4	-	-	9.9	
37-1	b	Cu	10.1	
37-2	b	Cu	10.1	
37-3	b,c	Cu	10.6	

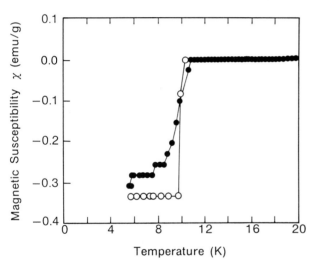

Fig. 2. Temperature dependence of magnetic suscpetibility in $(BEDT-TTF)_2Cu(NCS)_2$ (●) with tensile stress and (○) without stress

formula (7) in [5] and assumed that the relevant vibrational modes and their interaction strengths are the same with TTF as adopted by *Yamaji* [5]. Further, to fit the T_c in the absence of the tensile strain to that of free samples, 10 K, we took the values of the electron–acoustic phonon coupling constant and the Coulomb energy to be 0.21 eV and 0.41 eV, respectively. The resultant T_c increases monotonically with decreasing band width (Fig. 4). The arrows indicate the samples in the elongated state. It is obvious that the increase in T_c is not consistent with the calculation for the tensile strain in the b-c plane.

Table 2. Results of magnetic susceptibility measurements of $(BEDT-TTF)_2Cu(NCS)_2$

Sample	Direction of Elongation	Holder	Transition Temperature (K)	$\triangle T_C$ (K)
(BEDT-TTFh$_8$)$_2$Cu(NCS)$_2$	b	Cu	10.7±0.4	0.8
	b	Al	10.7±0.4	0.8
	b,c	Cu	9.9±0.2	-0.1
	b,c	Quartz Glass	10.1±0.2	0.2
	-		9.8±0.3	
	-		10.1±0.2	
(BEDT-TTFd$_8$)$_2$Cu(NCS)$_2$	b	Cu	10.9±0.2	0.4
	c	Cu	10.6±0.2	0.1
	b,c	Cu	10.4±0.2	-0.1
	-		10.8±0.2	
	-		10.2±0.2	

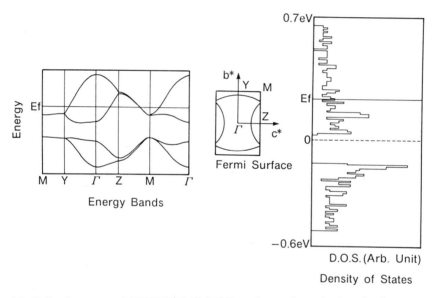

Fig. 3. Band structure of $(BEDT-TTF)_2Cu(NCS)_2$ under tensile strain along b-axis

As a reason for this discrepancy we conjecture that the interlayer interaction, as well as the intralayer interaction, is important for T_c. By applying uniaxial tensile stress, the interlayer distance along the a-axis is inevitably reduced, but its effect is not taken into account for the calculation of T_c.

The band structure is dominated by the two-dimensional arrangement of BEDT-TTF molecules, in that the interlayer interaction plays a minor role. To explain the remarkable effect of the interlayer interaction, we should take account of additional

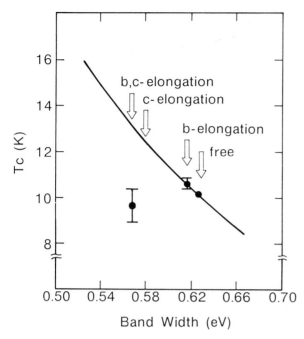

Fig. 4. The band width dependence of T_c in $(BEDT\text{-}TTF)_2Cu(NCS)_2$; • indicates the mean value of the increase in T_c, and the error bar shows the maximum and the minimum change in T_c

mechanisms, such as the lattice softness in relation to the hydrogen-like bonding as proposed by *Whangbo* et al. [7] and the enhancement of superconductivity either by anisotropic pairing or by two-dimensionality.

Next we focus our attention on the enhancement by the tensile strain of the resistance maximum appearing around 90 K. The temperature dependence exhibits nonmetallic behavior down to 90 K and then it turns metallic, forming a pronounced maximum. This maximum has attracted much attention since it resembles the resistivity of the heavy-electron metals. *Urayama* et al. assert that it is caused by defects in crystals [8], while *Parker* et al. suggest that it is a structural transition to a higher resistivity state in the cooling process [9]. However, from the results of the extraordinary enhancement of the resistivity by the tensile stress, we consider that the characteristics reflect the somewhat more intrinsic nature of the conduction electrons in this material.

To explain the resistance maximum, the possibility of the contribution of small polarons is pointed out by *Yamaji* [10], who speculated that the resistivity increases with decreasing temperature due to the formation of polarons through a strong electron–phonon interaction. The material turns metallic when the polarons exhibit coherent motion at low temperature. According to *Holstein* [11], the small polaron exhibits a resistance maximum since it develops a hopping motion above a certain characteristic temperature T^*, where thermal excitation enhances the conductivity, while it undergoes coherent motion. Below T^*, the thermal excitation scatters the polarons. In the hopping region, the height of the barrier which the polarons must overcome may be modified by the externally applied stress. Thus, the Holstein model may explain the temperature dependence of the resistivity, but it has the shortcoming that the theory is applicable only when the polarons are dilute. In the present system, the interaction among the polarons or the many-body effects should be significant.

In conclusion, the superconducting transition temperature T_c increases with elongation of the crystal. The increase in T_c is smaller than the value estimated by the effective volume and band calculation. The small value of the increase in T_c suggests that interlayer interactions have a significant effect on T_c, as well as intralayer interactions. The maximum value of the ratio of the temperature-dependent resistivity to the room-temperature resistivity is extraordinarily enhanced by elongation. It is possibly explained by a small polaron mechanism.

Acknowledgements. We are indebted to Prof. Gunzi Saito of the Institute for Solid State Physics for instructing us in the preparation method of κ-(BEDT-TTF)$_2$-Cu(NCS)$_2$. We are grateful to Dr. Takehiko Mori of the Institute for Molecular Science for the band calculation program.

References

1. G. Saito, H. Urayama, H. Yamochi, K. Oshima: Synth. Meth. **27**, A331 (1988)
2. M. Tokumoto, K. Murata, H. Bando, H. Anzai, K. Kajimura, T. Ishiguro: Physica **143B**, 338 (1986)
3. G. Saito, H. Urayama: Solid State Phys. **23**, 198 (1988) in Japanese
4. Y. Ueba, T. Mishima, H. Kusuhara, K. Tada: These proceedings
5. K. Yamaji: Solid State Commun. **61**, 413 (1987)
6. T. Mori, A. Kobayashi, Y. Sasaki, H. Kobayashi, G. Saito, H. Inokuchi: Bull. Chem. Soc. Jpn. **57**, 627 (1984)
7. M.-H. Whangbo, J.M. Williams, A.J. Schults, T.J. Emge, M.A. Beno: J. Am. Chem. Soc. **109**, 90 (1987)
8. H. Urayama, H. Yamochi, G. Saito, K. Nozawa, T. Sugano, M. Kinoshita, S. Sato, K. Oshima, A. Kawamoto, J. Tanaka: Chem. Lett. 55 (1988)
9. I.D. Parker, R.H. Friend, M. Kurmoo, P. Day, C. Lenoir, and P. Batail: J. Phys. Condens. Matter **1**, 4479 (1989)
10. K. Yamaji: Synth. Met. **27**, A115 (1988)
11. T. Holstein: Annals Phys. **8**, 325 (1959)

Highly Correlated Fermi Liquids in the High-T_c Organic Conductor κ-(BEDT-TTF)$_2$Cu(NCS)$_2$

N. Toyota[1], E.W. Fenton[2], T. Sasaki[1], and M. Tachiki[1]

[1] Institute for Materials Research, Tohoku University, Katahira 2-1-1, Sendai 980, Japan
[2] National Research Council of Canada, Ottawa, Canada K1A 0R6

Abstract We present evidence of many-body renormalizations of the cyclotron mass and the lifetime of electrons moving in the basal plane of a high-T_c organic conductor, κ-(BEDT-TTF)$_2$Cu(NCS)$_2$. It is concluded that this dimerized material is a highly correlated Fermi liquid satisfying Luttinger's theorem. The importance of the on-site and/or the nearest neighbor Coulomb correlations in the metallic state is pointed out. The temperature-dependent electrical resistance exhibiting a broad peak around 100 K is discussed in terms of fine-scaled quasiparticle states near the Fermi level.

During the last decade there have been great developments in synthesizing new classes of organic conductor [1]. One of them was the discovery of the high-T_c (10-11 K) superconductivity in κ-phased (BEDT-TTF)$_2$Cu(NCS)$_2$, by Prof. Saito's group [2]. This salt contains two dimerized pairs of (BEDT-TTF) donors in a monoclinic unit cell. The dimers stack in a a-c plane nearly orthogonal to each other, forming a conducting layer which is sandwiched by insulating anion sheets of [Cu-NCS)$_2$]$^-$[3]. Besides the high-T_c, this material exhibits several characteristic properties such as an electrical resistance peak around 100 K [4], the puzzling giant peak of the ^{1}H-NMR relaxation rate below T_c [5], and a large upper critical field H_{c2} [4,6,7]

Recently a consensus has been reached on the Fermi surface (FS) of this material, first observed in Shubnikov-de Haas oscillations by Oshima et al. [8] and followed by Toyota et al. [9], Pratt et al. [10] and Müller et al. [11]. The obtained FS, which is considered to be a slightly warped cylinder, occupies 16-20 % of the area of the basal plane of the first Brillouin zone. This FS is assigned to be a lens-like closed orbit centered at the Z-point in the Brillouin zone, which has been predicted by the extended Hückel method using the tight-binding approximation [8]. This assignment is further supported by the observation of the magnetic breakdown [11], which is explained to occur across the small gap at the Z-M zone boundary between the lens and open sheets. We might conclude, therefore, that the band-structure calculations well reproduce the SdH-derived FS as a first approximation.

However, a discrepancy between SdH experiments [8-10] and calculations [8] is clearly found in the cyclotron mass m_c of the conduction electrons moving in the basal planes. The observed $m_c = (3.4-4.0)m_0$ (m_0 is the free electron mass) is much heavier than $m_c = (0.5-1.0)\,m_0$ expected from the dispersion relations of the band-structure calculations. This discrepancy is also recognized in the density-of-states calculations [12] which reveal too wide a bandwidth of 0.6-0.7 eV due to the upper two split bands. The corresponding band mass is again estimated to be $\leq 1\, m_0$.

In our previous paper [13], we pointed out that just this discrepancy in m_c and the agreement of the FS-topology between SdH observations and band-structure calculations could be direct evidence for many-body renormalizations of the cyclotron mass and the electron lifetime τ in " heavy-mass" organic conductors as κ-(BEDT-TTF)$_2$Cu(NCS)$_2$ (treated in this paper) and the high-T_c phased β$_H$-(BEDT-TTF)$_2$I$_3$ [14]. These materials are highly correlated Fermi liquids which satisfy Luttinger's theorem [15]: the volume of the FS in momentum space is unaffected by the many-body interactions, but the effective mass is enhanced. The essential point of the renormalization theory is that any many-body interaction necessarily leads to quasi-particles with renormalized mass $m = Zm_b$ and lifetime $\tau = Z\tau_b$, where Z is a renormalization constant defined as $1-\partial\Sigma/\partial(i\omega) > 1$ ($\Sigma(\omega)$ is a self-energy), m_b is the band mass and τ_b the unrenormalized (bare) lifetime. [See ref. 13.]

If we take the maximum band mass $m_b \sim 1 m_0$, we find $Z \approx 3\text{-}4$. It is interesting to note that this estimation might be consistent with that obtained by Mori and Inokuchi [16]. They succeeded in fitting the rather complicated temperature dependence of the thermoelectric power of the present material by assuming the effective bandwidth near the Fermi level to be a factor of almost five narrower than that of their original band structure calculations [17]. For this large Z, we have proposed that the Coulomb correlations of the extended Hubbard model expressed as follows might be important [13]:

$$H = \sum_{i \neq j,\sigma} t_{ij}(a^+_{i\sigma} a_{j\sigma} + a^+_{j\sigma} a_{j\sigma}) + \sum_i U n_{i\uparrow} n_{i\downarrow} + \sum_{<i,j>} V_{ij} n_i n_j \qquad (1)$$

where $n_{i\sigma} = a^+_{i\sigma} a_{i\sigma}$ and $n_i = n_{i\uparrow} + n_{i\downarrow}$. The first term represents the kinetic energy of electrons in a tight-binding form with the transfer integral t_{ij}, being proportional to the overlaps between the molecular orbital wave functions on sites i and j. The second term means the on-site Coulomb interaction with U and the last term the nearest neighbor Coulomb interaction with V_{ij} where, in the present salt, each intradimer molecule might be a dominant neighboring site.

Recently Pratt et al. [10] have pointed out the importance of the Coulomb correlations U and V, which were estimated to be ~1.3 eV and 0.2-0.5 eV, respectively. This estimation is based on the possible interpretation of the optical conductivity peaks [18] as resulting from the interband transitions across V. Taking into account these values, it is likely that the quasiparticle states in the vicinity of the Fermi level might be expected to be fine-scaled in energy and strongly temperature-dependent.

This might be responsible for the broad peak of the resistivity, as follows. First, as assumed using the one-dimensional model [10], the quasiparticle subbands which are split by V open a gap of $V - W$ (>0, being assumed at room temperature), where V and W are of comparable magnitude, resulting in a narrow gap. Secondly, we should take into account the expected broadening of W on cooling, since the transfer integral t_{ij} should become larger due to the lattice contraction on cooling. Thirdly, we might include the effect of electron-phonon interactions via the thermal-vibration-induced fluctuation of the transfer integral: $\delta t_{ij} = (\partial t_{ij}/\partial r_{ij}) \delta r_{ij}$ where δr_{ij} is approximated at high temperatures to be $(2k_BT/C)^{1/2}$ (C is an elastic constant).[19] We thus expect that these three competing effects could explain the broad peak of the resistivity, although the quantitative comparison is open to further studies. The enhancement and suppression of the resistivity peak observed, respectively, under elongational [12] and compressional [20] strains, might be consistent with the above picture. Because, by controlling the lattice volume and hence the transfer integrals, the former (the latter) strain qualitatively widens (narrows) the energy gap. It is also noted that the quasiparticle states might be quite sensitive to impurities which give a band-tail via electron-impurity scatterings. Thus impurity-induced subband tails should easily make vague such a small gap around room temperature. This effect might explain the resistivity peaks which are considered to be sensitive to the impurity and/or lattice defects [21].

In the present paper, we have shown that high-T_c κ-(BEDT-TTF)$_2$Cu(NCS)$_2$ is a highly-correlated, heavy-mass metal and the resistivity peak might be attributed to the fine-scaled quasiparticle states near the Fermi level. As pointed out [13], the high-T_c phase of β_H-(BEDT-TTF)$_2$I$_3$ might also be renormalized, while its low-T_c phase with an incommensurate superstructure might not be. Although the reason is not clear at present, the high-T_c organic salts seem to be necessarily correlated systems. This suggests that a possible understanding of superconductivity in organic salts may ultimately be achieved on the basis of a strong-correlated electron system, as is now being attempted for heavy-fermion and oxide systems.[22]

Acknowledgement ---- We thank Dr. Pratt for sending us their preprint and valuable discussions at the Conference. This work was supported in part by a Grant-in-Aid for Special Project Research from the Ministry of Education, Science and Culture of Japan.

References

[1] For recent Proceedings, Synthetic Metals **27 A-D** and these Proceedings
[2] H. Urayama, H. Yamochi, G. Saito, K. Nozawa, T. Sugano, M. Kinoshita, S. Sato, K. Oshima, A. Kawamoto and J. Tanaka, Chem. Lett. **1988** (1988) 55
[3] H. Urayama, H. Yamochi, G. Saito, S. Saito, A. Kawamoto, J. Tanaka, T. Mori, Y. Maruyama and H. Inokuchi, Chem. Lett. **1988** (1988) 463
[4] K. Oshima, H. Urayama, H. Yamochi and G. Saito, J. Phys. Soc. Jpn. **57** (1988) 730

[5] K. Takahashi, T. Tokiwa, K. Kanoda, H. Urayama, H. Yamochi and G. Saito, Synth. Met. 27 (1988) A319
[6] K. Murata, Y. Honda, H. Anzai, M. Tokumoto, K. Takahashi, N. Kinoshita, T. Ishiguro, N. Toyota, T. Sasaki and Y. Muto, Synth. Met. 27 (1988) A341
[7] K. Oshima, R. C. Yu, P. M. Chaikin, H. Urayama, H. Yamochi and G. Saito, this Proceedings.
[8] K. Oshima, T. Mori, H. Inokuchi, H. Urayama, H. Yamochi and G. Saito, Phys. Rev. B38 (1988) 938
[9] N. Toyota, T. Sasaki, K. Murata, Y. Honda, M. Tokumoto, H. Bando, N. Kinoshita, H Anzai, T. Ishiguro and Y. Muto, J. Phys. Soc. Jpn. 57 (1988) 2616
[10] F. L. Pratt, J. Singleton, M. Kurmoo, S. J. R. M. Spermon, W. Hayes and P. Day, to be published in J. Phys. Condens. Matter and these Proceedings
[11] M.Müller, C. -P. Heidmann, A. Lerf, R. Sieburger and K. Andres, these Proceedings
[12] M. Kusuhara, Y. Sakata, Y. Ueba, K. Tada, M. Kaji and T. Ishiguro, these Proceedings
[13] N. Toyota, E. W. Fenton, T. Sasaki and M. Tachiki, Solid State Commun. 72 (1989) 859
[14] W. Kang, G. Montambaux, J. R. Cooper, D. Jerome, P. Batail and C. Lenoir, Phys. Rev. Lett. 62 (1989) 2559
[15] J. M. Luttinger, Phys. Rev. 121 (1961) 1251
[16] T. Mori and H. Inokuchi, J. Phys. Soc. Jpn. 57 (1988) 3674
[17] In ref. [16], the γ-value of the linear term in specific heats was estimated to be 100 mJ/mole·K^2 (being surprisingly large), using the data: S. Katsumoto, S. Kobayashi, H. Urayama, H. Yamochi and G. Saito, J. Phys. Soc. Jpn. 57 (1988) 3672
[18] T. Sugano, K. Nozawa, H. Hayashi, K. Nishikida, K. Terui, T. Fukasawa, H. Takenouchi, S. Mino, H. Urayama, H. Yamochi, G. Saito and M. Kinoshita, Synth. Met. 27 (1988) A325
[19] N. Toyota, P. Koorevaar, J. v. d. Berg, P. H. Kes, J. A. Mydosh, T. Shishido, Y. Saito, N. Kuroda, K. Ukei, T. Sasaki and T. Fukuda, J. Phys: Condens. Matter 1 (1989) 3721
[20] K. Murata, M. Tokumoto, H. Anzai, Y. Honda, N. Kinoshita, T. Ishiguro, N. Toyota, T. Sasaki and Y. Muto, Synth. Met. 27 (1988) A263
[21] A. Ugawa, G. Ojima, K. Yakushi and H. Kuroda, Synth. Met. 27 (1988) A445
[22] E. W. Fenton, Proc. Int. Seminar on High Temperature Superconductivity, Dubna, USSR (1989) (to be published, World Scientific Publ.) and Naukova Dymka (to be published)

Electronic Properties of $(BEDT-TTF)_3Cl_2 \cdot 2H_2O$

S.D. Obertelli[1], I.R. Marsden[1], R.H. Friend[1], M. Kurmoo[2], M.J. Rosseinsky[2], P. Day[2], F.L. Pratt[3], and W. Hayes[3]

[1]Cavendish Laboratory, Madingley Road, Cambridge CB3 0HE, UK
[2]Inorganic Chemistry Laboratory, South Parks Road, Oxford OX1 3QR, UK
[3]Clarendon Laboratory, Parks Road, Oxford OX1 3PU, UK

We report measurements of the optical, magnetic and transport properties of this 3:2 BEDT-TTF salt. There is clear evidence from the magnetic susceptibility and thermopower for the opening of a gap at the Fermi energy below about 100K. Though this salt shows semimetallic behaviour, the magnetic susceptibility shows a large Pauli conduction electron susceptibility, of 7×10^{-4} emu/mole formula unit.

1. Introduction

$(BEDT-TTF)_3Cl_2.2H_2O$ has been synthesised electrochemically by several groups [1-3] under a range of conditions, some intended for synthesis of other salts [1,2], others for the synthesis of the chloride salt [3]. The structure shows the usual separation of BEDT-TTF sheets which are organised as well defined stacks along the a axis in the ac plane with two inequivalent stacks and three inequivalent donor molecules on each stack. The anions form sheets between the donor sheets, with hydrogen bonding between the water molecules and chloride ions. The unit cell thus contains 2 formula units, with 6 BEDT-TTF molecules and 4 chloride counterions. The band calculation of Mori and Inokuchi [2] shows that the strongest interactions within the donor sheet are inter-stack in the c direction, and that there is little anisotropy within the ac plane. With a total of six donor molecule bands, and a charge transfer of 2/3 electron, a total of four are filled and two empty. The band calculation shows that there is overlap between bands at the Fermi energy, so that the material is a semimetal with electron and hole pockets in the Fermi surface.

This salt shows metallic properties at room temperature, but shows a gradual transition to an insulating state below about 100 K. Under pressure it has been found to undergo a transition from this non-metallic state to a superconducting state. The critical pressure is about 10 kbar, and critical temperatures of up to 4 K are found [3,4]. It is of particular interest because it is a relatively rare example of a superconductor with a 2/3 band filling.

We report here measurements of the electronic properties in the normal state. We find clear evidence that the onset of the insulating state is associated with the opening of an energy gap at the Fermi level.

2. Results and Discussion

Optical reflectivity spectra are shown in figure 1 for the two in-plane polarisations, parallel and perpendicular to the long (c) axis. These spectra can be fitted to a single carrier Drude model and the fitted parameters are given in table 1.

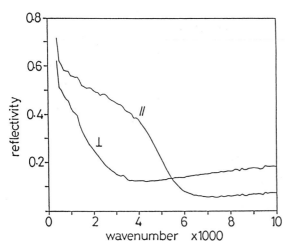

Figure 1 Reflectivity polarised parallel and perpendicular to the c axis.

Table 1 Fitted parameters for Drude reflectivity

	$\omega_p/\sqrt{\varepsilon_c}$ (eV)	γ (eV)	ε_c	ω_p (eV)
//	0.73	0.26	9.9	2.3
$\perp$	0.33	0.30	6.3	0.82

These values can be used to give the optical conductivity, $\sigma = \varepsilon_0 \omega_p^2/\gamma$, and we find values of $\sigma_{//} = 2700$ S/cm and $\sigma_\perp = 300$ S/cm. These values are a little higher than measured DC, but the anisotropy of 9:1 is in close agreement with the value of 7:1 reported by Mori and Inokuchi [2]. The plasma frequency, ω_p can be used to derive the optical effective mass, $m^* = n\, e^2/\varepsilon_0\omega_p^2$ and we find values of 0.46 and 3.6 m_e for the parallel and perpendicular directions. We estimate the parallel bandwidth, $4t_{//} = \pi\hbar\rho/m^*d^2\sin(\pi\rho/2)$ to be 1.1 eV Using the analysis of Tajima et al.[5] to estimate the anisotropy of the transfer integrals we obtain $4t_\perp = 0.5$ eV. At the photon energies used here the semimetallic character of this material is not directly probed since the bulk of the contribution to the reflectivity comes from interband transitions.

The excess molar magnetic susceptibility (χ) of the chloride salt is shown in figure 2. The data have been corrected for a small Curie tail corresponding to a concentration of 3000 ppm s = 1/2 spins. A core diamagnetism of value -685×10^{-6} emu/mole formula unit has been subtracted, to force χ to zero at low temperatures. This is close to the value of -649×10^{-6} emu/mole formula unit calculated from Pascal's constants. The salt clearly undergoes a phase transition at 100 K, as seen by the fall in χ at this temperature and the peak in $d\chi/dT$ as shown in the inset to the figure. The value of χ_{Pauli} in the metallic phase, of some 7×10^{-4} emu/mole formula unit is high. There are 3 donor molecules per formula unit, so a comparable value for a 2:1 salt is 4.7×10^{-4} emu/mole. This value is typical for the β phase and κ phase superconducting 2:1 salts [6,7] and as previously noted, considerably higher than calculated for a non-interacting electron gas using the optically-determined bandwidths. The

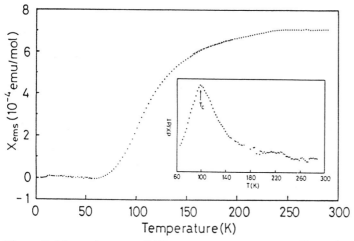

Figure 2 Magnetic susceptibility, χ versus temperature. Inset shows dχ/dT.

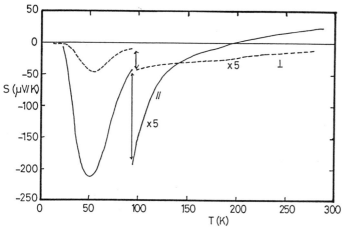

Figure 3 Thermopower parallel and perpendicular to the long axis of the crystal

high value of χ here is not inconsistent with the semimetallic bandstructure since for a two-dimensional system the density of states is high close to the band edge.

Measurements of the thermopower in one symmetry direction were reported by Mori and Inokuchi [2]. We have resolved the two components in the two symmetry directions in the ac plane. We find that the thermopower measured by Mori and Inokuchi is equivalent to the value measured along the long axis of the crystals used here, and is positive at high temperatures, changing sign at around 200 K and reaching a maximum value of -210 µV/K at around 50 K, before falling towards zero at lower temperatures. In the direction transverse to this the thermopower is always negative. It increases more slowly at low temperatures, reaching a peak value of about -45 µV/K at 50 K before also falling towards zero. The room temperature values of the thermopower are fully consistent with the semimetallic bandstructure. The low temperature increase however, indicates that the density of states at the Fermi energy, $N(E_F)$ is reduced below 100 K. The return of the thermopower towards zero below 50 K, together with a resistivity which though rising, does not diverge towards infinity, indicates that the density of states at the Fermi energy though low remains finite.

In summary, we see clear evidence that there is a phase transition at around 100 K and that this reduces $N(E_F)$ to a low, though finite level at low temperatures. There is some indication for the role of disorder in this material [3]. We consider that the effect of pressure is to reduce the effect of this transition on the conduction electrons, so that in the high pressure phase, the high density of states at the Fermi energy that is indicated from the large paramagnetic susceptibility of the conduction electrons at room temperature and at ambient pressure, is preserved at low temperatures. It seems therefore that a very high value of $N(E_F)$ is a prerequisite for superconductivity in both 2:1 and 3:2 BEDT-TTF salts.

References

1. M. J. Rosseinsky, M. Kurmoo, D. R. Talham, P. Day, D. Chasseau and D. Watkin, J. Chem. Soc. Chem. Commun. 88 (1988).
2. T. Mori and H. Inokuchi, Chem. Lett. 1657 (1987).
3. M. Kurmoo, M. J. Rosseinsky, P. Day, P. Auban, W. Kang, D. Jérome and P. Batail, Synthetic Metals **27**, A425 (1988).
4. T. Mori and H. Inokuchi, Solid State Commun. **64**, 335 (1987).
5. H. Tajima, H. Kanbara, K. Yakushi, H. Kuroda, G. Saito and T. Mori, Synthetic Metals **25**, 323 (1988)
6. D. R. Talham, M. Kurmoo, P. Day, S. D. Obertelli, I. D. Parker and R. H. Friend, J. Phys. C**19**, L383 (1986)
7. S. Klotz, J. S. Schilling, S. Gärtner and D. Schweitzer, Solid State Commun. **67**, 981 (1988).

Part V

**BEDT-TTF Family:
Fermiology and
Related Subjects**

Galvanomagnetic Properties of the Organic Metals β-(ET)$_2$X: Magnetoresistance and Shubnikov–de Haas Oscillations

V.N. Laukhin[1], *M.V. Kartsovnik*[1], *S.I. Pesotskii*[1], *I.F. Schegolev*[2], *and P.A. Kononovich*[2]

[1]Institute of Chemical Physics, USSR Academy of Sciences,
 SU-142432 Chernogolovka MD, USSR
[2]Institute of Solid State Physics, USSR Academy of Sciences,
 SU-142432 Chernogolovka MD, USSR

Abstract. Shubnikov–de Haas oscillations have been measured in single crystals of the organic metals β-(ET)$_2$IBr$_2$ and β-(ET)$_2$I$_3$, as well as the angle and field dependences of the magnetoresistance. The classical part of the magnetoresistance exhibits strong angular oscillations on rotation of the field in the ac^* and $b'c^*$ planes. A possible shape of the Fermi surface is discussed.

The magnetoresistance angle and field dependences in high-quality single crystals of β-(ET)$_2$IBr$_2$ have been measured. Shubnikov–de Haas (SdH) oscillations have been observed in a wide interval of angles φ between the magnetic field and c^* direction ($c^* \perp ab$ plane). Analogous measurements have been also carried out in single crystals of β_L-(ET)$_2$I$_3$.

The angle dependence of the resistance along the a direction, R_a, for the field lying in the $b'c^*$ plane ($b' \parallel [a \times c^*]$) is shown in Fig. 1. Analogous dependences are observed for R_a and R_c for the field in the ac^* plane [1].

The magnetoresistance angle dependences exhibit strong oscillations, which are not SdH ones since their period does not depend on the magnetic field, which affects only the magnitudes of the peaks.

The angle dependence of the resistance along the c^*-axis, R_c, for the field in the ab plane is shown in Fig. 2. It is seen that the transverse magnetoresistance anisotropy in the ab plane is surprisingly high. The minimum of the magnetoresistance corresponds to the field perpendicular to the a-axis, and the maximum to the field nearly perpendicular to the stack direction, $a + b$.

With the field out of the ab plane, two types of SdH oscillations are observed: (i) slow ones [1,2] with a typical frequency 0.5 MG existing in an angle interval about $\pm 10°$ around the c-axis, and (ii) fast oscillations [2,3] with the minimum frequency 40 MG observable in a wide angle interval and exhibiting characteristic beats (Fig. 3) in some field directions. The beats reveal the presence of two types of oscillations with nearly equal frequencies and amplitudes.

The angle dependence of the amplitude of the fast SdH oscillations exhibits two big maxima [3] for the angles corresponding to the first local resistance maxima in the angle dependence of Fig. 1.

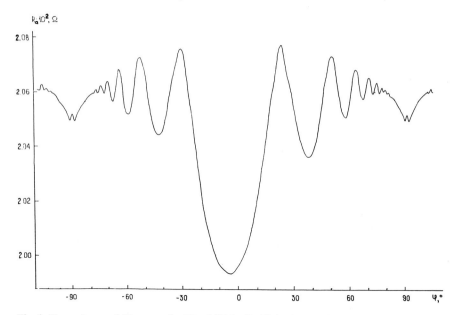

Fig. 1. Dependence of R_a on φ for $H = 15\,\text{T}$ in the $b'c^*$ plane and $T = 1.45\,\text{K}$

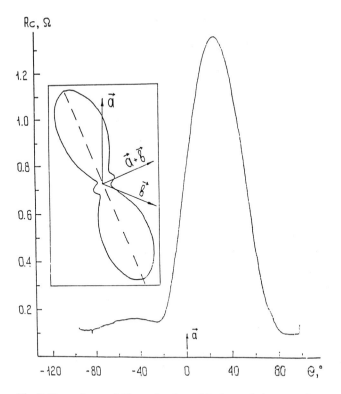

Fig. 2. Dependence of R_c on θ, where θ is the angle between the a-axis and the field $H = 15\,\text{T}$ in the ab plane, $T = 1.45\,\text{K}$. Inset shows the same dependence in polar coordinates

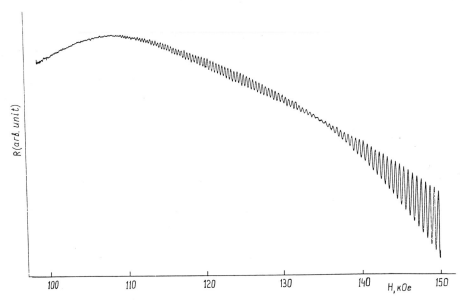

Fig. 3. Dependence of R_a on field in the $b'c^*$ plane at $\varphi \simeq 11°$ and $T = 1.45$ K

The angle dependences of the frequency of the fast SdH oscillations reveal that this type of oscillation is associated with a branch of the Fermi surface (FS) in the form of a not too warped cylinder extending along the c^* direction [2, 3]. The high anisotropy of the resistivity in the ac^* plane at $H = 0$, characteristic of β-(ET)$_2$X, is further evidence for this statement, as well as the smallness of the SdH oscillation beat frequency.

The oscillation frequencies at $H \parallel c^*$ correspond to the cylinder cross-section areas of 50% and 0.6% of the Brillouin zone area in the ab plane for the fast and slow oscillations, respectively [1–3]. The evaluation of the cyclotron masses results in $m_f \simeq 4.5 m_e$ for the fast oscillations and $m_s \simeq 0.5 m_e$ for the slow ones.

The strong angle oscillations of the magnetoresistance may be due to the presence of so-called self-crossing trajectories [4], which act as traps for electrons, giving rise to an increase in resistivity. They may result from warping of the FS cylinder, however, it is unlikely that the warping would be strong enough to provide a marked effect at relatively small angles between the field and the c^*-axis. Another possibility is associated with necks joining the FS cylinders in neighbouring unit cells of reciprocal space. The high anisotropy of the magnetoresistance in the ab plane reveals the shape of the FS cylinder to be far from circular. Therefore, taking into account the large value of its cross-sectional area, amounting to 50% of the Brillouin zone cross-section, some intersections of the Fermi surface with the Brillouin zone boundaries are very likely.

It may be easily seen that in all these cases, regardless of the way of forming the self-crossing trajectories, the angular oscillations in the classical part of the

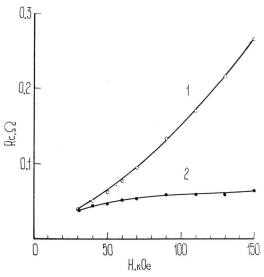

Fig. 4. 1: Field dependence of R_c in the first local peak at $\varphi = 17°$. 2: Field dependence of the peak "pedestal" at the same φ (see text)

magnetoresistance should be periodic in $\tan\varphi$ [3]. In Fig. 4, curve 1 shows the field dependence of R_c for $\varphi = 17°$ corresponding to the first local peak [1] in the $R_c(\varphi)$ dependence for $H \perp b'$. Curve 2 shows the field behaviour for the resistance of a peak "pedestal" whose value has been determined by an envelope of the resistance minima at $\varphi = 17°$. The difference in the behaviour of these curves reveals the existence of two carrier groups for this field direction, one with closed and the other with open or self-crossed trajectories. The difference in the behaviour of the two curves reveals that two carrier groups contribute to the magnetoresistance for this field direction, trajectories being closed for one group, and open or self-crossed for the other.

The genesis of the slow SdH oscillations is not quite clear, to date. They may arise from another branch of the Fermi surface or may result from some extreme cross-sections of the necks, if they do exist.

The obtained results allow us to construct possible Fermi surfaces of β-(ET)$_2$IBr$_2$. One of variants is shown in Fig. 5.

It is interesting to compare the magnetoresistance behaviour in β-(ET)$_2$IBr$_2$ and β_L-(ET)$_2$I$_3$. The angular oscillations characteristic of β-(ET)$_2$IBr$_2$ have not been observed for the latter metal [5]. There are also no fast SdH oscillations in the $R_a(H)$ dependences, so that only slow oscillations are present with a typical frequency of the order of 1 MG.

The absence of the fast SdH oscillations may be explained by the reduced perfection of β_L-(ET)$_2$I$_3$ crystals in comparison with crystals of β-(ET)$_2$IBr$_2$ [5]. It is not quite clear whether the lower crystal quality may completely suppress the rather strong angular oscillations in the classical part of the magnetoresistivity. It is not excluded that the absence of the latter is associated with a profound difference in the Fermi surface topology of the two metals.

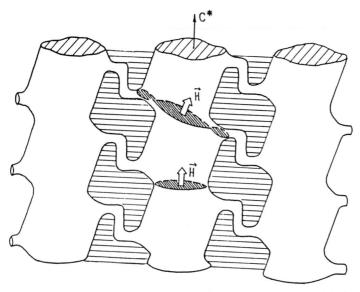

Fig. 5. Possible view of the Fermi surface of β-(ET)$_2$IBr$_2$ (not to scale). An extreme orbit for $H \parallel c^*$ and a self-crossed orbit are shown

Acknowledgements. We thank N.E. Alekseevskii and T. Palewsky for support, E.B. Yagubskii and E.E. Laukhina for the crystal synthesis, and V.G. Peschanskii for useful discussions.

References

1. M.V. Kartsovnik, P.A. Kononovich, V.N. Laukhin, I.F. Schegolev: Pis'ma Zh. Exp. Teor. Fiz. **48**, 498 (1988)
2. M.V. Kartsovnik, V.N. Laukhin, V.I. Nijankovskii, A.A. Ignat'ev: Pis'ma Zh. Exp. Teor. Fiz. **47**, 302 (1988)
3. I.F. Schegolev, P.A. Kononovich, V.N. Laukhin, M.V. Kartsovnik: Phys. Scr. 1989 (in press)
4. I.M. Lifshiz, M.Ya. Azbel, M.I. Kaganov: *Electron Theory of Metals* (Moscow, 1971), in Russian
5. M.V. Kartsovnik, P.A. Kononovich, V.N. Laukhin, S.I. Pesotskii, I.F. Schegolev: Pis'ma Zh. Exp. Teor. Fiz. **49**, 453 (1989)

The Fermi Surface in the Organic Superconductor β-(BEDT-TTF)$_2$IBr$_2$

T. Sasaki[1], N. Toyota[1], T. Fukase[1], K. Murata[2], M. Tokumoto[2], and H. Anzai[2]

[1]Institute for Materials Research, Tohoku University, Katahira 2-1-1, Aoba, Sendai 980, Japan
[2]Electrotechnical Laboratory, Umezono 1-1-4, Tsukuba 305, Japan

Abstract. The angular dependence of the magnetoresistance and the Shubnikov-de Haas (SdH) effect have been measured on an organic superconductor β-(BEDT-TTF)$_2$IBr$_2$ at ambient pressure. The magnetoresistance takes a minimum near the c^*-direction and exhibits weak oscillations as a function of the angle. The angular dependence of the SdH frequency is rather complicated, so that it cannot be explained by a simple cylindrical Fermi surface. We propose a Fermi surface model based on the experimental results, which is constructed by the Brillouin zone folding of a simple ellipsoidal Fermi surface along the k_c-axis.

About thirty organic superconductors have been found with combinations of various cation and anion molecules. The superconducting transition temperature T_c has been rising and at present, κ-(BEDT-TTF)$_2$Cu(NCS)$_2$ has the highest T_c of about 10K among organic materials. In spite of considerable work, however, some characteristic phenomena in organic superconductors, such as pressure, lattice modulation and low dimensionality effects on the superconductivity, are not fully understood. Knowledge of the Fermi surface is helpful in solving these problems. Recently magnetic quantum oscillations have been observed to obtain direct information about the Fermi surface in the (BEDT-TTF)$_2$X series with X=I$_3$ [1][2][3], IBr$_2$ [1][2][4][5], AuI$_2$ [6] in the β phase, X=AuBr$_2$ in the β" phase [7] and X=Cu(NCS)$_2$ in the κ phase [8][2].

In this paper, we present detailed experimental results of the angular dependence of the magnetoresistance and the Shubnikov-de Haas (SdH) effect in β-(BEDT-TTF)$_2$IBr$_2$. Based on these results, we propose a Fermi surface model.

The single crystals of β-(BEDT-TTF)$_2$IBr$_2$ were grown by a conventional electrochemical oxidation method. Gold films were evaporated on the crystal and then gold wires (10μmφ) were attached with gold paint. Resistance measurements were carried out at ambient pressure by a standard four-terminal method using dc or ac (31.3 Hz) current along the crystallographic a-axis. The sample was attached to a gear assembly in an adiabatic cell, which was able to rotate in the field and the accuracy of the angle was ±1°.

The sample was cooled slowly from room temperature to liquid helium temperature for about 12 hours. The superconducting transition temperature was 2.3 K at the mid-point of the resistance transition. The residual resistivity ratio ($RRR \equiv \rho(300K)/\rho(3K)$) was about 1500. Figure 1 shows the angular dependence of the magnetoresistance at 1.7K in the b'-c^* plane. The magnetoresistances take a minimum near the c^*-direction in both the b'-c^* and

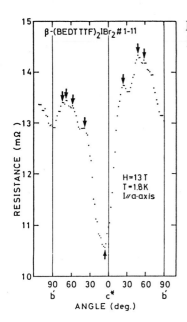

Fig. 1 Angular dependence of the magnetoresistance in the b'-c^* plane.

c^*-a planes. The angular dependence of the magnetoresistance in the b'-c^* plane exhibits weak oscillations as a function of the angle, which are superposed on the background magnetoresistance, but those oscillations cannot be seen in the c^*-a plane. Peaks of the angular oscillation are periodic in $\tan\theta$, where θ is the angle between the resistance minimum direction and the applied field. The angular variation of the magnetoresistance and the angle of peaks are slightly asymmetric. This asymmetry might be relevant to the low symmetric crystal structure (triclinic) of β-(BEDT-TTF)$_2$IBr$_2$. Similar angular oscillation was observed more clearly for the same material by Kartsovnik et al. [5] A recent theory proposed by Yamaji can explain these phenomena by the vanishing of the distribution width of the area of the orbits on the Fermi surface in the quasi-two-dimensional system. [9] Using this theory, the cross-sectional area of the Fermi surface is estimated to be about 30% of the area of the first Brillouin zone (S_{BZ}).

The oscillatory magnetoresistance was measured with applied fields in the b'-c^* and c^*-a planes. Figure 2(a) shows a typical example of the SdH oscillations as a function of the inverse of the field, where the monotonous background is subtracted. This oscillatory wave pattern with somewhat distorted beats is analyzed by means of a fast Fourier transform (FFT) method.

The obtained frequency power spectrum is shown in Fig. 2(b), where three independent peaks are found. Each frequency F_i is connected with the extremal cross-sectional area S_i of the Fermi surface normal to the field: $S_i = (2\pi e/\hbar c) F_i$ $= 9.545 \times 10^{11} F_i$ (cm^{-2}), where F is units of T.

Figure 3 shows the angular dependence of the peak frequency in the spectra. This angular dependence is rather complicated. However, some characteristic features are seen in this figure. First, in both planes, dominant frequencies with strong amplitudes are concentrated in $(1.5-2.1) \times 10^3$ T. This frequency range corresponds to the cross-sectional area of $20-28\%$ of S_{BZ}. Second, in the b'-c^*

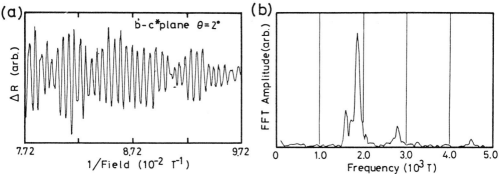

Fig. 2 (a) Typical Shubnikov-de Haas oscillations. (b) Frequency spectrum corresponding to Fig. 2(a). Some peaks obviously exist in the spectrum.

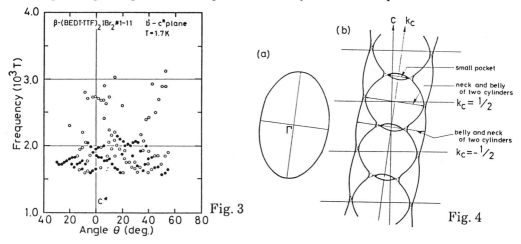

Fig. 3 Angular dependence of the Shubnikov-de Haas frequencies in the b'-c* plane. ● and ○ show strong and relatively weak peaks in frequency spectra, respectively.

Fig. 4 Fermi surface model. (a) A simple ellipsoid centered at the Γ point. (b) Fermi surfaces reduced by zone folding at the boundary along the k_c-direction.

plane, there seems to exist a branch which follows the orientation dependence for a nearly cylindrical Fermi surface given by $F=F_0/\cos\theta$, where F_0 is the frequency for the minimum cross-sectional area. It seems that this branch has $F_0=1.8\times10^3$ T at $\theta=0°$. Third, in the b'-c* plane, there seems to exist a branch with higher frequency of 2.7×10^3 T, centered only near $\theta=0°$. In the c*-a plane, we do not find the second and the third features. This is because the oscillation amplitudes are weaker than those in the b'-c* plane. We always measure the magnetoresistance in the transverse condition in the b'-c* plane, while, in the c*-a plane, except at $\theta=0°$, in between the transverse and the longitudinal condition.

Based on these features, we consider the Fermi surface model as follows. Figure 4(a) shows a single band of an ellipsoid of revolution, centered at the Γ point. The volume enclosed by this ellipsoid is assumed to be one-half of the first Brillouin zone. This assumption is made by requiring that one hole carrier exists in the unit cell and then the conduction band is half filled.

This ellipsoid elongated along the k_c-direction crosses the zone boundary, and then, the Fermi surface is reconstructed by zone folding. Figure 4(b) shows the thus obtained Fermi surface in a repeated zone scheme. It consists of three parts: two of them are slightly and considerably warped cylinders extending along the k_c-direction, the last one is a small lens-like pocket at the Γ point. It should be noted that the small pocket expected in this model might be consistent with the pockets in the β-type BEDT-TTF salts observed by other groups [4][6]. The angular dependence of the magnetoresistance can be explained by the cylindrical Fermi surface in terms of Yamaji's theory. This fact is not inconsistent with the present model, because the open orbit along the k_c-direction, which conserves the character of the cylinder, gives a dominant contribution to the magnetoresistance.

For β-(BEDT-TTF)$_2$IBr$_2$, Kartsovnik et al. reported two kinds of branches, one is of low frequency corresponding to 0.7% of S_{BZ}, and the other is of high frequency corresponding to 50% of S_{BZ}, and then the $\cos\theta$ dependence of the latter high frequency branch means a nearly cylindrical Fermi surface. The reason for the differences between their and our results is not known. A possible cause of the differences may be concerned with the quality of the sample. The value of RRR is 2000−3000 and T_c is 2.8K for the sample used by Kartsovnik et al., while about 1500 and 2.3K for our sample. Our sample was examined by X-ray diffraction down to liquid helium temperature and identified as the β-phase.

We would like to thank Prof. S. Kagoshima for studying the X-ray diffractions. This work was supported in part by a Grant-in-Aid for Special Project Research from the Ministry of Education, Science and Culture of Japan.

References

[1] K. Murata, N. Toyota, Y. Honda, T. Sasaki, M. Tokumoto, H. Bando, H. Anzai, Y. Muto and T. Ishiguro: J. Phys. Soc. Jpn. 57 (1988) 1540.
[2] N. Toyota, T. Sasaki, K. Murata, Y. Honda, M. Tokumoto, H. Bando, N. Kinoshita, H. Anzai, T. Ishiguro and Y. Muto: J. Phys. Soc. Jpn. 57 (1988) 2616.
[3] W.Kang, G. Montambaux, J. R. Cooper, D. Jérome, P. Batail and C. Lenoir: Phys. Rev. Lett. 62 (1989) 2559.
[4] M. V. Kartsovnik, V. N. Laukhin, V. N. Nijankovskii and A. A. Ignatiev:Pis'ma Zh. Eksp. Teor. Fiz. 47 (1988) 302 [JETP Lett. 47 (1988)363].
[5] M. V. Kartsovnik, P. A. Kononovich, V. N. Laukhin and I. F. Shchegolev: Pis'ma Zh. Eksp. Teor. Fiz. 48 (1988) 498 [JETP Lett. 48 (1988) 541].
[6] I. D. Parker, D. D. Pigram, R. H. Friend, M. Kurmoo and P. Day: Synth. Metals 27 (1988) A387.
[7] F. L. Pratt, A. J.Fisher, W. Hayes, J. Singleton, S. J. R. M. Spermon, M. Kurmoo and P. Day: Phys. Rev. Lett. 61 (1988)2721
[8] K. Oshima, T. Mori, H. Inokuchi, H. Urayama, H. Yamochi and G. Saito:Phys. Rev. B38 (1988) 938.
[9] K. Yamaji: J. Phys. Soc. Jpn. 58 (1989) 1520.

On the Electronic Properties of $ET_2Cu(NCS)_2$ as well as of Some New Organic Salts

H. Müller, C.-P. Heidmann, A. Lerf, W. Biberacher, R. Sieburger, and K. Andres

Walter Meißner Institute for Low Temperature Research,
D-8046 Garching, Fed. Rep. of Germany

Samples of $ET_2Cu(NCS)_2$ prepared by two different methods show differences in transport and superconducting properties, but show similar Shubnikov - de Haas oscillations. Possible mechanisms of conduction and superconduction for this compound are discussed.
The properties of new compounds analogous to the superconductor $ET_2Cu(NCS)_2$, namely $ET_2Ag_3(SCN)_4$ and $ET_2Hg_x(SCN)_4$, as well as some salts of substituted triphenylene systems are also discussed.

$ET_2Cu(NCS)_2$: NORMAL AND SC PROPERTIES

Crystals of $ET_2Cu(NCS)_2$ were synthesized in two ways: First following the recipe of Urayama /1/ (samples B) and secondly by our own method /2/ (samples A), which differs from /1/ by the molar concentration and the solvent for electrolysis: We used ET (1 mmol/l), CuSCN (36 mmol/l), KSCN (25 mmol/l) and dibenzo-18-crown-6 (10 mmol/l) in methylene chloride. With platinum electrodes and a current density of 1.5 $\mu A/cm^2$, conglomerates of thin black platelet-like crystals were obtained after 11 days. The morphology of samples B was somewhat different (elongated black hexagons) and the crystals showed flatter and more perfect surfaces. The resistivity versus temperature behaviour of both types of samples is shown in Fig. 1. Sample B pretty much reproduces the behaviour reported by Oshima et al. /3/, namely first an increase in resistivity on cooling, a maximum at 90 K followed by a rapid decrease down to the SC transition temperature (with a SC onset at T_c = 10.3 K). This increase is much smaller in sample A, which shows a slightly lower transition temperature of T_c = 9.4 K. In the structure work of Oshima et al. /3/ it is shown that the unit cell contains two ET-dimers tilted nearly 90° against each other and forming a checkerboard-like pattern in the plane of the ET-molecules. In principle, therefore, this double-dimer could lead to an insulating state, if the 2:1 stoichiometry between donors and anions is taken into account (two holes per unit cell). This is somewhat similar to the case of α-ET_2I_3, which also contains four ET molecules per unit cell and which in fact shows a metal-insulator transition at 136 K. Indeed, the resistivity of sample B shows activated behaviour between room temperature and 170 K, corresponding to an energy gap of 51 meV. The Seebeck coefficients, which are positive for thermal gradients in the c-direction and negative in the b-direction are also of rather large magnitude ($\sim$ 30 $\mu V/K$) /4/ and are more typical of a semiconducting or semimetallic state. Below 90 K, a transition from semiconducting to metallic behaviour occurs, which is rather unusual and normally requires a structure change. It is interesting to note in this context the anomalous <u>negative</u> thermal expansion in the a-direction (i.e. normal to the plane of packing of ET-molecules), that has been observed by Urayama et al. /1/. It in turn causes a correspondingly larger positive thermal expansion especially in the c-direction and thus a denser packing of ET-molecules on cooling, which might lead to a stabilization of the metallic

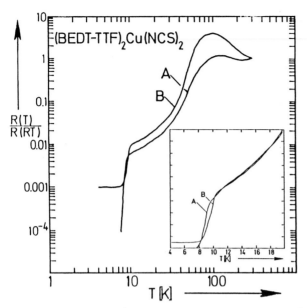

Fig. 1: Resistivity versus temperature and difference in SC transition temperature (insert) of samples A and B.

state at lower temperatures. If superconductivity arises from modulation of the Π overlap (and thus of the band structure) due to the libron-motion of the ET molecules, then the pairing interaction will be strongest for a critical Π overlap. This critical overlap seems to be close to the one which separates the metallic from the semiconducting state.

$ET_2Cu(NCS)_2$: SHUBNIKOV - DE HAAS OSCILLATIONS

These experiments were done in the High Field Laboratory of the Max Planck Institute in Grenoble, in fields up to 24 T and at temperatures down to 0.3 K. The crystals were mounted in a holder such that they could be tilted (during the experiments) around their b- and c-axis, the initial field orientation always being along the a*-axis (normal to the b-c plane of conduction) /5/. The resistivity was always measured along the b-direction. In both the A- and B-samples, SdH-oscillations were seen in fields above 12 T. The SdH-frequency was similar in both A- and B-samples, amounting to 597 $\pm$ 7 T. This is in agreement with the observations of Pratt et al. /6/, but somewhat smaller than the frequency of 666 T observed by Oshima et al. /3/ and would amount to a Fermi surface cross section of 16 % of the first Brillouin zone. The oscillation amplitude for the B-samples was considerably larger, indicating a higher crystal quality. When tilting the field away from the a*-axis by an angle Θ in any direction, we observe a $(\cos \Theta)^{-1}$ dependence of the SdH frequency, which confirms the high degree of two-dimensionality of the conduction band in this crystal. This is shown in Fig. 2 for tilts around the b-axis as well as around the c-axis (insert of Fig. 2). A field sweep on a B-sample in the highest field range (plotted versus 1/H) is shown in Fig. 3. Above B=22 T there is evidence for a superimposed second periodicity. This is seen most clearly in a Fourier analysis of the data (Fig. 4). The Fourier analysis shows that besides the ground frequency of 597 T and its higher harmonics (1194 T, 1800 T), there are new frequencies at

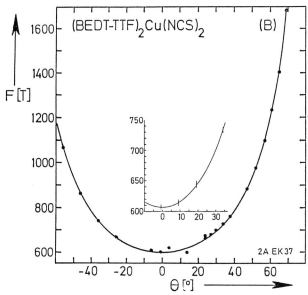

Fig. 2: SdH-oscillation frequency as a function of tilt angle for tilts around the b-axis (solid dots) and around the c-axis (insert). The solid lines are the $(\cos \theta)^{-1}$ functions.

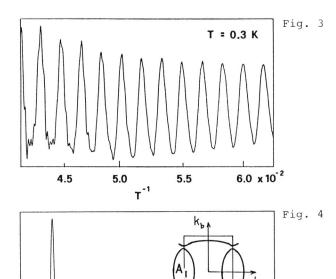

Figs. 3 and 4: SdH-oscillations in sample B (plotted versus 1/H) and their Fourier spectrum.

2650 T + n × 597 T (n = 0, 1, 2, 3). We tentatively interpret this as a magnetic breakdown effect, which connects a hole and an electron Fermi surface and which thus creates a new (and larger) Fermi surface cross section. The observed frequencies are qualitatively consistent with the Fermi surface structure calculated by Mori et al. /4/ (see insert of Fig. 4), the area A_2 amounting to 71 % of the first Brillouin zone.

$ET_2Ag_3(SCN)_4$

Electrocrystallization with the $Ag(SCN)_2^-$ anion in various solvents produces, in contrast to $Cu(SCN)_2^-$, a compound with the complex anion stoichiometry $ET_2Ag_3(SCN)_4$. The crystals were black, rhombic and of typical size 1.8 × 1.0 × 0,5 mm³. X-ray studies revealed an orthorhombic structure with cell parameters a = 4.204 (1) Å, b = 11.597 (2) Å and c = 40.125 (9) Å, with a possible superstructure doubling the a-axis. Conductivity measurements, carried out both at ambient pressure and at 2.6 kbar, show semiconducting behaviour, with energy gaps of 185 meV down to 215 K and 230 meV below this temperature (at p = 0). At p = 2.6 kbar, these gaps are reduced to 167 and 210 meV respectively, indicating that a metallic state might require pressures in excess of 30 kbar. The semiconducting behaviour presumably arises from a less dense packing of the ET molecules due to the larger anion size.

$ET_2Hg_x(SCN)_4$

Electrolysis of a solution of ET and $(n-Bu_4N)_2Hg(SCN)_4$ in trichloroethane or a mixture (4:1) of chlorobenzene and acetonitrile produced black, thin and extremely fragile needles. When using the latter solvent, a stoichiometry x = 1.4 - 1.5 is derived from analytical data. Conductivity data again show semiconducting behaviour below room temperature, the energy gap being rather small (85 meV).

SEARCH FOR NEW DONORS

We have investigated the donor properties of substituted triphenylene systems, namely benzotrithiophene, hexamethoxytriphenylene and tris(-methylenedioxy)-triphenylene. In spite of rather high redox-potentials (+1450 mV, +1180 mV and +1290 mV respectively vs SCE (solvent: CH_2Cl_2); first oxidation potentials), we have been able to grow 2:1 salts of each compound with various inorganic anions. Fragility and smallness of crystals isolated, however, made resistivity measurements impossible. Recently, we have been able to synthesize the promising sulphur-based π donors (I) and (II)

I: n=1   II: n=3

with redox-potentials of +750 mV/1250 mV vs SCE for (I) and +620/1150 mV vs SCE for (II) (first and second oxidation potentials, solvent: CH_2Cl_2). A detailed synthesis and results of electrocrystallization experiments and conductivity data will be published elsewhere.

REFERENCES

1. H. Urayama, H. Yamochi, G. Saito, S. Sato, A. Kawamoto, J. Tanaka, T. Mori, Y. Maruyama and H. Inokuchi, Chem. Letters 1988, 463.
2. H. Müller, C.-P. Heidmann, H. Fuchs, A. Lerf, K. Andres, R. Sieburger and J. S. Schilling, Nato ASI, Spetses (Greece), Plenum Press, 1989, to be published.
3. K. Oshima, H. Yamochi, H. Urayama and G. Saito, Physica C 153, 1148 (1988)
4. T. Mori, this conference.
5. C.-P. Heidmann, W. Biberacher, H. Müller, W. Joss and K. Andres, Nato ASI, Spetses (Greece), Plenum Press 1989, to be published.
6. F. L. Pratt, J. Singleton, M. Kurmoo, S.J.R.M. Spermon, W. Hayes and P. Day, this conference.

Fermi Surface and Band Structure of κ-(BEDT-TTF)$_2$Cu(NCS)$_2$

F.L. Pratt[1], *J. Singleton*[2], *M. Kurmoo*[3], *S.J.R.M. Spermon*[2], *W. Hayes*[1], *and P. Day*[3]

[1]Clarendon Laboratory, University of Oxford, Oxford OX1 3PU, UK
[2]High Field Magnet Laboratory and Research Institute for Materials,
  University of Nijmegen, 6522 ED Nijmegen, The Netherlands
[3]Inorganic Chemistry Laboratory, University of Oxford, Oxford OX1 3QR, UK

Abstract. Shubnikov–de Haas measurements of Fermi surface parameters in κ-(BEDT-TTF)$_2$Cu(NCS)$_2$ are reported. Magnetoresistance oscillations are observed with fundamental field 596±2 T in transverse geometry and 645±3 T in longitudinal geometry. Associating these different periodicities with the two extremal areas of a quasi–two–dimensional Fermi surface with weak modulation in the interplane direction leads to an estimate of 1.6±0.2 meV for the inter–plane bandwidth. Comparison of the experimental results with the proposed band structure indicates an enhancement of the effective mass; this is attributed to the effect of the short range electron correlation.

1. Introduction

The κ-phase of the BEDT-TTF (ET) series of organic conductors is particularly interesting as κ-ET$_2$Cu(NCS)$_2$ is the highest T_c organic superconductor and also the crystal structure and measured properties of the κ-phase indicate a greater degree of two–dimensionality than with the other phases. Besides giving the highest T_c for organics, the ET salts are the first family of organic metals in which closed sections of the Fermi surface have been studied by means of the Shubnikov–de Haas (SdH) effect. SdH oscillations in the magnetoresistance have now been measured in several β phase materials [1-4], β''-ET$_2$AuBr$_2$ [5], κ-ET$_2$Cu(NCS)$_2$ [6,7] and ET$_2$KHg(SCN)$_4$ [8]. The reported SdH measurements for the β-phase materials show complex behaviour and several different oscillation periods and beats have been observed by different groups. The κ-phase by comparison appears more straightforward with the observation of just one SdH period in the earlier work [6]. We report here SdH measurements on κ-ET$_2$Cu(NCS)$_2$ in which we have observed two different SdH periods, depending on the measurement configuration, and we discuss the band structure and Fermi surface in conjunction with these measurements. A fuller account of this work will be found elsewhere [9].

2. Experimental Details

Crystals of κ-ET$_2$Cu(NCS)$_2$ were obtained by electrocrystallisation of ET in distilled 1,1,2-TCE using TBA SCN / Cu SCN as the source for the

Cu(NCS)$_2^-$ anion. The ESR linewidth was 60 G and the midpoint of the superconducting transition was 9.7 K. X-ray measurements confirmed the previously reported κ-phase crystal structure [10].

Electrical contacts to the samples were made using gold wire attached with silver paint to evaporated gold pads. Both four-in-line and square contact arrangements were used. Resistance measurement was by the low frequency AC method. Magnetic fields up to 20 T were applied perpendicular to the bc plane of the crystals and the samples were cooled in a helium dilution refrigerator.

3. Experimental Results and Discussion

Oscillations in the magnetoresistance become observable above 9 T for temperatures below 1 K. Both Fourier analysis and direct fitting of the oscillations were used to estimate Fermi surface parameters from the data. For longitudinal geometry (j||a*) a single series of SdH oscillations is observed. The fundamental field for these oscillations is 645±3 T which is consistent with the previous longitudinal measurements of Oshima et al.[8]. For transverse geometry (j⊥a*) a significantly different SdH frequency is observed (Fig.1). The fundamental field in this case is 596±2 T which corresponds to a Fermi surface area 9% smaller than in longitudinal configuration. This lower frequency has also been observed by Müller et al.[7]. From the temperature dependence of these oscillations we derive an effective mass of 3.6±0.2 m_e and from the field dependence of the

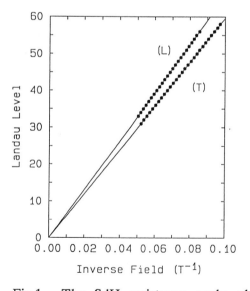

Fig.1 The SdH resistance peaks plotted against reciprocal field at 0.2 K with B||a*. Longitudinal configuration j||B (L) compared with transverse configuration j⊥B (T). The straight lines indicate the fitted parameters.

oscillation amplitude we estimate T_D = 0.4±0.1 K; both these parameters are consistent with the previous longitudinal results.

Some modulation of the area of the quasi–two-dimensional Fermi surface is expected as k varies in the inter–plane (a^*) direction and a possible explanation for the observation of the two SdH frequencies is that they correspond to the maximum and minimum areas. It is not entirely clear why longitudinal and transverse geometry are sensitive to different extrema but differences in the details of the scattering processes at the two extrema may be responsible e.g. the anisotropic scattering expected from anion disorder. Following this interpretation, t_a, the inter–plane transfer integral may be estimated as follows. We take the Fermi surface to be given by $E_F = E_{bc}(k_a) - 2 t_a \cos k_a a$, where E_F is the Fermi energy and E_{bc} is the energy in the bc plane. The inter–plane bandwidth is then given by $4 t_a = \hbar e \Delta B_F / m^*_{cyc}$ where ΔB_F is the difference in fundamental fields between the two extrema [9]. Taking ΔB_F to be 49 T and m^*_{cyc} to be 3.6 m_e we get $4 t_a$ = 1.6±0.2 meV. By comparison, we note that in β_H-ET$_2$I$_3$ a beat frequency of 37T was seen with $m^*_{cyc} \sim 4 m_e$ [2]. This implies an inter–plane bandwidth of ~ 1.1 meV, which is comparable to that we estimate for the Cu(NCS)$_2$ salt.

An extended Hückel band structure calculation for this material predicts a single closed orbit with an area of about 20% of the zone [6]. This orbital area corresponds quite well to the experimental value (16–17%). We have calculated the cyclotron effective mass expected for this orbit as m^*_{cyc}= 1.0±0.1 m_e whereas the experimental value is 3.6 m_e. This indicates a mass enhancement and narrowing of the bands by a factor of ~ 3.6. A similar reduction in bandwidth has been shown to be consistent with the thermopower [11]. The band narrowing may be understood as the result of short–range electron correlations, of which the most important are expected to be the on-site correlation energy U and the intra–dimer correlation energy V_D.

5. Conclusion

Studies of the SdH oscillations in κ-ET$_2$Cu(NCS)$_2$ show that whilst the size of the closed orbit on the Fermi surface is reasonably consistent with the calculated band structure, the measured effective mass is significantly enhanced over the calculated value. This enhancement is proposed to result from the short–range electron correlations which are believed to be comparable to the electron hopping bandwidth and consequently must be taken into account in calculations of the electronic band structure. Two different SdH periods have been observed in our experiments, one in longitudinal geometry and the other in transverse geometry. We have interpreted these different periods as corresponding to the maximum and minimum cross-sectional areas of the tubular closed section of Fermi surface. It is hoped that further experiments will clarify the origin of the different SdH frequencies and allow a more detailed study of the Fermi surface.

References

[1] K. Murata, N. Toyota, Y. Honda, T. Sasaki, M. Tokumoto, H. Bando, H. Anzai, Y. Muto and T. Ishiguro, *J. Phys. Soc. Jpn.* 57, 1540 (1988); N. Toyota, T. Sasaki, K. Murata, Y. Honda, M. Tokumoto, H. Bando, N. Kinoshita, H. Anzai, T. Ishiguro and Y. Muto, *J. Phys. Soc. Jpn.* 57, 2616 (1988); see also papers of K. Murata et al and T. Sasaki et al. in these proceedings.

[2] W. Kang, G. Montambaux, J.R. Cooper, D. Jerome, P. Batail and C. Lenoir, *Phys. Rev. Lett.* 62, 2559 (1989).

[3] M.V. Kartsovnik, V.N. Laukhin, V.I. Nizhankovskii and A.A. Ignat'ev, *JETP Lett.* 47, 363 (1988); M.V. Kartsovnik, P.A. Kononovich, V.N. Laukhin and I.F. Shchegolev, *JETP Lett.* 48, 541 (1988); also V.N. Laukhin et al (these proceedings).

[4] I.D. Parker, D.D. Pigram, R.H. Friend, M. Kurmoo and P. Day, *Synth. Met.* 27, A387 (1988).

[5] F.L. Pratt, A.J. Fisher, W. Hayes, J. Singleton, S.J.R.M. Spermon, M. Kurmoo and P. Day, *Phys. Rev. Lett.* 61, 2721 (1988).

[6] K. Oshima, T. Mori, H. Inokuchi, H. Uruyama, H. Yamochi and G. Saito, *Phys. Rev.* B38, 938 (1988).

[7] H. Müller et al (these proceedings).

[8] T. Osada et al (these proceedings).

[9] F.L. Pratt, J. Singleton, M. Kurmoo, S.J.R.M. Spermon, W. Hayes and P. Day, submitted to *J. Phys: CM*.

[10] H. Urayama, G. Yamochi, G. Saito, T. Sugano, M. Kinoshita, T. Inabe, T. Mori, Y. Maruyama and H. Inokuchi, *Chem. Lett.* 1988, 1057 (1988).

[11] T. Mori and H. Inokuchi, *J. Phys. Soc. Jpn.* 57, 3674 (1988).

Fermi Surface and Thermoelectric Power of Two-Dimensional Organic Conductors

T. Mori and H. Inokuchi

Institute for Molecular Science, Myodaiji, Okazaki 444, Japan

Abstract. Electronic band structures and Fermi surface of organic conductors are discussed on the basis of the single-band tight-binding approximation. The band structures of $(BEDT-TTF)_2Cu_5I_6$ and $(BEDT-TTF)_4Hg_3Cl_8$ (BEDT-TTF: bis-(ethylenedithio)tetrathiafulvalene) are calculated according to this approximation. There is also an attempt made to interpret the anisotropy of thermoelectronic power of two-dimensional organic conductors.

1. Introduction

Electronic band structures and Fermi surface of organic conductors have been widely calculated on the basis of a single-band tight-binding approximation. In particular, the discovery of two-dimensional BEDT-TTF conductors enhanced the necessity of the band calculation due to their large variety of electronic structures. Though the results of the calculation have been, at the earlier stage of the investigation, compared only with the anisotropy of various physical properties, the recent experiments of the Shubnikov-de Haas (SdH) oscillations have afforded a much stricter touchstone of the calculation. Here we present (1) the band structure of $(BEDT-TTF)_2Cu_5I_6$ and $(BEDT-TTF)_4Hg_3Cl_8$, (2) a band-theoretical attempt to reproduce the anisotropy of thermoelectric power, and (3) comments on the agreement of the calculations with the SdH experiments.

2. Band Structure of $(BEDT-TTF)_2Cu_5I_6$

The band structure of $(BEDT-TTF)_2Cu_5I_6$, whose preparation and structure has been reported by Shibaeva et al.[1], is shown in Fig. 1. The interactions p1, p2, and p3 (Fig. 1(a)) form a two-dimensional network of the BEDT-TTF molecules, similarly to other BEDT-TTF salts. The band structure (Fig. 1(b)), however, has an open Fermi surface perpendicular to the a axis. This is due to the uniform arrangement of the BEDT-TTF molecu-

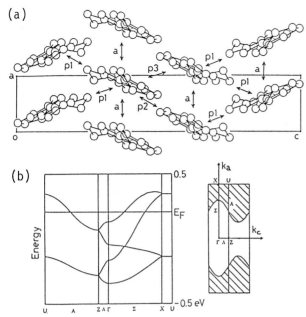

Fig. 1. (a) Donor arrangement and (b) band structure of (BEDT-TTF)$_2$Cu$_5$I$_6$. Overlap integrals are a=5.3, p1=-7.5, p2=-7.7, and p3=16.4 (x10^{-3}).

les along the a axis. Though the Fermi surface is open, this complex does not undergo a Peierls transition (metal down to 1.3 K [2]), because the Fermi surface is seriously distorted due to the comparable magnitude of the "intercolumnar" interactions. This salt is probably the first BEDT-TTF metal which has an open Fermi surface along the "stacking" direction. It will also be noteworthy that the gross feature of the band structure (Fig. 1(b)) closely resembles β-(BEDT-TTF)$_2$PF$_6$ due to the resemblance of their symmetry properties [3]. However, the latter salt is one-dimensional along the direction perpendicular to the "stacking" direction; its one-dimensionality has been established from the observation of the two-fold modulated structure which appears below the metal-insulator transition temperature [4].

The organic superconductors (BEDT-TTF)$_4$Hg$_3$Cl$_8$ and (BEDT-TTF)$_4$Hg$_3$Br$_8$, also reported by Shibaeve et al. [5], have κ-type donor arrangement, and their band structures are basically identical to κ-(BEDT-TTF)$_2$I$_3$ [6].

3. Thermoelectric Power

Thermoelectric power of BEDT-TTF conductors has been theoretically calculated on the basis of the single-band model [7].

This calculation affords a good interpretation of the observed anisotropy and temperature dependence.

4. Fermiology

The above-mentioned band calculations as well as the anisotropy of the various physical properties have established the two-dimensionality of the BEDT-TTF conductors. The validity of the calculated Fermi surface should be, however, tested by such experiments as the SdH oscillations. Table 1 lists the single-band calculations of BEDT-TTF conductors which show metallic conduction down to low temperatures. The fundamental frequencies of clear oscillations observed in κ-(BEDT-TTF)$_2$-Cu(NCS)$_2$ [10] and β_H-(BEDT-TTF)$_2$I$_3$ [16] correspond with the cross section of the Fermi surface expected from the calculations. Many other experiments also basically agree with the calculations [17]. The observed slow oscillations of β''-(BEDT-TTF)$_2$AuBr$_2$ are essentially consistent with the relatively small closed piece of the calculated Fermi surface [18]. The observed oscillation of (BEDT-TTF)$_2$KHg(SCN)$_4$ [12] also agrees with the exsistence of the closed orbit in the calculated band structure. In conclusion, the single-band calculation affords, in spite of its drastic simplification, a good starting point for understanding the Fermi surface.

Table 1. Single-band calculations of metallic BEDT-TTF conductors.

Compounds	S_{FS}/S_{BZ}	Ref.
β-(BEDT-TTF)$_2$I$_3$	0.50	[8]
θ-(BEDT-TTF)$_2$I$_3$	Close	[9]
κ-(BEDT-TTF)$_2$I$_3$	Close	[6]
κ-(BEDT-TTF)$_2$Cu(NCS)$_2$	0.18	[10]
(BEDT-TTF)$_4$Hg$_3$Cl$_8$	(0.22)	
β''-(BEDT-TTF)$_2$AuBr$_2$	0.043	[11]
(BEDT-TTF)$_2$KHg(SCN)$_4$	0.20	[12]
(BEDT-TTF)$_2$ReO$_4$	Open+Close	[13]
(BEDT-TTF)$_2$Cu$_5$I$_6$	Open	
(BEDT-TTF)$_2$ClO$_4$(TCE)$_{0.5}$	Semimetal	[14]
(BEDT-TTF)$_3$Cl$_2$(H$_2$O)$_2$	Semimetal	[15]

References

[1] R.P.Shibaeva and R.M.Lobkovskaya, Krystallografiya 33, 408 (1988).
[2] L.I.Buravov, A.V.Zvarykina, M.V.Kartsovnik, N.D.Kushch, V.N.Laukin, R.M.Lobkovskaya, V.A. Merzhanov, L.N.Fedutin, R.P. Shibaeva, and E.B. Yagubskii, Zh. Eksp. Teor. Fiz. 92, 594 (1987).
[3] T.Mori, A.Kobayashi, Y. Sasaki, R. Kato, and H.Kobayashi, Solid State Commun. 53, 627 (1985).
[4] H.Kobayashi, T.Mori, R.Kato, A.Kobayashi, Y.Sasaki, G. Saito, and H.Inokuchi, Chem. Lett. 1983, 581 (1983).
[5] R.P.Shibaeva and L.P.Rozenberg, Krystallografiya 33, 1402 (1988).
[6] R.Kato, H.Kobayashi, A.Kobayashi, S. Moriyama, Y.Nishio, K.Kajita, and W.Sasaki, Chem. Lett. 1987, 507 (1987).
[7] T.Mori and H.Inokuchi, J. Phys. Soc. Jpn. 57, 3674 (1988).
[8] T.Mori, A.Kobayashi, Y.Sasaki, H.Kobayashi, G.Saito, and H.Inokuchi, Chem. Lett. 1984, 957 (1984); M.H.Whangbo, J.M. Williams, P.C.W.Leung, M.A.Beno, T.J.Emge, H.H.Wang, K.D. Carlson, and G.W.Crabtree, J. Am. Chem. Soc. 107, 5815 (1985).
[9] H.Kobayashi, R.Kato, A.Kobayashi, Y.Nishio, K.Kajita, and W.Sasaki, Chem. Lett. 1986, 833 (1986).
[10] K.Oshima, T.Mori, H.Inokuchi, H.Urayama, H.Yamochi, and G.Saito, Phys. Rev. B 38, 938 (1988); K.D.Carlson et al. Inorg. Chem. 27, 965 and 2904 (1988).
[11] T.Mori, F.Sakai, G.Saito, and H.Inokuchi, Chem. Lett. 1986, 1037 (1986); H.Kobayashi, R. Kato, and A.Kobayashi, Synth. Metals 19, 623 (1987).
[12] T.Osada, R.Yagi, S.Kagoshima, N.Miura, M.Oshima, and G.Saito, this proceedings.
[13] M.H.Whangbo, M.A.Beno, P.C.W.Leung, T.J. Emge, H.H.Wang, and J.M.Williams, Solid State Commun. 59, 813 (1986).
[14] T.Mori, A.Kobayashi, Y.Sasaki, H.Kobayashi, G.Saito, and H.Inokuchi, Bull. Chem. Soc. Jpn., 57, 627 (1984).
[15] T.Mori and H.Inokuchi Chem. Lett. 1987, 1657 (1987).
[16] W.Kang, G.Montambaux, J.R.Cooper, D.Jérome, P.Batail, and C.Lenoir, Phys. Rev. Lett. 61, 2559 (1989).
[17] See the papers of K.Yamaji, V.N.Laukin et al., H.Müller et al., F.L. Pratt et al., K.Murata et al., and T. Sasaki et al. in these proceedings.
[18] F.A.Pratt, W.Hayes, A.J.Fisher, J.Singleton S.J.R.M. Spermon, M.Kurmoo, and P.Day, Synth. Metals, 29, F667 (1989).

Self-Consistent Band Structure and Fermi Surface for β-(BEDT-TTF)$_2$I$_3$

J. Kübler[1] and C.B. Sommers[2]

[1] Institut für Festkörperphysik, Technische Hochschule Darmstadt, Hochschulstr. 6, D-6100 Darmstadt, Fed. Rep. of Germany
[2] Laboratoire de Physique des Solides, Bât. 510, F-91401 Orsay, France

Abstract. Results of self-consistent electronic band-structure calculations for β-(BEDT-TTF)$_2$I$_3$ are presented. The calculations are based on the local-density-functional approximation and the ASW-algorithm. Energy bands and the topology of the Fermi surface are determined and compared with results of older calculations by us and by Mori et al. and with recent experimental Fermi-surface data. By means of partial density of states the local composition of the states at the Fermi energy is elucidated.

1. Introduction

Recently we reported results for the band structure and the Fermi surface of β-(BEDT-TTF)$_2$I$_3$ from ab initio calculations [1]. Since our findings were preliminary - suffering from rather crude approximations -and since new measurements appeared [2], we felt a critical reassessment of our old results was necessary and, therefore, considerably refined our calculational procedure. We report our new results here comparing them with older ones [1], [3] and those of Mori et al. [4] as well as with the experimental data on the Fermi surface by Kang et al. [2].

Let us begin by remarking on the theoretical basis of our calculations. There is first density-functional theory in its local approximation (LDA) [5]; it supplies effective-single-particle Schrödinger equations for the ground-state properties of the complicated many-particle problem at hand. One must solve a self-consistent-field problem for wave functions $\varphi_k(\vec{r})$ which through

$$n(\vec{r}) = \sum_{k=1}^{N} |\varphi_k(\vec{r})|^2$$

determine the electron-density $n(\vec{r})$. But the effective potential that appears in the Schrödinger equation is a well-defined function of the density $n(\vec{r})$ [6]. Thus an iterative procedure is mandatory for solving this problem by means of an efficient method. Before describing the latter briefly, we must comment on the physical significance of the eigenvalues ϵ_k of the Schrödinger equation.

Unlike Hartree-Fock theory, there is no Koopmans's theorem in density-functional theory (DFT), thus the ϵ_k are merely Lagrange parameters of the variational procedure needed to derive the Schrödinger equation from the total energy [5]. The one-particle energies, ϵ_k, are not quasi-particle energies. Still, one can prove that the Fermi energy, i.e. the energy of the highest occupied state, is an exact quantity of DFT [5]. Although this fact and Luttinger's theorem are not enough to ensure that the Fermi surface is given correctly by the DFT there is plenty of empirical evidence that the Fermi surfaces calculated in the LDA approximate real Fermi surfaces astonishingly well. In fact, much more severe are the approximations necessary to solve the Schrödinger equation, particularly for systems like β-(BEDT-TTF)$_2$I$_3$.

In our approach the wave functions and the charge density are obtained using the augmented spherical wave method (ASW) [7]. It essentially is a combination-of-atomic-orbitals method. The unit cell is partitioned completely into slightly overlapping atomic spheres (Wigner-Seitz spheres) in each of which the potential is assumed to be spherically symmetric. These approximations are not without problems for the system at hand which is not closed-packed at all. To fill space completely one must center potentials with no nuclear charge in between the two $C_{10}H_8S_8$-molecules and also the I-ions. These so-called empty spheres are placed emperically and they will be empty only in the beginning of the calculation acquiring charge in the course of the self-consistency cycles.

The crystal structure of β-(BEDT-TTF)$_2$I$_3$ is as given by Leung et al. [11]. We distinguish α-, β- and γ-C, as well as α- and β-S; all H-atoms are assumed to be equivalent, all three I-atoms inequivalent. Thus we abandoned one severe approximation made in our previous calculation [1], where we treated all C-atoms as equivalent, similarly with the S-atoms. The interstitial space was filled with 32 empty spheres as described above.

2. Results and Discussion

In Fig. 1 the band structure near the Fermi energy is shown. In contrast to our old results [1] (those given in ref. [3] were ill-converged) the Fermi energy, E_F, is now located in antibonding states thus giving hole-states at E_F. The two narrow bands at the bottom originate from the center C-atoms. The other four bands are not easily localizable (but see the discussion of the partial density-of-states below). We can construct the intersection of the Fermi surface with the basal plane from Fig. 1 and show it in Fig. 2. The surface is not closed connecting to the other zones through the point Y. But Fig. 1 at Y shows that a tiny shift of E_F suffices to cause a closure of the surface. Thus our results cannot decide about this extremal orbit. - It is interesting to compare these results with those of Mori et al. [4] obtained with a Hückel-molecular-orbital calculation. Except for the number of bands at E_F (we see four, Mori et al. have two) the upper-most bands are very similar. Thus our previous remark [1] claiming that Hückel-molecular-orbital calculations cannot describe the bands at E_F can no longer be maintained.

Fig. 3 gives the band structure in a plane parallel to the basal plane half-way between it and the top of the Brillouin-zone. The corresponding intersection of the Fermi surface with this plane is shown in Fig. 4 and is now seen to be closed. It

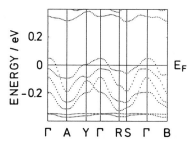

Fig. 1 Band structure of β-(BEDT-TTF)$_2$I$_3$ in the basal plane.

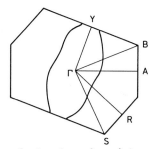

Fig. 2 Fermi-surface intersection with basal plane.

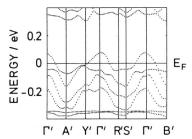

Fig. 3 As Fig. 1 but in plane half-way between basal plane and top plane.

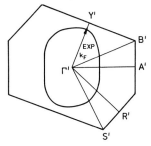

Fig. 4 Fermi-surface intersection with plane used in Fig. 3.

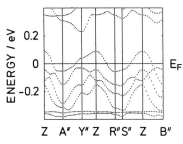
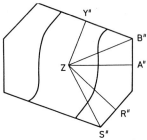

Fig. 5 As Fig. 1 but in top plane of the Brillouin-zone.

Fig. 6 Fermi-surface intersection with plane used in Fig. 5.

is in good agreement with an experimental cross-section given by Kang et al. [2] indicated by k_F^{EXP} in Fig. 4, and an average effective mass ratio is calculated as $m^*/m \simeq 3$.

Fig. 5, finally, gives the band structure in a plane parallel to the basal plane but now passing through the point Z in the top of the Brillouin-zone. The corresponding intersection of the Fermi surface with this plane is shown in Fig. 6; it is seen to be open again. But unlike the Fermi-surface cut with the basal plane this connection to the other zones around point Y" is quite certain, see Fig. 5 at Y". Thus the Fermi surface is warped strongly. It obviously gives rise to another extremal orbit perpendicular to the basal plane between Y" and Y if the connection in Fig. 3 is real or another, larger one, if the connection at Y is not. In the latter case we estimate $\pi k_F^2 \simeq 0.3 \times 10^{15}$ cm^{-2} which is roughly 10% of the dominant extremal area of $\pi k_F^2 = 3.56 \times 10^{15}$ cm^{-2} given by Kang et al. [2] for the extremal area in (or parallel) to the basal plane.

An attempt to characterize the local composition of the states at the Fermi energy, E_F, can be made by means of the partial density of states shown in Fig. 7. Here the height of each column above the atom indicated below gives the state density, $N(E_F)$, for this atom (for both spins), I_F denotes one of the I-atoms nearly in the face of the unit cell, I the corner atom and + the dominant interstitial charges. The total state density calculated is $N_{tot}(E_F) = 16$ eV^{-1} per cell which is probably accurate to within ± 6 eV^{-1}. An experimental value was discussed in ref. [1] and we felt ~ 8 eV^{-1} per cell not unreasonable. The interstitial charge contributes approximately 18% to the total value of $N(E_F)$.

An important question concerns the stability of the present results with respect to changes of the crucial partitioning of space into atomic spheres and changes of the total volume. Since these calculations are rather lengthy they are still ongoing. But we do have results for a case where the volume of the unit cell was decreased uniformly by $\sim 4\%$. The effect is rather enormous and in one important asspect unexpected. Although the Fermi energy remains within the four bands depicted in Fig. 1, and those become slightly distorted, as we would have expected, the low lying C-bands at - 3.5 eV are "dissolved" and mixed up with two other bands previously not below E_F, which we did not expect. This has the drastic effect that the Fermi energy now cuts through the lower lying of the four bands near E_F whereas before it was in the upper-most one. The changes in the Fermi surface are correspondingly large. The resulting changes in charge transfer are only moderate and the density-of-state profile shown in Fig. 7 retaining its shape, but the values for all I-atoms grow by more than a factor of 3. It remains to be seen by experimental results under hydrostatic pressure, especially for the Fermi surface, whether these effects are an artifact of our approximations or whether they are real.

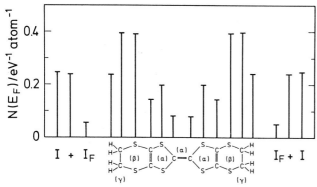

Fig. 7 Partial density of states at E_F per atom (for both spins) for atoms shown below, I corner iodine, I_F near-face iodine, + interstitial.

3. Summary

In closing we summarize that we have calculated band structure and Fermi surface for β-(BEDT-TTF)$_2$I$_3$ for the experimental volume at ambient pressure and for a slightly decreased volume. Our calculations are self-consistent and are based on the local-density approximations. Our results at ambient pressure can explain the recently [2] measured Fermi surface extremal cross-section, but the effect of pressure is enormous. Further calculations are needed (and in progress) to verify our theoretical findings.

Acknowledgement. The work of one of us (J.K.) was supported by the DFG through Sonderforschungsbereich 252 Darmstadt, Frankfurt, Mainz. One of us (C.B.S.) wishes to thank IBM for the loan of the 6150-5080 graphics workstation which was supported by a University-IBM research agreement; as well as a computational grant from the C.N.R.S. (CIRCE).

References

[1] J. Kübler, M. Weger, and C.B. Sommers, Sol. State Comm. 62, 801 (1987)

[2] W. Kang, G. Montambaux, J.R. Cooper, D. Jérome, P. Batail and C. Lenoir, Phys. Rev. Lett. 62, 2559 (1989)

[3] M. Weger, J. Kübler, and D. Schweitzer, in: Novel Superconductivity, S.A. Wolf and V.Z. Kresin eds. (Plenum, New York 1987), p. 149

[4] T. Mori, A. Kobayashi, Y. Sasaki, H. Kobayashi, G. Saito, and H. Inokuchi, Chem. Lett. 1984, 957 (1984)

[5] W. Kohn and P. Vashishta, in: Theory of the Inhomogeneous Electron Gas, S. Lundqvist and N.H. March eds. (Plenum, New York 1983), p. 79

[6] L. Hedin, B.I. Lundqvist, J. Phys. C4, 2064 (1976)

[7] A.R. Williams, J. Kübler, and C.D. Gelatt, Phys. Rev. B19, 6094 (1979)

[8] P.C.W. Leung, T.J. Emge, M.A. Beno, H.H. Wang, J.M. Williams, V. Petricek, and P. Coppens, J. Am. Chem. Soc. 107, 6184 (1985)

Anomalous Magneto-oscillation in θ-Type Crystals of (BEDT-TTF)$_2$I$_3$

K. Kajita[1], Y. Nishio[1], T. Takahashi[1], W. Sasaki[1], R. Kato[1], H. Kobayashi[1], A. Kobayashi[2], and Y. Iye[3]

[1]Faculty of Science, Toho University, Miyama 2-2-1, Funabashi 274, Japan
[2]Faculty of Science, University of Tokyo, Hongo 7-3-1, Tokyo 113, Japan
[3]ISSP, University of Tokyo, Roppongi, Tokyo 106, Japan

A new type of oscillatory magneto-transport phenomenon has been observed in θ-type crystals of (BEDT-TTF)$_2$I$_3$. A large oscillation appears in the magnetoresistance when the constant magnetic field is rotated in a plane containing the direction normal to the two-dimensional conducting plane (c*-axis) and an axis in the conducting plane. The oscillation does not depend on the angle between the electric current and the magnetic field, but depends on the magnetic field direction versus the crystal axis. For the magnetic field in the ac*-plane, the position of the bottoms of the oscillation is expressed as $\tan(\theta_{min}) = \alpha \cdot n$ ($n=0,1,2,3,\ldots$), where angle θ is measured from the c*-axis. The amplitude of the oscillation is determined by the magnetic field component normal to the two-dimensional plane.

1. Introduction

A θ-type crystal of (BEDT-TTF)$_2$I$_3$ is an organic crystal having metallic character down to low temperatures and undergoing a superconducting transition at about 3.6K [1]. Figure 1 gives the arrangement of BEDT-TTF molecules in a conducting plane and the Fermi surface which has been calculated by the tight binding approximation [2]. Conducting carriers are holes on the BEDT-TTF molecules, which has a two-dimensional nature. The low-temperature behavior of the resistivity is shown in Fig.2.

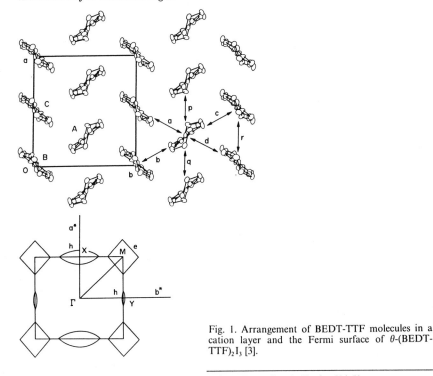

Fig. 1. Arrangement of BEDT-TTF molecules in a cation layer and the Fermi surface of θ-(BEDT-TTF)$_2$I$_3$ [3].

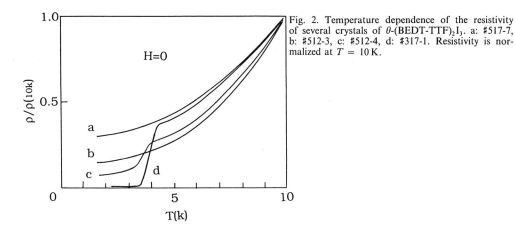

Fig. 2. Temperature dependence of the resistivity of several crystals of θ-(BEDT-TTF)$_2$I$_3$. a: #517-7, b: #512-3, c: #512-4, d: #317-1. Resistivity is normalized at $T = 10$ K.

At an early stage of the investigation, we noticed that the magnetoresistance of this material in the normal state is anomalous [3]. The magnetoresistance seems not to be determined by the Lorentz force. First, it is larger than expected from the Lorentz force. The most important is that the effect is independent of the angle between the magnetic field and electric current; it depends only on the magnetic field direction. The magnetoresistance is largest when the field is applied in the a-axis, which is in the two-dimensional plane, and it is smallest when the field is normal to the conducting plane. In the course of extending the investigation of this effect into the higher magnetic field region, we encountered a new type oscillation in the magnetoresistance effect.

2. Experiment

Figure 3 gives the resistance at several magnetic fields plotted against the angle between the magnetic field and the c*-axis. In low magnetic fields below 3T, we have sinusoidal curves with the minimum and the maximum at $\theta=0$ ($H//c^*$) and $\theta=90$ ($H//a$), respectively. Such a low field magnetoresistance is the one we have mentioned in the introduction. We call this "background". Above 3T, there appears oscillation on top of the background. In Fig.4, the same type of experiments done in the bc*-crystal plane are depicted. In this plane, the background component is very large and the oscillation is not so clear as that in Fig.3. Nevertheless, the existence of the oscillation is doubtless.

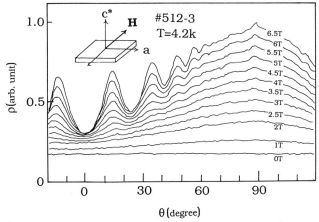

Fig.3 Resistivity ρ plotted against the magnetic field direction. Magnetic field is rotated in the ac^*-plane.

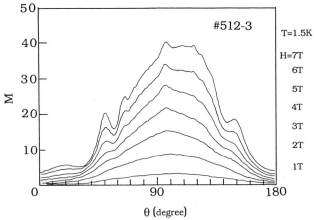

Fig.4 Magnetoresistance M plotted against the magnetic field direction. Magnetic field is rotated in the bc^*-plane.

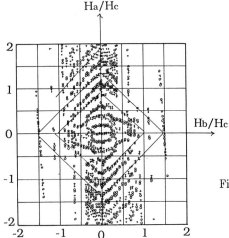

Fig.5 Positions of the bottoms (crosses) and peaks (circles) of the oscillation of the magnetoresistance.

The positions of the oscillation peaks and bottoms are independent of the magnetic field strength, which excludes the possibility that this is due to the Schubnikov-de Haas effect. The character of the oscillation period is well represented in the figure in which we plot the magnetoresistance against the tangent of the angle θ. We get the expression for the angle of the bottoms as $\tan(\theta_{min}) = \alpha \cdot n$ ($n=0,1,2,3,....$), where the parameter α, which gives the spacing of the bottoms, is about 0.39.

The experiments are extended to the magnetic field in the 4π direction. The results are plotted in Fig.5. In this figure the positions of the peaks and bottoms of the oscillation are plotted in the map of H_a/H_c and H_b/H_c. The solid lines are guides to the eye giving the positions of the bottom of the oscillation with the index n=2,3 and 4. We obtain an expression giving the bottom positions as $aH_a/H_c + bH_b/H_c = n$ ($n=0,1,2,3,....$).

Another important feature of the oscillation is that the amplitude is determined by the field component normal to the two-dimensional plane, H_c. This is demonstrated in Fig.6 which gives the oscillation of the magnetoresistance as a function of the magnetic field component normal to the two dimensional plane. The experimental measurements done in several different magnetic fields are confined in the same envelope, indicating the importance of the field component H_c in determining the oscillation amplitude.

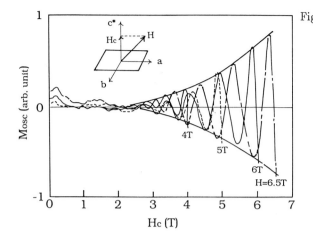

Fig.6 Oscillatory part M_{osc} of the magnetoresistance is plotted against the magnetic field component Hc $(=H\cos(\theta))$. Curves for $H=6.5T, 6T, 5T, 4T$ are plotted simultaneously. The envelope shown by the solid line is a guide for the eye.

3. A Possible Mechanism

The same type of oscillation has been found in other organic conductors, in β-(BEDT-TTF)$_2$I$_2$Br [5] and recently in (BEDT-TTF)$_2$KHg(SCN)$_4$ [6]. Thus, we have to consider that this is common to some organic conductors.

Yamaji [7] has proposed a mechanism which may give the magnetoresistance oscillation. His theory is based on the fact that the systems in which the oscillation is observed are two dimensional with a slight three dimensionality. In a magnetic field, the motion of the carrier in the k-space is given by the intersection of the Fermi surface and the plane normal to the magnetic field. For general directions of the magnetic field, there is a distribution in the area of the intersections because of the existence of the three dimensionality. Yamaji has shown that for several special magnetic field directions, the area becomes independent of the positions of the cross sections. In that direction of the magnetic field, the application of the magnetic fields enhances the two dimensionality of the system and that gives rise to an increase of the magnetoresistance. For the geometrical positions of the bottoms and peaks of the oscillations, the theory of Yamaji seems to explain the experiment. The problem remaining is how to explain the very large amplitude of the oscillation observed in the experiment.

References

[1] K.Kajita, Y.Nishio, S.Moriyama, W.Sasaki, R.Kato, H.Kobayashi and A.Kobayashi, Solid State Commun. 64 , 1279(1987).
[2] A.Kobayashi, R.Kato, H.Kobayashi, S.Moriyama, Y.Nishio, K.Kajita and W.Sasaki, Chem. Lett. , 2017 (1986).
[3] K.Kajita, Y.Nishio, T.Takahashi, W.Sasaki, R.Kato, H.Kobayashi and A.Kobayashi, Solid State Commun. 70 , 1181 (1989).
[4] K.Kajita, Y.Nishio, T.Takahashi, W.Sasaki, R.Kato, H.Kobayashi , A.Kobayashi and Y.Iye, Solid state Commun. 70 , 1189 (1989).
[5] M.V.Kartsovnik, P.A.Kononovich, V.N.Laukhin, I.F.Schegolev, JETP Lett. 48 , 541 (1988).
[6] T.Osada, R.Yagi, S.Kagoshima, N.Miura, M.Oshima and G.Saito , in this issue,
[7] K.Yamaji , J. Phys. Soc. Jpn. 58 , 1520 (1989).

Nearly Complete Quantization in Quasi-Two-Dimensional Organic Superconductors

K. Yamaji

Electrotechnical Laboratory, 1-1-4 Umezono, Tsukuba 305, Japan

Abstract. The new angle-dependent oscillation of magnetoresistance in β- and θ-(ET)$_2$X is shown to arise from a nearly complete discretization of Landau levels due to the disappearance of the dependence on wave number, which specifies the level together with the Landau quantum number, in the vicinity of the Fermi energy. This is shown by a semiclassical argument to occur at special angles for a weakly corrugated cylinder form of Fermi surface. Here the magnetoresistance makes peaks. Theoretical values of the angles are in good agreement with the observed ones. This result gives support to the validity of the tight-bonding bands based on a single HOMO of the ET molecule for ET superconductors. We treat also the case of the general form of in-plane Fermi surface and compare the results with experiments. These findings open a new method of Fermiology.

1. Introduction

Recent advance in magnetotransport studies on (ET)$_2$X superconductors is remarkable (here ET stands for BEDT-TTF) [1-5]. Results are going to verify the tight-binding bands calculated on the basis of the t_{1u}-type HOMO by Mori et al. [6]. This is important by itself but even more so from the viewpoint of the controversy on the mechanism of superconductivity in the ET superconductors. For example, this type of HOMO with the electronic charge concentrated in the central part of the TTF skeleton has strong couplings with intramolecular vibrations, which are asserted to provide the major contribution to the BCS-type electron-electron attraction in all superconductors composed of TTF derivatives [7].

The new effect of angle-dependent oscillation of magnetoresistance found in β-(ET)$_2$IBr$_2$ by Kartsovnik et al. [3] and in θ-(ET)$_2$I$_3$ by Kajita et al. [4] is another touchstone for the tight-binding bands. In this report it is shown to be well understandable on the basis of such bands, namely in the case of β-(ET)$_2$IBr$_2$ on the basis of the tight-binding band calculated by Mori et al., which consists of a single HOMO on each molecule and has a closed Fermi surface in the conducting plane directions. The peaks of the magnetoresistance appear at special angles, i.e. two-dimensionalization angles, for which all Landau levels lose the dependence on wave number in the direction of the field so that they are quantized into discrete levels, each of which has a three-dimensional degeneracy in the neighborhood of the Fermi energy ε_F [8]. The peak angles in the case of the general form of Fermi surface in the conducting plane is also derived. These results open a new method to determine the form of the Fermi surface by means of magnetoresistance measurements.

2. Case of Corrugated Cylinder Form of Fermi Surface

On the basis of the Shubnikov-de Haas effect observed by Kartsovnik et al. [3] and Mori et al.'s tight-binding band [6], we first assume the following simplified band:

$$\varepsilon_k = \hbar^2(k_x^2 + k_y^2)/2m - 2t \cdot \cos(ck_z), \tag{1}$$

where t is a small transfer energy between conducting layers. The Fermi surface in the conducting plane makes a circle with a radius nearly equal to k_F. k_F is defined by $\hbar^2 k_F^2/2m = \varepsilon_F$.

When the magnetic field H is applied perpendicularly to the conducting plane, we get Landau levels specified by integer n and wave number k_z as $E(n, k_z) = \hbar\omega(n+1/2) - 2t \cdot \cos(ck_z)$ with the cyclotron frequency $\omega_c = eH/mc_l$ (c_l being the light velocity). Due to the k_z-dependence, Landau levels with a specific n value make a continuous band with a width equal to 4t. When we incline the magnetic field by φ from the k_z-direction to the k_x direction, this width vanishes at special angles according to the following semiclassical argument. Trajectories of the semiclassical closed orbits are given by the inter-

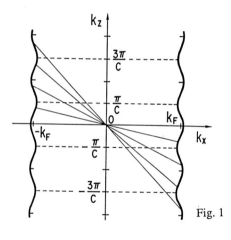
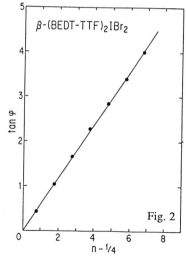

Fig. 1. Semiclassical-orbital planes satisfying (3) are shown by oblique lines. Here the special angle φ corresponds to the angle between the oblique and horizontal lines

Fig. 2. Values of $\tan\varphi$ at peaks of the magnetoresistance (curve 1 in Fig. 1 [3]) versus $n-1/4$ where n numbers the peaks from the $\varphi = 0$ side

section of the constant energy surface and the planes perpendicular to the magnetic field defined by $k_x\sin\varphi + k_z\cos\varphi = k_z^{(0)}\cos\varphi$; here $k_z^{(0)}$ denotes the point of intersection of the plane with the k_z-axis. The area S_k of the orbit specifically on the Fermi surface is obtained as

$$S_k = [\pi k_F^2 + 4\pi mt\cdot\cos(ck_z^{(0)})J_0(ck_F\tan\varphi) + O(t^2)]/\cos\varphi. \tag{2}$$

When the Bessel function $J_0(z)$ vanishes, the second term in the square bracket vanishes for any value of $k_z^{(0)}$. Since $J_0(z) \approx (2/\pi z)^{1/2}\cos(z-\pi/4)$, this occurs when φ satisfies

$$ck_F\tan\varphi \approx \pi(n - 1/4), \quad n = 1, 2, 3, \ldots. \tag{3}$$

These correspond to the angles between the oblique and horizontal lines in Fig.1. This means that for such special angles, the Landau levels lose the $k_z^{(0)}$ dependence in the neighborhood of the Fermi energy. Therefore, the Landau band width due to the $k_z^{(0)}$ dependence vanishes and the Landau levels are completely discretized with each discrete level being degenerate with a number proportional to the total site number. In other words, the system is two-dimensionalized at these angles. Although the correction term of the order of t^2/ε_F hinders the complete vanishing of the $k_z^{(0)}$ dependence, the quantization is almost complete since t^2/ε_F is very small.

For such angles, first the amplitude of the Shubnikov-de Haas oscillation must be maximized. Second, as is known in the theory of the quantum Hall effect [9], the two-dimensionalized degenerate Landau level must be broadened into a band by impurities, and since for almost all values of the magnetic field the Fermi energy must be on a non-conducting level away from the center of an impurity-broadened band, the coarse-grain-averaged value of the magnetoresistance must be enhanced especially at very low temperatures. Therefore, these angles should correspond to the peaks of the angle-dependent magnetoresistance.

This statement is verified by Fig. 2 plotted from the Russian data for β-(ET)$_2$IBr$_2$ [3]. The data points in the $\tan\varphi$ vs $n-1/4$ plane make a straight line, as expected, and its slope 0.596 is close to the theoretical value $\pi/ck_F \approx 0.62$. Furthermore, they report that the amplitude of the Shubnikov-de Haas oscillation takes a local maximum for these angles. This success of the cylindrical model for the Fermi surface clears away the question raised by Kübler et al. about the validity of the tight-binding bands for β-(ET)$_2$X [10]. The $\tan\varphi$ vs $n-1/4$ plot for the data of Kajita et al. for θ-(ET)$_2$I$_3$ [4] also makes a straight line. They clearly show that the peak angles do not depend on the strength of H nor on the temperature in accord with the theory.

The condition for the oscillation to appear is given by $t^2/\varepsilon_F < e\hbar H/mc_1 < 4t$. The lower bound of H comes from the condition that the correction of the order of t^2/ε_F does not break the three-dimensional quantization. The upper bound comes from the condition that the Landau bands overlap for most angles. For β-$(ET)_2IBr_2$ if we set $m = 5m_0$, $H = 15$ T, then we get $1\ k_B < t < 148\ k_B$. For θ-$(ET)_2I_3$, insufficient data are available for an estimate. In the latter material the layers are dimerized in the normal direction. This makes the effective value of t smaller. This may be why the condition for the appearance of the angle-dependent oscillation is less stringent in this material. The role of the Zeeman energy is remembered here. The difference of the in-plane Fermi surface areas for both spins is about 1 % of the Brillouin zone. This is about the order of magnitude related with the observed beat and slow oscillation of the magnetoresistance, although they have not yet been successfully explained.

3. Case of General Form of Quasi-Two-Dimensional Fermi Surface

In real systems the special angles should depend on the form of the Fermi surface in the conducting plane and on the direction in which the field is inclined. In fact we see data indicating the latter dependence for β- and θ-$(ET)_2X$ salts. Thus, we need the treatment of the general case. When it is elliptic, because of the anisotropic effective mass m_x and m_y, the angles are given by (3) with k_F suitably transformed for the inclination direction. In particular cases of the x- and y- directions the relevant k_F value is k_F itself in these directions.

When a more general quasi-two-dimensional band is given by $\varepsilon_\mathbf{k} = g(k,\theta) - 2t\cdot\cos(ck_z)$, where (k, θ) is the two-dimensional polar coordinate for $\mathbf{k}$, and the in-plane Fermi surface for $t = 0$ takes a general form $k = f(\theta)$, the area of semiclassical orbits treated above is modified to the following expression:

$$S_k = S_k(t=0) + 2t\cdot\cos(ck_z^{(0)})\cdot I(\varphi,\varphi_0) + O(t^2), \tag{4}$$

$$I(\varphi,\varphi_0) = \int_0^{2\pi} d\theta \cos[cf(\theta)\tan\varphi\cos(\theta-\varphi_0)]/\left[\frac{\partial g(k,\theta)}{\partial k}\right], \tag{5}$$

where $S_k(t=0)$ is the zeroth order term in t, and φ_0 is the angle of the inclination plane relative to the k_x-axis. Again the correction of the order of t disappears when the integral $I(\varphi,\varphi_0)$ vanishes.

In the case of the anisotropic quadratic $\mathbf{k}$ dependence in the conducting plane, we get the diagram in Fig.3 in which the value of the $\tan\varphi$ for which $I(\varphi,\varphi_0)$ vanishes is plotted at the polar angle equal to φ_0.

In the case of the tight-binding band proposed by Mori et al. for β-$(ET)_2I_3$ with a set of transfer energy values [11] adjusted by fitting to Hennig's thermopower data, we get Fig. 4. The trajectories of $I(\varphi,\varphi_0) = 0$ are not necessarily circular around the origin. The slopes of the plot of $\tan\varphi$ vs n, when the field is inclined in the ac*- and b'c*-planes, are deviated from the line defined by (3) to the gentler and steeper side, respectively. This is in agreement with the data reported by Schegolev et al. [12] and Sasaki et al. [13] showing such deviations, although the jump in the calculated result is not observed. This result indicates the appropriateness of the new version of the tight-binding band for β-$(ET)_2I_3$.

Concerning the result on θ-$(ET)_2I_3$, a band calculation suggests the Fermi surface consisting of three cylinders [14]. The peaking angle diagram for the lens-form of the Fermi surface is similar to that for the elliptic form. The diagram for the diamond form gives complicated forms. These are not close to the observed data obtained by Kajita et al. [15]. We may need more band calculations for θ-$(ET)_2I_3$.

Concerning κ-$(ET)_2Cu(NCS)_2$, the Shubnikov-de Haas oscillation takes a maximum amplitude for $\varphi = 18°$ when the field is inclined in the ab-plane [16]. This is in fair agreement with the above argument, if we use the known form of the closed piece of the Fermi surface [1]. Preliminary data of magnetoresistance indicate peaks in the range of $\varphi = 30°$ to $50°$ when the field is inclined in the ac-plane [17]. These angle values are in fair agreement with the calculated band structure.

When data for the whole region of inclination angle φ_0 become available in these systems, we could determine the bands by adjusting the band parameters through fitting the calculated diagram to experimental data. Therefore, the present effect opens a new method of Fermiology.

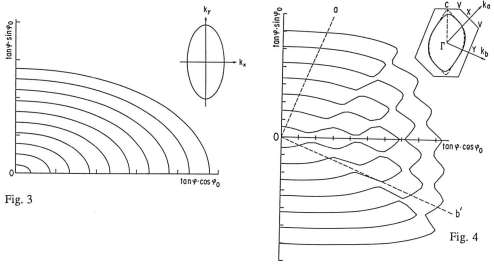

Fig. 3. Values of $\tan\varphi$ satisfying $I(\varphi,\varphi_0) = 0$ are plotted at the polar angle equal to φ_0 for the case of an elliptic Fermi surface shown in the inset

Fig. 4. Same diagram as Fig. 3 for the case of the tight-binding band of Mori et al. [11], the Fermi surface of which is shown by a continuous curve in the inset

References

1. K. Oshima, T. Mori, H. Inokuchi, H. Urayama, H. Yamochi, G. Saito: Phys. Rev. B **38**, 938 (1988)
2. K. Murata, N. Toyota, Y. Honda, T. Sasaki, M. Tokumoto, H. Bando, H. Anzai, Y. Muto, T. Ishiguro: J. Phys. Soc. Jpn. **57**, 1540 (1988)
3. M.V. Kartsovnik, P.A. Kononovich, V.N. Laukhin, I.F. Shchegolev: JETP Lett. **48**, 541 (1988)
4. K. Kajita, Y. Nishio, T. Takahashi, W. Sasaki, R. Kato, H. Kobayashi, A. Kobayashi, Y. Iye: Solid State Commun. **70**, 1189 (1989)
5. W. Kang, G. Montambaux, J.R. Cooper, D. Jérome, P. Batail, C. Lenoir: Phys. Rev. Lett. **62**, 2559 (1989)
6. T. Mori, A. Kobayashi, Y. Sasaki, H. Kobayashi, G. Saito, H. Inokuchi: Chem. Lett. 957 (1984)
7. K. Yamaji: Solid State Commun. **61**, 413 (1987)
8. K. Yamaji: J. Phys. Soc. Jpn. **58**, 1520 (1989)
9. H. Aoki, T. Ando: Surface Science **170**, 249 (1986)
10. J. Kübler, M. Weger, C.B. Sommers: Solid State Commun. **62**, 801 (1987)
11. T. Mori, H. Inokuchi: J. Phys. Soc. Jpn. **57**, 3674 (1988)
12. I.F. Schegolev, P.A. Kononovich, V.N. Laukhin, M.V. Kartsovnik: to be published in Chemica Scripta (Proc. Gen. Meeting of CMD of EPS in Nice, 1989)
13. T. Sasaki, N. Toyoda et al. obtained oscillatory magnetoresistance data in the b'a*-plane.
14. H. Kobayashi, R. Kato, A. Kobayashi, T. Mori, H. Inokuchi, Y. Nishio, K. Kajita, W. Sasaki: Synth. Metals **27**, A289 (1988)
15. K. Kajita: Parity (in Japanese) **4**, No. 7, 52 (1989)
16. K. Oshima, T. Mori, H. Inokuchi, H. Urayama, H. Yamochi, G. Saito: Synth. Metals **27**, A165 (1988)
17. M. Tokumoto: private communication

High-Field Magnetotransport in the Organic Conductor (BEDT-TTF)$_2$KHg(SCN)$_4$

T. Osada[1], R. Yagi[1], S. Kagoshima[1], N. Miura[2], M. Oshima[2,*], and G. Saito[2,**]

[1]Department of Pure and Applied Sciences, University of Tokyo,
3-8-1 Komaba, Meguro-ku, Tokyo 153, Japan
[2]Institute for Solid State Physics, University of Tokyo,
7-22-1 Roppongi, Minato-ku, Tokyo 106, Japan
*On leave from Japan Carlit Co. Ltd., Shibukawa, Gunma 377, Japan
**Present address: Department of Chemistry, Kyoto University,
Sakyo-ku, Kyoto 606, Japan

Abstract. The "fermiology" of the organic conductor (BEDT-TTF)$_2$KHg(SCN)$_4$ is discussed based on the experimental results of the Shubnikov-de Haas effect and the angle-dependent resistance oscillations. The possible origins of the anomalous kink structure found in the high-field magnetoresistance are also discussed. We present the quantum-mechanical version of Yamaji's model for the angle-dependent oscillations.

(BEDT-TTF)$_2$KHg(SCN)$_4$ is a novel organic conductor synthesized as a modification of (BEDT-TTF)$_2$Cu(SCN)$_2$ [1]. It shows the metallic behavior down to 0.5K without any superconducting or metal-insulator transitions. It has a layered structure consisting of the conducting BEDT-TTF layers and the polymeric anion layers whose thickness is rather large compared to other BEDT-TTF compounds. Therefore, the strong two-dimensionality (2D) is expected because of its weak interlayer coupling.

In this paper, we discuss the electronic structure of this system based on the results of the magnetotransport experiments. The high-field magnetotransport measurements were carried out with an a.c. technique under pulsed magnetic fields up to 40T. The angle-dependence of the magnetoresistance was measured under static fields up to 12T using a rotating holder [2].

Figure 1 shows a typical example of the transverse magneto-resistance recordings under magnetic fields perpendicular to the conducting plane (ac-plane). The magnetoresistance exhibits remarkable field-dependence in this arrangement: (i) negative slope above 10T, (ii) sharp kink structure around 22.5T, and (iii) large enhancement of Shubnikov-de Haas (SdH) effect above the kink field. The kink field shows no explicit temperature-dependence. Figure 2 shows the angle-dependence of magneto-resistance at several magnetic fields. The field direction is tilted from the normal (b*-axis) of the conducting plane. As shown by arrows in Fig.2, the angle-dependent quantum oscillations similar to those observed in θ-(BEDT-TTF)$_2$I$_3$ [3] and β-(BEDT-TTF)$_2$IBr$_2$ [4] are found. The dip position of the angle-dependent oscillations is periodic against the tangent of the tilted angle, and this period depends on the direction to which the magnetic field is tilted from the normal [2].

First, we discuss the Fermi surface (FS) of (BEDT-TTF)$_2$KHg(SCN)$_4$ based on the experimental results. The SdH period $\Delta(1/B)=0.0015T^{-1}$ leads to the FS cross-sectional area: 16% of the 2D Brillouin zone (BZ). The temperature-dependence and the field-dependence of the SdH amplitude give the cyclotron mass: $m_c=1.4m_0$ and the Dingle temperature: $T_D=4.0K$, respectively. The period of the angle-dependent oscillation gives the information about the Fermi wave number along the special direction

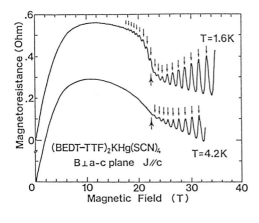

Fig.1 The transverse magnetoresistance under the fields perpendicular to the conducting plane.

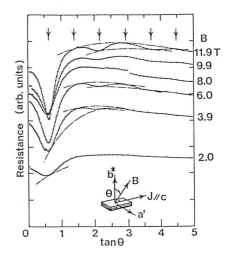

Fig.2 The angle-dependence of the magnetoresistance.

according to the Yamaji's model [5]. Employing Yamaji's formula $\Delta(\tan\theta) = \pi/bk_F$, we get the Fermi wave numbers: $k_{Fa} \sim 0.1 \text{Å}^{-1}$ along the a-axis and $k_{Fc} \sim 0.05 \text{Å}^{-1}$ along the c-axis. The presence of the angle-dependent oscillations is evidence of the existence of the weakly warped cylindrical FS in this compound. Thus, the experimental results suggest the weakly warped cylindrical FS whose cross-section is 16% of the 2D BZ and strongly anisotropic in the 2D k-space. These results are qualitatively consistent with the results of Mori's tight-binding band calculation [6].

Next, we discuss the origin of the high-field kink structure. Absence of the temperature-dependence of the kink field suggests that the origin of this structure is not ascribed to the many-body effect but to the single-particle effect. The magnetic breakdown effect is ruled out as an origin because the magnetic field range used in this work is too low to cause the magnetic breakdown [2]. Vanishing of an overlap of the Landau sub-bands might be a possible mechanism for the kink structure. When the Landau level spacing (cyclotron energy) $\hbar\omega_c = \hbar eB/m_c c$ becomes larger than the Landau sub-band width 4t (t: interlayer transfer matrix), the energy spectrum is discretized to a set of Landau sub-bands separated by gaps, so that the SdH amplitude is largely enhanced. If we ascribe the kink field to this critical field, we can estimate the transfer matrix as t=0.5meV, which is a reasonable value close to that estimated from the beating of the giant SdH effect in β_H-(BEDT-TTF)$_2$I$_3$ [7].

Finally, we consider Yamaji's model for the angle-dependent oscillations. Yamaji treated the magnetic energy spectrum of quasi-2D system with a warped cylindrical FS by a semiclassical method, and found the fact that the Landau sub-band width around the Fermi level oscillates against the tilted angle of the magnetic field from the normal of the 2D plane. Yamaji claimed that this band width oscillation is the origin of the observed angle-dependent conductivity oscillations. However, it was not well explained how the conductivity change is caused by the band width change. Yamaji's model is very similar to that for the new type of magnetic quantum oscillations discovered recently in the 2D electron gas formed in a GaAs/Al$_x$Ga$_{1-x}$As heterostructure with the weak lateral periodic potential (lateral superlattice) [8,9]. We can rewrite Yamaji's semiclassical model as a quantum-mechanical one referring to the model for the new type of the magnetic oscillations observed in the lateral superlattice. We consider the following band model describing the quasi-2D system,

$$E(k) = \hbar^2(k_x^2 + k_y^2)/2m^* - 2t \cdot \cos(c^* k_z), \tag{1}$$

where c^* is the interlayer spacing. We calculated the energy dispersion when the magnetic field $B = (0, B\sin\theta, B\cos\theta)$ tilted from the normal of the 2D plane (z-direction) by θ, using the effective mass approximation. Choosing a gauge $A = (0, Bx\cos\theta, -Bx\sin\theta)$, the effective mass equation is deduced into the one-dimensional eigenvalue equation:

$$-d^2 f(x^*)/dx^{*2} + B^{*2} x^{*2} \cos^2\theta \, f(x^*) - 2t^* \cos(B^* x^* \sin\theta - c^* k_z^{(0)}) f(x^*) = E^* f(x^*), \tag{2}$$

where $l^2 = \hbar c/eB$, $x^* = (x + l^2 k_y/\cos\theta)/c^*$, $B^* = (c^*/l)^2$, $t^* = t/(\hbar^2/2m^* c^{*2})$, $E^* = E/(\hbar^2/2m^* c^{*2})$ and $k_z^{(0)} = k_B/\cos\theta$ (k_B is the wave number along the field direction). Figure 3 shows the calculated dispersion of the Landau sub-bands against $k_z^{(0)}$ for several angles in the absence of scatterers. We see that several sub-bands have very weak dispersion, in other words, very large band mass along the field direction. We propose that the conductivity oscillation results from the change of the sub-band mass around the Fermi level. We should note that the oscillations could appear by this mechanism even at low fields where SdH effect disappears. An example of the energy spectrum as a function of $\tan\theta$ is shown in Fig.4. It is seen that the sub-band width oscillates periodically against $\tan\theta$ when the Fermi level is fixed. Though this angle-dependent oscillation and the magnetic oscillation observed in the lateral superlattice are different from each other in appearance, the essential mechanism seems to be common in these two new phenomena.

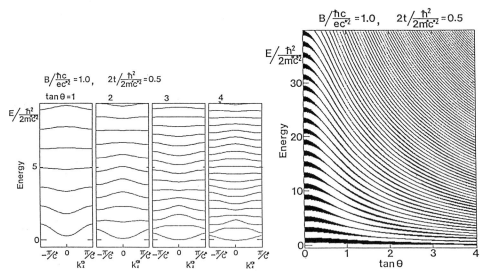

Fig.3 Energy dispersion of the Landau sub-bands under tilted magnetic fields.

Fig.4 Energy spectrum of the quasi-2D system as a function of the direction of the magnetic field with a fixed strength.

References

1. M.Oshima, H.Mori, G.Saito, and K.Oshima: Chem. Lett., 1159 (1989).
2. T.Osada, R.Yagi, A.Kawasumi, S.Kagoshima, N.Miura, M.Oshima, and G.Saito: to be published.
3. K.Kajita, Y.Nishio, T.Takahashi, W.Sasaki, R.Kato, H.Kobayashi, A.Kobayashi, and Y.Iye: Solid State Commun. 70, 1189 (1989).
4 M.V.Kartsovnik, P.A.Kononovich, V.N.Laukhin, and I.F.Shchegolev: JETP Lett. 48, 541 (1988).
5. K.Yamaji: J. Phys. Soc. Jpn. 58, 1520 (1989).
6. T.Mori and H.Inokuchi: in this proceedings.
7. W.Kang, G.Montambaux, J.R.Cooper, D.Jérome, P.Batail, and C.Lenoir: Phys. Rev. Lett. 62, 2559 (1989).
8. R.R.Gerhardts, D.Weiss, and K.von Klitzing: Phys. Rev. Lett. 62, 1173 (1989).
9. R.W.Winkler, J.P.Kotthaus, and K.Ploog: Phys. Rev. Lett. 62, 1177 (1989).

Electronic Properties in (BEDT-TTF)$_2$X: Magnetoresistance and Hall Effect

K. Murata, M. Ishibashi*, Y. Honda*, T. Komazaki**, M. Tokumoto, N. Kinoshita, and H. Anzai

Electrotechnical Laboratory, Tsukuba, Ibaraki 305, Japan
*Visiting student from Tokyo Metropolitan University
**Visiting student from University of Tsukuba

Abstract. The Shubnikov de Haas effect under pressure of 1.8 kbar (T_c=6.8 K) and Hall effect under ambient pressure of β-(BEDT-TTF)$_2$I$_3$ are presented. The Fermi surface area perpendicular roughly to the c^*-axis is found to be 56.2 % of the first Brillouin zone. Beat-like phenomena and the undulation in the magnetoresistance oscillation are observed. The Dingle temperature and the cyclotron effective mass are 0.5 K and 5.93 m_0, respectively. The Hall coefficient at ambient pressure is found to be positive and independent of temperature except for a stepwise decrease by 8 % at 175 K and a sudden decrease below 20 K, which suggests the existence of a new phase between the metallic and the superconducting phases in the pressure-temperature phase diagram. We speculate on the close relation between the superconductivity and this new "phase" at low temperature.

Introduction

We report in the first part of this paper our recent results on the Shubnikov de Haas effect study under pressure of 1.8 kbar where T_c is 6.8 K of the high T_c state of β-(BEDT-TTF)$_2$I$_3$. In the second part, in association with the Fermi surface and with the low temperature phase of this material, a Hall effect study of the same material at ambient pressure is presented. The Hall effect measurements are discussed in more detail elsewhere[1].

I. The Shubnikov de Haas effect under 1.8 kbar.

I-1. Experiment

In the Shubnikov de Haas experiment, the samples are grown by the usual method of electrochemical oxidation. Two samples are located in the pressure cell with the a-b-plane perpendicular to the magnetic field. The a- and b- dimensions of the samples are 0.03 x 2, and 0.07 x 1.1 mm^2, respectively. The resistances of the samples at ambient pressure at room temperature are 45 ohm and 7.7 ohm, respectively. The four terminal electrical contacts are made by using 10 μm gold wires in the gold-evaporated spots, which are aligned in line, on the c-surfaces of the samples. The contact resistances are 30 ohm. The two samples exhibit identical behavior in the pressure and the temperature dependences of resistivity. The superconducting transition temperature, T_c, for both samples is 6.8 K. The apparent resistivity ratio between 296 K and 8 K(just above T_c) is 450 - 550.*   Although the resistivity behaviors down to low temperature are very similar, the dc magnetoresistances at 4.2 K for $H//c*$ are slightly different between the two samples, i.e. one is saturating and the other slightly increasing above H_{c2}. The difference may be due to the slight difference in the orientation of the samples to the magnetic field. The Shubnikov de Haas effect is studied using an ac current of 42 µA at 19 Hz. Temperature is measured at the mixing chamber, where magnetic field is nulled. The studied temperature is 0.19 K, 0.50 K and 0.73 K. Above 1 K, the magnetoresistance oscillations are obscured. The magnetic field is swept up to 13.5 Tesla. Above 7.5 Tesla, pronounced magnetoresistance oscillations are observed.

I-2. Results for SdH Effect

Figure 1 shows the ac magnetoresistance with $I//a$ and $H//c*$ configuration at 0.19 K. The peak to peak amplitude of the oscillatory part of the magnetoresistance, which is demonstrated by the width of the curve in Fig. 1, is 2.8 % of the normal state resistance. The beat like phenomena and the undulating behavior are observed:; the analysis has not yet been completed. Figure 2 shows the oscillatory part of the magnetoresistance against inverse magnetic field. The fundamental frequency shows the two-dimensional Fermi surface area of $(3.93 \pm 0.02) \times 10^{15}$ cm^{-2} which corresponds to 56.2 % of the first Brillouin zone, where the value of the area of the first Brillouin zone of 6.989×10^{15} cm^{-2} is used, neglecting the thermal and the pressure contractions.**   If these effects are taken into account and with no tilt in the sample mounting in the pressure cell, the value of 56.2 % is modified to 53.3 %. The Dingle temperature and the cyclotron effective mass are estimated to be 0.5 K and 5.93 m_0, respectively. These values can be compared with those obtained in samples which are under pressure only during the cooling process, with which a high T_c state is realized[4]. The noticeable difference be-

---

*With our clamp cell and the pressure medium, a pressure deficit of 1.5 kbar between room temperature and helium temperature should be taken into account. In this experiment, we started with 3.3 kbar at room temperature and achieved 1.8 kbar at low temperature. Considering the decrease in resistivity during the pressurization, the true resistivity ratio can be estimated to be 850 - 1030.

**The thermal contraction between room temperature and low temperature may be as large as 4 kbar. Considering the compression of the 5.8(=4 + 1.8) kbar, the expansion of the Brillouin zone due to this effect is about 1.054 from the compressibility data.[3]

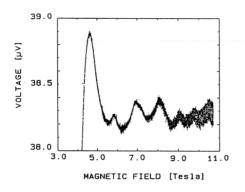

Fig. 1. The ac magnetoresistance behavior of β-(BEDT-TTF)$_2$I$_3$ under a pressure of 1.8 kbar (T_c=6.8 K) in a field perpendicular to the two-dimensional a-b plane at 180 mK.

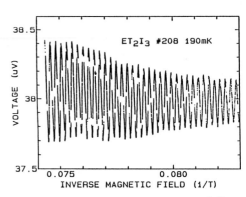

Fig. 2. The magnetoresistance of β-(BEDT-TTF)$_2$I$_3$ as a function of inverse magnetic field under a pressure of 1.8 kbar (T_c=6.8 K) in a field perpendicular to the two-dimensional a-b plane at 180 mK.

tween our results and those of Kang et al. is in the amplitude of the oscillation compared with the normal state resistance. Detailed discussions will be given in the separate paper including the pressure dependence of the SdH effect.

II. Hall Effect at P=0

II-1. Experiment

The Hall voltage contacts for this experiment are arranged on both sides along the b'-axis. The orientations of the dc current and the magnetic field are along the a- and c^*- axes, respectively. The field dependence of the Hall voltage is linear up to the studied field of 5 Tesla.

II-2. Results and Discussions of the Hall Effect at P=0

Figure 3 shows the Hall coefficient as a function of temperature. Except for the sudden change at 175 K and below 20 K, the Hall coefficient is constant with temperature, which demonstrates that this material can be described as a degenerate metal. On the other hand, the carrier of this material is thought to be holes originating from the transfer of one charge from two BEDT-TTF molecules to one I$_3$. By this consideration of chemistry, the number of carriers is not expected to change with the slight change in structure that occurs at 175 K. However, the stepwise decrease in R_H at 175 K indicates that the band structure is not simple. Rather, the introduction of more than one band or an anisotropic band is required. We note that the carrier number derived

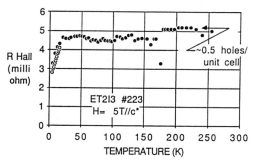

Fig. 3. The Hall coefficient as a function of temperature of β-(BEDT-TTF)$_2$I$_3$ at ambient pressure. The orientation of the current and magnetic fields are along the a- and c^*-axes, respectively. (After ref. [1])

by the relation $R_H = 1/nec$, suggests a smaller number than 1 carrier per unit cell.

The decrease of R_H below 20 K must be very important in terms of superconductivity. The value of T_c of this material close to ambient pressure is 1 K or 8 K depending on the presence or absence of the superstructure that usually appears below 175 K. It has been believed generally that only metallic and superconducting phases are present. (In TMTSF salts, SDW phase is observed in addition to those phases.). We note, however, that the proton NMR relaxation experiments[5,6] show an increase in the relaxation rate below about 20 K, where an increase in the apparent carrier number is observed. We suggest that between 20 K and 8 K a new phase exists, which may be necessary for superconductivity to occur.

References

1. K. Murata, M. Ishibashi, Y. Honda, M. Tokumoto, N. Kinoshita and H. Anzai: J. Phys. Soc. Jpn. **58** (1989) 3469.
2. V. F. Kaminskii, T.G. Prokhova, R.P. Shibaeva, E. B. Yagubskii, JETP Lett. **39** (1984) 17.
3. H. Tanino, K. Kato, M. Tokumoto, H.Anzai and G. Saito: J. Phys. Soc. Jpn. **54** (1985) 2390.
4. W. Kang, G. Montambeaux, J. R. Cooper, D. Jérome, P. Batail, C. Lenoir: Phys. Rev. Lett. **62** (1989) 2559.
5. Y. Maniwa, T. Takahashi, M. Takigawa, G. Saito, K. Murata, M. Tokumoto and H. Anzai: Jpn. J. Appl. Phys. Suppl. **26-3** (1987) 1361.
6. F. Creuzet, C. Bourbonnais, D. Jérome, D. Schweitzer, H. J. Keller: Euro. Phys. Lett. **1** (1986) 467.

Part VI

**DMET Salts
and Their Families**

Physical Properties and Crystal Structures of DMET Superconductors and Conductors

K. Kikuchi[1], *K. Murata*[2], *K. Saito*[1], *K. Kobayashi*[3], *and I. Ikemoto*[1]

[1]Department of Chemistry, Faculty of Science,
Tokyo Metropolitan University, Tokyo, Japan
[2]Electrotechnical Laboratory, Umezono, Tsukuba, Ibaraki 305, Japan
[3]Department of Chemistry, College of Arts and Sciences,
University of Tokyo, Komaba, Meguro-ku, Tokyo 153, Japan

Abstract. The physical properties and crystal structures of $(DMET)_2X$ salts were overviewed. $(DMET)_2X$ were found to be classified into five groups depending upon the shape of the anions. Among them, seven salts were discovered to show superconductivity. We compared the distances of short intercolumnar atomic contacts of some DMET superconductors with each other to consider the relation between superconductivity and the degree of two-dimensionality.

1. Introduction.

DMET is the first unsymmetrical donor that produces superconductors. The unsymmetrical donor is expected to have the following two possibilities: One is that it is expected to link two families of mother symmetrical donors. That is, the family of the donor AB is expected to link the familes of AA and BB together. In the case of the DMET family, it is expected to link the BEDT-TTF and the TMTSF families. This possibility is realized; our investigation revealed that some DMET salts are similar to TMTSF salts and some are similar to BEDT-TTF salts [1-5].

The second possibility is that there may be some unique properties characteristic of unsymmetrical donors. Their salts are likely to have κ-type structure because the unsymmetrical donors seem to be dimerized. In fact we found that some DMET salts have κ-type structure [3]. But we cannot find properties truly characteristic of unsymmetrical donors. Up to now we have found seven superconductors and many conductors in the DMET family [1-3,6,7]. In this report we overview the physical properties and crystal structures of DMET salts.

2. Physical properties of DMET salts.

The temperature dependence of the resistivity of typical DMET salts is shown in Fig. 1 [6]. You can see that the salts with anions of different shapes show different temperature dependences. The salts with octahedral anions show semiconductive behavior even at room temperature though they have high conductivity at room temperature. They also show an antiferromagnetic transition at 20 K [10].

The salts with tetrahedral anions have a broad metal-insulator transition. They also show an AF transition at 20 K [10]. Although the temperature of the metal-insulator transition depends upon the anions, the temperature of the magnetic transition is almost the same.

The DMET salts with linear anions show three different types of temperature dependence of resistivity. At ambient pressure the $Au(CN)_2$ salt shows a resistance upturn at 28 K. Under 5 kbar the upturn disappears and a superconducting transition appears at 0.80 K [6]. The ESR study revealed that the metal-insulator transition at ambient pressure is due to the formation of SDWs [8-10]. The T-P phase diagram of the $Au(CN)_2$ salt is similar to that of TMTSF salts, but the pressure range where SDWs and superconductivity are observed

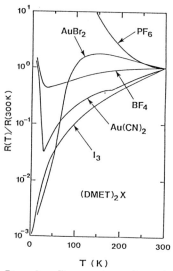

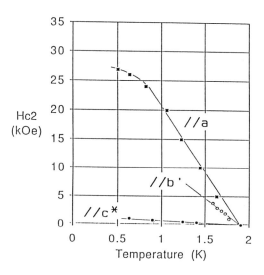

Fig. 1. Temperature dependence of resistivities of five typical DMET salts.

Fig. 2. Temperature dependence of critical field of $(DMET)_2AuBr_2$.

is very large [11]. AuI_2 and $AuCl_2$ salts show a similar temperature dependence of ambient resistivity.

The I_3 salt exhibits no resistance upturn at ambient pressure and shows superconductivity at 0.47 K [2]. This temperature dependence is similar to that of β-(BEDT-TTF)$_2$X salts. IBr_2, I_2Br, SCN, and $AuBr_2$ salts also show no resistance upturn. The IBr_2 salt shows a superconducting transition at 0.58 K. But the other three salts show no superconductivity. In the case of I_2Br and SCN salts the absence of superconductivity may be related to the orientation of the anions.

The rhombus-like $AuBr_2$ salt shows another type of temperature dependence of resistivity [3]. It shows a resistance maximum at 180 or 150 K. Such a resistance maximum is also observed in κ-(BEDT-TTF)$_2$Cu(NCS)$_2$. Some of them undergo a superconducting transition at 1.9 K at ambient pressure. Another did not exhibit superconductivity at pressures below 1.5 kbar. Figure 2 gives the temperature dependence of the critical field of the superconductivity of rhombus-like $AuBr_2$ salts. This suggests that rhombus-like $AuBr_2$ is a two-dimensional material. In summary, DMET salts with anions of different shape show different temperature dependence of resistivity and can be classified into five groups.

3. Crystal structures of DMET salts.

In the PF_6 salt, a columnar structure is observed as in TMTSF salts [12]. However the normal to the DMET molecular plane is largely tilted from the stacking axis and some dimerization is observed. This dimerization may be related to the semiconductive behavior of resistivity.

The BF_4 salt has two columns which are almost perpendicular to each other [12]. This structure is very peculiar and a similar structure is reported only in the ClO_4 salt of DMET-TTF. One stack is similar to the one in the PF_6 salt but little dimerization is observed. Another stack is similar to those of TMTSF salts and little dimerization is also observed. Little dimerization is consistent with metallic behavior. In this salt, there is only one short contact in one column, but the electrical property is expected to be two-dimensional because of the presence of two columns. Indeed anisotropy of resistivity is about two.

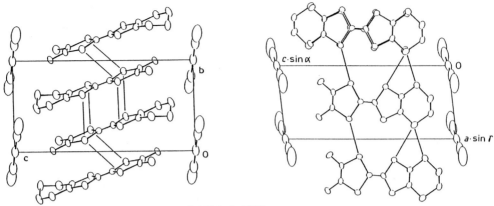

Fig. 3. Crystal structure of (DMET)$_2$Au(CN)$_2$

Figure 3 shows the crystal structure of the Au(CN)$_2$ salt [12,13]. A column similar to the one in the PF$_6$ salt is observed, but little dimerization is observed. In this salt, there are one short Se-Se and two S-S contacts between stacks. Therefore the Au(CN)$_2$ salt has some two-dimensional character. The AuI$_2$ salt also has a similar structure.

The structure of the I$_3$ salt is very similar to that of the Au(CN)$_2$ salt [12]. But their electrical properties are different from each other. That is, the (DMET)$_2$Au(CN)$_2$ salt is similar to TMTSF salts, the (DMET)$_2$I$_3$ salt is similar to β-BEDT-TTF salts. The difference in properties between TMTSF and BEDT-TTF salts has generally been thought to be caused by a difference in dimensionality. TMTSF salts have less two-dimensionality than BEDT-TTF salts, so they have SDW and superconducting phases. On the other hand, the electronic systems of BEDT-TTF salts are simple, being metallic and superconducting. Table 1 shows the distances of short interstack contacts of DMET superconductors which have a structure similar to that of the Au(CN)$_2$ salt. The distances in I$_3$ and IBr$_2$ salts are shorter than the corresponding distances in Au(CN)$_2$ and AuI$_2$ salts. This suggests that Au(CN)$_2$ and AuI$_2$ salts have less two-dimensionality than I$_3$ and IBr$_2$ salts. This fact is consistent with their electrical properties. That is, the less-two-dimensional Au(CN)$_2$ and AuI$_2$ salts have SDW and superconducting phases and the more-two-dimensional I$_3$ and IBr$_2$ salts have a simple electronic system.

Figure 4 shows the crystal structure of rhombus-like AuBr$_2$. The structure is not columnar in contrast to other DMET salts [3]. It consists of sheets of dimers of DMET molecules and AuBr$_2$ anions, so this salt is a two-dimensional material. This structure is very similar to that of κ-(BEDT-TTF)$_2$Cu(NCS)$_2$. This is consistent with their similarity of electrical properties. In conclusion, the salts in each group have the crystal structures characteristic of each group and reflected in thier electrical properties.

Table 1. Comparison of distances of short intercolumnar atomic contacts in DMET superconductors.

Compound	Se-Se / Å	S-S / Å	S-S / Å
(DMET)$_2$Au(CN)$_2$	3.884	3.580	3.672
(DMET)$_2$AuI$_2$	3.832	3.552	3.611
(DMET)$_2$I$_3$	3.784	3.504	3.604
(DMET)$_2$IBr$_2$	3.774	3.494	3.588

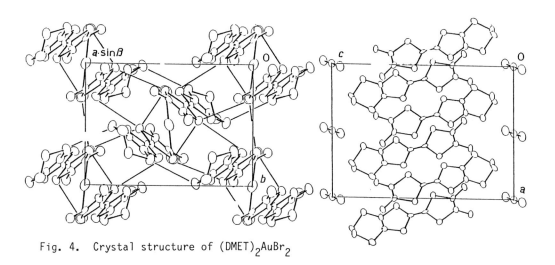

Fig. 4. Crystal structure of (DMET)$_2$AuBr$_2$

References

1. K. Kikuchi, K. Murata, Y. Honda, T. Namiki, K. Saito, K. Kobayashi, T. Ishiguro, and I. Ikemoto, J. Phys. Soc. Jpn., 56, 3436(1987).
2. K. Kikuchi, K. Murata, Y. Honda, T. Namiki, K. Saito, K. Kobayashi, T. Ishiguro, and I. Ikemoto, J. Phys. Soc. Jpn., 56, 4241(1987).
3. K. Kikuchi, Y. Honda, Y. Ishikawa, K. Saito, I. Ikemoto, K. Murata, H. Anzai, T. Ishiguro, and K.Kobayashi, Solid State Commun., 66, 405(1988).
4. K. Murata, K. Kikuchi, T. Takahashi, K. Kobayashi, Y. Honda, K. Saito, K. Kanoda, T. Tokiwa, H. Anzai, T. Ishiguro, and I. Ikemoto, J. Mol. Electron., 4, 173(1988).
5. K. Kikuchi, K. Saito, I. Ikemoto, K. Murata, T. Ishiguro, and K. Kobayashi, Synth. Metals, 27, B269(1988).
6. K. Kikuchi, M. Kikuchi, T. Namiki, K. Saito, I. Ikemoto, K. Murata, T. Ishiguro, and K. Kobayashi, Chem. Lett., 931(1987).
7. K. Kikuchi, K. Murata, Y. Honda, T. Namiki, K. Saito, K. Kobayashi, T. Ishiguro, and I. Ikemoto, J. Phys. Soc. Jpn., 56, 2627(1987).
8. K. Kanoda, T. Takahashi, T. Tokiwa, K. Kikuchi, K. Saito, I. Ikemoto, and K. Kobayashi, Phys. Rev. B38, 39(1988).
9. Y. Nogami, M. Tanaka, S. Kagashima, K. Kikuchi, K. Saito, I. Ikemoto, and K. Kobayashi, J. Phys. Soc. Jpn., 56, 3738(1987).
10. K. Kanoda, T. Takahashi, K. Kikuchi, K. Saito, I. Ikemoto, and K. Kobayashi, Phys. Rev. B39, 3996(1989).
11. Y. Honda, K. Murata, K. Kikuchi, K. Saito, I. Ikemoto, and K. Kobayashi, Solid State Commun., in press.
12. K. Kikuchi, Y. Ishikawa, K. Saito, I. Ikemoto and K. Kobayashi, Synth. Metals, 27, B391(1988).
13. K. Kikuchi, Y. Ishikawa, K. Saito, I. Ikemoto and K. Kobayashi, Acta Cryst., C44, 466(1988).

Superconductivity and Spin Density Waves in (DMET)$_2$Au(CN)$_2$

K. Murata[1], K. Kikuchi[2], Y. Honda[2], T. Komazaki[1,], K. Saito[2], K. Kobayashi[3], and I. Ikemoto[2]*

[1]Electrotechnical Laboratory, Tsukuba, Ibaraki 305, Japan
[2]Department of Chemistry, Faculty of Science, Tokyo Metropolitan University, Fukazawa, Setagaya-ku, Tokyo 158, Japan
[3]Department of Chemistry, College of Arts and Sciences, University of Tokyo, Komaba, Meguro-ku, Tokyo 153, Japan
*Visiting student from Tsukuba University

Abstract. The Hall effect of (DMET)$_2$Au(CN)$_2$, in which both SDW and superconductivity are observed, is presented. On decreasing temperature, the Hall coefficient, R_H, changes sign from positive to negative at a temperature much higher than that of the SDW transition. We show that the variation of the Hall coefficient must be caused by a SDW fluctuation. The problem of the coexistence of SDW and superconductivity is discussed.

1. Introduction

The family of (DMET)$_2$X salts presents a variety of crystals which show electronic properties similar to TMTTF-, TMTSF-, and BEDT-TTF- salts depending on the counter anions, X [1]. It is thus suitable to study this family to understand the continuity between the preceding families. Among the (DMET)$_2$X salts, (DMET)$_2$Au(CN)$_2$ is metallic at high temperature and exhibits spin density wave (SDW) and superconductivity like (TMTSF)$_2$PF$_6$. However, (DMET)$_2$Au(CN)$_2$ is different from (TMTSF)$_2$PF$_6$ in two respects [2]. One is the wide pressure range (~ 3 kbar) where both superconductivity and SDW are observed. Thus, it is expected that this material will be a typical example to study the real coexistence of superconductivity and SDW at the same temperature and pressure. The other feature is the saturation in resistivity below the temperature of the SDW, T_{SDW}. In contrast, in (TMTSF)$_2$PF$_6$, according to our measurement, resistivity is well expressed by $\rho = \exp(-2\Delta/k_B T)$ with $2\Delta = 24$ K for T_{SDW} of 12.5 K between the temperatures 0.1 and 0.9 of T_{SDW}. The ratio of $2\Delta/k_B T_{SDW}$ is then 2, which is close to the mean field value within a factor of 2.

In this paper we focus on the properties which have not been commonly observed and discussed in (TMTSF)$_2$PF$_6$: i) The resistivity saturation below T_{SDW}. ii) Sign reversal in R_H at a temperature much higher than T_{SDW}. iii) The wide pressure range where superconductivity and SDW are both observed.

2. Experiment

Samples are prepared by a conventional electrochemical method. In the Hall measurement, the current contacts are arranged at both ends of the long and most conductive a-axis to make a uniform current. The contacts for the Hall voltage are located on both sides along the b-axis. The Hall voltage is taken by the difference between the values at the Hall voltage terminals of +5 and -5 Tesla. The current and field dependences are checked at low temperature. No significant field dependence is observed in R_H. DC currents are used in the measurement by eliminating the thermoelectric power. The polarity of the Hall signal is checked by measuring the known GaAs semiconductor.

3. Results and Discussions

Figure 1 shows the Hall coefficient at low temperature. It is obvious that the carrier concentration is constant below 10 K, which confirms the resistivity saturation below T_{SDW}, which is about 22-26 K. The polarity of the carrier is found to be that of electrons. The carriers that dominate at high temperature are expected to be holes, since the conducting path must be realized in the donor stacks of DMET, which is supported by thermoelectric power [3]. Therefore, a positive Hall coefficient is expected at high temperature.

Figure 2 shows the temperature dependence of the Hall coefficient. The dominant carriers at high temperature are proved to be positive and consistent with the charge transfer from two DMET molecules to one $Au(CN)_2$ anion. The sign reversal of the Hall coefficient takes place around 40 K, which is much higher than T_{SDW}. Further symptoms of variation of R_H appear around 75 K.

Figure 3 shows the plot of the position of the sign reversal in the pressure-temperature phase diagram. We notice that the phenomena of sign re-

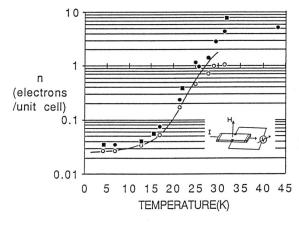

Fig. 1 Temperature dependence of Hall coefficient of $(DMET)_2Au(CN)_2$ at low temperature. Different symbols denote different samples.

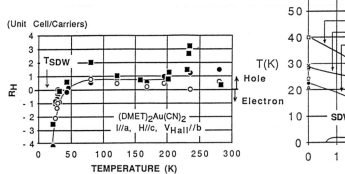
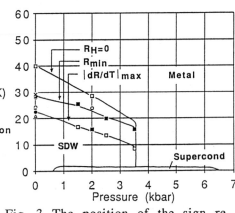

Fig. 2 Temperature dependence of Hall coefficient of (DMET)$_2$Au(CN)$_2$ in the temperature range below room temperature.

Fig. 3 The position of the sign reversal in the Hall coefficient of (DMET)$_2$Au(CN)$_2$ in the pressure-temperature phase diagram.

versal takes place only at pressures where SDW is present at low temperature. We then conclude that the variation and sign reversal in R_H are closely related to SDW, and may be a fluctuation of SDW. In (TMTSF)$_2$PF$_6$, the sign reversal appears at the boundary of the SDW and metal in the pressure-temperature phase diagram[4] or in the temperature-magnetic field phase diagram[5], which is in the case of field-induced SDW.

It is interesting to study whether or not superconductivity below 3.5 kbar is realized using the electrons at the pocket of the Fermi surface, which means a real coexistence of SDW and superconductivity. To solve this problem, we considered the following. Above 3.5 kbar, superconductivity is realized from the metallic part, where $R_H > 0$. The Hall coefficient is independent of field strength so that the R_H can be considered to be positive even in the low field where superconductivity is present. The superconductivity seems to be the same through 3.5 kbar since no significant change in the superconducting transition temperature is observed through 3.5 kbar. Then we considered that the superconductivity is realized by the carriers where R_H is positive even below 3.5 kbar. With this consideration, we are tempted to conclude that the superconductivity below 3.5 kbar is *not* realized by the electrons at the pockets with $R_H < 0$. Since the SDW is specified by negative R_H in this material, we may conclude that the real coexistence of SDW and superconductivity is not realized. The difficulty in this discussion is that the transition temperature of the SDW is about 25 times larger than that of superconductivity. The condensation energy of the SDW must be much larger than that of superconductivity. According to our consideration, a big reconstruction in the electronic state should take place at the transition temperature of superconductivity below 3.5 kbar even with a big difference in the condensation energy of the two states. From the experimental point of view, since the presence and absence of the phases against pressure is very reproducible, the existence of the wide pressure range where both superconductivity and the SDW are present is a fact that should be taken into account when we discuss the properties of this material, irrespective of the above consideration.

References

1. K. Murata, K. Kikuchi, T. Takahashi, K. Kobayashi, Y. Honda, K. Saito, K.Kanoda, T. Tokiwa, H. Anzai, T.Ishiguro and I.Ikemoto: J. Mol. Electronics, **4**, 173 (1988).
2. Y. Honda, K. Murata, K. Kikuchi, K. Saito, I. Ikemoto and K. Kobayashi: Solid State Commun. **71**, 1087 (1989).
3. K. Saito, H. Kamio, Y. Honda, K.Kikuchi, K. Kobayashi and I. Ikemoto: to be published in J. Phys. Soc. Jpn. **58**, No. 11(1989).
4. Unpublished data of K. Murata *et al.*
5. J. F. Kwak, J.E. Schirber, P.M. Chaikin, J. M. Williams, H. H.Wang and L. Y. Chiang: Phys. Rev. Lett. **56**, 972 (1989). Also, M. Ribault, D. Jérome, J. Tuchendler, C. Weyl, and K. Bechgaard: Mol. Cryst. Liq. Cryst. **119**, 91 (1989)

Phase Transition of the Organic Metal (DMET)₂Au(CN)₂ at 180 K

K. Saito[1], K. Kikuchi[1], K. Kobayashi[2], and I. Ikemoto[1]

[1]Department of Chemistry, Faculty of Science, Tokyo Metropolitan University, Fukazawa, Setagaya-ku, Tokyo 158, Japan

[2]Department of Chemistry, College of Arts and Sciences, University of Tokyo, Komaba, Meguro-ku, Tokyo 153, Japan

Abstract. The temperature dependence of the heat capacity was determined by AC calorimetry. A small thermal anomaly was detected between 160 and 190 K, where anomalous behaviors have previously been reported in resistivity, ESR linewidth and lattice constants. A small, continuous anomaly also appeared in thermoelectric power at 180 K. The nature of the transition is discussed on the basis of the results. A possible mechanism of the phase transition which is consistent with all the experimental observations is proposed. The temperature dependence of an order-parameter associated with the transition is deduced from that of the thermoelectric power.

1. Introduction

$(DMET)_2Au(CN)_2$ is the first member [1] of organic superconductors based on an unsymmetrical donor. The crystal structure of $(DMET)_2Au(CN)_2$ has been determined at room temperature under the normal pressure [2] and is shown in Fig. 1. The DMET molecules stack with an alternate orientation along the b-axis, which is the most conductive and crystal-growth axis. There exist two types of overlap mode.

$(DMET)_2Au(CN)_2$ is metallic down to 28 K under the normal pressure [1]. The X-ray [3] and the magnetic resonance [4] studies revealed the transition is due to the formation of SDW. The transition is suppressed by pressure of 5 kbar and the superconductivity appears below 0.80 K [1]. The T-P phase diagram has recently been determined through the electrical measurements [5].

Murata et al. [6] pointed out the existence of a small anomaly (dip) in the resistivity curve at about 180 K and suggested that the anomaly resulted from some phase transition. According to so-called Elliott mechanism, a similar anomaly can be recognized in the linewidth of ESR [7]. An anomaly was also reported for

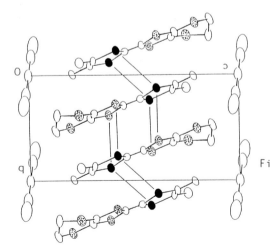

Fig. 1. Crystal structure of $(DMET)_2Au(CN)_2$. ●; Se, ◉; S. Solid lines indicate shorter interatomic contacts than the sum of van der Waals radii.

lattice constants [8]. Although the anomalies at high temperatures may have no direct correlation on the low-temperature properties such as superconductivity, it is desired to understand fully physical properties of organic conductors because the structural design is one of the most important points in developing new organic superconductors. In this paper, the results of heat capacity and thermoelectric power measurements are described. The mechanism of the transition is discussed.

2. Experimental

AC calorimetry was performed on two samples having a similar size (2 x 0.5 x 0.02 mm^3) using the AC calorimeter described previously [9] and consistent results were obtained. The measurement frequency was about 3 Hz and the magnitude of the temperature oscillation about 10 mK. The absolute value of heat capacity was not determined because of the difficulty in calibrating the heat absorbed.

The size of the sample used for thermoelectric power measurements was 1.4 x 0.2 x 0.3 mm^3. The measurements were carried out along the crystal-growth axis using the apparatus described previously [10]. An electrical contact was established through a thin gold wire (10 μm in diameter) with carbon paste, and a thermal contact through the wire and by a mechanical contact without adhesive (paste) in order to avoid the sample breakage. The primary data were corrected using the results which were obtained above ca. 200 K before sample breakage when directly contacted with paste.

3. Results and Discussion

The results of AC calorimetry in one typical run are plotted in Fig. 2. A very broad hump can be recognized between 160 and 190 K. Thus the anomalous behaviors of $(DMET)_2Au(CN)_2$ [6-8] are considered to be accompanied with a thermodynamic phase transition.

Although AC calorimetry is impossible for detecting latent heat due to a first-order phase transition unless one uses special techniques, the anomaly detected in this study is smooth within 0.2 per cent and, consequently, is very likely of a higher-order transition. The anomalous portion of heat capacity (excess heat capacity) is only 0.8 per cent of the total heat capacity of the sample when we draw a smooth interpolating curve as a normal portion as shown in Fig. 2. Assuming the value of the heat capacity at 200 K to be $5 \cdot 10^2$ $J \cdot K^{-1} \cdot mol^{-1}$, the roughly-estimated upper limits of the enthalpy and the entropy of transition are obtained as $1 \cdot 10^2$ $J \cdot mol^{-1}$ and 0.5 $J \cdot K^{-1} \cdot mol^{-1}$, respectively, by graphical integration of

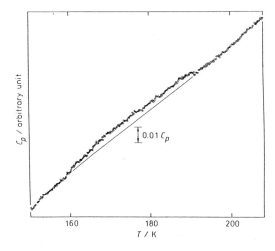

Fig. 2. Temperature dependence of the heat capacity of $(DMET)_2Au(CN)_2$.

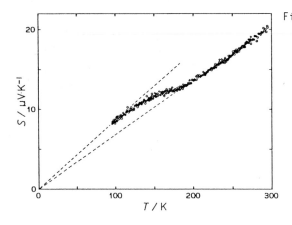

Fig. 3. Absolute thermoelectric power of $(DMET)_2Au(CN)_2$ measured in the cooling (circles) and the heating (squares) directions.

the excess heat capacity. The value of the entropy of transition is much smaller than $R \cdot \ln 2$ ($\simeq 5.8$ J·K^{-1}·mol^{-1}). Hence, it is hard to imagine some structural (positional) ordering as a possible mechanism of the transition. No disorder was observed in the crystallographic study [2].

The absolute thermoelectric power of $(DMET)_2Au(CN)_2$ is shown in Fig. 3. The positive thermoelectric power implies the hole-like character of the conduction carrier. The phase transition is detected as an inflection of the curve at about 180 K. The good agreement between the data obtained in the cooling (circles) and the heating (squares) directions strongly supports the higher-order nature of the transition; the transition is of the second kind in the Landau's classification.

Above the phase transition temperature, the thermoelectric power depends linearly on temperature. A straight line can be drawn as shown in Fig. 3. It passes through the origin. The slope is $6.8 \cdot 10^{-2}$ µV·K^{-2}. Assuming a simple sinusoidal band, the magnitude of transfer-integral t is estimated as 0.25 eV. In the following, the thermoelectric power of the high-temperature phase (including its extrapolation) is designated as S_H. Below the transition temperature, the magnitude of t was not determined as the temperature dependence of the thermoelectric power is not linear as seen in Fig. 3.

It was reported [5] that the SDW phase appeared in the pressure range where the (high-temperature) phase transition took place, and that the SDW and the high-temperature transitions simultaneously disappeared on applying pressure. These facts suggest the importance of electron-phonon coupling in the high-temperature transition. It is noted that no superlattice nor change in space group was detected through the transition [3,8]. Taking into account the crystal structure and the importance of electron-phonon coupling, the following mechanism may be considered: The crystal is electronically dimerized below the transition, i.e. the electronic system is characterized with two transfer-integrals. On the other hand, two transfer-integrals are effectively the same in magnitude above the transition. The non-dimerized state is stabilized through electron-phonon coupling. The fact that the zero-wavevector modulation in $(DMET)_2Au(CN)_2$ is equivalent to that of $4k_F$ will also play an important role.

The above model of the phase transition is only a possible, not exclusive, one, but seems consistent with all the experimental data: The narrowing of the band width implies the enhanced degree of localization of electrons, resulting in the upward anomaly in the resistivity and ESR linewidth on passing the transition temperature in the cooling direction. The model predicts the increase in thermoelectric power on cooling as observed in the experiments. The phase transition is of higher-order and displacive in nature.

Now, we try to deduce the temperature dependence of an order-parameter associated with the transition. It is anticipated that a rigid band structure with metallic nature is established at sufficiently low temperatures. In such a low temperature region, the thermoelectric power should recover a linear temperature dependence. Such a straight line is tentatively drawn in Fig. 3. The virtual

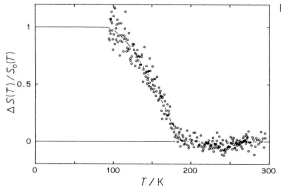

Fig. 4. Temperature dependence of the order-parameter. For the meaning of the symbols, see the text.

thermoelectric power below the transition temperature is designated as S_L, in the following. The following quantity seems to fulfill the requirement for an order-parameter, i.e., that is finite below a transition point but zero above it:

$$\Delta S(T) \cdot S_0(T)^{-1} = [S(T) - S_H(T)] \cdot [S_L(T) - S_H(T)]^{-1}.$$

This quantity is plotted in Fig. 4. The existence of the phase transition is evident. The location of the transition point is determined as (180 ± 5) K. The temperature dependence of the quantity seems reasonable as that of an order-parameter. The dependence can be expressed as $(T_{trs}-T)^{0.65\pm0.05}$. The large uncertainty in the exponent results from that in the transition temperature. The exponent will have some relation to the critical exponent β for the primary order-parameter (electric dimerization in our model). However the value is too large to be regarded as β though half of the value may be acceptable.

This work was partially supported by Grant-in-Aid for Scientific Research (No. 61430003), for Encouragement of Young Scientists (No. 63740267), and for Specially Promoted Research (No. 63060004) from the Ministry of Education, Science and Culture.

References

1 K. Kikuchi, M. Kikuchi, T. Namiki, K. Saito, I. Ikemoto, K. Murata, T. Ishiguro and K. Kobayashi, Chem. Lett. 551(1988).
2 K. Kikuchi, Y. Ishikawa, K. Saito, I. Ikemoto and K. Kobayashi, Acta Crystallogr., Sect. C 44, 466(1988).
3 Y. Nogami, M. Tanaka, S. Kagoshima, K. Kikuchi, K. Saito, I. Ikemoto and K. Kobayashi, J. Phys. Soc. Jpn. 56, 3783(1987).
4 K. Kanoda, T. Takahashi, T. Tokiwa, K. Kikuchi, K. Saito, I. Ikemoto and K. Kobayashi, Phys. Rev. B 38, 39(1988).
5 Y. Honda, K. Murata, K. Kikuchi, K. Saito, I. Ikemoto and K. Kobayashi, Solid State Commun. in press.
6 K. Murata, K. Kikuchi, T. Takahashi, K. Kobayashi, Y. Honda, K. Saito, K. Kanoda, T. Tokiwa, H. Anzai, T. Ishiguro and I. Ikemoto, J. Mol. Electron. 4, 173(1988).
7 K. Kanoda, T. Takahashi, K. Kikuchi, K. Saito, I. Ikemoto and K. Kobayashi, Synth. Metals 27, B385(1988).
8 S. Kagoshima and Y. Nogami, Synth. Metals 27, A299(1988).
9 K. Saito, H. Kamio, K. Kikuchi, K. Kobayashi and I. Ikemoto, J. Phys.: Condensed Matt. in press.
10 K. Saito, H. Kamio, K. Kikuchi, K. Kobayashi and I. Ikemoto, J. Phys. Soc. Jpn. 58, (1989).

Antiferromagnetic Transitions in (DMET)$_2$X and (DMPT)$_2$X

K. Kanoda[1], S. Okui[1], T. Takahashi[1], K. Kikuchi[2], K. Saito[2], I. Ikemoto[2], and K. Kobayashi[3]

[1]Department of Physics, Gakushuin University, Mejiro, Toshima-ku, Tokyo 171, Japan
[2]Department of Chemistry, Tokyo Metropolitan University, Fukazawa, Setagaya-ku, Tokyo 158, Japan
[3]Department of Chemistry, College of Arts and Sciences, University of Tokyo, Komaba, Meguro-ku, Tokyo 153, Japan

Abstract. The magnetism of the two families of organic conductors, (DMET)$_2$X and (DMPT)$_2$X, was investigated by means of the electron paramagnetic resonance. Several DMET salts were found to have antiferromagnetic ground states. The transition temperatures are 20-25 K and the spin susceptibility above the transition is similar for all the DMET salts studied here. Two DMPT salts were also found to undergo antiferromagnetic transitions. However, the transition temperature, the critical behavior and the absolute value of spin susceptibility were different from those of the DMET salts. The electronic properties of these two families are discussed comparatively.

The existence of antiferromagnetic ground states and the interplay with superconductivity in the TMTTF and TMTSF families shed light on the importance of low dimensionality and electron correlation in the organic conductors. Electron paramagnetic resonance (EPR) is a useful microscopic probe for the electronic states and has revealed antiferromagnetic aspects in these families [1]. In the present work, the EPR method was applied to the two families of (DMET)$_2$X [X= PF$_6$, SbF$_6$, BF$_4$, ClO$_4$, Au(CN)$_2$, I$_3$] and (DMPT)$_2$X [X= AsF$_6$, PF$_6$]. (Parts of the results on the DMET salts were already published [2-4].)

The DMET family exhibits a rich variety of transport properties ranging from semiconductor to superconductor [5]. The typical temperature dependence of resistivity is summarized in the inset of Fig. 1. (DMET)$_2$PF$_6$ is a semiconductor, at least, below room temperature [6]; (DMET)$_2$BF$_4$ exhibits a metal-insulator transition at 37 K [6]; (DMET)$_2$Au(CN)$_2$ also exhibits a metal-insulator transition at 25 K at ambient pressure but a superconducting transition occurs at 1 K under pressure [7]; (DMET)$_2$I$_3$ is an ambient-pressure superconductor with a T_c of 0.5 K [8]. All of the above salts have the same columnar structure of DMET molecules.

The DMPT salts are recently synthesized compounds. The DMPT molecule is a derivative of the DMET molecule; the ethylene group of the latter is replaced with propylene. In the present study, AsF$_6$ and PF$_6$ salts have been investigated. Both salts show a semiconducting behavior below room temperature, as (DMET)$_2$PF$_6$ does. The crystal structure is also quite similar to that of (DMET)$_2$PF$_6$.

The EPR measurements were performed in the temperature range between 10 and 300 K, using an X-band ESR spectrometer. In this report, we focus on the spin susceptibility and the linewidth.

Figure 1 shows the results on (DMET)$_2$X [X= PF$_6$, BF$_4$, Au(CN)$_2$ and I$_3$]. The overall temperature dependence and the magnitude of the susceptibility are essentially the same. χ_{spin} is 2.0-2.5 × 10^{-4} emu/mol at room temperature. It decreases gradually with decreasing temperature and is saturated at the value of 1.3-1.6 × 10^{-4} emu/mol. This is in strong contrast to the variety of the resistivity behav-

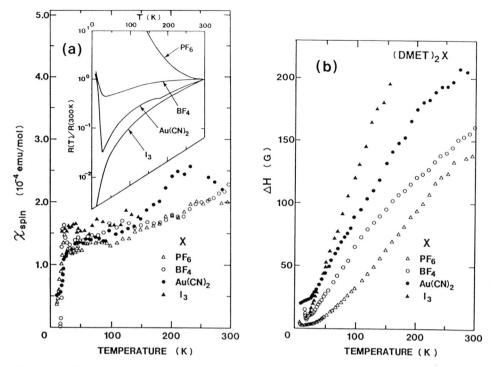

Fig. 1 (a) Spin susceptibility and (b) EPR linewidth for typical salts in the DMET family. Inset shows the temperature dependence of the resistivity for these salts.

ior. This fact implies that the same mechanism governs the magnetism, whether the system is metallic or semiconducting.

An explanation of these paradoxical behaviors is given in the context of correlated electrons in a one-dimensional quarter-filled band proposed by Emery et al. [9] for TMTTF and TMTSF salts. The $(DMET)_2X$ family is also believed to have the same electronic structure. They considered that dimerization of molecules along the stacking axis produces the so-called $4k_F$ potential for charge carriers. $4k_F$ electron-electron Umklapp scatterings lead to a correlation gap in the charge degree of freedom. On the contrary, the spin degree of freedom is free from the correlation gap, so that no anomaly in the spin susceptibility is expected. This picture applies in a pure one-dimensional system, where the degree of molecular dimerization is an important parameter. Now what about the role of dimensionality? It seems natural to consider that interstack couplings reduce the correlation gap, because the Fermi surface gets to warp and the $4k_F$ Umklapp scatterings become less effective.

The variation of the conductivity in the DMET family should be explained in terms of the correlation gap, that is, the gap should be the largest in the PF_6 salt and become smaller in order of the BF_4, $Au(CN)_2$, and I_3 salts. If this variation comes from the dimensionality of the conduction band, it is consistent with the variation of the linewidth shown in Fig. 1(b), on the assumption of the Elliott mechanism responsible for the EPR linewidth [10]. Indeed, in the $AuBr_2$ salt, which has a two dimensional electronic state, the linewidth is far greater than in any salt given in Fig. 1(b) [11]. In summary, the similarity in susceptibility and the variation in conductivity

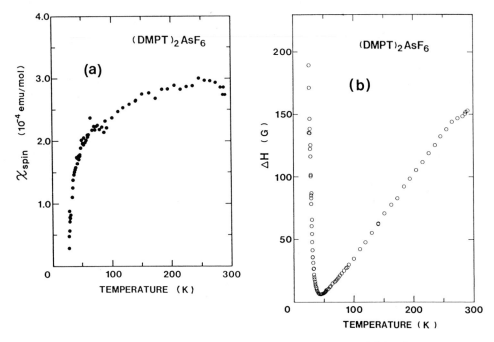

Fig. 2 (a) Spin susceptibility and (b) EPR linewidth for $(DMPT)_2AsF_6$.

come from the electron correlation and the dimensional variation in this family. Intrachain parameters such as the band width and the on-site Coulomb energy do not differ so much among these salts.

For the PF_6, BF_4 and $Au(CN)_2$ salts, the susceptibility decreases abruptly below 20-25 K, indicating some kind of ordering. A clear line broadening below 20 K in the BF_4 salt is indicative of an antiferromagnetic transition. The $Au(CN)_2$ salt shows no line broadening although a kink is visible in the transition region. Our subsequent ^1H NMR studies for this salt have shown clear antiferromagnetic natures of the transition; a large enhancement of nuclear spin-lattice relaxation rate around the transition and an appreciable broadening of the ^1H NMR line below 20 K [3]. In the PF_6 salt, the EPR line broadening is very small and only starts well below the transition in the susceptibility. Thus we have no confidence of antiferromagnetic transition at present. A spin-Peierls transition is not ruled out, considering the strong one-dimensional character in this salt. In addition to the three salts, mentioned above, SbF_6 and ClO_4 salts have been found to undergo antiferromagnetic transitions at 20 K through an abrupt drop of susceptibility and a clear line broadening (not shown in the figure) [12]: Many DMET salts have antiferromagnetic ground states with almost the same transition temperature.

Now the results of the DMPT salts are given. The spin susceptibility and the linewidth for $(DMPT)_2AsF_6$ are shown in Fig. 2. The susceptibility is 3×10^{-4} emu/mol at room temperature and weakly temperature dependent like the DMET salts. Below 50 K, χ_{spin} decreases rapidly and vanishes around 25-29 K. The linewidth decreases monotonously with temperature down to 45 K, and then starts to diverge, associated with the disappearance of the absorption line. The clear divergence evidences an antiferromagnetic nature of the transition.

The difference in the EPR properties, the spin susceptibility and the linewidth, between the $(DMPT)_2AsF_6$ and $(DMPT)_2PF_6$ was negligible. Both salts have antiferromagnetic ground states.

Since the measurements on the DMPT family have been done only for the two salts, general comparison between the two families is not attainable yet. It seems, however, worthwhile to compare $(DMET)_2PF_6$ and $(DMPT)_2AsF_6$ [or $(DMPT)_2PF_6$], with the same crystal structure. The similarities in the experimental results are summarized as; (a) the semiconducting behavior in the conductivity, (b) the weak temperature dependence of the susceptibility and (c) the absolute value and the temperature dependence of the line width above 50 K. The dissimilarities, on the other hand, are summarized as follows; (d) the absolute value of the spin susceptibility is greater in $(DMPT)_2AsF_6$ than in $(DMET)_2PF_6$; (e) the transition temperature is a little higher in the DMPT salt; (f) the line broadening in the DMPT salt starts at much higher temperature than in the DMET.

(a) and (b) imply that the overall picture of correlated electrons for the DMET salts is applicable to the DMPT salts; both salts have large correlation gaps associated with low dimensionality and/or large dimerization. The lattice parameter along the stacking of $(DMPT)_2AsF_6$ is longer by $\sim 2\%$ than that of $(DMET)_2PF_6$ [13], so that the band width of the former is expected to be narrower than that of the latter. The larger spin susceptibility of the DMPT salt, (d), is explained as due to the narrower band width.

(f) implies that precursor effects of the antiferromagnetic transition are more pronounced in DMPT than in DMET, while the transition temperatures do not differ so much, (e). This strongly suggests that the DMPT salt is more one-dimensional than the DMET. Now taking account of (c), one has to assume a larger spin-orbit coupling and/or a larger orbital scattering rate in the DMPT. Moreover, since the transition temperature should be given by the on-site Coulomb energy and the interstack (exchange) coupling, (e) requires a stronger on-site Coulomb in the DMPT. Thus far we do not have enough information to check these considerations, but the present EPR data should be crucial for further systematic analyses of these families.

References

1. H.J. Pedersen, J.C. Scott, and K. Bechgaard, Phys. Rev. B **24**, 5014 (1981).
2. K. Kanoda, T. Takahashi, K. Kikuchi, K. Saito, I. Ikemoto, and K. Kobayashi, Synth. Met. **27**, B385 (1989).
3. K. Kanoda, T. Takahashi, T. Tokiwa, K. Kikuchi, K. Saito, I. Ikemoto, and K. Kobayashi, Phys. Rev. B **38**, 39 (1988).
4. K. Kanoda, T. Takahashi, K. Kikuchi, K. Saito, I. Ikemoto, and K. Kobayashi, Phys. Rev. B **39**, 3996 (1989).
5. K. Murata, K. Kikuchi, T. Takahashi, K. Kobayashi, Y. Honda, K. Saito, K. Kanoda, T. Tokiwa, H. Anzai, T. Ishiguro, and I. Ikemoto, J. Mol. Elect. **4**, 173 (1988).
6. K. Kikuchi, I. Ikemoto, and K. Kobayashi, Synth. Met. **19**, 551 (1986).
7. K. Kikuchi, M. Kikuchi, T. Namiki, K. Saito, I. Ikemoto, K. Murata, T. Ishiguro, and K. Kobayashi, Chem. Lett. **1987**, 931 (1987).
8. K. Kikuchi, K. Murata, Y. Honda, T. Namiki, K. Saito, K. Kobayashi, T. Ishiguro, and I. Ikemoto, J. Phys. Soc. Jpn. **56**, 3436 (1987).
9. V.J. Emery, R. Bruinsma, and S. Barisic, Phys. Rev. Lett. **48**, 1039 (1982).
10. R.J. Elliott, Phys. Rev. **96**, 266 (1954).

11. K. Kanoda, T. Takahashi, K. Kikuchi, K. Saito, Y. Honda, I. Ikemoto, K. Kobayashi, K. Murata, and H. Anzai, Solid State Commun. **69**, 415 (1989).
12. K. Kanoda, S. Okui, T. Takahashi, K. Kikuchi, K. Saito, I. Ikemoto, and K. Kobayashi, (to be published).
13. T. Mochizuki, K. Kikuchi, G. Saito, and I. Ikemoto, (unpublished).

Conducting and Superconducting Salts Based on MDTTTF, EDTTTF, VDTTTF, EDTDSDTF, MDSTTF, BMDTTTF, Pd(dmit)$_2$, and Ni(dcit)$_2$

G.C. Papavassiliou[1], G. Mousdis[1], V. Kakoussis[1], A. Terzis[2], A. Hountas[2], B. Hilti[3], C.W. Mayer[3], and J.S. Zambounis[3]

[1] Theoretical and Physical Chemistry Institute, National Hellenic Research Foundation, 48, Vassileos Constantinou Ave., Athens 116 35, Greece
[2] Institute of Materials Science "Democritos" N.R.C., Ag. Parascevi Attikis, Athens 153 10, Greece
[3] Central Research Laboratories, CIBA-GEIGY AG, CH-4004 Basel, Switzerland

Abstract. Conducting and superconducting salts based on methylenedithiotetrathiafulvalene (MDTTTF), ethylenedithiotetrathiafulvalene (EDTTTF), vinylenedithiotetrathiafulvalene (VDTTTF), ethylenedithiodiselenadithiafulvalene(EDTDSDTF), methylenediselenotetrathiafulvalene(MDSTTF), bis(methylenedithio)tetrathiafulvalene(BMDTTTF), bis(4,5-dimercapto-1,3-dithiole-2-thione) palladate(Pd[dmit]$_2$), and bis(3,4-dimercapto-5-cyanoisothiazole) nickelate (Ni[dcit]$_2$) have been prepared and studied.

Introduction

Recently, a number of conducting and superconducting salts based on unsymmetrical donor molecules as well as on metal 1,2-dithiolenes have been prepared and studied (see [1-12], and refs.therein). In this paper the preparations, crystal structures and physical properties of some salts based on the compounds (1)-(8) (: π-donors or π-acceptors) are described

(1): (MDTTTF)   (2): (EDTTTF)   (3): (VDTTTF)

(4):(EDTDSDTF)   (5):(MDSTTF)   (6):(BMDTTF)

(7): Pd(dmit)$_2$ (8): Ni(dcit)$_2$

Experimental

Compounds (1)-(5) have been prepared from the corresponding 2-oxo-1,3-dithioles or selenium analogs and 4,5-bis(methylcarboxy)-1,3-dithiole-2-thione or selenium analog by a two-step sequence:coupling via triethylphosphite and demethoxycarboxylation with LiBr in hexamethylphosphoramide[1,2,6]. Compound (6) has been prepared by coupling of 4,5-methylenedithio-1,3-dithiole-2-one via triethyl phosphite [6]. The half-wave oxidation potentials ($E^1_{1/2}, E^2_{1/2}$) [13] of (1), (2),(5), (6) have values 535, 915; 545, 970; 490, 870; 595, 880 mV, respectively. These values are intermediate between those of TTF(470, 940 mV) and BEDTTTF (600, 980 MV) [13].Similar results are expected for the compounds (3), (4), etc. [4], [6]. $(Me_4N)_2Pd(dmit)_2$ and $(Bu_4N)_2Ni(dcit)_2$ have been prepared from the corresponding disodium salts of the ligands (dmit, dcit) after treatment with $Pd(NH_3)_4(NO_3)_2$ and $NiCl_2$ in presence of Me_4NBr and Bu_4NBr, respectively. From the compounds (1)-(8) a number of conducting salts have been prepared mainly by electrooxidation methods.$(Me_4N)_xPd(dmit)_2$ has been prepared by oxidation of aceton solutions of $(Me_4N)_2Pd(dmit)_2$ as follows:

$$(Me_4N)_2Pd(dmit)_2 \xrightarrow{I} (Me_4N)_1Pd(dmit)_2 \xrightarrow{air} (Me_4N)_xPd(dmit)_2 \text{ (needles)}$$

Results and Discussion

The results of conductivity measurements in a number of salts (1)-(12) are summarized in Fig.1 and Fig.2. (1): $(MDTTTF)_2AuI_2$ (orth., Pbmn [3,7]) is metallic (σ_{RT}=12-36 S/cm) and becomes a superconductor at low temperature (Tc=5K under 1bar, Tc=3K under 1.5 kbar) [3,8,14]. (2):$(EDTTTF)_2IBr_2$ (tricl., P$\bar{1}$ [7]) remains metallic down to 1.35 K σ_{RT}=150-550 S/cm) [3,12]. Same results have been obtained for (3): $(EDTDSDTF)_2 IBr_2$ (tricl., P$\bar{1}$); (σ_{RT}=1050-1660 S/cm). (4):$(EDTTTF)_2AuI_2$ (orth. F222 [7]) is metallic (σ_{RT}=500 S/cm) down to low temperature MIT ≈ 200 K under 1 bar [5]; MIT ≈ 20 K under 4.8 kbar). (5): $(EDTTTF)_2AuBr_2$ (monoch., C 2/m [7]) is metallic (σ_{RT}=230-330

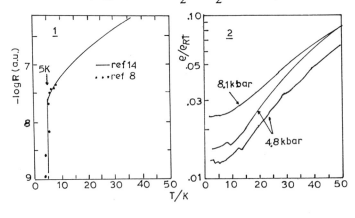

Fig.1. Temperature dependence of resistance and normalized resistivity of (1) and (2) respectively.

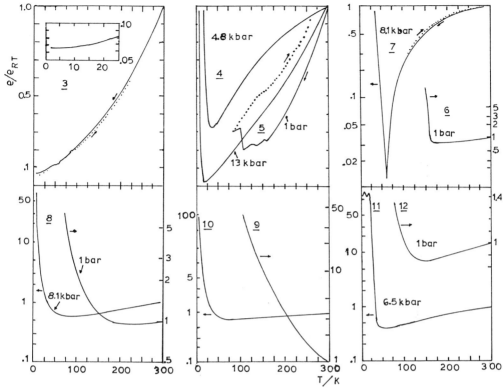

Fig.2. Temperature dependence of normalized resistivity of (3)-(12).

S/cm; MIT ≈ 140 K under 1 bar; MIT ≈ 15 K under 13 kbar) (see also [9]). (6): (EDTTTF)$_2$Ag(CN)$_2$ is metallic (σ_{RT}= 2-50 S/cm; MIT ≈ 150 K under 1bar). (7): (EDTTTF)$_2$SbF$_6$ is metallic (σ_{RT}= 20-120 S/cm; MIT ≈ 200 K under 1 bar, MIT ≈ 70 K under 8.1 kbar). (8):(VDTTTF)$_2$IBr$_2$ is metallic (σ_{RT}=50-714 S/cm; MIT ≈ 200 K under 1 bar, MIT ≈ 50 K under 8.1 kbar). (9): (MDTTTF)$_2$AuBr$_2$ (triecl., P$\bar{1}$) is semiconducting (σ_{RT}=205S/cm). (10): (MDSTTF)$_2$AuI$_2$ is metallic (σ_{RT}=305 S/cm; MIT ≈ 50K under 1 bar). (11):(Me$_4$N)$_x$Pd(dmit)$_2$ (needles) is metallic (σ_{RT}= 28 S/cm; MIT ≈ room temp. under 1bar, MIT ≈ 20 K under 6.5 kbar) (see also [15]). (12): (BMDTTTF)$_x$Ni(dcit)$_2$ (plates) is metallic (σ_{RT}= 20-40 S/cm; MIT ≈ 120K under 1 bar).

The salts (2) and (3) are isostructural with the β-(BEDTTTF)$_2$X superconductors [7,16] and have a residual conductivity $\sigma_{1.35}$>10000 S/cm. According to the treatment reported in [17] these crystals could be candidates for superconductivity at lower temperature. The temperature dependence of the resistivity of (4)-(8) and (10-(12) indicate that these salts could be superconductors under pressure-values higher than those reported above.

References

1. G.C.Papavassiliou,,in Proc. NATO-ASI on Lower-Dim. Systems and Mol.Electronics, Spetses Island, Greece, 12-23 June, 1989, Ed.R.M.Metzger, Plenum, in press.
2. G.C.Papavassiliou, Pure, Appl.Chem., in press.
3. G.C.Papavassiliou, G.A.Mousdis, J.S.Zambounis, A.Terzis, A.Hountas, B.Hilti, C.W.Mayer, and J.Pfeiffer, Synth.Metals **27**, B379 (1988).
4. G.C.Papavassiliou, G.A.Mousdis, S.Y.Yiannopoulos, V.C.Kakoussis and J.S.Zambounis, Synth.Metals **27**, B373 (1988).
5. A.Terzis, A.Hountas, A.E.Underhill, A.Clark, B.Kaye, B.Hilti, C.Mayer, J.Pheiffer, S.Y.Yiannopoulos, G.Mousdis and G.C.Papavassiliou, Synth.Metals **27**, B97 (1988).
6. G.C.Papavassiliou, V.C.Kakoussis, G.A.Mousdis and J.S.Zambounis, Chem.Scripta, in press (1989).
7. A.Hountas, A.Terzis, G.C.Papavassiliou, B.Hilti, M.Burkle, C.W.Mayer and J.S.Zambounis, Acta Cryst. in press (1989) and refs. [2,11] cited therein.
8. A.M.Kini, M.A.Beno, D.Son, H.H.Wang, K.D.Carlson, L.C.Porter, U.Welp, B.A.Vogt, J.M.Williams, D.Jung, M.Evain, M.-H.Whangbo, O.L.Overmyer, and J.E.Schirber, Sol.St.Commun. **69**, 503(1989).
9. T.Mori and H.Inochuchi, Sol.St.Commun. **70**, 823 (1989).
10. R.Kato, H.Kobajashi and A.Kobajashi, Chem.Lett. 781(1989), and in press.
11. H.Nakano, K.Miyawaki, Y.Shirota, S.Harada, and N.Kasai, Bull.Chem.Soc.Jpn **62**, 2604 (1969); Y.Honda, K.Murata, K.Kikuchi, K.Saito, I.Ikemoto, and K.Kobayashi, Sol.St.Commun., in press.
12. See refs. [37-41, 49-51, 61-99] cited in ref.1 here.
13. Donor :7×10^{-4} M, Bu_4NPF_6:0.025 M in benzonitril at $23°C$ vs.SCE at a Pt-disc electrode (anode); scam rate 100 mV/s; see also H.Tatemitsu, J.Chem.Soc. Chem. Commun. 106 (1985).
14. M.Freund, J.L.Olsen, B.Hilti, C.W.Mayer and J.S.Zambounis, presented at EPS meeting, Nice, France March 1989; B.Hilti, A.Terzis, G.C.Papavassiliou to be published.
15. A.Kobayashi, H.Kim, Y.Sasaki, Synth.Metals **27**, B339 (1988).
16. T.Ishiguro, Physica C **153-155**, 1055(1988).
17. M.Tokamoto, H.Anzai, K.Murata, K.Kajimura and T.Ishiguro, Synth.Metals **27**, A251 (1988).

Part VII

Crystal and Electronic Structures

Structural Instabilities and Electronic Structures of Some Organic Conductors and Superconductors

S. Ravy[1], E. Canadell[2], and J.P. Pouget[1]

[1]Laboratoire de Physique des Solides (CNRS LA 2), Bât. 510,
Université Paris-Sud, F-91405 Orsay, France
[2]Laboratoire de Chimie Théorique (CNRS UA 506), Bât. 490,
Université Paris-Sud, F-91405 Orsay, France

Abstract. Recent X-ray diffuse scattering investigations of organic conductors and superconductors are reviewed. They consist in the observation of diffuse streaks in the 10 K organic superconductor (BEDT-TTF)$_2$Cu(NCS)$_2$, ascribed to a $2k_F$ wave of faults stabilized by the open portions of the Fermi surface. Charge density wave (CDW) instabilities are observed in the X [M(dmit)$_2$]$_2$ molecular superconductors (X = TTF, M = Pd or Ni) and conductor (X = Cs, M = Pd), and related to their conduction band structure involving respectively either both the HOMO and LUMO or an antibonding combination of the HOMO of the M(dmit)$_2$.

1. (BEDT-TTF)$_2$Cu(NCS)$_2$

This 10 K ambient pressure superconductor shows a short range modulation which consists of diffuse streaks along $\mathbf{a}^*$, with the reduced wave vector $\mathbf{q}=0.26(2)\ \mathbf{b}^*$ (figure 1). This modulation has been analysed[1] in terms of a periodic succession of faults stabilized by the one-dimensional (1D) open portions of the Fermi surface[2],

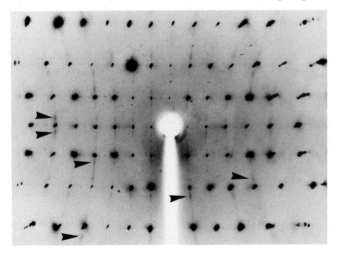

Figure 1. X-ray pattern from (BEDT-TTF)$_2$Cu(NCS)$_2$ at 23 K. The black arrows point towards the diffuse scattering. The $\mathbf{b}^*$ and $\mathbf{c}^*$ directions are vertical and horizontal respectively.

since $\mathbf{b}^*$-$\mathbf{q}$ corresponds to its average $2k_f$ wave vector. We suggest that the faults are the result of a cis to trans defect of coordination of the copper atoms in the Cu(NCS)$_2$ polymeric chains, implying a change in the relative orientation of neighbouring BEDT-TTF dimers. Since this diffuse scattering does not vary significantly in temperature, it is unlikely to drive the semiconductor-metal crossover observed around 90 K at ambient pressure in transport measurements[3].

2. TTF- [M(dmit)$_2$]$_2$; M= Ni or Pd

An X-ray diffuse scattering study[4] of these molecular conductors reveals 1 D structural fluctuations at several critical wave vectors (figure 2). They have been assigned[5] to CDW instabilities resulting from a complex 1 D band structure including, in addition to the TTF-donor band, several conduction bands built from the HOMO's and the LUMO's of the acceptor (figure 3).

In the Pd derivative, two main 1 D instabilities are observed at the reduced wave vector $\mathbf{q_1}$=0.5 $\mathbf{b}^*$ and $\mathbf{q_2}$=0.31 $\mathbf{b}^*$ (figure 2a). The associated diffuse scatterings originate from the Pd(dmit)$_2$ stacks, as shown by structure factor calculation. They can be respectively ascribed to $2k_f$ CDW instabilities within each bunch of LUMO and HOMO bands (figure 3a), which drive two successive smeared structural transitions at T_1=150 K and T_2=105 K, the metal-insulator transition occurring at T_2[6]. At low temperatures additional scatterings of weaker intensity are observed at $2\mathbf{q_2}$ and $\mathbf{q_1}\pm\mathbf{q_2}$. These data are consistent with a charge transfer of 0.76 or 1 electron per TTF molecule[5].

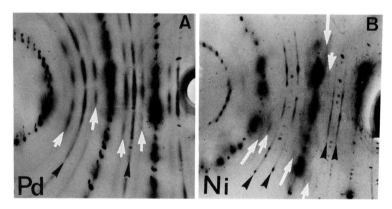

Figure 2. X-ray diffuse scattering pattern from :
A) TTF [Pd(dmit)$_2$]$_2$ at about 160K. The black and white arrows point towards the $\mathbf{q_1}$ and $\mathbf{q_2}$ scatterings,
B) TTF[Ni(dmit)$_2$]$_2$ at about 30K. The black arrows point towards the $\mathbf{q_1}$ scattering, and the short (long) white arrows towards the $\mathbf{q_2}$($\mathbf{q_3}$) scatterings. The 1D direction, $\mathbf{b}$, is horizontal.

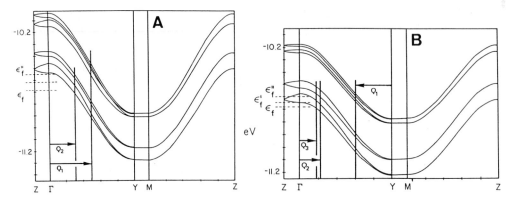

Figure 3. Band structure for the M(dmit)2 slabs in TTF[M(dmit)2]2 (A : M =Pd; B : M = Ni) - from ref. [5]. Half the 1D critical wavevectors ($Q_i = q_i/2$) measured by X-ray diffuse scattering are indicated.

At ambient pressure, the Ni derivative exhibits only one main 1 D structural instability at the reduced wave vector q_1=0.40(2) b^* and two quasi 1 D diffuse scatterings of weaker intensity at q_2=0.22(3) b^* and q_3=0.18(3) b^* (figure 2b); within experimental errors $q_1 = q_2 + q_3$. The q_1 instability leads to a structural transition at about T_1=40 K[4], which does not significantly affect the transport properties[7]. It could be associated with the bunch of the LUMO bands of the acceptor, since b^*-q_1 corresponds to an average nesting wave vector, as shown in figure 3b. The assignment of the q_2 and q_3 diffuse scattering is still uncertain but they could be associated with $2k_f$ instabilities in the HOMO bands of the Ni(dmit)2. They could be involved in the 10 K metal-semimetal transition recently detected by conductivity measurements at ambient pressure[7]. These structural results are compatible with a charge transfer of about 0.8 electron per TTF molecule.

3. Cs[Pd(dmit)2]2

This salt crystallises in the $C_{2/c}$ space group where layers of Cs and dimers of Pd(dmit)2 alternates along c. It undergoes a metal insulator phase transition at about 60 K[8], which is accompanied by a structural distortion. As shown by the X-ray pattern of figure 4, pretransitional fluctuations condense at T_c= 56.5 ±1 K into satellite reflections at the reduced wave vector q_c=(1,0.48,0). Additional reflections at q = (1,0,0) are also observed below T_c.

Because of the strong dimerization in the Pd(dmit)2 slabs, the antibonding combination of the HOMO's has a higher energy than the bonding combination of the LUMO's.The former level will receive an electron given by the cesium atom and form a half-filled conduction band. Interdimer interactions lead to a Fermi

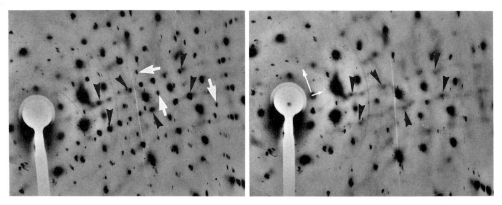

Figure 4. X-ray patterns from Cs(Pd(dmit)$_2$)$_2$ below (left) and above (right) T_c. The black arrows point towards the q_c satellite reflections and their pretransitional fluctuations. The white arrows show the q satellite reflections. The **a*** and **b*** vectors are indicated left by small and long arrows respectively.

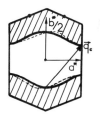

Figure 5. Superimposition of the Fermi surface of the partially filled band of the two slabs of Pd(dmit)$_2$ related by glide plane symmetry and its nesting wave vector q_c.

surface unexpectedly open in the stacking direction **a** (a similar situation occurs in δ-TTF[Pd(dmit)$_2$]$_2$[5]). The total Fermi surface, superimposition of those of the two Pd(dmit)$_2$ slabs related by a glide plane symmetry, is shown in figure 5. q_c is its nesting wave vector. This shows that a CDW instability drives the metal-insulator transition of Cs[Pd(dmit)$_2$]. A full investigation of the physical properties of this salt will be published elsewhere.

This works has benefited from collaboration with P. BATAIL, L. BROSSARD, P. CASSOUX and A. E. UNDERHILL. Fruitful discussions with D. JEROME and R. MORET are also acknowledged.

References

1. S. Ravy, J.P. Pouget, C. Lenoir and P. Batail, Solid State Comm., 73, 37 (1990)
2. K. Oshima, T. Mori, H. Inokuchi, H. Urayama, H. Yamochi and G. Saito, Phys. Rev. B 37, 938 (1988)
3. H. Urayama, H. Yamochi, G. Saito, K. Nozawa, T. Sugano, M. Kinoshida, S. Sato, K. Oshima, A. Kawamoto and J. Tanaka, Chem. Lett., 1988, 55 (1988)

4. S. Ravy, J.P. Pouget, L. Valade and J.P. Legros, Europhys. Lett. 9, 391 (1989)
5. E. Canadell, I.E.I. Rachidi, S. Ravy, J.P. Pouget, L. Brossard and J.P. Legros, J. Phys. France (1st October 1989)
6. L. Brossard, M. Ribault, L. Valade and P. Cassoux. J. Phys. France, 50, 1521 (1989)
7. L. Brossard, M. Ribault, L. Valade and P. Cassoux : C.R. Acad. Sc. Paris, in press.
8. R.A. Clark and A.E. Underhill, Synthetic Metals 27, B 515 (1988).

Structural and Physical Properties of (BEDT-TTF)$_2$[KHg(SCN)$_4$]

M. Oshima^1, H. Mori^2, G. Saito^3, and K. Oshima^4

^{1}Japan Carlit Co. Ltd., 2470 Handa, Shibukawa, Gunma 377, Japan
^{2}International Superconductivity Technology Center, 2-4-1 Mutsuno, Atsuta, Nagoya 456, Japan
^{3}Institute for Solid State Physics, University of Tokyo, 7-22-1 Roppongi, Minato-ku, Tokyo 106, Japan [1]
^{4}Department of Physics, Okayama University, 3-1-1 Tsushimanaka, Okayama 700, Japan

Abstract. The K^+ containing salt (BEDT-TTF)$_2$[KHg(SCN)$_4$] was grown by the electrochemical oxidation of BEDT-TTF with Hg(SCN)$_2$, KSCN, and 18-crown-6 ether from a mixed solvent of 1,1,2-trichloroethane and ethanol. The crystals were metallic down to 1.5 K with various temperature dependences. The crystal structures, electronic structures and magnetic properties of this salt are described.

1. Introduction

(BEDT-TTF)$_2$[KHg(SCN)$_4$] is one of the modifications of BEDT-TTF salts of Hg(II) thiocyanate. This salt has a unique crystal structure and shows metallic behavior down to low temperatures [2]. In this paper we report the crystal growth, electrical conductivity, magnetic property, and electronic and crystal structures of this salt.

2. Crystal Growth and Electrical Properties

A few kinds of BEDT-TTF salts with mercury thiocyanate were obtained by electrochemical oxidation of BEDT-TTF. By the combination of Hg(SCN)$_2$, KSCN, and 18-crown-6 ether as a supporting electrolyte and CH$_2$Cl$_2$ as a solvent, long black crystals of a composition of (BEDT-TTF)$_3$[Hg(SCN)$_3$] (Fig. 1a) were obtained which were metallic down to about 180 K where a sharp metal-insulator transition occurred. By the combination of the same electrolyte and 1,1,2-trichloroethane (TCE), we were not able to get crystals. The black thick plates of (BEDT-TTF)$_2$[KHg(SCN)$_4$] (Fig. 1b) were obtained when a mixed solvent of TCE and 10 vol% absolute ethanol was used in the electrocrystallization. Furthermore NH$_4^+$ containing crystals (Fig. 1c) were grown by changing KSCN to NH$_4$SCN in the electrolyte by using the same mixed solvent. The NH$_4^+$ containing salt is (BEDT-TTF)$_2$[NH$_4$(SCN)$_4$].

Very recently Müller et al. reported other modifications from TCE or a mixture of chlorobenzene/acetonitrile by using TBA$_2$Hg(SCN)$_4$ as a supporting electrolyte [3]. They obtained black needles and proposed the compositions as (BEDT-TTF)$_4$[Hg(SCN)$_4$](TCE) obtained from TCE and (BEDT-TTF)$_2$[Hg$_x$(SCN)$_4$, x=1.4-1.5] obtained from the mixed solvent. They mentioned that no conductivity data were available for the former due to the poor sample quality and the latter was semiconductive with room temperature conductivity of 3 S/cm.

The temperature dependences of the electrical resistivities of (BEDT-TTF)$_2$[KHg(SCN)$_4$] (abbreviated hereafter to K salt) and (BEDT-

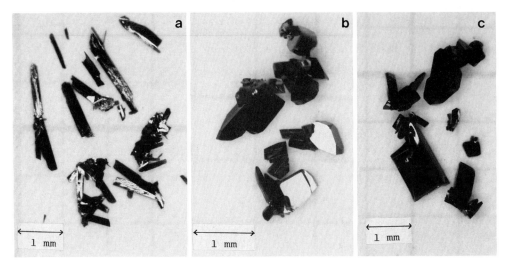

Fig. 1. Single crystals of (a) (BEDT-TTF)$_3$[Hg(SCN)$_3$], (b) (BEDT-TTF)$_2$[KHg(SCN)$_4$] and (c) (BEDT-TTF)$_2$[NH$_4$Hg(SCN)$_4$].

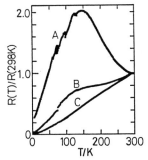

Fig. 2. Temperature dependences of normalized resistivity of (BEDT-TTF)$_2$[KHg(SCN)$_4$].

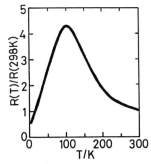

Fig. 3. Temperature dependence of normalized resistivity of (BEDT-TTF)$_2$[NH$_4$Hg(SCN)$_4$].

TTF)$_2$[NH$_4$Hg(SCN)$_4$] (NH$_4$ salt) were depicted in Figs. 2 and 3, and the room temperature conductivities are 20-100 S/cm and 23 S/cm, respectively. The temperature dependences of the K salt vary from sample to sample as shown in Fig. 2. An enhancement of resistivities was observed above 100 K in Type A or B and no anormaly was detected in Type C at higher temperatures, and all the samples were metallic down to 1.5 K. The NH4 salt was also metallic down to 1.5 K with an anomaly at around 100 K but the sample dependences were not investigated extensively in this case (only one sample was measured up to now).

3. Crystal and Electronic Structures and Magnetic Properties of K Salt

The crystallographic parameters of the K salt were summarized in ref. 2. Along the b-axis the conducting BEDT-TTF layer and the insulating

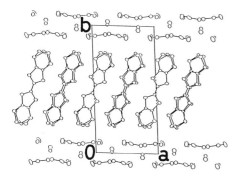

Fig. 4. Crystal structure of (BEDT-TTF)$_2$[KHg(SCN)$_4$] at 298 K along the c-axis.

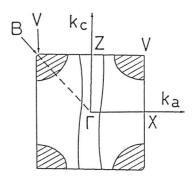

Fig. 5. Fermi surface of (BEDT-TTF)$_2$[KHg(SCN)$_4$] calculated by tight-binding method. Shaded regions indicated hole-like parts.[4]

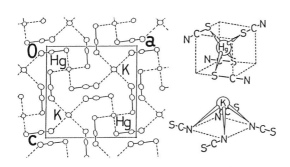

Fig. 6. Triple-sheet structure of anion KHg(SCN)$_4$ and schematic structures of its component clusters.

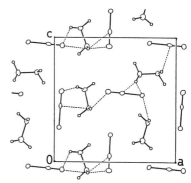

Fig. 7. Short contacts between the SCN groups of anion and the terminal ethylene groups of BEDT-TTF are shown by dashed lines.

KHg(SCN)$_4$ layer repeat alternately (Fig. 4). The ac- plane is the two-dimensional conducting plane of BEDT-TTF molecules which stack along the c-axis and have a number of short S..S atomic contacts along the a-axis. These features give a two-dimensional Fermi surface (Fig. 5) calculated by Mori based on the crystal structure at room temperature [4], which consists of hole pockets at the corner of the Brillouin zone and a one-dimensional surface along ZΓ. Shubnikov-de Haas signals were observed by Osada et al. with an oscillation period of 0.0015T^{-1} which corresponds to the area of the extremal orbit of 16.5 % of the 1st Brillouin zone [5]. This is in good agreement with the area of the hole-pockets in Fig. 5 ; 19 % of the 1st Brillouin zone. The details of the Shubnikov-de Haas effect in this salt are described in these proceedings.

Not only the BEDT-TTF molecules but also the anions form a two-dimensional network along the ac-plane (Fig. 6). Every SCN is linked to K$^+$ and Hg^{2+} as is shown in Fig. 6 schematically. As a result infinite chains of ..SCN..K..NCS..Hg..SCN.. spread two-dimensionally in the ac-plane to form a thick (6.8 Å) network of anion polymers. Several short anion..ethylene contacts were observed at low temperature (104 K, Fig. 7) such as S(anion)..H(2.79-2.96 Å),

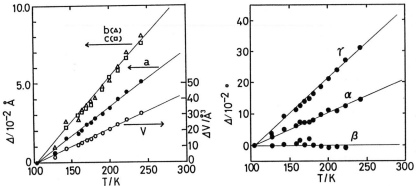

Fig. 8. Temperature dependences of lattice parameters of $(BEDT-TTF)_2[KHg(SCN)_4]$.

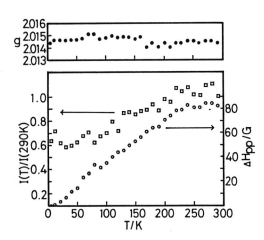

Fig. 9. Temperature dependences of the EPR intensity, the g-value and the peak-to-peak linewidth for a single crystal of $(BEDT-TTF)_2[KHg(SCN)_4]$.

C(anion)..H (2.68-2.90 Å) and N(anion)..H (2.44-2.68 Å). Fig. 8 presents the temperature dependences of lattice parameters. The unit cell volume and lattice constants except β decreased monotonically with temperatures down to 104 K.

The crystal structures of the NH_4 salt were solved which is isostructural to the K salt; $P\bar{1}$, a=10.091(1), b=20.595(2), c=9.963(1) Å, α=103.65(1), β=90.53(1), γ=93.30(1)°, V=2008.1(3) Å^3, Z=2. The details will be reported elsewhere.

A preliminary result of EPR measurments on single crystals is shown in Fig. 9 where the external magnetic field is approximately parallel to the b-axis. The g value at 290 K is 2.0144 as usually observed in BEDT-TTF salts. The g values are almost constant down to 4.2 K with small scattering which indicates that the molecular orientation of BEDT-TTF does not change substantially with respect to the crystal axis. The linewidth at 290 K is 82.5 G and remained constant down to 230 K then decreased monotonically with temperature. The EPR intensity also showed some constant region down to around 200 K and decreased below it slightly, but did not show rapid change down to low temperatures. All these EPR data are consistent with the metallic nature of this compound. More precise experiments on the K and NH_4 salts are under way.

The authors wish to thank Drs. M.Kinoshita and T.Sugano, and Mr. K.Nozawa for measurement of EPR spectra. This work was partly supported by the Grant-in-aid for Scientific Research from the Ministry of Education, Science and Culture.

References

1. Present address; Department of Chemistry, Faculty of Science, Kyoto University, Sakyoku, Kyoto 606, Japan.
2. M.Oshima, H.Mori, G.Saito and K.Oshima, Chem. Lett., **1989**, 1159.
3. H.Muller, C.P.Heidmann, H.Fuchs, A.Lerf, K.Andres, R.Sieburger and J.S.Schilling, to be published in "Lower-Dimensional Systems and Molecular Electronics" R.M.Metzger ed. NATO ASI, 1989.
4. T.Mori, private communication.
5. T.Osada, R.Yagi, S.Kagoshima, N.Miura, M.Oshima and G.Saito, these proceedings.

Importance of Weak Hydrogen Bonding C-H···Donor and C-H···Anion Interactions in Governing the Structural Properties of Organic Donors BEDT-TTF and BEDO-TTF and Their Charge-Transfer Salts

M.-H. Whangbo[1], D. Jung[1], J. Ren[1], M. Evain[1], J.J. Novoa[2], F. Mota[2], S. Alvarez[2], J.M. Williams[3], M.A. Beno[3], A.M. Kini[3], H.H. Wang[3], and J.R. Ferraro[3]

[1]Department of Chemistry, North Carolina State University, Raleigh, NC 27695, USA
[2]Facultat de Quimica, Universitat de Barcelona, E-08028-Barcelona, Spain
[3]Chemistry and Materials Science Divisions, Argonne National Laboratory, Argonne, IL 60439, USA

Abstract. On the basis of ab initio SCF-MO/MP2 and semi-empirical MNDO SCF-MO calculations, we examined the structural characteristics of neutral organic donor molecules BEDT-TTF and BEDO-TTF and their charge-transfer salts. The donor···donor and donor···anion interactions associated with the C-H bonds of the donor molecules are found to play a crucial role in governing the packing patterns of the donor molecules, which, in turn greatly influence the physical properties of the donor molecule salts.

1. Introduction

Organic donor molecules BEDT-TTF (**1a**) and its analogs form charge transfer-salts with a variety of monovalent anions X^- [1]. Except for a few cases, these salts have a layered structure in which layers of donor molecules alternate with layers of anions. The π-framework of the donor is typically inclined to the anion layer, so that the C-H bonds at both ends of a donor molecule make short contacts with the anions. The electronic properties of the donor salts are essentially determined by the packing patterns of their donor molecule layers. In synthesizing new organic salt metals, therefore, it is important to appreciate how donor packing patterns are affected by donor···donor and donor···anion interactions. Over the past years it has been recognized [1-8] that crystal structures of donor salts, and thus their donor-layer packing patterns, are strongly controlled by donor···donor and donor···anion interactions involving the C-H bonds of the donor molecules. These weak hydrogen bonding C-H···donor and C-H···anion interactions provide a key to understanding the anion- and pressure-dependence of the superconducting transition temperatures of β-(BEDT-TTF)$_2$X ($X^- = I_3^-$, AuI_2^-, IBr_2^-) [3,4], the thermal conversion of α-(BEDT-TTF)$_2$I$_3$ to α$_t$-(BEDT-TTF)$_2$I$_3$ [5], the occurrence of two different C-C-H bending modes in the polarized reflectance spectrum of κ-(BEDT-TTF)$_2$Cu(NCS)$_2$ [6], and the differences in the structural properties of various κ-phase organic donor salts [7, 8]. In this report, we discuss the structural characteristics of neutral BEDT-

BEDT-TTF   BEDO-TTF
  1a   1b

TTF and BEDO-TTF (**1b**) solids and their charge-transfer salts on the basis of ab initio SCF-MO/MP2 [9] and MNDO SCF-MO [10] calculations on model systems.

2. Neutral Donor Solid

The C-H···donor contacts shorter than the van der Waals radii sum present in neutral BEDT-TTF solid [11] are shown by dashed lines in **2**, which reveals that the C-H bonds of a donor make short contacts with the S and C (double bond carbon) atoms of its adjacent donor molecules. Aside from these C-H···donor contacts, there are two important structural features to note in **2**: (a) The π-framework of BEDT-TTF is bent as depicted in **3** ($\theta \cong 15°$), and (b) one six-membered ring of a donor molecule has a staggered conformation **4a**, but the other ring has a conformation resembling the eclipsed conformation **4d**.

3. Ring Conformation and π-Framework Bending

The observation (b) is striking since steric strain is expected for the eclipsed conformation of **4d**. To estimate the energy necessary to convert **4a** to **4d**, ab initio SCF-MO/MP2 calculations were performed on the conformations **4a**-**4d** of model ring **5** by employing the 3-21G basis set. (**4b** has only one methylene carbon atom on the π-plane, while **4c** has both methylene carbon atoms above the π-plane with one at half the height of the other.) Our calculations show that **4a** is the only minimum energy conformation, and that **4b**, **4c**, and **4d** are less stable than **4a** by 0.61, 2.94, and 5.67 kcal/mol, respectively. Thus, the conformational change of the six membered ring does not require much energy as long as the ring avoids the eclipsed conformation **4d**.

The observation (a) is also found for neutral BEDO-TTF (**1b**) solid [12]. To estimate the energy required for the π-framework bending, MNDO SCF-MO calculations were carried out on BEDT-TTF^0 and BEDT-TTF$^+$ as a function of the bending angle θ (See **3**). The π-framework is calculated to be most stable when it is planar (θ=0°) for both BEDT-TTF^0 and BEDT-TTF$^+$. For θ=5°, 10°, and 15°, BEDT-TTF^0 (BEDT-TTF$^+$) becomes less stable by 0.19 (0.46), 0.78 (1.85), and 1.81 (4.21) kcal/mol, respectively. Thus the bending potential of the π-framework is rather soft for both BEDT-TTF^0 and BEDT-TTF$^+$.

4. C-H···Donor Interactions

As described above, the π-framework bending and the adoption of an eclipsed-like ring conformation are destabilizing in nature, though not strongly. These structural changes lead to numerous C-H···donor contacts between donor molecules of the neutral donor solid. To estimate the stabilization energies of various C-H···donor interactions, we performed ab intio SCF-MO/MP2 calculations on H$_3$C-H···OH$_2$ and H$_3$C-H···CH$_2$=CH$_2$ by employing the 6-311+G** basis set, and on H$_3$C-H···SH$_2$ and H$_3$C-H···SeH$_2$ by employing the 6-311G** basis set with the pseudo-potential core for Se [13]. The stabilization energies of the C-H···Y (Y=O, S, Se, C) interactions calculated from these model compounds are 1.08, 0.46, 0.35, and 0.20 kcal/mol at their calculated optimum H···Y distances of 3.65, 3.96, 4.34, and 2.74 Å for Y=O, S, Se, and C, respectively. All of these C-H···Y interactions are weak, but are certainly stabilizing in nature. Thus the π-framework bending and the adoption of an eclipsed-like ring conformation in neutral BEDT-TTF solid give rise to energy stabilization via the numerous C-H···donor interactions they bring about. The stabilization energy of the C-H···S interaction is about twice that of the C-H···C interaction, but is only about half that of the C-H···O interaction. Thus the donor-packing pattern of neutral BEDO-TTF solid is, as expected, different from that of neutral BEDT-TTF solid. For example, within a layer of BEDO-TTF molecules in the neutral BEDO-TTF solid, the donor molecules make optimum use of the C-H···O contacts as illustrated in **6**.

6

5. C-H···Anion Interactions

In contrast to the case of the neutral donor solids, the π-frameworks of the donor molecules are planar in their charge-transfer salts. This finding can be easily understood if the C-H···anion interactions are more stabilizing than the C-H···donor interactions, because more short C-H···anion contacts can be made when the donor π-framework is planar. To estimate the stabilization energies of the C-H···X$^-$ (X$^-$=I$_3^-$, IBr$_2^-$, ICl$_2^-$) interactions; we carried out ab initio SCF-MO/MP2 calculations on H$_3$C-H···X$^-$ by using the 3-21+G** basis set with the pseudo-potential cores for I and Br [13]. The stabilization energies of the C-H···X$^-$ interactions are calculated to be 1.13, 1.27, and 1.58 kcal/mol at their calculated optimum H···X$^-$ distances of 3.50, 3.14, and 2.85 Å for X$^-$ = I-I-I$^-$, Br-I-Br$^-$, and Cl-I-Cl$^-$, respectively. The stabilization energies of the C-H···anion interactions are much greater than those of the C-H···donor interactions, C-H···S and C-H···C, as anticipated.

The C-H···donor interaction, C-H···O, is only slightly weaker than the C-H···anion interactions, which may well result in charge-transfer salts with very different properties for these two similar donor molecules. Therefore, the crystal packing patterns of BEDO-TTF salts should not be dominated by the C-H···anion interactions, and hence should differ from those of BEDT-TTF salts. For example, in each donor stack of (BEDO-TTF)$_2$AuBr$_2$ [14], donor molecules slip in the direction perpendicular to the central C=C bond (See **7**) instead of slipping along the C=C bond as found for most BEDT-TTF salts. The BEDO-TTF molecules make short C-H···O contacts within each donor stack (**8**) and between adjacent donor stacks (**9**).

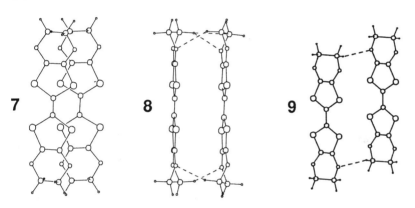

6. Concluding Remarks

For BEDT-TTF and BEDO-TTF molecules, the weak hydrogen bonding stabilization energies of the C-H...donor interactions are calculated to be much smaller than those of the C-H···anion interactions, except for the C-H···donor interaction C-H···O, which is comparable in magnitude to the C-H···anion interactions. Thus, the crystal packing patterns of BEDO-TTF salts should be determined primarily as a compromise between the C-H···O and the C-H···anion interactions, and therefore would be expected to be somewhat different from those of the BEDT-TTF salts. The donor···donor and donor···anion interactions involving the C-H bonds of the donor molecules are crucial factors to consider in understanding the structural and electronic properties of organic donor salts and derived synthetic metals.

Acknowledgment. Work at North Carolina State University and at Argonne National Laboratory is supported by the U. S. Department of Energy, Office of Basic Energy Sciences, Division of Materials Sciences under Grant DE-FG05-86ER45259 and Contract W-31-109-ENG-38, respectively. Work at Universitat de Barcelona was supported by CICYT Grant PB86-0272.

References

1. J. M. Williams, H. H. Wang, T. J. Emge, U. Geiser, M. A. Beno, P. C. W. Leung, K. D. Carlson, R. J. Thorn, A. J. Schultz, and M. -H. Whangbo, *Prog. Inorg. Chem.*, **35**, 51 (1987).
2. M. A. Beno, G. S. Blackman, P. C. W. Leung, and J. M. Williams, *Solid State Commun.*, **48**, 99 (1983).
3. M. -H. Whangbo, J. M. Williams, A. J. Schultz, T. J. Emge, and M. A. Beno, *J. Am. Chem. Soc.*, **109**, 90 (1987).
4. M. -H. Whangbo, J. M. Williams, A. J. Schultz, and M. A. Beno, *Organic and Inorganic Low-Dimensional Crystalline Materials*, P. Delhaes and M. Drillon, eds., Plenum, New York, 1987, p 333.
5. H. H. Wang, J. R. Ferraro, K. D. Carlson, L. K. Montgomery, U. Geiser, J. R. Whitworth, J. A. Schlueter, J. M. Williams, S. Hill, M. -H. Whangbo, M. Evain, and J. J. Novoa, *Inorg. Chem.*, **28**, 2267 (1989).
6. J. R. Ferraro, H. H. Wang, U. Geiser, A. M. Kini, M. A. Beno, J. M. Williams, S. Hill, M. -H. Whangbo, and M. Evain, *Solid State Commun.*, **68**, 917 (1988).
7. A. M. Kini, M. A. Beno, D. Son, H. H. Wang, K. D. Carlson, L. C. Porter, U. Welp, B. A. Vogt, J. M. Williams, D. Jung, M. Evain, M. -H. Whangbo, D. L. Overmyer, and J. E. Schirber, *Solid State Commun.*, **69**, 501 (1989).
8. D. Jung, M. Evain, J. J. Novoa, M. -H. Whangbo, M. A. Beno, A. M. Kini, A. J. Schultz, J. M. Williams, and P. J. Nigrey, *Inorg. Chem.*, in press.
9. *GAUSSIAN 86*, M. J. Frish, J. S. Binkley, H. B. Schlegel, K. Raghavachari, C. F. Melius, R. L. Martin, J. J. P. Stewart, F. W. Bobrowicz, C. M. Rohlfing, L. R. Kahn, D. J. Defrees, R. Seeger, R. A. Whiteside, D. J. Fox, E. M. Fleuder, and J. A. Pople, Carnegie-Mellon Quantum Chemistry Publishing Unit, Pittsburgh, PA, 1984.
10. M. J. S. Dewar and W. Thiel, *J. Am. Chem. Soc.*, **99**, 4907 (1977).
11. H. Kobayashi, A. Kobayashi, Y. Sasaki, G. Saito, and H. Inokuchi, *Bull. Chem. Soc. Jpn.*, **59**, 301 (1986).
12. T. Suzuki, H. Yamochi, G. Srdanov, K. Hinkelmann, and F. Wudl, *J. Am. Chem. Soc.*, **111**, 3108 (1989).
13. W. R. Wadt and P. J. Hay, *J. Chem. Phys.* **82**, 284 (1985).
14. M. A. Beno, A. M. Kini, U. Geiser, H. H. Wang, K. D. Carlson, and J. M. Williams, *this volume*.

Electronic Structure of β-(BEDT-TTF)$_2$I$_3$ Studied by Positron Annihilation

S. Tanigawa[1], P.K. Tseng[2], T. Kurihara[1], K.Y. Chang[2], K. Watanabe[1], T. Kubota[1], M. Tokumoto[3], N. Kinoshita[3], and H. Anzai[3]

[1] Institute of Materials Science, University of Tsukuba, Tsukuba, Ibaraki 305, Japan
[2] Physics Department, National Taiwan University, Taipei, Taiwan, Rep. of China
[3] Electrotechnical Laboratory, Umezono, Tsukuba, Ibaraki 305, Japan

Abstract. The electron momentum distribution of β-(BEDT-TTF)$_2$I$_3$ at 23K and 300K was studied by positron annihilation. One electron cylinder with an almost elliptic cross section running along $\vec{c}*$ axis was found at 23K. At 300K, this cylinder was distorted.

1. Introduction

β-(BEDT-TTF)$_2$I$_3$ is an organic metal with superconductivity. It has two superconducting states; one is the low-T_c state with T_c about 1.1-1.6K at ambient pressure, the other is the high T_c state with T_c about 7.5K under the slight pressure of 1.2kbar[1]. It is believed that the low-T_c state is associated with the incommensurate lattice modulation which takes place around 175K under ambient pressure[2]. Information about the effect the lattice modulation on the electronic structure of the crystal is helpful for the understanding of the mechanism of superconductivity. In this paper, we will report a study of the electronic structure of β-(BEDT-TTF)$_2$I$_3$.

The electronic and crystal structure of the title compound has two-dimensional character. BEDT-TTF molecules stack loosely and form a column along the [110] direction. The interstack S-S interactions form a 'corrugated sheet' network lying principally on the $\vec{a}$-$\vec{b}$ plane. Electrical conduction probably occurs through this network. The linear I$_3$ anions are encapsulated in an ethylene group H-atom cavity and form an anionic sheet. The anionic sheet receives an electron from the BEDT-TTF molecular sheet to leave a conduction hole in the network. The alternation of the insulating anionic sheet and the conducting cationic-radical sheet forms the layer-structure of the organic metal. According to the measurement of conductivity at room temperature and coherence length by Tokumoto et al. one finds that $\sigma_a:\sigma_b':\sigma_{c^*}=260:620:1$, $\xi_a(0):\xi_b'(0):\xi_{c^*}=21:20:1$, respectively[1]. The anisotropy in conductivity and coherence length reflects the two-dimensional electronic character of β-(BEDT-TTF)$_2$I$_3$. The evidence for the existence of a two-dimensional Fermi surface is drawn from the Shubnikov-de Haas experiment performed by Murata et al.[3]

2D-ACAR measurements can provide information about the electron momentum distribution in metals. In this paper, we will show the result of the application of two-dimensional angular correlation measurement of annihilation radiations (2D-ACAR) to β-(BEDT-TTF)$_2$I$_3$.

2. 2D-ACAR Method

2D-ACAR spectrum $N(p_x,p_y)$ is the projection of photon pair momentum density on the p_x-p_y plane in the extended scheme[4],

$$N(p_x, p_y) = \int \rho(\vec{p}) \, dp_z \tag{1}$$

where $\rho(\vec{p})$ is the probability that the electron-positron system annihilates into a photon pair with total momentum $\vec{p}$. According to the independent particle model[4], $\rho(\vec{p})$ is expressed by

$$\rho(\vec{p}) = \text{const.} \sum_{b,\vec{k}}^{occ} \left| \int d\vec{r} \, \Psi_{b,\vec{k}}(\vec{r}) \, \Phi_p(\vec{r}) \exp(-i\vec{p}\cdot\vec{r}) \right|^2, \tag{2}$$

where $\Psi_{b,\vec{k}}(\vec{r})$ is the wavefunction of an electron, $\Phi_p(\vec{r})$ the wavefunction of a thermalized positron and the summation is over all occupied states. If we measure 2D-ACAR spectra for many different projections, we can obtain a full three-dimensional density $\rho(\vec{p})$ by means of an image reconstruction method [5] in the same way as medical CT scanning techniques. The information about the Fermi surface can be extracted by the folding procedure proposed by Lock, Crisp and West [6] (LCW folding). The LCW folding is a periodical superposition of $\rho(\vec{p})$ on every reciprocal lattice point. That is,

$$\rho(\vec{k}) = \sum_{\vec{G}_i} \rho(\vec{p} + \vec{G}_i), \tag{3}$$

where $\vec{G}_i$ is the i-th reciprocal lattice vector and $\vec{k}$ is defined within the first Brillouin zone. By the Bloch theorem, $\rho(\vec{k})$ can be given by

$$\rho(\vec{k}) = \text{const.} \sum_{b,\vec{k}}^{occ} \theta(E_F - E_{b,\vec{k}}) \int_{cell} d\vec{r} \, |\Psi_{b,\vec{k}}(\vec{r})|^2 |\Phi_p(\vec{r})|^2. \tag{4}$$

Therefore, $\rho(\vec{k})$ should have breaks corresponding to the integral in Eq.(4) between inside and outside of the Fermi surface. If the density of positrons can be assumed to be uniform in space, the integral in Eq.(4) is equal to unity for occupied electron states. Even in the general case where the density of positrons is not uniform, the $\vec{k}$ dependence of the integral still seems to be not so large that the breaks in $\rho(\vec{k})$ can indicate the location of the Fermi surface in $\vec{k}$ space.

3. Experiment

The 2D-ACAR spectra for a single crystal of β-(BEDT-TTF)$_2$I$_3$ were measured at both 23K and 300K. The injected direction of positrons was parallel to the $\vec{c}*$ axis. The first integration direction was parallel to the $\vec{a}$ axis. Then the crystal was rotated around the $\vec{c}*$ axis by a 10° step for another spectrum. We got 18 2D-ACAR spectra in total. From the set of 2D-ACAR spectra, a full 3D density $\rho(\vec{k})$ was determined.

4. Results and Discussion

Figs. 1(a), 1(b) and 1(c) show the cross sections of $\rho(\vec{k})$ with different values of k_3 at T=23K. In figs. 1(a) and 1(b), the distributions look the same around the Γ and Z points. In fig. 1(b), the distribution around the origin is slightly different from the other two. However, an almost elliptic distribution is found around the origin in all three figs. From figs. 2 and 3, one finds that the $\rho(\vec{k})$ look quite different along the a* and b* axes. But the maxima of the distributions both occur at zero wave number. This indicates that the crystal has an electron-like band structure around zero wave number and one may imagine an almost elliptic electron cylinder running along the c* direction. This result is similar to the theoretical calculation by Kübler et al.[7] Moreover, the

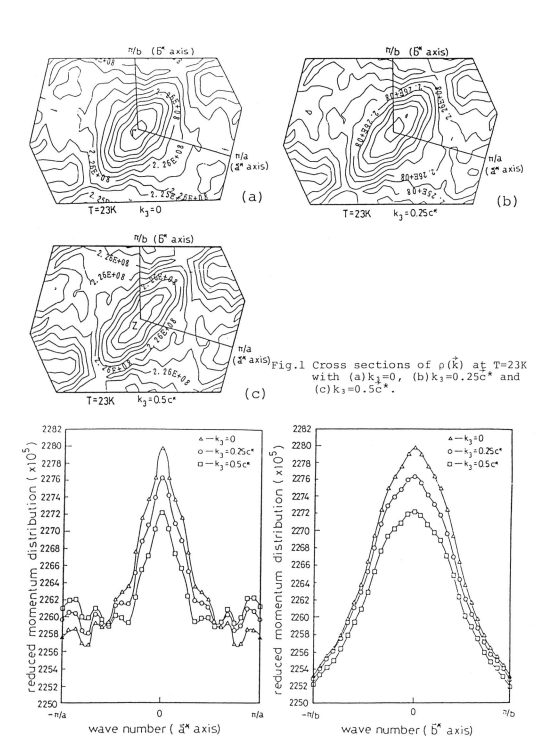

Fig.1 Cross sections of $\rho(\vec{k})$ at T=23K with (a) $k_3=0$, (b) $k_3=0.25c^*$ and (c) $k_3=0.5c^*$.

Fig.2 Reduced momentum density along $\vec{a}^*$ axis at T=23K.

Fig.3 Reduced momentum density along $\vec{b}^*$ axis at T=23K.

curves have almost the same shape but without overlapping around zero wave number. This indicates that $\rho(\vec{k})$ has a slight dependence on k_3.

Figs. 4, 5 and 6 show the corresponding results at T=300K. In fig. 4, the cross sections at $k_3=0$ and $k_3=0.5\vec{c}*$ have no great difference and elliptic distributions are found around zero wave number. The result in fig. 4(b) somehow surprises us. The elliptic distribution shifts away from the origin. In fig. 5, the curves for different k_3 overlap around the zero wave number. It suggests that the distributions are independent of k_3 along the $\vec{a}*$ direction. In fig. 6, the shape of the distributions is almost the same except for the one corresponding to $k_3=0.25\vec{c}*$. One finds that the maximum of this distribution shifts and has a more complicated dependence on wave number. From these figures, one may discover that the electron cylinder which we find at low temperature is distorted. It hints at a more complicated electron momentum distribution at room temperture.

The result shown above is very preliminary work on the organic superconductor. We hope that our reserch may stimulate further theoretical calculation, including of positron wavefunctions.

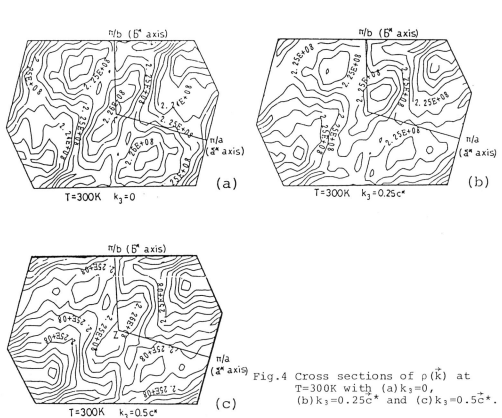

Fig.4 Cross sections of $\rho(\vec{k})$ at T=300K with (a) $k_3=0$, (b) $k_3=0.25\vec{c}*$ and (c) $k_3=0.5\vec{c}*$.

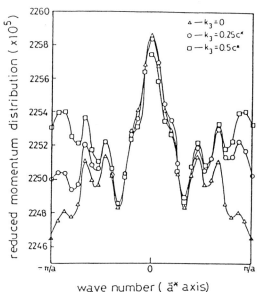

Fig.5 Reduced momentum density along $\vec{a}^*$ axis at T=300K.

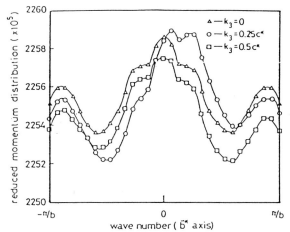

Fig.6 Reduced momentum density along $\vec{b}^*$ axis at T=300K.

References

1. M. Tokumoto, Res. of the Electrotechnical Lab., No.892, 1 (1988).
2. T. J. Emge, P. C. S. Leung, M. A. Beno, A. J. Schultz, H. H. Wang, L. M. Sowa and J. M. Williams, Phys. Rev. B$\underline{30}$, 6789 (1984).
3. K. Murata, M. Toyota, Y. Honda, T. Sasaki, M. Tokumoto, H. Bando, H. Anzai, Y. Muto and T. Ishiguri, J. Phys. Soc. Jpn. $\underline{57}$, 1540(1988).
4. S. Berko, Positron Solid-State Physics (North-Holland, 1$\overline{98}$3) 64.
5. R. Suzuki, M. Osawa, S. Tanigawa, M. Matsumoto and N. Shiotani, J. Phys. Soc. Jpn. $\underline{58}$, 3251 (1989).
6. D. G. Loch, V. H. C. Crisp and R. N. West, J. Phys. F$\underline{3}$, 561 (1973)
7. J. Kübler, M. Weger and C. B. Sommers, Solid State Commun. $\underline{62}$, 801 (1987).

Pressure Dependence of the Transport Properties of κ-(BEDT-TTF)$_2$Cu(NCS)$_2$

I.D. Parker[1], *R.H. Friend*[1], *M. Kurmoo*[2], *P. Day*[2], *C. Lenoir*[3], *and P. Batail*[3]

[1]Cavendish Laboratory, Madingley Road, Cambridge CB3 0HE, UK
[2]Inorganic Chemistry Laboratory, South Parks Road, Oxford OX1 3QR, UK
[3]Laboratoire de Physique des Solides, Bât. 510, F-91405 Orsay, France

Abstract The molecular superconductor κ-(BEDT-TTF)$_2$Cu(NCS)$_2$ shows a very high resistivity at ambient pressure ($\approx 10^{-1}$ Ωcm at room temperature) which rises on cooling to reach a maximum near 90K. We find that with the application of pressures of up to 2.5kbar this peak remains strong and is progressively sharpened, but that above 2.5 kbar the size of the resistive peak falls very quickly. In our crystals the resistively-measured T_c is 7 K at ambient pressure and remains constant up to 3 kbar, before falling steeply at -1.8 K/kbar above this pressure.

1. Introduction

κ-(BEDT-TTF)$_2$Cu(NCS)$_2$ shows a superconducting transition temperature at up to 12 K [1,2], the highest transition temperature of any organic superconductor yet discovered. Its normal state conductivity shows a curious temperature dependence. On cooling, the resistivity decreases to a shallow minimum near 250K, before increasing to reach a maximum near 110K. The ratio ρ_{max}/ρ_{300K} appears to be sample dependent, and has been reported to be as high as 6 [1], but more generally, and as in this work, is found to be nearer 2 [3-6]. Below 110K the resistivity decreases steeply to the superconducting transition which has been reported between about 7 and 12 K [1-6]. Furthermore, the room temperature conductivity in the *bc* plane is very low for an organic metal (10-20 S/cm). We report here a study of the transport properties under high pressure [7]. We have made our measurements under well-controlled conditions [8] with particular emphasis on the pressure regime below 4 kbar, which is the region in which we find very considerable changes in the normal state properties.

2. Results

The samples were grown by the method of Uramaya et al. [1]. Crystals show an EPR linewidth of about 60 Oe with the applied DC field parallel to the *a* axis at room temperature, and lattice parameters $a = 13.121(2)$ Å, $b = 8.441(2)$ Å, $c = 16.231(2)$ Å, $\beta = 110.32(1)°$, $V = 1685.7$ Å^3. The high pressure system used here has been described previously [8]. Pentane was used as the pressure-transmitting medium in a beryllium-copper pressure cell in communication with a large intensifier/pressure reservoir, which remains at room temperature, via a stainless steel capillary tube. This system allows accurate control of pressure down to the freezing point of the pentane (120 K at ambient pressure), and shows little pressure variation below the freezing. It does not, therefore, suffer from the considerable pressure losses with cooling usually encountered with "clamp" cells, and provides a particularly convenient system for the present set of measurements.

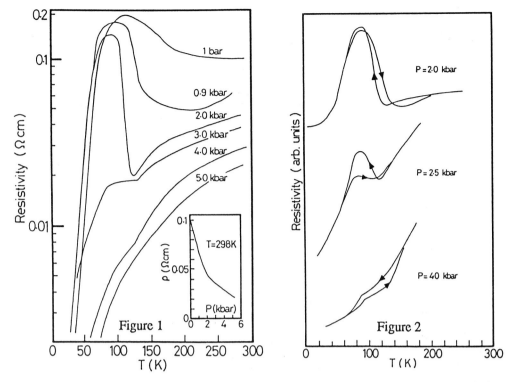

Figure 1 The temperature dependence of resistivity at various pressures. Data are shown for cooling. The inset shows the pressure dependence of the resistivity at room temperature.

Figure 2 Hysteresis in the resistivity around the resistance anomaly at various pressures. Typical cooling and warming rates were in the range 0.3 to 1 K/minute

Resistivity as a function of temperature at several pressures is shown in figure 1. The ambient pressure results are similar to those reported by other groups, with a room temperature resistivity of around 10 S/cm, and a broad maximum in resistivity near 110 K. Below 100K, the resistance drops sharply and the onset of superconductivity is observed near 8K. The ratio $R_{(300K)}/R_{(10K)} \approx 270$. Temperature cycling shows no evidence for hysteresis in the resistivity at ambient pressure. Under pressure, the room temperature conductivity increases sharply, initially at 30% per kbar, as shown in the inset to figure 1. The temperature dependence of the resistivity for several pressures is also shown in figure 1. At the two lower pressures indicated, of 0.9 and 2.0 kbar, the resistivity shows the same general form as at ambient pressure, with an initial fall with decreasing temperature, followed by an increase to a maximum and a further fall at lower temperatures. However, in comparison with the ambient pressure behaviour, the resistivity anomaly is much more strongly pronounced, having a sharp transition from the 'metallic' behaviour at higher temperatures to the anomalous regime. Thus, for example, at 2.0 kbar the sample is metallic down to 122K, but the resistance rises by a factor of 7 to a sharp peak at 95K. Furthermore, the resistance of the sample around the resistive peak shows substantial hysteresis between cooling and warming data, as indicated in figure 2. Within this pressure regime T_c remains constant. However, with further increase of the pressure the resistive anomaly is rapidly weakened. At 3.0 kbar the peak in resistivity is completely suppressed, and only a slight anomaly remains to indicate its position. It appears that the temperature at which it is observed is little affected by pressure, though it is increasingly difficult to detect it as pressure is further raised, and at 5.0 kbar all traces of the anomaly are removed.

For pressure below 2.5 kbar, the superconducting transition temperature, T_c, remains constant, with an onset at 8.0K, half height resistance at 7.0K and zero resistance at 6.0K. Above 2.5 kbar, T_c drops rapidly, at a rate of -1.8K/kbar. It should be noted that the pressure is non-hydrostatic since the sample is encased in frozen pentane, and as such, it is exposed to shear stresses due to the differential contraction of the sample and frozen pressure medium.

3. Discussion

The anomaly in the resistivity in $(BEDT-TTF)_2Cu(SCN)_2$ cannot easily be associated with a structural phase transition. There is considerable evidence that the carrier concentration changes little on cooling, notably (i) the weakly temperature independent paramagnetic susceptibility for temperatures above 90K (χ_p is proportional to the density of states at the Fermi level), though there is a fall below 90 K [9,10], (ii) the absence of an anomaly in the thermopower corresponding to the resistance anomaly [11], (iii) the fact that R_H remains small [7], and (iv) the observation that the Fermi surface calculated at room temperature corresponds well with the low temperature SdH data [12]. Thus it appears that there is no reorganisation of the Fermi surface which leads to a large reduction in the density of states at the Fermi energy.

Recent X-ray diffuse scattering measurements [13], do show the presence of a temperature independent superlattice, tentatively assigned to an incommensurate series of 'line-defects' within the anion layers, and which does match the wavevector spanning the Fermi surface. This match may be driven by Fermi surface nesting, as modelled for incommensurate mass density waves in various metal alloys. If this coupling is effective, then the variation in the form of the Fermi surface with lattice parameter, due either to thermal contraction with cooling or to the application of high pressure, may cause the superlattice potential to be brought in or out of effective coupling with the conduction electron system. Evidence in support of this is provided by the large increase in the resistance maximum observed under conditions of uniaxial strain [14].

The pressure dependence of the superconducting transition temperature, T_c, is different here from that reported by other groups. Specifically, we find that there is a correlation between the plateau in T_c up to 3 kbar and the presence of the large resistive anomaly seen clearly in figure 1. We have discussed these results elsewhere[7].

References

1. H.Urayama, H.Yamochi, G.Saito, K.Nozawa, T.Sugano, M.Kinoshita, S.Sato, K.Oshima, A.Kawamoto and J.Tanaka : Chem. Lett. 55 (1988).
2. H. Mori, H. Yamochi, G. Saito and K. Oshima, these proceedings.
3. K.Oshima, H.Urayama, H.Yamochi and G.Saito : J. Phys. Soc. Japan **57**, 730 (1988).

4. S.Gärtner, E.Gogu, I.Heinen, H.J.Keller, T.Klutz and D.Schweitzer : Solid State Comm. **65**, 1531 (1988)
5. A.Ugawa, G.Ojima, K.Yakushi and H.Kuroda : Synthetic Metals **27**, A445 (1988).
6. H. Veith, C-P Heidmann, H. Muller, H. P. Fritz, K. Andres and H. Fuchs, Synthetic Metals **27**, A361 (1988).
7. I. D. Parker, R. H. Friend, M. Kurmoo, P. Day, C. Lenoir and P. Batail, J. Phys. CM **1**, 4479 (1989)
8. D. R. P. Guy and R. H. Friend, J. Phys. E (Scientific Instruments) **19**, 430 (1986)
9. K.Nozawa, T.Sugano, H.Urayama, H.Yamochi, G.Saito and M.Kinoshita, Chem. Lett. 617 (1988)
10. S. Klotz, J. S. Schilling, S. Gärtner and D. Schweitzer, Solid State Commun. **67**, 981 (1988).
11. H.Urayama, H.Yamochi, G.Saito, T.Sugano, M.Kinoshita, T.Inabe, T.Mori, Y.Maruyama and H.Inokuchi : Chem. Lett. 1057 (1988)
12. K.Oshima, T.Mori, H.Inokuchi, H.Urayama, H.Yamochi and G.Saito, Phys. Rev. **B38**, 938 (1988).
13. S. Ravy, J-P. Pouget and R. Moret, these proceedings
14. H. Kusuhara, Y. Sakata, Y. Ueba, K. Tada, M. Kaji and T. Ishiguro, these proceedings.

Anomalous Transport Behavior in κ-(BEDT-TTF)$_2$Cu(NCS)$_2$

K. Oshima[1], R.C. Yu[2], P.M. Chaikin[2], H. Urayama[3], H. Yamochi[3], and G. Saito[3]

[1]Okayama University, 3-1-1 Tsushima-naka, Okayama 700, Japan
[2]Dept. of Physics, Princeton University, Princeton, NJ 08544, USA
[3]Institute for Solid State Physics, University of Tokyo, Roppongi, Tokyo 113, Japan

Abstract. Anomalous transport behavior is reported in the title compound including superconducting critical fields, temperature dependence of the normal resistance, and negative magnetoresistance. The experimental condition for the parallel critical field has been extended up to 24 T and down to .5 K. The anomalous critical field behavior at low temperatures is confirmed in the different configurations. The negative magnetoresistance and the anomalous temperature dependence of the normal resistance in the magnetic field have been observed. The results are discussed in relation to the magnetic impurity effect.

1. Introduction

κ-(BEDT-TTF)$_2$Cu(NCS)$_2$ is the superconductor which has the highest critical temperature in the organic materials to date (10.4 K).[1] We have shown that the electronic structure can be understood using the simple free electron model as confirmed by the Shubnikov-de Haas (SdH) effect.[2] But the superconducting behavior is anomalous, and still to be explained. As this system contains Cu atoms, we should be careful about the possible effect of the magnetic interaction. In this paper, we present new experimental results and discussions about the magnetic field effect on transport properties.

2. Critical Field

The curious behavior of the upper critical field has been observed and already reported.[3] The parallel critical field ($H_{c2\|}$) becomes very high due to the 2-dimensional character. Therefore we extended the measurement to higher fields using the Bitter magnet at the National Magnet Laboratory (U.S.A.). The newly obtained data are shown in Fig.1. They are consistent with the earlier data at low fields as shown in the figure. A peculiar upward curvature has been noticed in the perpendicular critical field ($H_{c2\perp}$). One of our new observations is a non-saturating behavior of $H_{c2\|}$ at the lowest temperatures. Unfortunately the full superconducting-to-normal transition could not be observed below 1 K due to insufficient field, but the critical field seems to exceed the Pauli limit. We know that the lower critical field (H_{c1}) is also anomalous as shown in Fig.1. Therefore all of the critical fields seem to show an anomalous behavior.

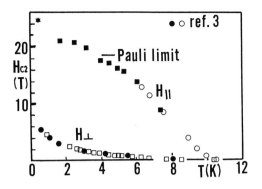
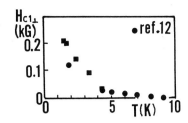

Fig. 1. Temperature dependence of the upper and lower critical fields.

3. Negative Magnetoresistance

The existence of peculiar negative magnetoresistance has been known from the very beginning of the study of this material.[2] Newly measured results are shown in Fig.2a. Recently, several articles have reported negative magnetoresistance behavior in the organic materials [4, 5] One of the proposed mechanisms is weak localization, and the other is internal boundary scattering. The negative magnetoresistance in our case has been confirmed to exist irrespective of current directions when the magnetic field is perpendicular to the two-dimensional crystal plane. Therefore our results cannot be understood by the boundary scattering mechanism. The strength of our negative magnetoresistance seems too large to be explained by the localization effect. The SdH effect revealed the scattering rate at the lowest temperature to be lower than 10^{12} sec^{-1}(Dingle temperature < 1 K), therefore it seems that the system is far from the localization condition.[6]

4. Normal Resistance in the Field

The temperature dependence of the sample resistance in the magnetic field is shown in Fig.2b. One of the striking features is the resistance minimum in the low field conditions, which reminds us of the Kondo effect in the transition

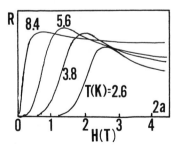
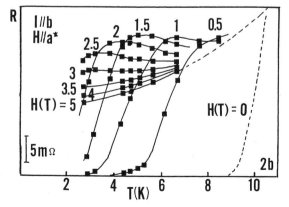

Fig. 2. Negative magnetoresistance and anomalous normal resistance in the field.

metal alloys. We notice a lnT-like temperature dependence at low fields, though the range is restricted by the superconducting transition. Another possibility might be the SDW transition near Tc. But in this case, we cannot expect the suppression of the resistance anomaly due to the magnetic field. We also notice the temperature dependence in high fields. It is nearly expressed by the T^2 law except for a residual part. This temperature dependence can also be seen above Tc in zero field.[1]

5. Magnetic Impurity Effect

We should next discuss the Kondo effect. If we assume a magnetic scattering contribution to exist in the resistivity, the negative magnetoresistance can be understood as due to the suppression of the magnetic scattering rate at higher magnetic fields through the polarization of the impurity magnetic moment. If this is the case, we should identify the origin of impurity. Possible evidence for the existence of a magnetic impurity would be an E.S.R. signal observed even in the superconducting state.[9] Another interesting result which seems to be related to magnetism is the N.M.R. relaxation enhancement observed around 5 K.[8] The temperature corresponds to the temperatures where critical fields start to show an anomaly. Therefore we prepared purified samples to exclude possible magnetic impurities. The transition temperature and the transition width are actually dependent on the sample preparation process.[7] We could obtain a sample with a sharp superconducting transition width from the purified batch. The transition temperature also seems to depend on the quality of the crystal. We are not sure that the external impurity is the only deciding factor for the transition temperature. We still need a systematic study to confirm the impurity effect. We have observed the critical temperature difference between deuterated and protonated crystals.[10] We should reconfirm such a difference using actually purified samples. The systematic study is really difficult as H.Kusuhara et al. showed by demonstrating the possibility of raising the critical temperature by elongating crystals.[11] We sometimes experience a double step superconducting transition with a high transition temperature. They are not usually reproducible. Therefore we should be careful about the effect of stress on samples. The E.S.R. result seems to suggest the existence of BEDT radicals in the crystal. [9] Therefore we should also be careful about crystal imperfections. If an electron can localize at the sulfur atom it can be a candidate for the E.S.R. signal and for a magnetic scatterer. Lastly, we should mention the Kondo effect on the critical field. It can actually reproduce the anomalous critical field, if we assume a temperature-dependent pair-breaking mechanism by spin-flip scattering. But it seems that this is not the case, as we would have to assume a much greater transition temperature suppression due to impurities than we have observed. The anomalous behavior of the critical field is probably related to the flux pinning mechanism, as can be suspected from the temperature dependence of the diamagnetism.[12] Below 4 K, the flux pinning seems stronger than that above. We suspect that magnetic interaction is responsible for the flux pinning mechanism. As we have seen, the parallel critical field behavior does not exclude the possibility of non-singlet superconductivity. There exists a proposal to understand this system as a highly correlated system.[13, 14] The temperature dependence of normal resistance is consistent with such a viewpoint.

In summary, we have noticed an important role of impurities in the physical properties of our system. Though the systematic study is very cumbersome, the physics contained in our system is fundamental and similar to the problems proposed in the high Tc and heavy-fermion superconductors.

One of the authors (K.O.) would like to thank Princeton University for hospitality during his visit to the department of Physics. We would also like to thank the Francis Bitter National Magnet Laboratory.

References

[1] H.Urayama et al., Chem.Lett.**1988** 55
[2] K.Oshima et al., Phys.Rev.B**38** 938 (1988)
[3] K.Oshima et al., J.Phys.Soc.Jpn.**57**,730 (1988)
[4] J.P.Ulmet et al., Solid State Commun.**67**,145 (1988)
[5] F.L.Pratt et al., Phys. Rev. Lett.**61**, 2721 (1988)
[6] Y.Hasegawa and H.Fukuyama, J.Phys.Soc.Jpn.**55** 3717 (1986)
[7] H.Mori, et al., these proceedings.
[8] T.Takahashi et al., Physica C153-155,487 (1988)
[9] H.Urayama et al., Chem.Lett.**1988** 1057
[10] K.Oshima et al., Synthetic Metals,**27**,A473 (1988)
[11] H.Kusuhara et al., these proceedings.
[12] M.Tokumoto et.al.,Synthetic Metals,**27**,A305 (1988)
[13] K.Kanoda et al., these proceedings.
[14] N.Toyota et al., these proceedings.

STM Study of (BEDT-TTF)$_2$Cu(NCS)$_2$ Surface

M. Yoshimura[1], N. Ara[1], M. Kageshima[1], R. Shioda[1], A. Kawazu[1],
H. Shigekawa[2], H. Mori[3] [1], M. Oshima[3] [2], H. Yamochi[3], and G. Saito[3]

[1]Department of Applied Physics, University of Tokyo,
 Bunkyo-ku, Tokyo 113, Japan
[2]Institute of Materials Science, University of Tsukuba,
 Tsukuba Science City 305, Japan
[3]Institute for Solid State Physics, University of Tokyo,
 Roppongi, Minato-ku, Tokyo 106, Japan

   The surface structure of an organic superconductor (BEDT-TTF)$_2$Cu(NCS)$_2$ was observed by means of a scanning tunneling microscope (STM) in air and at room temperature. Observed periodic structures were in good agreement with the two dimensional BEDT-TTF molecular structure projected onto its crystal b-c plane. Strong tunneling voltage dependence of STM image was also observed.

1. Introduction

(BEDT-TTF)$_2$Cu(NCS)$_2$ crystal has a monoclinic structure which consists of alternating stacking of BEDT-TTF molecular layers and Cu(NCS)$_2$ molecular layers (Fig.1, a=16.248Å, b=8.440Å, c=13.124Å, β=110.30 degree, V=1688.0Å^3)[3]. Through the electron transfer from the BEDT-TTF layer to the Cu(NCS)$_2$ layer, the former has two dimensional

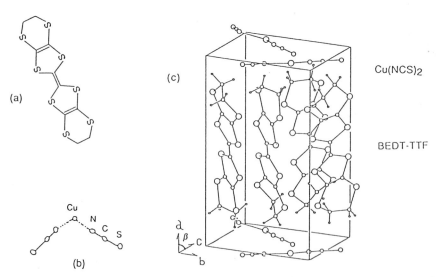

Fig.1 (a) Molecular structure of BEDT-TTF. (b) Molecular structure of Cu(NCS)$_2$. (c) Crystal structure of (BEDT-TTF)$_2$Cu(NCS)$_2$ determined by X-ray diffraction, which has a monoclinic structure[3].

conductivity in its crystal b-c plane, while the latter becomes insulating. This crystal shows superconductivity at Tc=10.4K under ambient pressure. In order to reveal the microscopic mechanism of this high-Tc organic superconductor, we must have knowledge of the atomically resolved electronic and geometric structure of the crystal.

2. Experimental Results and Discussion

STM observations were performed over the crystal b-c plane in air and at room temperature. A platinum-iridium tip was used to prove the surface. Figure 2 shows a gray scale image obtained by the constant current mode (66mV tip to sample, 4.6nA, scan area is 60x60-$Å^2$). The voltage dependence of tunneling current indicates that the surface has a metallic property (images were the same on both signs of the bias voltage), and the periodic structure in Fig.2 is expected to correspond to the surface electronic structure of BEDT-TTF layers[4].

The arrangement of BEDT-TTF molecules, projected onto the b-c plane, is shown in Fig.3[5]. When the crystal axis determined by X-ray diffraction is taken into consideration two choices for the unit cell are possible (drawn in Fig.2). In addition, STM images showed a strong tunneling voltage dependence (Fig.4), which corresponds to the change of local distribution of electronic states according to their energy. In order to give a comprehensive explanation of the observed images, we shall have to perform further experiments and detailed calculations on the local density of states of this crystal.

The mechanism of superconductivity for organic materials[6] will be revealed by a systematic STM study of those materials. For comparison an STM image of (BEDT-TTF)$_2$KHg(SCN)$_4$ surface acquired over the a-c plane is shown in Fig.5 together with the crystal structure determined by X ray diffraction[7,8].

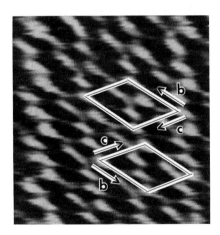

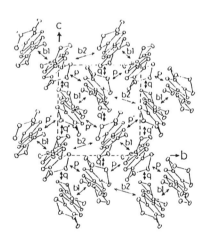

Fig.2 Gray scale image of (BEDT-TTF)$_2$Cu(NCS)$_2$ surface obtained by the constant current mode (66mV tip to sample, 4.6nA). Scan area is 60x60-$Å^2$. Two possible unit cells are drawn.

Fig.3 BEDT-TTF molecular structure projected onto the crystal b-c plane[5]. The unit cell is drawn.

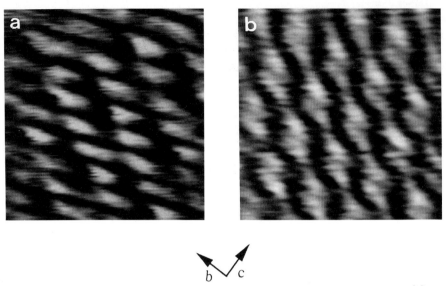

Fig.4 Tunneling voltage dependence of STM images (a) 180mV, 30x40-Å^2. (b) 300mV, 50x50-Å^2.

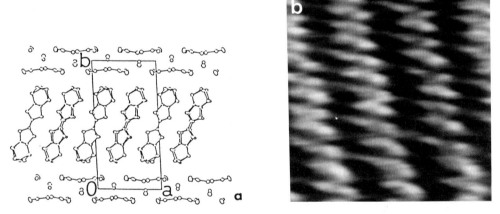

Fig.5 (a) Crystal structure of (BEDT-TTF)$_2$KHg(SCN)$_4$ determined by X-ray diffraction[7]. (b) Surface image obtained by STM (90mV, 1.01nA, 75x75-Å^2).

3. Conclusion

In summary, we obtained STM images of (BEDT-TTF)$_2$Cu(NCS)$_2$ surface structure. The observed images indicate that the surface is metallic and the periodic structure observed was in good agreement with the BEDT-TTF molecular structure projected onto the crystal b-c plane. STM images showed a strong tunneling voltage dependence.

It is a pleasure to acknowledge many helpful discussions with Prof. T.Mori, Institute for Molecular Science.

References

[1] Present address: International Superconductivity Technology Center, Atsuta, Nagoya 456
[2] On leave of absence from Japan Carlit Co., Ltd., Shibukawa Gunma 377
[3] H.Urayama, H. Yamochi, G. Saito, S. Sato, A. Kawamoto, J. Tanaka, T. Mori, Y. Maruyama, and H. Inokuchi, Chem.Lett.,1988(1988)463.
[4] M. Yoshimura, N. Ara, M. Kageshima, R. Shioda, A. Kawazu, H. Shigekawa, and S. Hyodo, J.Vac.Sci.Technol.A, to be published
[5] K. Oshima, T. Mori, H. Inokuchi, H. Urayama, H. Yamochi, and G. Saito Phys.Rev.B,38(1988)938
[6] K. Yamaji, Synthetic Metals, 27(1988)A115
[7] M. Oshima, H. Mori, G. Saito, and K. Oshima, Chem.Lett.,1989(1989)1159
[8] M. Yoshimura, N. Ara, M. Kageshima, R. Shioda, A. Kawazu, H. Shigekawa, M.Oshima, H. Yamochi, and G. Saito, Jpn.J.Appl.Phys., to be submitted

Crystal Growth and Properties of $(BEDT-TTF)_2Cu(NCS)_2$

Y. Ueba, T. Mishima, H. Kusuhara, and K. Tada

R&D Group, Sumitomo Electric Industries, 1-1-3 Shimaya,
Konohanaku, Osaka 554, Japan

Abstract. The crystal growth for $(BEDT-TTF)_2Cu(NCS)_2$ was examined by changing such factors as amount of alcohol added into a solvent 1,1,2-trichloroethane, temperature, electric current and electrode size. It was found that the addition of alcohol into the solvent enhances the rate of crystal growth significantly. A typical crystal size of $5 \times 1.5 \times 0.05$ mm^3 was obtained on 1 mm diameter Pt wire electrode in 1 to 2 weeks at a temperature of 20°C under a current of 0.5 μA. The largest crystal was about 10 mm long and 3 mm wide.

1. Introduction

An organic superconductor with a superconducting transition temperature higher than 10 K, $\kappa\text{-}(BEDT-TTF)_2Cu(NCS)_2$ was recently synthesized./1/ Chemical and physical properties of the compound have been extensively studied./2/ However, crystal growth of the compound has been reported on very little. The effect on the crystal growth of solvents, molecules like water added in a solvent, and temperature have been investigated. /3,4/ We studied the crystal growth for $(BEDT-TTF)_2Cu(NCS)_2$, changing factors such as amount of alcohol added into solvent, temperature, current, and electrode size.

2. Experimental

Starting Materials. BEDT-TTF used in this study was purchased from Tokyo Kasei Co. and purified by recrystallization from monochlorobenzene(m.p. 241°C). Copper(I)thiocyanate and 18-Crown-6 ether were purchased from Aldrich Chemical Co. and used without further purification. Potassium thiocyanate was obtained from Kanto Chemical and purified by recrystallization from ethanol and drying at 150°C for 1 hr and 200°C for 10 min. The solvent 1,1,2-trichloroethane was purified by washing with H_2SO_4, aq NaOH, and distilled water, drying over $CaCl_2$ and distilling just before use(b.p.113-113.5°C). Methanol(Kokusan Chemical, H_2O max.0.03%) and ethanol(Kanto Chemical, 99.5%) were used without further purification. Removal of water from ethanol was done by drying over $CaSO_4$ and by azeotropic distillation with dichloromethane.

Crystal Growth. Crystal growth of $(BEDT-TTF)_2Cu(NCS)_2$ was carried out using an electrochemical crystallization method. Air in glass cells was replaced with nitrogen gas before use. BEDT-TTF(30 mg), CuSCN(70 mg), KSCN(126 mg), and 18-Crown-6 (210 mg) were put into each cell under N_2 gas flow. One hundred ml of 1,1,2-trichloroethane with varied amounts of alcohol was added into each cell and stirred overnight under exclusion of light. After removal of the stirrer tip, each cell was kept still to precipitate a small amount of undisolved starting material. Platinum electrodes with Teflon holders were introduced into each cell under N_2 gas flow. The cells were placed in an incubator and kept at a constant temperature(10, 20, 30, or 50°C). Constant current(0.5, 1,5,

10, or 50 µA) was then applied. The size of the crystals was monitored during crystal growth. After a given period of time crystals were harvested and washed with methanol followed by drying under vacuum.

Property Measurements. Electrical resistivity was measured by the usual four terminal method using gold paste and 20 µm gold wire. Susceptibility was measured by using a SQUID susceptometer Hokusan HSM-2000. X-ray structural analysis was carried out with a Rigaku RASA 5R system.

3. Results and Discussion

The effects of alcohol in a solvent, 1,1,2-trichloroethane on crystal growth were studied and the results are shown in Tables 1 and 2. It is clear that the addition of methanol and ethanol in the solvent is effective to enhance both the nucleation and the rate of crystal growth of (BEDT-TTF)$_2$Cu(NCS)$_2$. Without alcohol, no crystals form or the nucleation occurs only after a few weeks. The addition of 1 to 2 wt% of ethanol gives the fastest rate and the largest crystals. A typical size of 5 x 1.5 x 0.05 mm^3 was obtained in a growth period of 1 to 2 weeks at a temperature of 20°C under a current of 0.5 µA. Figure 1 shows a photograph of crystals growing on a Pt wire electrode in a cell. Figure 2 shows a photograph of a typical crystal obtained.

The effects of temperature on crystal growth are shown in Table 3. At 10°C, crystals grow very slowly and small size crystals less than 1 mm long are obtained. Those crystals have also a platelet shape and the same superconducting property as κ-salt by susceptibility measurement. At 30°C, only a few crystals grow and at 50°C no crystals are obtained. The best temperature for crystal growth is 20°C.

The effects of electric current on crystal growth at 20°C are shown in Table 4. A range of current from 0.5 to 50 µA was examined. As current increases, nucleation tends to occur in shorter time and at many more points. At 10 µA a large crystal more than 10 mm long was obtained. Crystals grown under current weaker than 10 µA show superconductivity, however at 50 µA feathery crystals form and show no superconductivity.

The effects of electrode size are shown in Table 5. When Pt wire electrode with 2 mm diameter was used instead of 1 mm dia. electrode, the growth rate decreased and a number of small crystals were obtained. This may be a result of decreasing the current density. In order to obtain appropriately large size crystals the use of 1 mm dia. Pt wire electrode under the current of 0.5 µA is recommended.

Table 1. Effect of the addition of ethanol on the crystal growth of (BEDT-TTF)$_2$Cu(NCS)$_2$.

Amount of ethanol (wt%)	Time of growth (days)				
	1	2	5	7	14
0	none	none	nucleation	growing	0.2mm
0.5	-	-	1mm	1.5mm	1.7mm
1.0	nucleation	growing	3mm	5 mm	5 mm
2.0	nucleation	growing	4mm	5 mm	5 mm
5.0	none	nucleation	growing	0.2mm	0.5mm
10	none	nucleation	growing	0.1mm	0.5mm
20	-	nucleation	-	-	0.5mm

Length in the table shows the length of the largest crystal.
The conditions of growth: 20°C, 0.5µA, 1,1,2-trichloroethane.

Table 2. Effect of the addition of methanol on the crystal growth of $(BEDT-TTF)_2Cu(NCS)_2$.

Amount of methanol (wt%)	Time of growth (days)			
	1	4	7	14
0	none	none	growing	0.2mm
1.0	nucleation	0.8mm	1.1mm	1.6mm
2.0	nucleation	0.4mm	0.5mm	1.0mm
5.0	nucleation	0.2mm	0.2mm	0.6mm

Figure 1. A photograph of $(BEDT-TTF)_2Cu(NCS)_2$ crystals on 1 mm diameter Pt wire electrode in a cell.

Figure 2. A photograph of $(BEDT-TTF)_2Cu(NCS)_2$ crystal obtained which has a size of 6 x 3 x 0.1 mm^3.

Table 3. Effect of temperature on the crystal growth of $(BEDT-TTF)_2Cu(NCS)_2$.

Temperature (°C)	Time of growth (days)				
	1	2	5	7	14
10	nucleation	0.1mm	0.2mm	0.2mm	1mm
20	nucleation	growing	4 mm	5 mm	5mm
30	none	none	none	1 mm	-
50	none	none	none	none	none

The conditions of growth: 0.5μA, 2wt% ethanol in 1,1,2-trichloroethane.

Table 4. Effect of electric current on the crystal growth of $(BEDT-TTF)_2Cu(NCS)_2$.

Current (μA)	Time of growth (days)				
	1	2	5	7	14
0.5	nucleation	growing	4mm	5mm	5mm
1.0	1mm	1.5mm	1.5mm	2mm	3mm
5.0	2mm	-	2mm	3mm	-
10.0	2mm	-	-	10mm	-
50	1mm	2mm	-	-	-

The conditions of growth: 20°C, 2wt% ethanol in 1,1,2-trichloroethane.

Table 5. Effect of electrode size on the crystal growth of $(BEDT-TTF)_2Cu(NCS)_2$.

Electrode	Time of growth (days)					
	1	2	5	7	14	30
Pt wire(1mm dia.)	nucleation	3mm	4mm	5mm	6mm	10mm
Pt wire(2mm dia.)	nucleation	growing	growing	2mm	2mm	4mm

The conditions of growth; 20°C, 0.5μA, 2wt% ethanol in 1,1,2-trichloroethane.

A single crystal X-ray structural analysis shows that the crystal has κ-type structure. The crystallographic data, monoclinic, a=16.219 b=8.436, c=13.107, β=110.313, V=1681.8 $Å^3$, are in agreement with reported ones/1/ with the exception that a disorder in two of the ethylene groups was resolved(R=0.041).

The results of the resistivity and the susceptibility measurements are shown in Figures 3 and 4, respectively. Crystals used in these measurements are those obtained under optimum conditions. The Tc at onset was 9.9-11.1 K, which is consistent with the reported Tc for κ-$(BEDT-TTF)_2Cu(NCS)_2$. Superconducting transition occurs within 2 K. A resistivity peak at around 90 K which was seen in previous studies can be seen in crystals produced in this experiment. The resistivity ratio just above Tc in Fig.3 was one order of magnitude high compared to the reported value. The reason for high residual resistivity is not

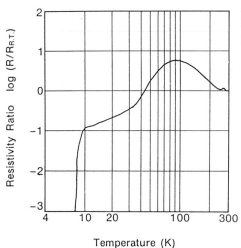

Figure 3. Temperature dependence of the resistivity of (BEDT-TTF)$_2$Cu(NCS)$_2$.

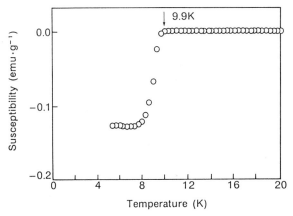

Figure 4. Temperature dependence of the susceptibility of (BEDT-TTF)$_2$Cu(NCS)$_2$.

clear at this moment. Investigation on the effects of purity of starting materials is underway.

4. Conclusions

(BEDT-TTF)$_2$Cu(NCS)$_2$ crystals grow faster with the addition of methanol or ethanol into a solvent, 1,1,2-trichloroethane. One to two weight percent addition of ethanol into the solvent gives the most effective result on the crystal growth of (BEDT-TTF)$_2$Cu(NCS)$_2$. Crystals with a typical size of 5 x 1.5 x 0.05 mm^3 are obtained in 1 to 2 weeks. The effects of temperature, electric current and electrode size on the crystal growth were also examined. The optimum condition is 20°C, 0.5μA, and 1 mm dia. Pt wire electrode.

Acknowledgement. We are greatly indebted to Professor Gunzi Saito for instructing us on the preparation method of $(BEDT-TTF)_2Cu(NCS)_2$ crystals and suggesting that we study the effects of the addition of alcohol on the crystal growth. We are grateful to Dr. Mike Extine and Rigaku for X-ray single crystal structural analysis.

References

1. H. Urayama, H. Yamochi, G. Saito, K. Nozawa, T. Sugano, M. Kinoshita, S. Sato, K. Oshima, A. Kawamoto, and J. Tanaka, Chem. Lett.,55-58, (1988).
2. G. Saito, H. Urayama, H. Yamochi, and K. Oshima, Synth. Met.,27,A331, (1988).
3. H. Anzai, M. Tokumoto, N. Kinoshita, K. Murata, K. Takahashi, T. Ishiguro, 2p-G4-1, Spring Meeting of The Physical Society of Japan 1988; H. Anzai, M. Tokumoto, T. Iwai, and T. Uchida, 28p-P-1, Spring Meeting of The Physical Society of Japan, 1989.
4. N. Kinoshita, K. Takahashi, K. Murata, M. Tokumoto and H. Anzai, Solid State Comm., 67(5), 465-470 (1988).

Synthesis, Crystal Structure and Properties of (BEDT-TTF)$_3$CuCl$_4$·H$_2$O

M. Kurmoo[1], T. Mallah[1], P. Day[1], I. Marsden[2], M. Allan[2], R.H. Friend[2], F.L. Pratt[3], W. Hayes[3], D. Chasseau[4], J. Gaultier[4], and G. Bravic[4]

[1]Inorganic Chemistry Laboratory, South Parks Road, Oxford OX1 3QR, UK
[2]Cavendish Laboratory, Madingley Road, Cambridge CB3 0HE, UK
[3]Clarendon Laboratory, University of Oxford, Oxford OX1 3PU, UK
[4]Laboratoire de Cristallographie et Physique Cristalline,
 Université de Bordeaux I, Cours de la Liberation, F-33405 Talence, France

Abstract. The synthesis of (BEDT-TTF)$_3$CuCl$_4$·H$_2$O and its crystal structure, electrical transport, magnetic and optical properties are reported. The structure consists of alternate layers of BEDT-TTF and of dimeric units of CuCl$_4$·H$_2$O. It shows metallic behaviour (10-300K) and an anisotropy of 1:7 within the plane. EPR shows two signals belonging to the two types of electrons, conduction electrons and electrons localised on the Cu, with very little interaction between them. The bandwidth is estimated to be 0.6 and 0.3 eV parallel and perpendicular to the a-c axis respectively.

1. Introduction

The current interest in charge transfer salts of the type D$_2$X with organic donors, D = TMTSF, BEDT-TTF, DIMET, and MDT-TTF, and with inorganic mono-anions, X, may be attributed to the observation of superconductivity for several of these salts [1]. The highest critical temperature is 11 K for κ-(BEDT-TTF)$_2$Cu(NCS)$_2$ [2]. The structure-property relationships of the various classes of compound have been drawn by several groups. In most cases, the anions used are non-magnetic inorganic mono-anions and the use of magnetic and di-anions is sparse. The few examples in the BEDT-TTF family are those of (BEDT-TTF)$_3$(MnCl$_4$)$_2$ [3], (BEDT-TTF)$_2$FeCl$_4$ and (BEDT-TTF)FeBr$_4$ [4], which are semiconductors, and the interaction of the electrons on the cations with those on the anions is minimal. We report here the synthesis of the first example of a conducting salt that contains a magnetic anion and its crystal structure [5], electrical, magnetic and optical properties.

2. Experimental Details

Hexagonal crystals of (BEDT-TTF)$_3$CuCl$_4$·H$_2$O were obtained by electrocrystallisation of BEDT-TTF in a 9:1 mixture of CH$_2$Cl$_2$ and CH$_3$CN saturated with [(CH$_3$)$_4$N]$_2$CuCl$_4$. A second crystal phase, which is a semiconductor, was also obtained under the same conditions [6]. The resistivity was measured by four probe low frequency AC method; contacts were made by gold wire attached with silver paint to evaporated gold pads. The magnetic properties were studied by EPR on oriented single crystals. Reflectance was measured on single crystals with the polarisation parallel and perpendicular to the long axis (a-c) of the crystal.

3. Experimental Results and Discussion

The crystal belongs to the triclinic system, $P\bar{1}$, a=16.634, b=16.225, c=8.980Å, α=90.72, β=93.24, γ=96.76°, V=2402Å³, Z=2. The crystal structure consists of stacks of BEDT-TTF parallel to the c-axis and forming layers in the ac-plane (Fig. 1a). These layers are interleaved with the inorganic anions and the water molecules. There are three independent BEDT-TTF and two crystallographically different stacks; one contains one type of BEDT-TTF and the other has an alternation of the remaining two. The stacks are arranged in an ABBABBA fashion. The BEDT-TTF within a layer are almost coplanar along the a-c axis with shorter S-S contacts than those along the stacks. The anions lie in planes parallel to the layers of the donors. Two $CuCl_4$ anions are bridged by two water molecules to form isolated units (Fig. 1b).

The oxidation state of the Cu is two as evidenced by the close bond length of the Cu-Cl of the title compound (2.25Å average) to that of $CuCl_2.6H_2O$ (2.29Å), and the flattened tetrahedral coordination geometry (Cl-Cu-Cl angle of 150°), which is widely observed in $CuCl_4^{2-}$ salts [7], and is a result of the Jahn-Teller distortion. From this stoichiometry and the Cu in 2+ state, this compound will have a 2/3 filled band, as for

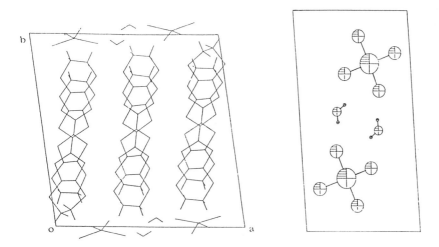

Figure 1: (a) Projection along c and (b) view of the anion dimer.

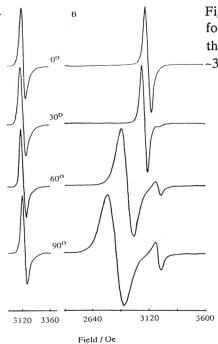

Figure 2: Variation of the EPR spectrum for rotation of the crystal about the b axis (A) and the a–c axis (B) at ~35 K.

$(BEDT-TTF)_3Cl_2 \cdot 2H_2O$ [8]. It is worth noting that within the donor layers the arrangement of the BEDT–TTF molecules of these two materials is almost identical.

The conductivity is anisotropic in the plane, 140 S cm^{-1} along the a–c axis and 20 S cm^{-1} perpendicular to that axis. It decreases on lowering temperature to a minimum at ~10K. Preliminary magneto-resistance measurements at 200 mK shows a moderate $\Delta R/R$ of +25% in a field of 20 T. The thermopower is constant (ca. 0 μV/K) down to 200 K and increases to a maximum of 11 μV/K at ~60K.

The EPR spectra were recorded as a function of orientation with respect to the static and microwave fields, and of temperature. The spectra are characterised by two signals at g~2 with intensity ratio of 1:6. On rotation of the crystal (Fig. 2), the g-value of the strong signal increases from to 2.05 to 2.29 and the linewidth increases from 40 to 140 G while the weak signal remains at g = 2.03 and the width at ca. 50 G. Thus they are assigned, by analogy to the g-value and linewidths of known $CuCl_4^{2-}$ salts and other BEDT–TTF compounds, to the Cu and the conduction electrons, respectively. On lowering temperature, the intensity of the Cu signal increases while the intensity ratio decreases to 1:40 at ~35 K, and the linewidth of both signals decrease by ca. 25% at 4 K. The spin susceptibility as estimated by the intensity of the Cu signal closely obeys a $T^{-0.8}$ power law. The interaction between the conduction electron and the localised spin on the Cu atoms must be very small, since in a bottleneck regime one would expect a large change in the g-value with temperature [9]. In the present case, the g-value changes at a rate of -4×10^{-5} / K. With the propagation of the microwave parallel to the a-c axis, a Dysonian lineshape is observed, which we have analysed using the flat plate geometry approximation of Feher and Kip [10] to give a microwave conductivity of 120 S cm^{-1} at room temperature and increasing at low temperature.

The optical reflectivity spectra are characterised by a Drude edge at ca. 2500 cm^{-1} with a superimposed weak band at 4000 cm^{-1}. The Drude-Lorentz fits of the data give a screened plasma frequency of 9400 cm^{-1} and width 2700 cm^{-1} for E $\parallel$ a-c and 8900 and 3400 cm^{-1} for E $\perp$ a-c. Kramers-Kronig analysis gives a conductivity along the a-c axis of ~600 S cm^{-1} at 400 cm^{-1}. The bandwidth is estimated to be 0.6 eV along the a-c axis and 0.3 eV perpendicular to a-c.

4. Conclusion

A highly conducting salt containing an inorganic anion carrying a magnetic moment has been prepared. The structure is closely related to the only 3:2 organic superconductor, $(BEDT-TTF)_3Cl_2 \cdot 2H_2O$. Good agreement on the conductivity is obtained between four probe, EPR and optical measurements.

References

[1] Proceedings of the International Conferences on Science and Technology of Synthetic Metals, *Synth. Met.*, 27-29 (1988).
[2] H.Urayama, H.Yamochi, G.Saito, S.Sato, A.Kawamoto, J.Tanaka, T.Mori, Y.Murayama and H.Inokuchi, *Chem. Lett.*, 463 (1988).
[3] T.Mori and H.Inokuchi, *Bull. Chem. Soc. Jpn.*, 61, 591 (1988).
[4] T.Mallah, C.Hollis, S.Bott, P.Day and M.Kurmoo, *Synth. Met.*, 27, A381 (1988).
[5] D.Chasseau, G.Bravic, J.Gaultier, M.Kurmoo, T.Mallah and P.Day, *Acta Cryst.*, submitted (1989).
[6] M.Kurmoo, unpublished.
[7] A.F.Wells in "Structural Inorganic Chemistry", (CP), 1136 (1984).
[8] M.Kurmoo, M.J.Rosseinsky, P.Day, P.Auban, W.Kang, D.Jérome and P.Batail, *Synth. Met.*, 27, A425 (1988); P.Day, M.Kurmoo, D.R.Talham, P.Day, D.Chasseau and D.Watkin, *J. Chem. Soc., Chem. Commun.*, 88 (1988).
[9] S.E.Barnes, *Adv. Phys.*, 30, 801 (1980).
[10] G.Feher and A.F.Kip, *Phys. Rev.*, 98, 337 (1955).

Electronic Properties of Charge Transfer Complexes of BEDT-TTF and Related Donors with Transition Metal Halides

T. Enoki[1], I. Tomomatsu[1], Y. Nakano[1], K. Suzuki[1], and G. Saito[2]

[1]Department of Chemistry, Faculty of Science, Tokyo Institute of Technology, Ookayama, Meguro-ku, Tokyo 152, Japan
[2]Institute for Solid State Physics, University of Tokyo, Roppongi, Minato-ku, Tokyo 106, Japan

Abstract. BEDT-TTF and C_1TET-TTF charge transfer complexes were synthesized with $FeCl_4^-$, $CoCl_4^{2-}$ and $CoBr_4^{2-}$ anions by electrocrystallization and characterized by electrical and ESR measurements. A semiconductive $(C_1TET-TTF)FeCl_4$ was found to have dimerized donors behaving as a singlet-triplet system. A metallic compound was obtained in BEDT-TTF complexes with $CoCl_4^{2-}$.

1. Introduction

Recently, organic superconductors have been intensively investigated in organic charge transfer complexes with BEDT-TTF, DMET and other donors. Most of the organic superconductors consist of nonmagnetic inorganic anions such as ClO_4^-, PF_6^-, I_3^-, IBr_2^-, etc. On the other hand, there have been quite a few studies on organic conductors with magnetic anions such as 3d transition metal halides[1-2]. In these organic conductors, magnetic moments of d-electrons on the anions may be able to interact with conduction electrons on the donor π-electronic systems, so that we expect cooperative phenomena induced by π-d interactions. From this point of view, we have been surveying organic conductors with magnetic anions to find good candidates for the conductors with π-d interactions. In this paper, we present synthesis and electrical and magnetic properties of organic charge transfer complexes of BEDT-TTF and C_1TET-TTF with 3d transition metal halides anions $FeCl_4^-$, $CoCl_4^{2-}$ and $CoBr_4^{2-}$.

BEDT-TTF   C₁TET-TTF

2. Experimental Results and Discussion

Samples of BEDT-TTF and C_1TET-TTF charge transfer complexes were prepared by means of electrocrystallization. Complexes with $FeCl_4^-$ anion were crystallized using a supporting electrolyte of $[(CH_3)_4N]FeCl_4$ in 1,1,2-trichloroethane or mixed solvent of 1,1,2-trichloroethane with 5% ethanol. Complexes with CoX_4^{2-} (X=Cl,Br) were obtained using $[(C_2H_5)_4N]_2CoX_4$ electrolyte in benzonitrile.

2.1 C_1TET-TTF with $FeCl_4^-$ anion

Two kinds of crystal forms were obtained for C_1TET-TTF complexes with $FeCl_4^-$; plate and needle crystals. X-ray crystal structure analysis for the plate crystal gave a monoclinic (C2/c) structure with the stoichiometry of $(C_1TET-TTF)FeCl_4$

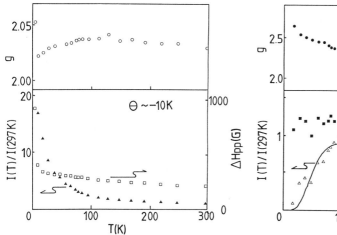

Fig.1. Temperature dependence of intensity I, g-value and linewidth ΔH_{pp} of Fe^{3+} ESR signals in $(C_1TET-TTF)-FeCl_4$ (needles).

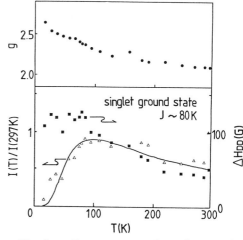

Fig.2. Temperature dependence of donor ESR signals in $(C_1TET-TTF)FeCl_4$ (platelet).

where a=26.421, b=11.777, c=14.023Å, β=103.9°, V=4235.4Å^3, Z=8 [3]. Four dimers of $C_1TET-TTF$ donors are involved in the unit cell. Both the crystals are semiconductive; Ea=0.60eV, ρ_{rt}=4.8x10^{6}Ωcm for the needle crystals and Ea=0.43eV, ρ_{rt}=4.4x10^{6}Ωcm for the plate crystals.

The powder ESR spectra for Fe^{3+} ions have g=2.02 and ΔH_{pp}=210G or 2.04 and 510G for the needle or the plate crystals, respectively. Temperature dependence of Fe^{3+} ESR spectra for the needle crystals are shown in Fig. 1. The intensity above about 10K shows Curie-Weiss behavior with the Weiss temperature of θ=-10K, suggesting the presence of antiferromagnetic interaction among Fe^{3+} ions. Below 10K, it starts deviating from the Curie-Weiss law. ΔH_{pp} has a slight increase with the lowering of temperature, and below about 10K, it abruptly shifts upward. The g-value also shows an anomaly below about 10K. The abrupt change in the ESR spectra below about 10K are considered to be associated with the development of antiferromagnetic ordering. For the plate crystal, the temperature dependence of Fe^{3+} ESR spectra have similar behavior to that of the needle crystals. In the $(C_1TET-TTF)FeCl_4$ plate crystals, ESR spectra associated with $C_1TET-TTF$ donors were also observed down to liquid helium temperature (g=2.08 and ΔH_{pp}=49G at room temperature). Temperature dependence is shown in Fig. 2 for the donor signal. The intensity increases with the lowering of temperature in the high temperature region above about 100K, while, below that temperature, it steeply decreases to zero as temperature is lowered. ΔH_{pp} increases with the lowering of temperature, then, below about 100K, it becomes constant around 110G. The g-value increases as temperature is lowered. Below about 100K, temperature dependence of the g-value is more enhanced. The temperature dependence of the intensity can be explained in terms of the singlet-triplet system with the exchange interaction of J=80K. As suggested by the X-ray analysis, the crystal has dimers of $C_1TET-TTF$ donors with a half-filled band. The singlet-triplet type behavior is considered to be associated with the dimer coupled with antiferromagnetic interactions. The temperature dependence of the g-value is too large to explain in terms of isolated coupled spins without magnetic environment. In $(C_1TET-TTF)FeCl_4$, there are Fe^{3+} spins which tend to be ordered antiferromagnetically. The internal field associated with the Fe^{3+} spins may contribute to the large change in the g-value of the donor spins in the low temperature domain.

2.2 BEDT-TTF with $CoCl_4^{2-}$ and $CoBr_4^{2-}$ anions

BEDT-TTF gives two kinds of crystal forms with $CoCl_4^{2-}$; needle and block crystals. The needle crystals, stoichiometry of which is expected to be 4:1 by EPMA, show metallic behavior at high temperatures ($\rho_{rt}=0.014\Omega cm$) as shown in Fig. 3, where temperature dependence of resistivities are exhibited for four crystals in the same batch. As temperature is lowered, the resistivities suggest a metal-insulator transition at a temperature ranging between 130K and 40K depending on crystal. In the sample (c), a broad hump appears around 150K, which is reminiscent of the resistivity behavior of other BEDT-TTF compounds[4,5]. The broad feature might be caused by the quenching of the molecular vibration of ethylene groups in BEDT-TTF donors. Temperature dependence of thermoelectric power was also investigated for a crystal in the same batch as that for the resistivity measurement. At high temperatures, thermoelectric power has negative values. As temperature is lowered, it increases gradually toward zero, reaches zero around 150K, and becomes positive below that temperature. It tends to be constant around 5μV/K below about 100K. The trend in thermoelectric power cannot be explained in term of a simple band structure based on BEDT-TTF donor, since the BEDT-TTF should have hole carriers through charge transfer to acceptors. A band calculation is necessary on the basis of crystal structure analysis to explain the detailed behavior of thermoelectric power. Anyway, the monotonous increase in thermoelectric power indicates that a metallic state survives at least above 60K. Powder ESR spectra were also investigated, where g=2.004 and $\Delta H=60G$ at room temperature in the donor signal. Temperature dependence of the ESR intensity suggests the presence of Pauli paramagnetism in the high temperature domain, while below about 100K, the contribution of Curie susceptibility covers the presence of Pauli paramagnetism at low temperatures. Among all the BEDT-TTF complexes with magnetic 3d transition metal ions ever investigated, the present crystal with $CoCl_4^{2-}$ is involved in a rare group with metallic nature[6,7]. An ESR signal associated with Co^{2+} ions appears around g~6 with Curie behavior at low temperatures below about 20K. The block crystal has different electronic nature from that of the needle metallic material. Resistivity behaves semiconductively with Ea=0.067eV and $\rho_{rt}=0.5\Omega cm$.

Plate crystals were obtained in BEDT-TTF complex with $CoBr_4^{2-}$ (4:1 stoichiometry was expected by EPMA). Resistivity measurement shows semiconductive behavior with Ea=0.08eV and $\rho_{rt}=0.5\Omega cm$. Moreover, the measurement suggests the presence of the first order phase transition around 180K with a hysteresis as shown in Fig. 4. The activation energy is not so influenced by the presence of the phase transition. The first order phase transition was confirmed by means of heat capacity measurement, where an anomaly associated with the phase transition was estimated to be 2% of the total heat capacity at transition point T_c. Pressure dependence of resistivity was investigated as shown in Fig. 4. The application of pressure makes the phase transition unclear, and above P=4kbar, there is no indication of the phase transition except a change in the slope of the logρ vs. 1/T curves. ρ_{rt} decreases by 75% as pressure increases from 1bar to 10kbar. The change in the activation energy as a function of pressure is estimated to be $dEa/dP=-6.5\times10^{-3}$

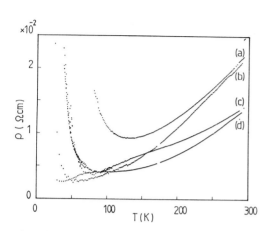

Fig. 3. Temperature dependence of resistivity for (BEDT-TTF)-CoCl$_4$ (needles): different samples (a)-(d).

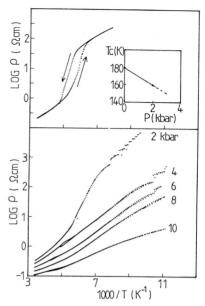

Fig. 4. Temperature dependence of resistivity for (BEDT-TTF)-CoBr$_4$ under various pressures.

eV/kbar and -9.5×10^{-3} eV/kbar for $T \ll T_c$ and $T \gg T_c$, respectively. Thus, the system is expected to be metallic without any activation energy above around P=15kbar.

References

1. M. Inoue, M. B. Inoue, C. Cruz-Vazguez, S. Roberts and Q. Fernando, Synth. Metals, 19, 641 (1987).
2. M. Lequan, R. M. Lequan, G. Maceno and P. Delhaes, J. Chem. Soc. Chem. Commun., 174 (1988).
3. Y. Nakano, K. Suzuki, T. Enoki, A. Otsuka, G. Saito and M. Konno, unpublished results.
4. G. Saito, T. Enoki and H. Inokuchi, Chem. Lett., 1345 (1982).
5. H. Urayama, H. Yamochi, G. Saito, K. Nozawa, T. Sugano, M. Kinoshita, S. Sato, K. Oshima, A. Kawamoto and J. Tanaka, Chem. Lett., 55 (1988).
6. M. Tanaka, A. Kawamoto, J. Tanaka, M. Sano, T. Enoki and H. Inokuchi, Bull. Chem. Soc. Jpn., 60, 2531 (1987).
7. M. Kurmoo, T. Mallah, P. Day, I. Marsden, M. Allen, R. H. Friend, F. L. Pratt, W. Hayes, D. Casseau, J. Gautier and G. Bravic, First ISSP International Symposium on the Physics and Chemistry of Organic Superconductors (These proceedings).

Structural Electrical and Magnetic Properties of the (BEDT-TTF)$_4$Ni(CN)$_4$ Complex

M. Tanaka[1], *H. Takeuchi*[1], *A. Kawamoto*[2], *J. Tanaka*[2], *T. Enoki*[2], *K. Suzuki*[3], *K. Imaeda*[4], *and H. Inokuchi*[4]

[1]Department of Chemistry, College of General Education, Nagoya University, Chikusa-ku, Nagoya 464-01, Japan
[2]Department of Chemistry, Faculty of Science, Nagoya University, Chikusa-ku, Nagoya 464-01, Japan
[3]Department of Chemistry, Faculty of Science, Tokyo Institute of Technology, Meguro-ku, Tokyo 152, Japan
[4]Institute for Molecular Science, Myodaiji, Okazaki 444, Japan

Abstract. The X-ray diffraction, the electrical conductivity and the ESR spectra of a single crystal of (BEDT-TTF)$_4$Ni(CN)$_4$ complex were measured. This crystal belongs to the triclinic system and exhibits a metal-insulator transition at 160K. The ESR signals show the presence of the [Ni^{III}(CN)$_4$]$^-$ anion in addition to the BEDT-TTF cation radical.

1. Introduction

Since the discovery of the superconductor (BEDT-TTF)$_2$Cu(NCS)$_2$ exhibiting complete superconductivity around 10K at ambient pressure[1], studies of many complexes with metal salts as the counter anion have been made. Such charge transfer complexes between BEDT-TTF and metal salts are interesting for their magnetic properties in addition to their electrical properties, because the metal ion can have an incompletely occupied electron shell. From this point of view, we investigated the magnetic and electronic properties of (BEDT-TTF)$_2$CuCl$_2$[2]. In the present paper, we report on the complex (BEDT-TTF)$_4$Ni(CN)$_4$.

2. Crystal structure

The crystal of (BEDT-TTF)$_4$Ni(CN)$_4$ was obtained by the electrochemical oxidation of the TCE solution containing BEDT-TTF and K$_2$Ni(CN)$_4$. The crystal data: triclinic, space group P$\bar{1}$, a=10.994(2), b=16.443(2), c=9.703(1) A, α=97.85(1)°, β=115.15(1)°, γ=95.94(1)°, Z=1. Intensities were measured on a Rigaku automated diffractometer with Cu Kα radiation. The number of the observed independent reflections was 4268. The structure was solved by the Monte-Carlo direct method with the aid of MULTAN78 program system[3] and refined by the full matrix least squares. The final R value was 0.067.

The crystal structure is shown in Fig. 1. The long molecular axis of BEDT-TTF is located almost parallel to the b axis and the BEDT-TTF molecules stack on each other along the [$\bar{2}$01] direction. The [Ni(CN)$_4$]$^-$ anion has a square planar structure and is located in the ac plane. The interstack contacts are longer than the the sum (3.70 A) of the Pauling's van der Waals radius of a sulfur atom. Several S-S contacts shorter than 3.70 A are observed along the [$\bar{2}$01] direction.

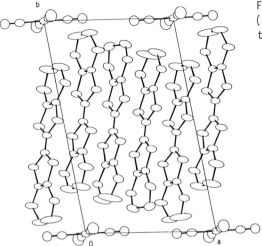

Figure 1. Crystal structure of $(BEDT-TTF)_4Ni(CN)_4$ projected on the ab plane.

3. Electrical conductivity

The temperature dependence of electrical conductivity of the single crystal is shown in Fig. 2. The room temperature conductivity was about 20 S/cm. The crystal was metallic between 230 and 300K and became semi-conductive with an energy gap of 100K below 160K after the temperature dependence of the conductivity was not observed between 160 and 230K. This kind of resistance behavior may be explained by the phase transition of the crystal between 160 and 230K.

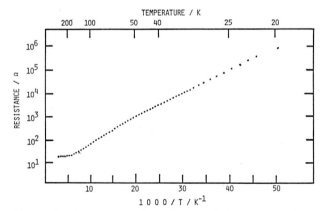

Figure 2. Temperature dependence of the a axis electrical conductivity of the single crystal of $(BEDT-TTF)_4Ni(CN)_4$

4. Magnetic properties

The ESR spectra of the single crystal were taken by a Varian E112 at X-band with a continuous helium flow cryostat and a rectangular microwave cavity. The crystal was rotated around the a axis and was measured over a temperature range of 4-300K.

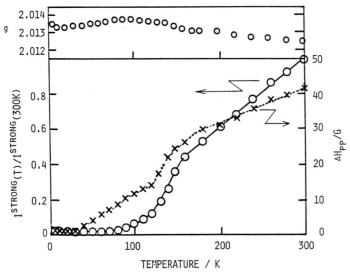

Figure 3. Temperature dependence of the g value, the linewidth and the absorption intensity of the ESR spectra of the strong band.

The ESR spectrum at 4K consists of the strong band at g=2.013, the weak band at g=2.056 and the satellite band at 2.004. The strong band is assigned to the unpaired π-electron of the BEDT-TTF cation radical[2]. The weak band can be due to the d electron of $[Ni^{III}(CN)_4]^-$ anion in a square planar arrangement[4].

The temperature dependence of the g value, the linewidth ΔH_{pp} and the absorption intensity of the ESR spectra are shown in Fig. 3.

The temperature dependence of the absorption intensity of the weak band can be depicted by the Curie law as shown in Fig. 4. The number of spins per $[Ni(CN)_4]^-$ anion is estimated to be 0.01 of the total Ni ions.

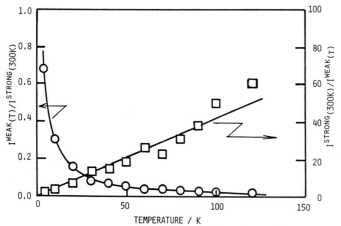

Figure 4. Temperature dependence of the absorption intensity of the ESR spectra of the weak band.

5. References

[1] H.Urayama, H.Yamochi, G.Saito, K.Nozawa, T.Sugano, M.kinoshita, S.Sato, K.Oshima, A.Kawamoto & J.Tanaka, Chem.Lett., 558(1988).
[2] M.Tanaka, A.Kawamoto, J.Tanaka, M.Sano, T.Enoki & H.Inokuchi, Bull.Chem.Soc.Jpn., 60, 2531(1987).
[3] P.Main, S.E.Hull, L.Lessinger, G.Germain, J.-P.Declercq & M.M.Woolfson, MULTAN78. A system of Computer programs for the Automatic Solution of Crystal Structures from X-ray Diffraction Data. Univs. of York, England and Louvain, Belgium.
[4] A.G.Lappin, C.K.Murray & D.W.Margerum, Inorg.Chem., 17, 1630(1978).

Crystal Structures and Electrical Conductivities of EDT-TTF Salts with TaF_6^-, AsF_6^-, PF_6^-, ReO_4^-, ClO_4^-, BF_4^-, $Au(CN)_2^-$ and $Ni(dmit)_2^{n-}$

A. Kobayashi[1], R. Kato[2], and H. Kobayashi[2]

[1]Department of Chemistry, Faculty of Science, University of Tokyo,
  Hongo, Bunkyo-ku, Tokyo 113, Japan
[2]Department of Chemistry, Faculty of Science, Toho University,
  Funabashi, Chiba 274, Japan

Abstract. Crystal and electronic structures of organic conductors, $(EDT-TTF)_2X$(EDT-TTF=ethylenedithiotetrathiafulvalene; $X=PF_6$, AsF_6, TaF_6, BF_4, ClO_4, ReO_4, $Au(CN)_2$), have been studied. The asymmetric EDT-TTF molecules stack alternately to form almost uniform columns. Two donor-acceptor type molecular conductors, α- and β-(EDT-TTF)[Ni(dmit)$_2$](dmit=4,5-dimercapto-1,3-dithiole-2-thione) have also been examined. The α form retains metallic behavior down to 1.5 K, exhibiting a small resistivity anomaly around 20 K. The β form is a typical one-dimensional charge transfer complex with a mixed column of EDT-TTF and Ni(dmit)$_2$.

1. Introduction

Up to now, five types of molecular metals are known to become superconductors: (1) quasi 1D Bechgaard salts $(TMTSF)_2X$ composed of organic donor TMTSF and closed shell inorganic anions X[1], (2) 2D $(BEDT-TTF)_2X$ salts[2] (3) $(DMET)_2X$ salts characterized by asymmetric donors[3] (4)$(MDT-TTF)_2X$ salts with asymmetric donors[4] (5) M(dmit)$_2$ compounds based on organic donor molecules and inorganic closed shell cations[5]. We have examined a new series of cation radical salts based on asymmetric π-donor EDT-TTF with TaF_6^-, AsF_6^-, PF_6^-, ReO_4^-, ClO_4^-, BF_4^-, $Au(CN)_2^-$ and $Ni(dmit)_2^{n-}$[6,7]. EDT-TTF and Ni(dmit)$_2$ have given two conducting 1:1 salts, α-, β-(EDT-TTF)-[Ni(dmit)$_2$].

2. $(EDT-TTF)_2X$ ($X=PF_6$, AsF_6, TaF_6, BF_4, ClO_4, ReO_4, $Au(CN)_2$)

EDT-TTF is a hybrid between TTF and BEDT-TTF molecules. EDT-TTF was obtained by cross-coupling of the appropriate 1,3-dithiole-2-thione and 1,3-dithiole-2-one in triethyl phosphite. Separation of the desired cross-coupling product from the symmetrical self-coupling co-products was accomplished by HPLC(silica gel, CS_2). The product distribution was TTF : EDT-TTF : BEDT-TTF=1:29:18. Cation radical salts of EDT-TTF were prepared by electrochemical oxidation in presence of (n-Bu)$_4$NX($X=PF_6$, AsF_6, TaF_6, BF_4, ClO_4, ReO_4, $Au(CN)_2$) in 1,1,2-trichloroethane using a constant current(1 μA). Intensity data were collected on a Rigaku automatic four-circle diffractometer with monochromated Mo-Kα radiation. The structures were determined by the direct method. Crystal data, electrical properties and final R values are listed in Table 1.

In all these compounds, the asymmetric EDT-TTF molecules stack alternately to form almost uniform columns. The EDT-TTF molecule is almost planar except the ethylenedithio fragment. Conformational disorder of the ethylene group was not observed in each case. These cation radical salts of EDT-TTF are classified into three almost isostructural groups.

(1) The triclinic group contains the $Au(CN)_2$, TaF_6, and AsF_6 (α form) salts. Molecular packing is similar to that of the Bechgaard salt, $(TMTSF)_2X$. Band energy calculations show a pseudo one-dimensional Fermi surface.

(2) The orthorhombic group includes AsF_6 (β form) and PF_6 salts. Four almost equivalent donor columns run along the b axis.

(3) The monoclinic group contains the salts with tetrahedral anions (BF_4^-, ClO_4^-, and ReO_4^-). They do not exhibit the orientational disorder.

Resistivity measurements showed that almost all these salts are highly conductive at room temperature but semiconductive at low temperature. Although band energy calculations of these three groups show quasi-one-dimensional metallic Fermi surfaces, narrow ESR signals suggest these are strong electron correlation systems.

Table 1. Crystal data and electrical properties of $(EDT-TTF)_2X$.

X	$Au(CN)_2$	TaF_6	AsF_6 (α)	AsF_6 (β)	PF_6	ReO_4	ClO_4	BF_4
S.G.	P$\bar{1}$	P$\bar{1}$	P$\bar{1}$	Pccn	Pccn	C2/c	C2/c	C2/c
a Å	14.752	15.213	14.855	28.303	28.014	29.720	29.283	29.133
b	7.225	7.049	7.043	7.120	7.122	7.185	7.145	7.134
c	6.388	6.463	6.457	12.706	12.650	12.416	12.417	12.379
α°	106.15	102.22	102.38					
β	101.56	102.57	96.15			110.56	111.54	111.85
γ	90.50	98.77	101.68					
Z	1	1	1	4	4	4	4	4
V Å³	639.2	646.7	638.1	2560.5	2523.9	2482.4	2416.5	2388.0
R	0.074	0.088	0.037	0.090	0.063	0.115	0.129	0.064
σ S cm^{-1}	5	100	50	30	100	10	100	10
T_{M-I}	semicon	57	50	ca. 50	42	150	ca. 100	170

3. $(EDT-TTF)[Ni(dmit)_2]$

Single crystals of α- and β-$(EDT-TTF)[Ni(dmit)_2]$ were electrochemically obtained using constant current(1 μA). The acetonitrile solution gave the α form as black hexagonal plates, and the 1,2-dichloroethane-acetonitrile(1:3) solution gave mainly the β form as black thin plates. Crystal data are: α form, triclinic, P$\bar{1}$, a=6.658(3), b=7.627(8), c=27.385(3) Å, α=93.23(3), β=91.43(3), γ=119.29(3)°, V=1208.6 Å³, Z=2. The β form, monoclinic, P2$_1$/c, a=

27.685(7), b=7.845(2), c=11.508(3) Å, β=101.33(2)°, V=2450.7 Å^3, Z=4. Intensity data were collected on an automatic four-circle diffractometer. The structures were solved by the direct method. The final R values are 0.096 for the α form and 0.082 for the β form.

3.1 α-(EDT-TTF)[Ni(dmit)$_2$]

The planar Ni(dmit)$_2$ molecules form a uniform column along the b axis with the interplanar distance of 3.52 Å. The crystal structure is shown in Fig. 1. The asymmetric EDT-TTF molecules stack alternately along the [110] direction. The interplanar distances are 3.62 Å and 3.61 Å. The mode of molecular overlapping is the "double bond over ring" type. The stacking direction of the EDT-TTF molecules ([110]) is different from that of the Ni(dmit)$_2$ molecules ([010]). The electrical resistivity of α form is shown in Fig. 2. The room temperature conductivity of the α form is about 100 S cm^{-1}. Upon cooling, the electrical resistivity exhibits a monotonous decrease down to 20 K. At 20 K, the resistivity begins to increase and has a peak around 14 K. It then decreases again to lower temperature, as a metal. At 15 K, X-ray photographs were taken, however no structural change was observed.

The reflection spectra of the α form suggest a small anisotropy of the electronic structure in the ab plane. The calculated Fermi surface consists of two quasi one-dimensional metallic bands, each of which is associated with Ni(dmit)$_2$ and EDT-TTF, respectively. The Fermi surface consists of two pairs of "warped" planes, which are not completely nested by a single modulation wave vector. This structural feature may contribute to the stabilization of the metallic state in the α form.

3.2 β-(EDT-TTF)[Ni(dmit)$_2$]

β form has a mixed-stacking structure. A structural feature is that there exist short intermolecular S...S distances (3.426-3.686 Å) based on the terminal thionyl S atoms between the Ni(dmit)$_2$ molecules aligned along the a axis. The room temperature conductivity is 100 Ω cm and temperature dependence is semiconductive.

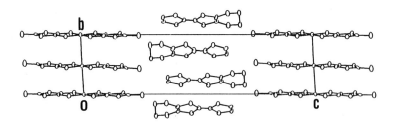

Fig. 1. Crystal structure of α-(EDT-TTF)[Ni(dmit)$_2$].

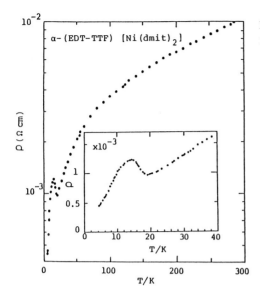

Fig. 2. Electrical resistivity of α-(EDT-TTF)[Ni(dmit)$_2$].

4. Acknowledgement

The authors acknowledge Prof. S. Kagoshima for his preliminary X-ray work on (EDT-TTF)[Ni(dmit)$_2$] at low temperature and also acknowledge Prof. T. Watanabe for ESR measurements.

References

[1] D. Jérome, A. Mazaud, M. Ribault, and K. Bechgaard, J. Phys. Lett. (Paris), 41, L95 (1980).
[2] R. P. Shibaeva et al., Mol. Cryst. Liq. Cryst., 119, 361 (1985); H. H. Wang, et al. Inorg. Chem., 24, 2465(1985); H. Kobayashi et al., Chem. Lett., 1986, 789; A. Kobayashi et al., ibid., 1987, 459; H. Urayama et al., ibid., 1988, 55.
[3] K. Kikuchi, T. Namiki, K. Saito, I. Ikemoto, K. Murata, T. Ishiguro, and K. Kobayashi, Chem. Lett., 1987, 931.
[4] G. C. Papavassiliou, A. Terzis, B. Hilti, C. W. Mayer, and J. Pfeiffer, Synthetic Metals, 27, B379 (1988).
[5] L. Brossard, M. Ribault, M. Bousseau, L. Valade, and P. Cassoux, C. R. Acad. Sc. Paris II 302, 205 (1986); A. Kobayashi, H. Kim, Y. Sasaki, R. Kato, H. Kobayashi, S. Moriyama, Y. Nishio, K. Kajita, and W. Sasaki, Chem. Lett., 1987, 1819.
[6] R. Kato, H. Kobayashi, and A. Kobayashi, Chem. Lett., 1989, 781.
[7] R. Kato, H. Kobayashi, and A. Kobayshi, Chem. Lett., 1989, 1839.

A New Transformable Cation-Radical Salt (EPT)$_2$I$_7$

V.E. Korotkov, R.P. Shibaeva, N.D. Kushch, and E.B. Yagubskii

Institute of Chemical Physics, USSR Academy of Sciences,
SU-142432 Chernogolovka, USSR

Abstract. A new cation-radical salt (EPT)$_2$I$_7$ has been synthesized. Its crystals are isostructural with the ε-(ET)$_2$I$_7$ crystals. On heating they lose their iodine and convert to (EPT)$_2$I$_3$ crystals.

One of the investigative approaches to the search for organic metals and superconductors based on cation-radical salts of bis(ethylenedithio)tetrathiafulvalene (ET) is its modification, the use of its symmetric and asymmetric derivatives. A rather large number of cation-radical salts based on such symmetric derivatives as MT, PT, BT, VT etc. with various anions have been obtained and characterized.

Some of them are organic metals. E.g., the first MT-based cation-radical salt of the 2:1 composition, (MT)$_2$Au(CN)$_2$, is an organic metal down to 80 K and its conducting cation-radical layer is of the k-type structure [1], characteristic of superconductors (ET)$_4$(Hg$_3$Cl$_8$) [2,3], (ET)$_4$(Hg$_3$Br$_8$) [4], k-(ET)$_2$I$_3$ [5], k-(ET)$_2$Cu(SCN)$_2$ [6]. At present several salts have been obtained on the base of the ET asymmetric derivatives: MET salts with ClO$_4$, PF$_6$, and ReO$_4$ anions [7], EPT.InI$_4$ [8], m- and t-(EPT)$_2$ICl$_2$ [9], one of them, m-(EPT)$_2$ICl$_2$, being isostructural with α'-(PT)$_2$ICl$_2$ [10].

Our paper reports on a new cation-radical salt based on ethylenedithiopropylenedithiotetrathiafulavalene, i.e., (EPT)$_2$I$_7$. The X-ray analysis of the salt has been undertaken and its properties studied.

The (EPT)$_2$I$_7$ crystals were obtained in electrochemical EPT oxidation (C = 2 x 10^{-3} M) on the Pt anode under the regime of constant current (I = 2.2 µa, j = 5 µa/cm^2) in benzonitrile. TBA.I$_3$ was used as an electrolyte with a concentration of 1x10^{-2} M. It should be noted that the main part of the crystals grew on the bottom and walls of the cell, the rest on the anode.

The crystal data for the (EPT)$_2$I$_7$ crystals are listed in the first column of Table 1. The experimental results, 2129 independent reflections with I $\geqslant$ 2σ (I), were obtained on the Syntex P$\bar{1}$ diffractometer (MoK$_\alpha$ radiation, graphite monochromator, (sin θ/λ)$_{max}$=

Table 1. The crystal data for isostructural cation-radical salts

Parameter	$(EPT)_2I_7$	ε-$(ET)_2I_7$ [11]	$(ET)_2(IBr_2)_2(C_2H_3Cl_3)_{0.5}$ [12]
a, Å	17.83(2)	13.974(9)	18.763(3)
b, Å	14.25(2)	13.40(3)	13.859(1)
c, Å	18.78(3)	18.77(3)	17.133(2)
α, β°	111.3(1)β	67.3(1)α	103.7(1)β
V, Å^3	4446	4210.5	4329.3
space group	$P2_1/a$	$P2_1/b$	$P2_1/c$
Z	4	4	4

0.586). The structure was solved by the heavy atom method and refined by the least squares method in the anisotropic approximation. The correction on the absorption was performed according to [13]. The R-factor was equal to 0.083. The atomic coordinates are summarized in Table 2.

Fig. 1 depicts the projection of the $(EPT)_2I_7$ crystal structure along the direction **b**. The presence of mixed cation-radical layers of $(EPT)_2I_3$ composition alternating with the I_8^{2-} anion layers along the **c** direction is characteristic of this structure. In the $(EPT)_2I_3$ layer the pairs of almost parallel EPT I-II (dihedral angle of 1.3°) form zig-zag bands along the direction **b**, thus forming a row of shortened intermolecular S...S contacts: S2...S8 (S8...S2) 3.52 Å (I-I$_i$), S4...S8 (S8...S4) 3.70 Å (I-I$_i$), S12...S16 (S16...S12) 3.67 Å (II-II$_i$), and S5...S11 3.69 Å (I-II$\bar{a}$), where the symmetry operations for I$_i$, II$_i$, and II$\bar{a}$ are (1-x, 2-y, $\bar{z}$), (1-x, 1-y, $\bar{z}$), and (x-1/2, 1$^1/_2$-y, z), respectively. The cation-radical EPT planes in the $(EPT)_2I_3$ layer are almost parallel to the I_3 anion (the angle of ~ 7°) and approximately perpendicular to the I_8^{2-} plane. They form an angle of ~ 6° with the [001] direction. The interatomic distances are I1-I2 2.909(4) and I1-I3 2.916(4) Å. The angle is equal to 176.6(1)$^\circ$.

In the anion layer the I_8^{2-} anions are located in the symmetry centers and have a Z-like form. They consist of two asymmetric I_3 anions with the distances I-I 2.803(8) and 2.938(7) Å and the angle of 175.9(3)$^\circ$ connected by the bridging molecule I_2 (I-I 2.87 Å). The interatomic distance I_3^- - I_2 is equal to 3.70(1) Å. The I_8^{2-} anion plane is almost parallel to the **ab** plane, the angle is ~ 10°.

It is noteworthy, that the $(EPT)_2I_7$ crystals are isostructural with the $\varepsilon-(ET)_2I_3(I_8)_{0.5}$ [11] and $(ET)_2(IBr_2)_2(C_2H_3Cl)_{0.5}$ [12,14] crystals, the positions of I_3 anions being occupied by the IBr_2 anions in the latter crystals and the place of the I_2 molecule is

Table 2. $(EPT)_2I_7$. The atomic coordinates for I, S ($x10^4$) and C ($x10^3$)

Atom	x/a	y/b	z/c
S1	3362(5)	7611(6)	901(4)
S2	4401(5)	9235(6)	1121(4)
S3	3387(5)	7518(7)	2467(5)
S4	4617(6)	9527(7)	2739(5)
S5	3065(5)	7642(6)	-921(4)
S6	4122(5(	9252(6)	-684(4)
S7	2630(6)	7570(8)	-2600(4)
S8	3961(6)	9418(8)	-2297(5)
C1	380(2)	838(2)	49(2)
C2	375(2)	807(2)	182(2)
C3	418(2)	887(3)	189(2)
C4	410(3)	805(4)	333(3)
C5	418(3)	894(4)	342(2)
C6	372(2)	841(2)	-26(2)
C7	318(2)	808(3)	-173(2)
C8	365(2)	884(3)	-160(2)
C9	206(2)	859(3)	-314(2)
C10	252(2)	913(3)	-355(2)
C11	307(3)	988(4)	-300(2)
S9	5906(5)	7919(6)	-44(5)
S10	4805(5)	6338(7)	-475(5)
S11	5982(6)	8245(7)	-1550(5)
S12	4662(7)	6318(8)	-2096(5)
S13	6073(5)	7765(7)	1668(5)
S14	4959(5)	6184(6)	1336(4)
S15	6401(7)	7789(7)	3333(5)
S16	5124(5)	5916(7)	2976(5)
C12	539(2)	708(2)	24(2)
C13	558(2)	760(2)	-106(2)
C14	503(2)	684(3)	-119(2)
C15	552(4)	779(3)	-246(3)
C16	542(2)	678(3)	-249(2)
C17	544(2)	697(2)	99(2)
C18	590(2)	730(3)	244(2)
C19	538(2)	652(3)	230(2)
C20	704(3)	683(3)	383(2)
C21	662(3)	615(3)	414(2)
C22	613(2)	542(3)	360(2)
I1	2964(1)	5068(1)	-181(1)
I2	2937(1)	5034(2)	-1739(1)
I3	3000(1)	5222(2)	1379(1)
I4	9165(2)	3566(3)	4550(1)
I5	7610(3)	3645(3)	4589(2)
I6	10796(2)	3340(6)	4531(2)
I7	10222(3)	945(6)	4902(2)

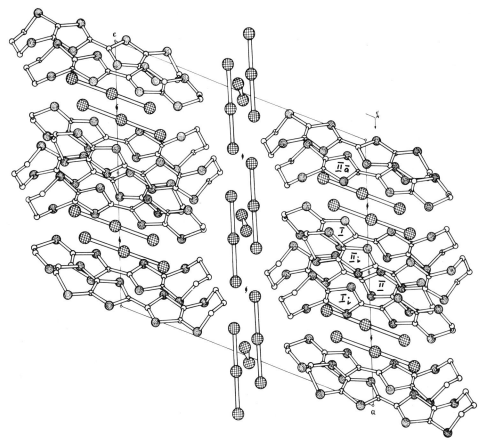

Fig. 1: The structure projection along the b direction.

occupied disorderedly by the molecule of trichloroethane. All these crystals are semiconductors in their initial state. The electronic zone structure of ε-$(ET)_2I_7$ [15] also showed that there was a large energy gap between the zones, *i.e.* the ε-phase crystals are typical semiconductors.

The room-temperature conductivity of the $(EPT)_2I_7$ crystals is 10^{-5} $ohm^{-1}cm^{-1}$. As shown earlier [16,17] on heating the ε-$(ET)_2I_7$ crystals lose their iodine and convert to β_H-$(ET)_2I_3$ with $T_c \sim 8$ K. The $(ET)_2(IBr_2)_2(C_2H_3Cl_3)_{0.5}$ crystals become organic metals on heating and retain their metallic state down to ~ 0.5 K [12,14].

Taking these facts into account we chose some $(EPT)_2I_7$ crystals which were placed into the argon flow for 4 hours at $90^{\circ}C$. The crystals' conductivity increased sharply up to 1-10 $ohm^{-1}cm^{-1}$, the conductivity dependence on temperature still being semiconducting. The X-ray analysis of these crystals showed that they had the $\sim$6-7 Å period along the direction of lengthening. Apparently, they converted to $(EPT)_2I_3$ crystals, which are isostructural with $\beta-(ET)_2I_3$ [17, 18] and $\beta-(PT)_2I_3$ [19], the former being superconductors, the latter semiconductors.

References

1. P.J. Nigrey, B. Morosin, J.E. Kwak, E.L. Venturini, R.J.Baughman: Synth. Met. **16**, 1 (1986).

2. R.N. Lyubovskaya, R.B. Lyubovskii, R.P. Shibaeva, M.Z. Aldoshina, L.M. Goldenberg, L.P. Rozenberg, M.L. Khidekel, Yu.P. Shulpyakov: Pis'ma v ZhETF **42**, 380(1985).

3. R.P. Shibaeva, L.P. Rozenberg: Kristallographiya **33**, 1402 (1988).

4. R.N.Lyubovskaya, E.I.Zhilyaeva, S.I. Pesotskii, R.B. Lyubovskii, L.O. Atovmyan, O.A. Dyachenko, T.G. Takhirov: Pis'ma v ZhETF **46**, 149 (1987).

5. A. Kobayashi, R. Sato, H. Kobayashi, S. Moriyama, K. Kajita, W. Sasaki: Chem. Lett., 459 (1987).

6. H. Urayama, H. Yamochi, G. Saito, K. Nozawa, T. Sugano, M. Kinoshita, S. Sato, K. Oshima, A. Kawamoto, J. Tanaka: Chem.Lett., 55 (1988).

7. M.A. Beno, U. Geiser, A.M. Kini, H.H. Wang, K.D. Carlson, M.M. Miller, T.J. Allen, J.A. Schulter, R.B. Proksch, J.M. Williams, Synth. Met. **27**, A209 (1988).

8. L.C.Porter, T.J. Allen, K.D. Carlson, M.Y. Chen, H.C.I. Kao, A.M. Kini, J.A. Schulter, H.H. Wang, J.M. Williams: Acta cryst. **C44**, 1712 (1988).

9. A.J. Schultz, U. Geiser, A.M. Kini, H.H. Wang, J. Schulter, C.S. Cariss, J.M. Williams: Synth. Met. **27**, A229 (1988).

10. R.P. Shibaeva, L.P. Rozenberg, M.A. Simonov, N.D. Kushch, E.B. Yagubskii: Kristallographiya **33**, 1156 (1988).

11. R.P.Shibaeva, R.M. Lobkovskaya, E.B. Yagubskii, E.E.Kostyuchenko: Kristallographiya **31**, 455 (1986).

12. H. Yamochi, H.Urayama, G. Saito, K. Oshima, A.Kawamoto, J.Tanaka: Chem. Lett., 1211 (1988).

13. N. Walker, D. Stuart: Acta cryst. **A39**, 158 (1983).

14. H. Yamochi, H.Urayama, G. Saito, K. Oshima, A.Kawamoto, J.Tanaka: Synth. Met. **27**, A485 (1988).

15. R. Kato, H. Kobayashi, A. Kobayashi, T. Mori, H. Inokuchi: Chem. Lett., 277 (1987).

16. V.A. Merzhanov, E.E. Kostyuchenko, V.N. Laukhin, R.M.Lobkovskaya, M.K. Makova, R.P. Shibaeva, E.B. Yagubskii: Pis'ma v ZhETF **41**, 146, (1985).

17. A.V. Zvarykina, P.A. Kononovich, V.N. Laukhin, V.N. Molchanov, S.I. Pesotskii, V.I. Simonov, R.P. Shibaeva, I.F. Shchegolev, E.B. Yagubskii: Pis'ma v ZhETF **43**, 257 (1986).

18. V.F. Kaminskii, T.G. Prokhorova, R.P. Shibaeva: Pis'ma v ZhETF **39**, 15 (1984).

19. H. Kobayashi, M. Takahashi, R. Kato, A, Kobayashi, W. Sasaki: Chem. Lett., 1335 (1984).

Microwave Conductivity of the Phthalocyanine and Dicyanoquinonediimine Salts

H. Yamakado[1], A. Ugawa[2], T. Ida[2], and K. Yakushi[2]

[1]Department of Structural Molecular Science, Graduate University for Advanced Studies, Myodaiji-cho, Okazaki, Aichi 444, Japan
[2]Institute for Molecular Science, Myodaiji-cho, Okazaki, Aichi 444, Japan

Abstract. The temperature dependence of microwave conductivity was measured on single crystals of phthalocyanine salts (NiPc(AsF$_6$)$_{0.5}$, PtPc(ClO$_4$)$_{0.5}$, CoPc(AsF$_6$)$_{<0.5}$), and dicyanoquinone-diimine (DCNQI) salts (Cu(Me$_2$-DCNQI)$_2$, Cu(MeBr-DCNQI)$_2$, Ag(Me$_2$-DCNQI)$_2$). Phthalocyanine salts exhibit a metal-insulator transition unlike the iodine salts of phthalocyanine. DCNQI salts show qualitatively the same behavior as studied by dc conductivity.

1. Introduction

Considerable effort has been devoted to a study of iodine salts of phthalocyanine. These materials show anomalous behavior of electron transport properties, e.g. a strong sample dependence and an absence of a Peierls transition in spite of their strong one-dimensional nature.[1-3] Although we have measured the dc electrical conductivity of electrochemically grown phthalocyanine salts, the resistance jumps prevented us from obtaining a reliable conductivity curve. In this paper we present the microwave conductivity results measured with a cavity perturbation method which enables the strain-free cooling of a sample crystal.

Some salts of dicyanoquinonediimine (DCNQI) exhibit a metal-insulator transition.[4] From the optical spectra of these materials, we pointed out the presence of an inter-chain interaction through the coordination bond between Cu and cyano groups of DCNQI, while alkali-metal salts of DCNQI have a one-dimensional character.[5] The difference between these two types may possibly be reflected in the microwave conductivity in its insulating phase.

2. Experimental

Single crystals of phthalocyanine and DCNQI salts were prepared using an electrochemical technique. The typical dimensions of the sample crystals are 2× 0.03× 0.03 mm³ and 5× 0.04× 0.04 mm³ for phthalocyanine and DCNQI salts, respectively. The microwave conductivity is measured by means of a cavity perturbation method.[6] The cavity resonator is a cylindrical type of TE_{011} mode at 9.4 GHz, the cavity being equipped with a liquid He cryostat. The sample crystal in the glass capillary is placed inside the cavity at the high electric field location in such a manner that the long axis of the crystal is parallel to the electric field. The microwave is sent to the cavity from a synthesized sweeper, HP8314B, and the reflected one is analyzed by a network analyzer, HP8757A. The resonant frequencies and the cavity Q-factors with and without a sample in the cavity are obtained from the reflectivity curves vs frequency. The operation and analysis are controlled by a microcomputer, PC9801-RX.

3. Results and Discussion

Figure 1 shows the temperature dependence of the microwave conductivity normalized to the room-temperature value of three phthalocyanine salts. The room-temperature values of conductivity are ca. 500 Scm⁻¹ for $NiPc(AsF_6)_{0.5}$ and $PtPc(ClO_4)_{0.5}$ and ca. 50 Scm⁻¹ for $CoPc(AsF_6)_{<0.5}$. The conductivity in the former two crystals reach a maximum around 50 K and 150 K, and then decrease on lowering the temperature. $CoPc(AsF_6)_{<0.5}$ shows a semiconductive behavior throughout the measured temperature range. These curves approximately follow straight lines in the intermediate temperature range. The activation energies are estimated to be 7 meV for $PtPc(ClO_4)_{0.5}$, and 21 meV and 8 meV for dc and microwave conductivity of $CoPc(AsF_6)_{<0.5}$, respectively.

The room-temperature conductivity of $NiPc(AsF_6)_{0.5}$ and $PtPc(ClO_4)_{0.5}$ are comparable with the iodine salts of NiPc and H_2Pc, which respectively exhibit conductivity peaks at 20-30 K and 15 K as well.[1,2] The difference between them is in the behavior of the low-temperature region; that is, the former have a drop in conductivity of more than 2 orders of magnitude whereas the iodine salts show little conductivity drop. Therefore the electrochemically prepared crystals appear to show a Peierls transition. However, the X-ray diffuse scattering characteristic of a Peierls transition has not been found down to 15 K in $NiPc(AsF_6)_{0.5}$ and $PtPc(ClO_4)_{0.5}$.[7] The temperature dependence of ESR intensity also shows no anomaly at the metal-insulator transition temperature.[8,9] In this sense, this transition is reminiscent

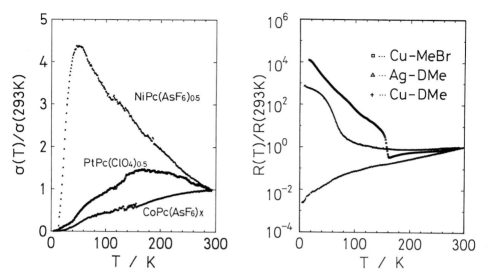

Fig. 1 left: Normalized microwave conductivities vs T for three phthalocyanine salts.

Fig. 2 right: Normalized microwave resistivities vs T for three DCNQI salts.

of (TMTTF)$_2$X, which has a strong electron-electron correlation.[10] However, the magnetic order like (TMTTF)$_2$X has not been found in phthalocyanine salts.

The electrical conductivity of CoPc(AsF$_6$)$_{<0.5}$ is qualitatively similar to that of CoPcI.[3] However, the details of temperature dependence are different, e.g. the conductivity of CoPcI linearly decreases with temperature. As mentioned before, the activation energies of the dc and microwave conductivities are remarkably different from each other, the reason for which is not clear at the moment. The semiconductive behavior of CoPc(AsF$_6$)$_{<0.5}$, although it is isostructural to PtPc(ClO$_4$)$_{0.5}$, is probably related to the d^7 configuration of Co^{2+}.

Figure 2 shows the temperature-dependent microwave conductivity of Cu(dimethyl-DCNQI)$_2$, Cu(2-methyl-5-bromo-DCNQI), and Ag(dimethyl-DCNQI)$_2$ normalized to the room-temperature value. The microwave conductivity of Cu(Me$_2$-DCNQI) progressively decreases down to 4.2 K following the relation $\rho \propto T^2$, which is exactly the same way as the dc conduction. Cu(MeBr-DCNQI)$_2$ shows a sharp metal-insulator transition at 160 K, the resistivity at low temperature being quite high. Ag(Me$_2$-DCNQI)$_2$ shows a gradual metal-insulator phase transition. The different natures of these phase transitions of Cu(MeBr-DCNQI)$_2$ and Ag(Me$_2$-DCNQI)$_2$ are probably related to the strong inter-chain interaction in the former and the one-dimensionality of the latter. In some one-dimensional materials, a huge dielectric constant is often observed in the insulating phase.[11] However, we could not find a large difference between them in the dielectric constant (10-100 below 10 K), the value of which is much smaller than a typical one-dimensional conductor, $K_{0.3}MoO_3$.[11] The small dielectric constant of $Ag(Me_2-DCNQI)_2$, although it is one-dimensional, is ascribed to the commensurate structure of the conduction band.

References

[1] J. Martinsen, S. M. Palmer, J. Tanaka, R. C. Greene, and B. M. Hoffman, Phys. Rev., B**30**, 6269 (1984).
[2] T. Inabe, T. J. Marks, R. L. Burton, J. W. Lyding, W. J. McCarthy, and C. R. Kannewurf, G. M. Reisner and F. H. Herbstein, Solid State Commun., **54**, 501 (1985).
[3] J. Martinsen, J. L. Stanton, R. L. Greene, J. Tanaka, B. M. Hoffman, and J. A. Ibers, J. Am. Chem. Soc., **107**, 6915 (1985).
[4] T. Mori, H. Inokuchi, A. Kobayashi, R. Kato, and H. Kobayashi, Phys. Rev., B**38**, 5913 (1988).
[5] K. Yakushi, G. Ojima, A. Ugawa, and H. Kuroda, Chem. Lett., 95 (1988).
[6] L. I. Buravov and I. F. Schegolev, Prib. Tekh. Eskp., **2**, 171 (1971).
[7] M. Tanaka and S. Kagoshima, private communication.
[8] K. Yakushi, H. Yamakado, M. Yoshitake, N. Kosugi, H. Kuroda, T. Sugano, M. Kinoshita, A. Kawamoto, and J. Tanaka, Bull. Chem. Soc. Jpn., **62**, 687 (1989).
[9] H. Yamakado, K. Yakushi, N. Kosugi, H. Kuroda, A. Kawamoto, J. Tanaka, T. Sugano, M. Kinoshita, S. Hino, Bull. Chem. Soc. Jpn., **62**, 2267 (1989).
[10] C. Coulon, S. S. P. Parkin, and R. Laversanne, Phys. Rev., B**31**, 3583 (1985).
[11] G. Mihály, T. W. Kim, and G. Grüner, Phys. Rev., B**39**, 13009 (1989).

Electron–Molecular Vibration Coupling in Organic Superconductors

T. Sugano and M. Kinoshita

Institute for Solid State Physics, University of Tokyo, Roppongi, Minato-ku, Tokyo 106, Japan

Abstract. Polarized reflectance spectra of some organic superconductors based on BEDT-TTF are examined over the range from infrared to near ultraviolet. Transition frequencies of the bands appearing in the infrared region are determined from the conductivity spectra obtained through a Kramers-Kronig transformation. Electron-molecular vibration (EMV) coupling constants and energies are estimated in terms of the dimer charge-oscillation model from the transition frequencies of the EMV coupled modes and the charge-transfer excitations. By use of the coupling constants and energies, some of the parameters describing superconducting state are evaluated and discussed on the basis of the BSC theory as well as Yamaji's theory for organic superconductivity.

1. Introduction

Electron-phonon interaction is a key parameter for discussing and understanding superconductivity in the materials of interest. In the reflectance spectra of the organic superconductors based on bis(ethylenedithio)tetrathiafulvalene (BEDT-TTF), several intense bands enhanced by the electron-molecular vibration (EMV) coupling are observed in the infrared region [1-3]. Since the EMV coupling has been suggested to be a possible mechanism of the electron-phonon interaction responsible for superconductivity [4], it is pertinent to discuss the relation between the EMV coupling and the superconducting nature of the BEDT-TTF salts.

In this paper, we summarize the EMV coupling constants and energies of the organic superconductors, protonated and deuterated κ-(BEDT-TTF)$_2$[Cu(NCS)$_2$] with the highest critical temperature T_c of $\sim$11 K known to date, and β-(BEDT-TTF)$_2$I$_3$ with T_c as high as $\sim$8 K in the high-T_c state.

2. Experimental

Polarized reflectance spectra of the BEDT-TTF salts were measured over the spectral range from 500 to 6000 cm^{-1} by using a Perkin-Elmer 1760 Fourier-transform infrared (FTIR) spectrometer equipped with a Spectra-Tech IR-PLAN$_{TM}$ microscope and a gold wire-grid polarizer and measured over the 4000 to 28000 cm^{-1} by using a home-made microspectrophotometer. Conductivity spectra were obtained through a Kramers-Kronig transformation of the reflectance spectra.

Transmittance spectra of the salts dispersed into potassium bromide pellets were recorded on a JACSO A-702 infrared spectrometer over the range from 200 to 4000 cm^{-1}.

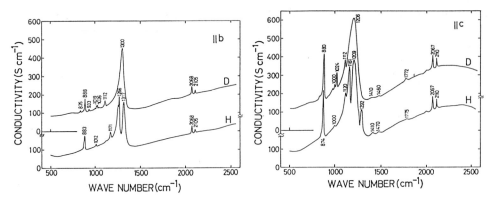

Fig. 1. Infrared conductivity spectra of protonated (H) and deuterated (D) κ-(BEDT-TTF)$_2$[Cu(NCS)$_2$] for the polarizations parallel to the b and c axes.

3. Results and discussion

To evaluate EMV coupling parameters, it is necessary to determine the frequencies of the intramolecular vibrational modes which couple with conduction electrons. In the case of the BEDT-TTF salts and complexes, the modes most strongly coupled with conduction electrons are the totally symmetric C=C and C-S stretching ones [1-4].

As was shown by us [1,2], a conductivity peak of the totally symmetric C=C stretching modes, enhanced by the EMV coupling, is superimposed by that of several C-C-H bending modes. It is therefore necessary to substitute deuterium in place of the hydrogen of the BEDT-TTF moiety to distinguish the C=C stretching peak from the C-C-H bending peaks.

Figure 1 shows the infrared conductivity spectra obtained by the Kramers-Kronig transformation for the reflectance spectra with light polarizations (E) parallel to the crystallographic b and c axes of the protonated (H) and deuterated (D) κ-(BEDT-TTF)$_2$[Cu(NCS)$_2$] salt. The broad conductivity peak near 1310 cm^{-1} in the E//b spectrum and near 1210 cm^{-1} in the E//c spectrum is hardly shifted by substituting deuterium for hydrogen, while the remaining sharp peaks at 1100-1300 cm^{-1} are significantly shifted toward the low frequency side. This clearly suggests that the broad peak is due to the EMV coupling, whereas the sharp peaks are related to the C-C-H bending modes. The EMV coupled frequencies Ω of the intramolecular vibrational modes thus determined are listed in Table 1.

On the basis of the assignments for κ-(BEDT-TTF)$_2$[Cu(NCS)$_2$], we have determined the EMV coupled frequencies of the intramolecular vibrational modes of the superconductor β-(BEDT-TTF)$_2$I$_3$ as well as the organic metal β"-(BEDT-TTF)$_2$AuBr$_2$. The infrared conductivity spectra of the salts are shown in Fig. 2. According to the discussion mentioned above, the intense peaks appearing at 1232 cm^{-1} in the spectrum of β-(BEDT-TTF)$_2$I$_3$ and at 1268 cm^{-1} in that of β"-(BEDT-TTF)$_2$AuBr$_2$ are attributed to the transition due to the totally symmetric C=C stretching mode induced by the EMV interaction.

Since the C-S stretching mode, coupled strongly with conduction electrons, appears in the 400-500 cm^{-1} region which was out of range for our reflectance measurements, we have examined transmittance spectra of κ-(BEDT-TTF)$_2$[Cu(NCS)$_2$] and β-(BEDT-TTF)$_2$I$_3$ in the range. Intense peaks are observed near 430 cm^{-1} in both salts. The frequencies Ω are listed in Table 1.

Table 1. Electron-molecular vibration coupling constant λ and energy g evaluated for the totally symmetric C=C and C-S stretching modes.

Compound	mode	ω_{CT}/cm^{-1}	ω/cm^{-1}	Ω/cm^{-1}	λ	g/meV
κ-(BEDT-TTF)$_2$[Cu(NCS)$_2$] Protonated salt	C=C					
	E//b	2910	1472	1317	0.15	70
	E//c	2390	1472	1209	0.20	74
	C-S	2650	501	435	0.25	50
Deuterated salt	C=C					
	E//b	2890	1477	1300	0.17	74
	E//c	2320	1477	1206	0.20	72
β-(BEDT-TTF)$_2$I$_3$	C=C	2450	1468	1230	0.19	73
	C-S	2450	501	428	0.26	49
β''-(BEDT-TTF)$_2$AuBr$_2$	C=C	2500	1470	1268	0.17	69

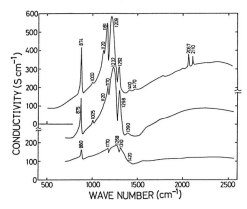

Fig. 2. Infrared conductivity spectra of protonated κ-(BEDT-TTF)$_2$[Cu(NCS)$_2$] (top), β-(BEDT-TTF)$_2$I$_3$ (middle), and β''-(BEDT-TTF)$_2$AuBr$_2$ (bottom).

We analyze the EMV coupling in terms of the dimer charge-oscillation model [3,5] because the BEDT-TTF molecules in the salts form a dimer as a repeating unit in the crystals [6,7]. In the model, the EMV coupling constant λ and energy g may be estimated by using the following equations,

$$(\omega^2 - \Omega^2)/\omega^2 = \lambda \omega_{CT}^2/(\omega_{CT}^2 - \omega^2) \tag{1}$$

and

$$g = (\omega \lambda \omega_{CT}/2)^{1/2}, \tag{2}$$

where ω is the bare frequency of the EMV coupled mode and ω_{CT} is the frequency of the charge-transfer transition in the dimer. We identify ω with the Raman frequency of the totally symmetric C=C stretching mode and ω_{CT} with the frequency of the electronic transition observed in the infrared region [2]. With the frequencies listed in Table 1, λ and g are obtained as is summarized also in Table 1. The constant λ is 0.15-0.20 for C=C and 0.25-0.26 for C-S. These values are very close to those of TTF salts [4].

If the energy g is assumed to be the electron-phonon interaction energy, the BCS theory in a weak-coupling limit predicts T_c as $T_c = 1.134\Theta_D\exp[-(gN_F)^{-1}]$, where Θ_D is the Debye temperature and N_F is

Table 2. Phonon frequency ω, coupling constant λ, screened Coulomb interaction term μ*, and critical temperature Tc. In parentheses are the observed values for the superconductors.

Compound	mode	$\hbar\omega/k_B$/K	λ	μ*	T_c/K
Protonated κ-(BEDT-TTF)$_2$[Cu(NCS)$_2$]	C=C	2120	0.20	0.013	11 (10.4)
	C-S	720	0.25	0.013	12 (10.4)
β-(BEDT-TTF)$_2$I$_3$	C=C	2110	0.19	0.030	5 (8.0)
	C-S	720	0.26	0.030	10 (8.0)

the density of states at the Fermi level. By combining the values g = 70 meV and N_F = 7.1 eV^{-1} spin^{-1} (f.u.)$^{-1}$ with T_c, Θ_D is obtained to be 68 K for κ-(BEDT-TTF)$_2$[Cu(NCS)$_2$]. This value is comparable to the value of Θ_D = 36 K obtained thermodynamically. Similar analysis for β-(BEDT-TTF)$_2$I$_3$ yields Θ_D = 49 K, being comparable to 52 K obtained thermodynamically.

The EMV coupling constant λ may be related to T_c in terms of Yamaji's theory using the following equation [4],

$$T_c = 1.134(\hbar\omega/k_B)\exp[-(\lambda - \mu^*)^{-1}], \qquad (3)$$

where k_B is the Boltzmann constant and μ* is the screened Coulomb interaction term. We have estimated μ* so as to reproduce observed T_c, since the effective Coulomb interaction in the conducting crystal would be sensitive to molecular arrangements. For β-(BEDT-TTF)$_2$I$_3$, μ* is estimated to be 0.030. This is about twice as large as 0.013, that estimated for κ-(BEDT-TTF)$_2$[Cu(NCS)$_2$]. Therefore, the Coulomb interaction in β-(BEDT-TTF)$_2$I$_3$ seems to be large compared with that in κ-(BEDT-TTF)$_2$[Cu(NCS)$_2$]. This difference between the Coulomb interactions in both salts is possibly due to the different molecular arrangements; the parallel array of the face-to-face BEDT-TTF molecular dimers in β-(BEDT-TTF)$_2$I$_3$ [7] and the perpendicular arrangement of the neighboring dimers in κ-(BEDT-TTF)$_2$[Cu(NCS)$_2$] [6]. The latter molecular arrangement brings about the quasi-two-dimensional nature of electronic properties. Since the EMV coupling constant λ is approximately the same for the conducting BEDT-TTF salts, the reduction of μ* could result in higher T_c. To confirm this, however, more elaborate estimation of the parameters is required.

References

1. T. Sugano, H. Hayashi, H. Takenouchi, K. Nishikida, H. Urayama, H. Yamochi, G. Saito, and M. Kinoshita, Phys. Rev. B, 37, 9100 (1988).
2. T. Sugano, H. Hayashi, M. Kinoshita, and K. Nishikida, Phys. Rev. B, 39, 11387 (1989).
3. M. Meneghetti, R. Bozio, and C. Pecile, J. Phys. (Paris), 47, 1377 (1986).
4. K. Yamaji, Solid State Commun., 61, 413 (1987).
5. R. Bozio, M. Meneghetti, and C. Pecile, J. Chem. Phys., 76, 5785 (1982).
6. H. Urayama, H. Yamochi, G. Saito, S. Sato, A. Kawamoto, J. Tanaka, T. Mori, Y. Maruyama, and H. Inokuchi, Chem. Lett., 459 (1988).
7. R. P. Shibaeva, V. F. Kaminskii, and V. K. Bel'skii, Kristallografiya, 29, 1089 (1984).

Dynamics of Charged Domain Walls in Semiconducting Charge Transfer Compounds

Y. Iwasa[1], *N. Watanabe*[1], *T. Koda*[1], *S. Koshihara*[2], *Y. Tokura*[2], *N. Iwasawa*[3], *and G. Saito*[3]

[1]Department of Applied Physics, University of Tokyo, Tokyo 113, Japan
[2]Department of Physics, University of Tokyo, Tokyo 113, Japan
[3]Institute for Solid State Physics, University of Tokyo, Tokyo 106, Japan

Abstract. One-dimensional molecular columns of many semiconducting charge transfer (CT) compounds are dimerized because of the bond-ordered wave (BOW) instability. Large dielectric constants and non-linear conductivity are commonly observed in these materials. Strong correlation between dielectric response and dc conductivity is understood in terms of the dynamics of domain-wall-type carriers characteristic of the dimerized stacks.

1. Introduction

A typical example of dimerized lattice is the 1/2-filled system of alkali-metal-TCNQ salts, which are semiconducting because of the large on-site Coulomb repulsion (Fig.1 (1)). If an alternating change occurs in the site energy of the adjacent molecules, the system becomes equivalent to the mixed-stack CT compounds with the ionic and dimerized ground state (Fig.1 (2)). These compounds are also semiconducting because of the site energy difference. In both systems, spin-Peierls-like transitions are observed and the lattice is dimerized at low temperatures.[1]

Some of the 1:2 TCNQ compounds show a $4k_F$ charge density wave (CDW)(Fig.1 (3)). A typical example is MEM-TCNQ$_2$, the lattice of which is dimerized at temperatures lower than 335K.

In all these systems, the ground state is doubly degenerated in terms of the lattice dimerization. In such a system, a domain-wall-like excitation (soliton) is thought to play an important role in the physical properties. This situation is quite similar to the BOW in the conjugated polymer trans-polyacetylene. In the polymer system, the existence of soliton has been confirmed experimentally,[2] but in the case of CT compounds, the solitonic nature of the low energy excitations is not well recognized yet. In order to clarify the nature of the charge excitation, we have measured the dielectric responses of CT crystals for which such solitonic excitations are expected to occur. It is reported here that anomalous dielectric properties are commonly observed in three groups of materials. A close correlation has been found between the dielectric and dc transport properties.

2. Dielectric response and dc conductivities

Samples were grown by the conventional solvent method, and the ac conductivity was measured in the frequency range $f(=\omega/2\pi)=$ 300Hz $\sim$ 10MHz. The dielectric constants were determined using the relation $\varepsilon_1 = \mathrm{Im}\sigma(\omega)/\omega$, $\varepsilon_2 = \{\mathrm{Re}\sigma(\omega) - \sigma(0)\}/\omega$. Electrical contacts were produced by painting silver paste onto single crystal samples. The dielectric constant for the stacking axis is one order of magnitude larger than that for the perpendicular direction. Typical dielectric response

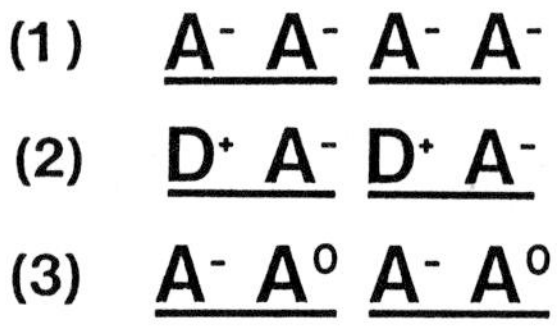

(1) A⁻ A⁻ A⁻ A⁻

(2) D⁺ A⁻ D⁺ A⁻

(3) A⁻ A⁰ A⁻ A⁰

Fig.1
Schematic representation of three kinds of dimerized stacks.
(1) 1:1 TCNQ salts
(2) mixed stack compounds
(3) 1:2 TCNQ salts

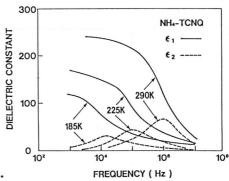

Fig.2
Dielectric dispersion for NH_4-TCNQ.

curves are shown in Fig.2 for NH_4-TCNQ which belongs to the group 1 in Fig.1. The real part ε_1 (solid line) is as large as several hundreds. It decreases slowly with frequency at low frequencies, but steeply decreases around the characteristic frequency $1/2\pi\tau$, where the imaginary part ε_2 (dashed line) shows a maximum. The observed frequency dependence of ε is approximately expressed by the Debye formula

$$\varepsilon(\omega) = \varepsilon(\infty) + \frac{\varepsilon(0) - \varepsilon(\infty)}{1 - i\omega\tau} \qquad (1)$$

An exact analysis should be done by the so-called generalized Debye formula, taking account of the distribution of the relaxation time τ.[3] At high temperatures, the dielectric constant at low frequencies is as large as several hundreds. As temperature is decreased, the relaxation frequency $1/2\pi\tau$ deceases. The temperature dependences of the τ value is expressed by the relation

$$\tau^{-1} = \tau_0^{-1} \exp(-E_\tau/kT). \qquad (2)$$

Here, E_τ is the activation energy, the physical meaning of which is mentioned below. As temperature decreases, the low frequency limit of dielectric constant $\varepsilon(0)$ decreases slightly. Such a dielectric response is more or less common to the three groups of materials shown in Fig.1. The observed Debye-like relaxation expressed in eq.(1) implies that there exist electric dipoles (oppositely charged bound pairs) which can expand and contract under an ac electric field. According to eq.(2), the motion of dipoles is hopping-type with the barrier height E_τ.

The dielectric constant ε is determined by the motion of the bound charges, while the dc conductivity σ by the motion of the unbound charges. There exists a strong relationship between the temperature dependences of ε and σ. All the materials investigated are semiconductors, where the temperature dependence of σ is approximately expressed by the activation-type relation $\sigma = \sigma_0 \exp(-E_\sigma/kT)$. E_σ is the activation energy of σ. In Fig.3, we show the correlation between the activation energies E_σ and E_τ for each material. Experimental points approximately fall on a straight dashed line, representing the relation $E_\sigma = E_\tau$.

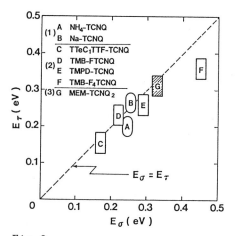
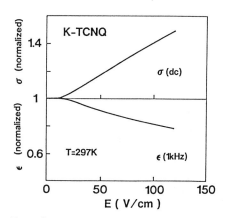

Fig.3
Relation between the activation energies E_σ and E_τ for several materials.

Fig.4
Electric field dependence of conductivity and dielectric constant for K-TCNQ.

This result indicates that the charged species responsible for the dielectric response and the dc transport are identical or very similar to each other in nature. In other words, both properties may be attributed to the same kind of oppositely charged pairs, which show a large dielectric response when they are bound and a dc conductivity when they are unbound. This property is very common not only in the BOW semiconductors but also in the mixed-stack CT compounds, the ground state of which is neutral.[4]

The strong correlation between the dc conductivity and the dielectric property is also confirmed by the observation of nonlinear conductivity. Figure 4 shows an electric field dependence of the dc conductivity σ and the dielectric constant ε_1 of K-TCNQ (Fig.1 (1)) at room temperature. As the electric field applied to the sample is increased, the σ increases and at the same time the ε_1 decreases. This nonlinearity is not attributed to the heating effect, since the ε_1 value would increase if temperature is raised as seen in Fig.2, in contrast with the decrease of the ε_1 under high electric fields. This result implies that the oppositely charged pairs which are bound by shallow potentials are dissociated and are driven into a free motion by the electric field. This tendency is observed in all materials we investigated. Especially, in the compound C in Fig.3, which shows the strongest nonlinearity, the dielectric constant ε_1 even becomes negative at high fields. The threshold electric field, where the nonlinearity appears, becomes lower when the activation energies E_σ, E_τ become smaller. There was also a general trend that the dielectric constant ε_1 at lower frequencies becomes larger when the two activation energies become small.

If the electric field is increased further, the conductivity increases significantly and shows a jump into a negative resistance regime. This effect is interpreted as due to a cooperative separation of bound charges.[5,6]

3. Domain wall dynamics

The characteristic dielectric and transport properties of BOW or dimerized semiconductors can be consistently interpreted in terms of the dynamics of solitons (domain walls of dimerized stacks). The lowest charge excitation in the one-dimensional BOW chain is theoretically predicted to be charged solitons.[7] Suppose that there exists a soliton pair. The region between the soliton pair is the domain having a different state of dimerization which is degenerated with that of the stack. These domains are excited thermally or introduced as native structure defects during the crystal growth process. In the BOW ground state, the dimerization of each stack is fixed so as to minimize the interchain interaction energy. Therefore, the average size of the domains of opposite dimerizations must not be so large. If the domain is confined by oppositely charged solitons which can move around, the size of the domain can be changed by applying an electric field. Therefore, this domain is regarded as a dipole which is responsible for the large dielectric response, but cannot contribute to the dc current. The observed large dielectric constant with Debye-type relaxation is interpreted in terms of the dynamics of the domain walls. The activation energy E_t, therefore, represents the barrier for the hopping motion of the charged solitons in this model.

On the other hand, if a domain wall pair is composed of charged soliton and spin soliton, this domain is similar to a polaron which can contribute to a dc current. Therefore, both the ac dielectric properties and dc transport properties are attributed to the same mechanism; the motion of the charged solitons. The strong correlation observed experimentally between dielectric constant and dc conductivity is reasonably understood in this manner. According to this model, the relation $E_\sigma = E_\tau$ (Fig.3) strongly indicates that the activation energy for dc conductivity is determined by a hopping process of the carrier rather than a carrier concentration.

From the nonlinear response of σ and ε shown in Fig.4, it is expected that a confined charge soliton pair is easily dissociated into unbound carriers under a high electric field. The energy barrier for this dissociation process is considered to be fairly small, since the nonlinearity is observed at comparatively low fields. (about 10V/cm in K-TCNQ at 297K)

It is worthwhile to note that the specific features, i.e., a huge dielectric constant and nonlinear conductivity, are also observed in the incommensurate CDW system in common.[8] In the latter system, the observed features are explained in terms of the CDW deformation and the depinning. However, in the strongly dimerized systems, the picture is quite different: the domain-wall-type excitations are playing the dominant role.

References

1. J.B.Torrance, in Low Dimensional Conductors and Superconductors, NATO ASI Series, edited by D.Jérome and L.G.Caron (Plenum, New York, 1987).
2. For a review, A.J.Heeger, S.Kivelson, J.R.Schrieffer, and W.-P. Su, Rev. Mod. Phys., 60,781 (1988).
3. S.Havriliak and S.Negami, J. Polym. Sci. C15, 99 (1966).
4. Y.Tokura, S.Koshihara, Y.Iwasa, H.Okamoto, T.Komatsu, T.Koda, N.Iwasawa, and G.Saito, Phys. Rev. Lett. 63, 2405 (1989).

5. Y.Tokura, H.Okamoto, T.Koda, T.Mitani, and G.Saito, Phys. Rev. B38, 2215 (1988).
6. Y.Iwasa, T.Koda, Y.Tokura, S.Koshihara, N.Iwasawa, and G.Saito, Phys. Rev. B39, 10441 (1989).
7. N.Nagaosa, J. Phys. Soc. Jpn., 55, 3754 (1986); 55, 3488 (1986).
8. R.J.Cava, R.M.Fleming, P.Littlewood, E.A.Rietman, L.F. Schneemeyer, and R.G.Dunn, Phys. Rev. B30, 3228 (1984).

The Effect of Pressure on the High Magnetic Field Electronic Phase Transition in Graphite

Y. Iye[1], C. Murayama[1], N. Mori[1], S. Yomo[2], J.T. Nicholls[3], and G. Dresselhaus[3]

[1]Institute for Solid State Physics, University of Tokyo, Roppongi, Minato-ku, Tokyo 106, Japan
[2]Department of Electronic and Information Engineering, Hokkaido Tokai University, Minami-ku, Sapporo 005, Japan
[3]Massachusetts Institute of Technology, Cambridge, MA 02139, USA

Abstract: The effect of hydrostatic pressure on the magnetic-field-induced electronic phase transition in graphite has been studied. The phase boundary shifts towards higher fields and lower temperatures with increasing pressure. To a first approximation, the pressure effect is accounted for by incorporating in the BCS-like expression for the transition temperature, the pressure dependences of the Fermi energy and the density of states, which enter through the graphite band parameter γ_2.

1. Introduction

Graphite occupies a unique position on the boundary between organic and inorganic systems. A magnetic-field-induced electronic phase transition in graphite has been studied by the present authors.[1-3] The features of the phase transition can be summarized as follows:

(1) The transition temperature T_c as a function of magnetic field B is empirically expressed as

$$T_c(B) = T^* exp(-B^*/B), \quad (1)$$

where T^* and B^* are fitting parameters.[1,2] For samples with a minimal concentration of ionized impurities, these parameters have been determined to be $T^* = 69\ K$ and $B^* = 104.7\ T$.[2] The above empirical relation is interpreted as a BCS-like expression for a pairing instability,

$$T_c(B) \sim 1.13\ \varepsilon_F\ exp(-1/N(\varepsilon_F)V), \quad (2)$$

by noting that the density of states at the Fermi level $N(\varepsilon_F)$ in high magnetic fields is proportional to B due to the Landau degeneracy factor.

(2) At the transition temperature $T_c(B)$, the conductivity shows a sudden decrease, indicating the development of a gap in the single particle excitation spectrum.[1,2]
(3) In the low temperature ordered phase, non-Ohmic transport reminiscent of collective transport in a charge density wave system has been observed.[3]
(4) For samples with a relatively high concentration of ionized impurities, the transition temperature at a given field is suppressed in such a way as to be attributable to the pair breaking effect of those impurities.[2]

Based on these experimental observations, we believe the nature of the low temperature phase to be a magnetic-field-induced charge density wave or a Wigner crystal state. In order to gain further insight into the nature of this electronic phase transition, we have carried out a high pressure experiment on this system. Since the graphite band parameters can be controlled by pressure, we can study the dependence of the phase transition on the band parameter values.

2. Experimental Method

Quasi-hydrostatic pressures up to $\sim 15\ kbar$ were attained in a 2.5 mm $dia.$ inner space of a teflon cell fitted in a 10 mm $o.d.$ copper-beryllium piston/cylinder clamp bomb. Because of the space limitation, we were not able to place a manometer in the cell together with the sample. The pressure values were therefore deduced from the relation between the clamped pressure and the resulting pressure at low temperatures, which was pre-calibrated for the cell by use of a lead manometer. The present experiments were done at two values of pressure, 5 $kbar$ and 10.5 $kbar$. The pressure cell was cooled in direct contact with the cryogenic helium-3 liquid. High magnetic fields up to 30.2 T were applied by use of a hybrid magnet at the Francis Bitter Magnet Laboratory, M.I.T. Resistance measurements were done by a standard d.c. method.

3. Experimental Results

Figure 1 shows traces of the magnetoresistance in the low field region for different pressure values. The effect of pressure on the Shubnikov-de Haas (SdH) oscillations is two-fold. First, the oscillatory features are shifted to higher fields with increasing pressure indicating an increase of the Fermi surface volume with pressure. Secondly, the amplitude of the SdH oscillations is significantly suppressed. The diminished SdH amplitude is partly due to an increase of the effective mass. But its more dominant cause presumably was introduction of dislocations during pressurization, since the SdH amplitude was not recovered when the same sample was brought back to ambient pressure. Such sample deterioration upon pressurization created particular difficulties for the present study, since the magnitude of the high field resistivity anomaly we were investigating critically depended on the sample quality.

Figure 2 shows high field portions of the magnetoresistance traces at several temperatures under 5 $kbar$. The onsets of resistivity anomaly are marked by arrows. The magnitude of the resistivity anomaly is much smaller than those typically observed at ambient pressure (see for example Fig. 1 of ref. 2). As the temperature is lowered, the magnitude of anomaly generally diminishes, for a reason that is not well understood. For the sample of Fig. 2, the resistivity anomaly was not well resolved for temperatures below $\sim 0.9\ K$ at this pressure.

The onset points of the resistivity anomaly are plotted in Fig. 3 for $p = 5\ kbar$ and 10.5 $kbar$ together with those for ambient pressure, obtained in earlier studies.[1,2]

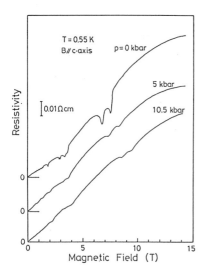

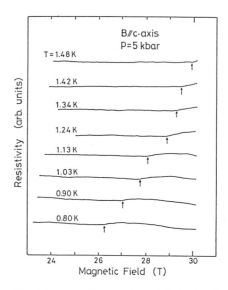

Fig.1 Traces of magnetoresistivity in the low field region showing the shift of the SdH oscillations with pressure. The ambient pressure data is for a different sample taken from the same batch of kish graphite as the one used for the high pressure experiment.

Fig.2 Traces of magnetoresistivity in the high field region showing the anomaly (indicated by arrows) associated with the phase transition. The magnitude of the resistivity anomaly is about one percent of the total resistivity.

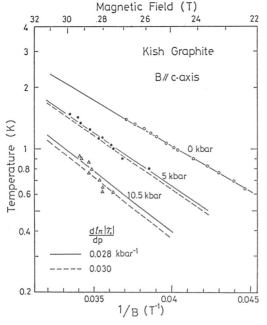

Fig.3 Onset points of the resisitivity anomaly at three different pressures plotted on a $\log T$ vs. $1/B$ scale. The lines represent the pressure dependence of the phase boundary according to Eq.(3). The solid and dashed lines correspond to $\frac{d \ln|\gamma_2|}{dP} = 0.028$ and $0.030\ kbar^{-1}$, respectively.

4. Discussion

The band structure of graphite near the Fermi level is appropriately described by the Slonczewski-Weiss-McClure parameters, γ_0, γ_1, γ_2, ... γ_5, and Δ.[4] Among these, the most important for the present problem is γ_2, where $2|\gamma_2|$ is the width of the π-band. The reported values for γ_2 lie between -0.0186 to $-0.0207 eV$. The application of pressure increases γ_2.[4] In the present pressure range, the increase is linear in pressure, $i.e.$ $\gamma_2(P) = (1+\alpha P)\gamma_2(0)$. The values for the logarithmic pressure derivative $\alpha \equiv \frac{d \ln|\gamma_2|}{dP}$ lie between $0.024\ kbar^{-1}$ and $0.043\ kbar^{-1}$.[4] Figure 4 schematically shows the lowest electron and hole Landau subbands and their density of states in the magnetic field range of our interest. The dashed and solid curves correspond to the situations at $P = 0$ and $10\ kbar$, respectively.

We now consider the pressure dependence of $T_c(B)$ expected from the functional form of Eq.(2). The preexponential factor is proportional to ε_F and hence to the π-band width $2|\gamma_2|$. The density of states at the Fermi level in the high field quantum limit is given by a product of the Landau degeneracy factor and the one-dimensional density of state of the lowest Landau subband. The latter factor is inversely proportional to the π-band width. Therefore, if we tentatively neglect a possible pressure dependence of the pairing interaction V and assume that the sole pressure dependence comes through γ_2, we obtain the following expression for the pressure dependence of $T_c(B)$

$$T_c(B, P) = T^*(1 + \alpha P)\ exp[-B^*(1 + \alpha P)/B], \qquad (3)$$

where T^* and B^* are the same as Eq. (1). The solid and dashed lines in Fig. 3 represent the phase boundaries as they are shifted by pressure according to Eq. (3). For the solid lines, the pressure coefficient was chosen as $\alpha = 0.028$, while for the dashed lines $\alpha = 0.030$. It is seen that good agreement can be obtained for a reasonable value of α.

Considering the simplicity of the model used here, which only takes into account the pressure-induced change in γ_2, we do not expect it to explain the full details of the pressure effect. In fact, the data points appear to deviate somewhat

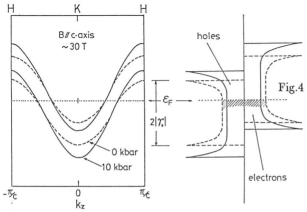

Fig.4 Schematic diagram of the dispersion along the k_z-axis of the lowest electron and hole Landau subbands at $B \sim 30\ T$. The dashed curves are for the ambient pressure, and the solid curves correspond to $P \sim 10\ kbar$. The figure on the right side shows the corresponding density of states profiles.

from the straight line behavior. It is possible that pressure dependences of the interelectron interaction potential responsible for the pairing and the basal plane effective mass play a role. Nonetheless, it is rather remarkable that the observed pressure effect can be accounted for, to a good first approximation, by simply considering the pressure-induced change in γ_2. This agreement gives additional support for the validity of Eq.(2) based on the BCS-like pairing instability as a model for the high magnetic field electronic phase transition in graphite.

References

[1] Y.Iye, P.M.Tedrow, G.Timp, M.Shayegan, M.S.Dresselhaus, G.Dresselhaus, A.Furukawa and S.Tanuma, Phys. Rev. **B25**, 5478 (1982).
[2] Y.Iye, L.E.McNeil, G.Dresselhaus, G.S.Boebinger and P.M.Berglund, *Proc. 17th Int. Conf. on Physics of Semiconductors*, eds. J.D.Chadi and W.A.Harris (Springer, New York, 1985) p.981.
[3] Y.Iye and G.Dresselhaus, Phys. Rev. Lett. **54**, 1182 (1985).
[4] N.B.Brandt, S.M.Chudinov and Ya.G.Ponomaev, *"Semimetals 1. Graphite and Its Compounds"* (North Holland, Amsterdam, 1988), and references therein

Ferro- and Antiferromagnetic Intermolecular Interactions of Organic Radicals, α-Nitronyl Nitroxides

K. Awaga, T. Inabe, U. Nagashima, and Y. Maruyama

Institute for Molecular Science, Myodaiji, Okazaki 444, Japan

Abstract. The temperature dependence of the magnetic susceptibilities for the three kinds of α-nitronyl nitroxides, 2-R-4,4,5,5-tetramethyl-4,5-dihydro-1H-imidazolyl-1-oxy 3-oxide (with R= phenyl (I), 3-nitrophenyl (II), and 4-nitrophenyl (III)) has been measured. It is found that the intermolecular spin interaction is ferromagnetic in the crystal of III, while it is antiferromagnetic in I or II. Furthermore, X-ray crystal analysis and MO calculation have been carried out on the nitroxide III, revealing a 2-D ferromagnetic network linked by the $N^{\delta+}...O^{\delta-}$ Coulomb attraction.

1. Introduction

There has recently been increasing interest in molecular ferromagnetism. In the 1960's, McConnell proposed [1] the resonance with the triplet charge-transfer (CT) state, for stabilizing the ferromagnetic (FM) intermolecular coupling, and it has been recently suggested [2,3] that the intramolecular exchange interaction (spin polarization effect) stabilizes the triplet CT state and results in FM coupling, based on the MO calculations. α-nitronyl nitroxide, a stable organic radical of S=1/2, has a quite interesting molecular structure from this point of view. The unpaired π-electron lies close to the non-bonding electrons (see Fig. 1), and therefore strong spin polarization caused by the n-π exchange interaction can be expected in this radical family [4].

In this report, we describe the magnetic properties of three α-nitronyl nitroxides, 2-R-4,4,5,5-tetramethyl-4,5-dihydro-1H-imidazolyl-1-oxy 3-oxide (with R= phenyl (I), 3-nitrophenyl (II), and 4-nitrophenyl (III), see Fig. 1). Furthermore, the magneto-structural correlation of III is discussed, based on the results of the X-ray crystal analysis and the MO calculation.

2. Experimental

The nitroxide radicals I, II and III were prepared by the reported method [5] and were purified by recrystallization from benzene or ether solution. The radical concentrations in them were estimated to be 100 % within experimental error from the Curie constants (see Table 1). The static magnetic susceptibility was measured under a

Table 1. The Curie and Weiss constants.

	C/emu K mol^{-1}	θ/K
I	0.375 ±0.05	-1.4 ±0.5
II	0.375	-0.5
III	0.374	+0.9

field of 1 T, by using a Faraday balance. The crystal data with $2<2\theta<60°$ were collected on a RIGAKU AFC-5 four-circle diffractometer with monochromatic Mo-K_α radiation.

3. Magnetic Properties [6]

The temperature, T, dependence of the paramagnetic susceptibilities χ_p of I, II and III was measured over the range from 2 to 250 K. χ_p is found to follow the Curie-Weiss law in this temperature range. The results are shown in Fig. 1, where $\chi_p T$ is plotted as a function of logarithms of T. $\chi_p T$ of the nitroxides I and II are almost constant in the high-temperature range and decrease monotonically below about 10 K, indicating the presence of weak antiferromagnetic (AFM) intermolecular interactions. On the other hand, the temperature dependence of χ_p of III is quite different from the others. $\chi_p T$ of III increases gradually with decreasing temperature, indicating FM coupling. The small humps commonly observed near 50 K could be due to adsorbed oxygen, whose amounts are very small. The Curie constants, C, and the Weiss constants, θ, of the three nitroxides, listed in Table 1, were determined from the reciprocal χ_p vs T plots (not shown). The FM interaction of III is considered to be comparable to the thermal energy of 1 K.

The FM coupling of III was firmly supported by the measurements of the field dependence of the magnetization at low temperatures [6]; the magnetization curves exhibited more rapid saturation than that of the S=1/2 spin entity. Through these magnetic measurements, anyhow, it is convincing that the magnetic coupling in these radicals changes from AFM to FM depending on the substituents, and consequently that the nitroxide III exhibits FM intermolecular interaction in contrast to the AFM interactions in I and II.

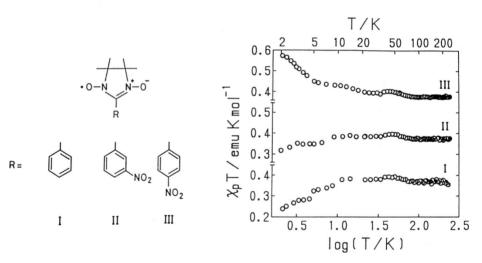

Fig. 1. The temperature dependence of the product $\chi_p T$ for I, II and III.

4. Magneto-Structural Correlation in the FM Nitroxide III [7]

The crystal of III is found to belong to the monoclinic Cc space group [a=10.960(3), b=19.350(3), c=8.257(3) Å, β=131.61(1)°, U=

1309.2(7) Å^3, Z=4, R(F)=0.0412 for 902 independent reflections with |F$_0$|>3σ(|F$_0$|)]. The nitroxide III has a twisted molecular structure, in spite of the fact that III can resonate with the planar quinonoid-structure. The phenyl ring is almost coplanar with the nitro group, but forms an angle of 50.3° with the plane of the nitronyl nitroxide, O-N-C-N-O. In nitroxide I, the corresponding dihedral-angle is about 30° [8,9], which is smaller than that in III. The molecular distortion in III could be ascribed to the effect of the intermolecular interaction, as is shown later.

The projection of the crystal structure of III along the b axis is shown in Fig. 2(a). There may occur weak intermolecular contacts between the O-atoms (O(1) and O(2)) in the NO groups and the N-atoms (N(3))in the NO$_2$ groups, and the nitroxide III forms a 2-D network by these contacts. Figure 2(b) shows the intermolecular conformation in this network, which is projected onto the ab plane. The N(3) is located at almost the mid point between O(1^i) and O(2^{ii}) in each of the neighboring radicals. The distances O(1^i)...N(3) and N(3)...O(2^{ii}) are 3.359(8) and 3.372(8) Å, respectively, and the angle O(1^i)...N(3)...O(2^{ii}) is 175.9°. The nitrophenyl plane is almost perpendicular to the O(1^i)...N(3)...O(2^{ii}) line.

A MO calculation has been carried out, by using an MNDO RHF-doublet method. It is found that the electronic polarization effect in III is so large that large positive and negative charges appear on N(3) and O-atoms, respectively. It is inferred from this calculation that this 2-D network is formed mainly by the Coulomb attraction between the negative charges on O(1^i) and O(2^{ii}), and the positive one on N(3). These intermolecular contacts probably result in the twisted molecular structure of III.

We consider the magnetic coupling in the 2-D network, referring to the results of the MO calculation. It is indicated that the unpaired electron occupying SOMO is localized on the side of the nitronyl nitroxide and has little population in the nitrophenyl ring, whereas NHOMO and NLUMO are distributed mainly in the nitrophenyl ring. NHOMO and NLUMO are defined as the highest

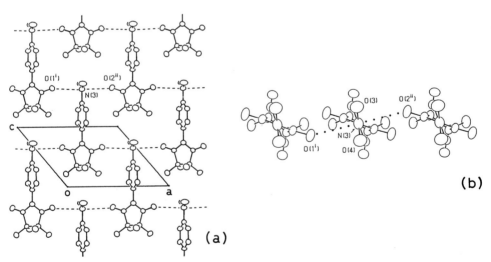

Fig. 2. (a) Projection of the structure of III along the b axis. Symmetry operations: (i) x, y, z+1; (ii) x+1, y, z+1. (b) Molecular arrangement of III in the 2-D network projected onto the ab plane.

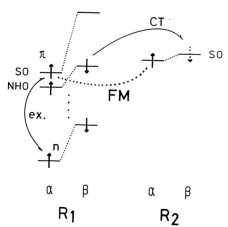

Fig. 3. FM coupling derived from the intramolecular exchange interaction and the intermolecular CT SO interaction.

doubly-occupied and the lowest unoccupied MO, respectively. At the contact points in the 2-D network, therefore, the intermolecular overlap between SOMO's could be much smaller than those between SOMO-NHOMO and/or between SOMO-NLUMO. For these, it could be concluded that the FM interaction of III originates mainly in the SOMO-NHOMO and/or SOMO-NLUMO CT interactions, in addition to the strong spin polarization effect. These features resemble those in the FM organic radical galvinoxyl [3]. Figure 3 shows schematically the mechanism of the FM coupling derived from the intramolecular exchange interaction and the intermolecular CT interaction. Details are given in Ref. [3]. The FM interaction in III is considered to work in this 2-D sheet through the N...O contacts.

Each of the 2-D sheets are connected by weak contacts, maybe hydrogen bondings, between O-atoms of NO and/or NO_2 groups and the H-atoms of phenyl rings. These contacts would also reflect the negative-charge polarization on the O-atoms. The nitroxide I, which has similar hydrogen bonding [9], shows an AFM interaction. Therefore, the interaction may not contribute to the FM coupling of III.

References

[1] H.M. McConnell: Proc. Robert A. Welch Found. Chem. Res. **11**, 144 (1967)
[2] K. Yamaguchi, T. Fueno, K. Nakasuji, and I. Murata: Chem. Lett. **1986**, 629
[3] K. Awaga, T. Sugano, and M. Kinoshita: Chem. Phys. Lett. **141**, 540 (1987)
[4] M.S. Davis, K. Morokuma, and R.W. Kreilick: J. Amer. Chem. Soc. **94**, 5588 (1972)
[5] E.F. Ullman, J.H. Osiecki, D.G.B. Boocock, and R. Darcy: J. Amer. Chem. Soc. **94**, 7049 (1972)
[6] K. Awaga and Y. Maruyama: Chem. Phys. Lett. **158**, 556 (1989); J. Chem. Phys. **91**, 2743 (1989)
[7] Kunio Awaga, Tamotsu Inabe, Umpei Nagashima, and Yusei Maruyama: J. Chem. Soc. Chem. Commun. **1989**, 1617
[8] W. Wang and S.F. Watkins: J. Chem. Soc. Chem. Commun. **1973**, 888
[9] Kunio Awaga, Tamotsu Inabe, and Yusei Maruyama: unpublished work

Part VIII

Structural Design of Organic Superconductors

Structure–Property Correlations in the Design of Organic Metals and Superconductors: An Overview

A.M. Kini[1], M.A. Beno[1], K.D. Carlson[1], J.R. Ferraro[1], U. Geiser[1], A.J. Schultz[1], H.H. Wang[1], J.M. Williams[1], and M.-H. Whangbo[2]

[1]Chemistry and Materials Science Divisions, Argonne National Laboratory, Argonne, IL 60439, USA
[2]Department of Chemistry, North Carolina State University, Raleigh, NC 27695, USA

> The submitted manuscript has been authored by a contractor of the U. S. Government under contract No. W-31-109-ENG-38. Accordingly, the U. S. Government retains a nonexclusive, royalty-free license to publish or reproduce the published form of this contribution, or allow others to do so, for U. S. Government purposes.

Abstract. Molecular structure and, more importantly, molecular packing in organic superconducting salts $(TMTSF)_2X$, $\beta\text{-}(BEDT\text{-}TTF)_2X$ and $\kappa\text{-}(BEDT\text{-}TTF)_2X$ will be examined in the context of deducing structure-property correlations in these systems. Such an approach has been instrumental in the discovery of superconductivity at 10.4 K in $\kappa\text{-}(BEDT\text{-}TTF)_2Cu(NCS)_2$, and it will continue to serve as an important tool in the rational design of new organic superconductors with even higher superconducting transition temperatures.

1. Introduction

Since the first discovery of superconductivity in 1979 in a cation-radical salt of the organic donor TMTSF, four other molecular systems — BEDT-TTF, $Ni(dmit)_2$, DMET and MDT-TTF — have now yielded superconducting solids. While there are about 30 superconductors known to date derived from these organic/metalloorganic precursors, BEDT-TTF has yielded the largest number (at least 10) of ambient pressure superconductors as well as the one with the highest transition temperature [T_c = 10.4 K in $\kappa\text{-}(BEDT\text{-}TTF)_2Cu(NCS)_2$]. This rapid progress in less than a decade is remarkable and it clearly raises the level of optimism for the discovery of "high-T_c" organic superconductors, particularly in view of the recent explosive progress made in the area of oxide superconductors.

2.1 Molecular and Structural Design

The organic conducting and superconducting solids are composed of "molecular building blocks", and their electronic bands are constructed from the overlap of molecular orbitals. Thus, a full understanding and control of the solid state properties rests primarily upon our ability to judiciously vary and control such parameters as molecular structure and molecular packing of these "building blocks" in the solid state. While the molecular design strategies for the preparation of electrically conducting solids are relatively well-understood [1], the structural design — our ability to induce molecules to pack in a desired and favorable (to achieve high conductivities and superconductivity) fashion — is still in its infancy. The preparation of a series of isostructural solids and the correlation of their properties with various identifiable molecular and structural parameters then becomes a crucial structural design strategy, from which we can gain important insights as an aid to the development of new materials. We present here an overview of such efforts, successful especially in the β-(BEDT-TTF)$_2$X series [2], which led to the discovery of the 10.4 K superconductor, κ-(BEDT-TTF)$_2$Cu(NCS)$_2$, with a distinctly different, nonstacking packing arrangement (the so-called κ-phase) of BEDT-TTF molecules. It is to be noted that no structure-property correlations yet exist for κ-phase salts. In the absence of such, we discuss some key molecular and structural parameters in known κ-phase salts and their relationship to the observed solid state properties. This analysis, we hope, will serve as a prelude to the development of useful structure-property correlations in the κ-phase superconductors.

2.2 (TMTSF)$_2$X and β-(BEDT-TTF)$_2$X Series

In the (TMTSF)$_2$X series of isostructural salts, the size of the anion dictates the packing density of TMTSF molecules. Thus, the inter-stack and intra-stack selenium-selenium network distances (critical parameters related to bandwidths and Fermi surfaces) have been correlated to the anion size by Williams et al. [3] and Kistenmacher [4]. This correlation can qualitatively rationalize the ambient pressure superconductivity in (TMTSF)$_2$ClO$_4$ while all other salts with anions larger than ClO$_4^-$ require applied pressure to become superconducting.

The cation-radical salts of BEDT-TTF, while being more complex due to the formation of multiple stoichiometries and structural phases (sometimes even within the same "apparently single" crystal [5]), for the same anion [2], are an ideal case in point to illustrate the intricate relationship between the structure (i.e., packing scheme of molecules) and the solid state properties. There are four (α, β, θ and κ) structurally unique phases known of BEDT-TTF-triiodide salts with a 2:1 stoichiometry. While the α-phase is a metal with a metal-insulator transition at 135 K, the three remaining phases are ambient pressure superconductors with T_c = 1.5 K (8 K, 0.5 kbar), 3.6 K and 3.6 K respectively. The different packing arrangements of BEDT-TTF molecules in these four salts are shown in Figure 1. The transfer integral (t) between two molecules is well-known to vary significantly depending upon the intermolecular separation and their relative orientation [6]. Thus, the electronic bandwidths and the shapes of the Fermi surfaces depend heavily on the packing arrangement of BEDT-TTF molecules.

The isostructural β-(BEDT-TTF)$_2$X superconductors, where $X^- $ = I$_3^-$, AuI$_2^-$, and IBr$_2^-$, with T_c = 1.5 K (8 K at 0.5 kbar), 4.9 K and 2.8 K respectively, provided the first opportunity to develop meaningful structure-property correlations in organic superconductors. These correlations (Figure 2) revealed a plausible way to prepare new superconductors with higher T_c's by simply going to longer (than I$_3^-$) linear anions. As evident from Figure 3, if the same structure is retained, the longer anions, which reside in a cavity created by –CH$_2$– units, will induce the BEDT-TTF molecules to pack less densely (than when X^- = I$_3^-$). This would result in narrower bandwidths (due to reduced orbital overlaps) and higher density of states at the Fermi level, and hence a higher T_c. The negative pressure dependance (–1K/kbar) of T_c's in β-(BEDT-TTF)$_2$X superconductors, i.e., suppression of T_c with applied pressure, also qualitatively lends support to this working hypothesis [7]. Realistically, however, there must be a limit to which this can be accomplished, since structural changes and dimensionality crossover (e.g., from 2-D to 1-D) may ensue at a certain longer anion length.

The superconducting (T_c = 10.4 K) cation-radical salt of BEDT-TTF with Cu(NCS)$_2^-$ counterion, first prepared by Japanese scientists [8] along a similar line of reasoning (in their case, the "effective volume" of BEDT-TTF), however, was shown to consist of not β-type but κ-type packing arrangement for BEDT-TTF molecules [9,10]. This unique packing motif does not consist of interacting stacks of molecules, as in β-(BEDT-TTF)$_2$X salts, but rather interacting dimers which are arranged nearly orthogonal to each other. This

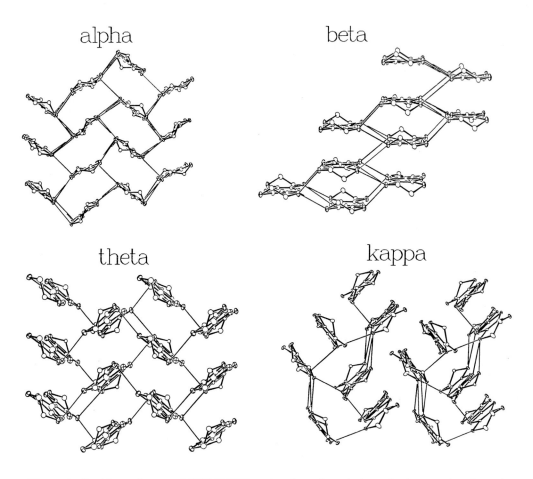

Figure 1. Packing schemes of BEDT-TTF molecules, viewed along the long molecular axis, in α, β, θ and κ phases of (BEDT-TTF)$_2$I$_3$.

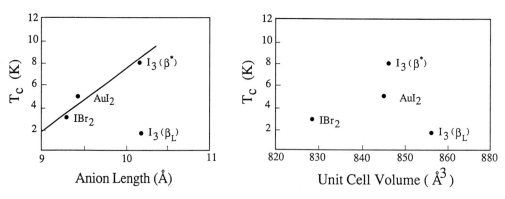

Figure 2. Plots of T_c vs. anion length (left) and T_c vs. unit cell volumes (right) in β-(BEDT-TTF)$_2$X salts, where X = I$_3$, AuI$_2$ and IBr$_2$.

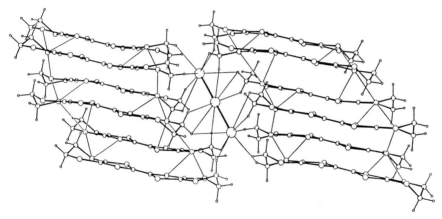

Figure 3. Structure of β-(BEDT-TTF)$_2$X salts from the perspective of the anion.

arrangement is still conducive to the formation of 2-dimensional S⋯S networks and the electronic band-structure calculations also reveal a 2-dimensional Fermi surface in this compound [10,11,12]. The 2-dimensional Fermi surface has also been recently verified experimentally by Shubnikov-de Haas oscillations observed in this salt [12,13]. Moreover, the anion in this case is not a linear, isolated species, but consists of a polymeric network involving tri-coordinated Cu(I). Thus, a new type of molecular packing arrangement has recently come to be recognized as a structure type that can also yield superconducting solids.

2.3 κ-(BEDT-TTF)$_2$X Series

The κ-type structure, first observed in a non-superconducting but metallic salt (BMDT-TTF)$_2$Au(CN)$_2$ [14], has now been found in superconducting salts derived from three other donor systems — BEDT-TTF, MDT-TTF and DMET (see Table 1). There are five BEDT-

Table 1. Intra-dimer Overlap Patterns, Intra-dimer Distances and Ethylene Endgroup Conformations in Known κ- or κ-like Salts

Compound	T_c	Overlap Pattern[*]	Intra-Dimer Distance, Å	Ethylene Conformation[#]
(BEDT-TTF)$_2$Cu(NCS)$_2$	10.4 K	B o R	3.35	S,S
(BEDT-TTF)$_2$I$_3$	3.6 K	B o R	3.35	E,E
(BEDT-TTF)$_4$Hg$_{3-\delta}$Cl$_8$	1.8 K (12 kbar) 5.3 K (29 kbar)	B o R	3.59[+]	S,S
(BEDT-TTF)$_2$Ag(CN)$_2$·H$_2$O	Non S.C. (Metal to 150 K)	B o R	3.7	E,S
(BMDT-TTF)$_2$Au(CN)$_2$	Non S.C. (T_{MI} 76 K)	B o B	3.6	--
(MDT-TTF)$_2$AuI$_2$	4.5 K	B o R	3.35	--
(DMET)$_2$AuBr$_2$	1.9 K	B o R	--	--

[*] B o R = Bond over Ring; B o B = Bond over Bond
[#] S = Staggered; E = Eclipsed
[+] At ambient pressure

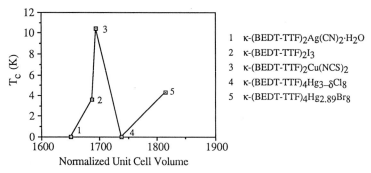

Figure 4. A plot of T_c vs. Normalized unit cell volumes of known κ-phase salts derived from BEDT-TTF.

TTF salts which are known to possess the κ-type structure, but no structure-property relationships, e.g., T_c vs. unit cell volume, have been elucidated yet (Figure 4). As can be seen in Table 1, κ-(BEDT-TTF)$_2$Cu(NCS)$_2$ and κ-(BEDT-TTF)$_2$I$_3$ [15] are ambient pressure superconductors, while κ-(BEDT-TTF)$_4$Hg$_{3-δ}$Cl$_8$ [16,17,18] requires applied pressure for superconductivity, and κ-(BEDT-TTF)$_2$Ag(CN)$_2$·H$_2$O [19] is metallic to 150 K. Yet another salt (BEDT-TTF)$_4$Hg$_{2.89}$Br$_8$, an ambient pressure superconductor (T_c = 4.3 K), is reported to be similar to κ-(BEDT-TTF)$_4$Hg$_{3-δ}$Cl$_8$, but its full structural details are presently not available [20,21]. Consequently, it will not be included in the following discussion.

We have examined intra-dimer overlap patterns (bond-over-ring or bond-over-bond), intra-dimer distances and conformations of ethylene end groups (eclipsed or staggered) of BEDT-TTF molecules in these κ-phase materials. This analysis has shed light on the question of why some κ-phase salts are superconducting, while some are not.

In the κ-(BEDT-TTF)$_2$X salts, the observed overlap pattern of the BEDT-TTF molecules forming the dimer is bond-over-ring type. This type of overlap is common to almost all segregated stack charge-transfer salts and is most likely due to inhomogeneous charge distributions on the radical-cation and radical-anion species [1,22]. The intra-dimer separations, however, show quite a variation: they are 3.35Å in both κ-(BEDT-TTF)$_2$-Cu(NCS)$_2$ and κ-(BEDT-TTF)$_2$I$_3$, and are 3.59 Å and ~ 3.7 Å, in κ-(BEDT-TTF)$_4$-Hg$_{3-δ}$Cl$_8$ and κ-(BEDT-TTF)$_2$Ag(CN)$_2$·H$_2$O, respectively. Since a smaller intra-dimer separation translates into a larger transfer integral and better electron delocalization, the absence of ambient pressure superconductivity in κ-(BEDT-TTF)$_4$Hg$_{3-δ}$Cl$_8$ and κ-(BEDT-TTF)$_2$Ag(CN)$_2$·H$_2$O can be reconciled with the larger intra-dimer distances. Furthermore, the superconducting transitions at 1.8 K and 5.3 K, under 12 kbar and 29 kbar pressure respectively, for κ-(BEDT-TTF)$_4$Hg$_{3-δ}$Cl$_8$ can also be understood in terms of intra-dimer distances most likely becoming smaller under applied pressure. A similar reasoning would then predict κ-(BEDT-TTF)$_2$Ag(CN)$_2$·H$_2$O to also become superconducting under pressure, but this compound has been reported to undergo a destructive phase transition (crystal shattering) when cooled below 150 K [23]. Therefore, a short intra-dimer distance (3.35 Å) appears to be critical for superconductivity.

Finally, some comments on the conformations of ethylene endgroups. Both the molecules forming the dimers in κ-(BEDT-TTF)$_2$Cu(NCS)$_2$ have ethylene groups in the staggered conformations, but they are both eclipsed in κ-(BEDT-TTF)$_2$I$_3$. The staggered conformation is the one found in the high-T_c (8 K) state of β-(BEDT-TTF)$_2$I$_3$ [24], which has been shown to result in "softer" anion-CH$_2$ contacts and hence a larger electron-phonon coupling constant (λ) than in other superconducting β-type salts with the eclipsed conformations [25]. Neutron diffraction studies on both κ-(BEDT-TTF)$_2$Cu(NCS)$_2$ and

κ-(BEDT-TTF)$_2$I$_3$ are needed to probe how the two different ethylene conformations are associated with the "hardness" or "softness" of anion-CH$_2$ contacts and to the observed T_c's in these salts. Difficulty in obtaining large, good quality single crystals suitable for such studies is the major stumbling block for clarifying this important question.

2.4 Other κ-Phase Salts

Besides BEDT-TTF, three other donors have also yielded cation-radical salts with κ-like structures. (BMDT-TTF)$_2$Au(CN)$_2$, first reported by Nigrey and coworkers [14], is the only metallic salt derived from BMDT-TTF thus far, and it undergoes a metal-semiconductor transition at 76 K. Application of pressure drives the transition to lower temperatures, but no superconductivity is observed up to 9000 psi (0.6 kbar) of pressure at or above 20 K [26]. Unlike BEDT-TTF salts, the dimer molecular pair in this salt has a rarely-observed bond-over-bond overlap pattern, and the intra-dimer separation is much larger (3.6 Å). Despite the larger separation, extended Hückel calculations reveal a much larger transfer integral in this dimer, owing to better orbital overlap attainable through the bond-over-bond overlap [11]. However, weaker dimer-dimer interactions and "hard" anion-CH$_2$ interactions are probably the reasons for the absence of superconductivity in this salt [11].

The ambient pressure superconductors derived from unsymmetrical donors, (MDT-TTF)$_2$AuI$_2$ [27,28] and (DMET)$_2$AuBr$_2$ [29], are quite interesting in that they have now clearly established that molecular symmetry is not essential for superconductivity. The overlap patterns in both these salts are bond-over-ring type, which is not due to the slipped packing pattern (as in BEDT-TTF salts) but rather due to the intrinsic asymmetry of the donor molecules. This may be an important consideration in the design of molecular dimers, which are the building blocks of κ-phase materials. In other words, unsymmetrical donors are possibly well-suited for the structural design of κ-phase materials. Coincidentally, the intra-dimer separation in (MDT-TTF)$_2$AuI$_2$ is 3.35 Å, same as that found in κ-(BEDT-TTF)$_2$Cu(NCS)$_2$ and κ-(BEDT-TTF)$_2$I$_3$!

It is noteworthy that ambient pressure superconductors with the highest T_c in the three donor systems, BEDT-TTF, MDT-TTF and DMET, all have the κ-type structures. The negative pressure derivative of T_c, (dT_c/dP), of κ-(BEDT-TTF)$_2$Cu(NCS)$_2$ is the largest (–3 K/kbar) recorded to date for any superconductor [30]. With the proper choice of anions, it is possible to enlarge the crystallographic unit cell in these salts to mimic the "negative pressure" effect and achieve higher superconducting transition temperatures. Thus, it is clear that κ-phase salts are very promising materials in the structural design of new superconductors.

3. Concluding Remarks

While the structure-property correlations in the κ-phase materials are still evolving at this point, it is our contention that structural subtleties discussed in the previous section have brought into focus their close relationship to the observed solid state properties. Obviously, more cation-radical salts of this structure type remain to be synthesized and studied, so that fully developed and useful structure-property correlations can be realized.

Finally, we consider the question of whether it is possible to "engineer" κ-type structures. Several approaches to build molecular dimers, the basic units of κ-phase materials, are deemed possible. As discussed earlier, unsymmetrical donors are good candidates in this regard. Secondly, transition metal-dithiolate chelate compounds are potential molecular components to build dimeric units, as shown by our recent work on Ni(dsit)$_2$ complexes [31]. Here, the square-pyramidal coordination of the transition metal ion by chalcogen atoms appears to favor a dimeric structure. There are numerous examples in the literature which illustrate this point. Synthesis of TTF-type compounds with a cyclophane-like structure is another way to engineer dimers, but this avenue is synthetically very challenging and laborious.

Inducing the potential molecular dimer units to pack in the desired κ-type structure is, however, not very straightforward. Molecular modelling studies may be a useful computational aid in this regard [22], but progress in this area has not been forthcoming because of the enormity of the problem beyond the dimer level. Resorting to trial-and-error methods using a variety of counterions, with a large dose of chemical "intuition", appears to be the only viable approach at this time. It may also be important to choose counterions which form "soft" donor-anion and acceptor-cation interactions, so that the "soft" phonon modes may give rise to strong electron-phonon coupling and hence higher superconducting transition temperatures.

Acknowledgement: Work at Argonne National Laboratory and North Carolina State University is supported by the U. S. Department of Energy, Office of Basic Energy Sciences, Division of Materials Sciences, under Contract W-31-109-ENG-38 and Grant DE-FG05-86ER45259, respectively.

References

1. D. O. Cowan: Proc. 4th Int'l Conf. New Aspects of Organic Chemistry, Kyoto, Japan, November 16–18, 1988, *New Aspects of Organic Chemistry* – I, ed. by Z. Yoshida, T. Shiba and Y. Ohshiro (Kodansha Ltd. Tokyo and VCH Verlagsgesellschaft, FRG 1989) In Press
2. J. M. Williams, H. H. Wang, T. J. Emge, U. Geiser, M. A. Beno, K. D. Carlson, R. J. Thorn, A. J. Schultz, M.-H. Whangbo: Prog. Inorg. Chem. 35, 51–218 (1987)
3. J. M. Williams, M. A. Beno, J. C. Sullivan, L. M. Banovetz, J. M. Braam, G. S. Blackman, C. D. Carlson, D. L. Greer, D. M. Loesing: J. Am. Chem. Soc. 105, 643–645 (1983)
4. T. J. Kistenmacher: Solid State Commun. 51, 275–279 (1984)
5. L. K. Montgomery, U. Geiser, H. H. Wang, M. A. Beno, A. J. Schultz, A. M. Kini, K. D. Carlson, J. M. Williams, J. R. Whitworth, B. D. Gates, C. S. Cariss, K. M. Donega, C. Wenz, W. K. Kwok, G. W. Crabtree: Synth. Metals 27, A195–A207 (1988)
6. S. S. P. Parkin, E. M. Engler, V. Y. Lee, R. R. Schumaker: Mol. Cryst. Liq. Cryst. 119, 375–387 (1985)
7. J. E. Schirber, L. J. Azevedo, J. F. Kwak, E. L. Venturini, P. C. W. Leung, M. A. Beno, H. H. Wang, J. M. Williams: Phys. Rev. B 33, 1987–1989 (1986)
8. H. Urayama, H. Yamochi, G. Saito, K. Nozawa, T. Sugano, M. Kinoshita, S. Sato, K. Oshima, A. Kawamoto, J. Tanaka: Chem. Lett. 55–58 (1988)
9. H. Urayama, H. Yamochi, G. Saito, S. Sato, A. Kawamoto, A. Tanaka, T. Mori, Y. Maruyama, H. Inokuchi: Chem. Lett. 463–466 (1988)
10. K. D. Carlson, U. Geiser, A. M. Kini, H. H. Wang, L. K. Montgomery, W. K. Kwok, M. A. Beno, J. M. Williams, C. S. Cariss, G. W. Crabtree, M.-H. Whangbo, M. Evain: Inorg. Chem. 27, 965–967 and 2904 (1988)
11. D. Jung, M. Evain, J. J. Novoa, M.-H. Whangbo, M. A. Beno, A. M. Kini, A. J. Schultz, J. M. Williams, P. J. Nigrey: Inorg. Chem. 28, 4516–4522 (1989)
12. K. Oshima, T. Mori, H. Inokuchi, H. Urayama, H. Yamochi, G. Saito: Phys. Rev. B 38, 938–941 (1988)
13. N. Toyota, T. Sasaki, K. Murata, Y. Honda, M. Tokumoto, H. Bando, N. Kinoshita, H. Anzai, T. Ishiguro, Y. Muto: J. Phys. Soc. Jpn. 57, 2616–2619 (1988)

14. P. J. Nigrey, B. Morosin, J. F. Kwak, E. L. Venturini, R. J. Baughman: Synth. Metals **16**, 1–15 (1986)
15. A. Kobayashi, R. Kato, H. Kobayashi, S. Moriyama, Y. Nishio, K. Kajita, W. Sasaki: Chem. Lett. 459–462 (1987)
16. R. N. Lyubovskaya, R. B. Lyubovskiĭ, R. P. Shibaeva, M. Z. Aldoshina, L. M. Gol'denberg, L. P. Rozenberg, M. L. Khidekel', Yu. F. Shul'pyakov: JETP Lett. **42**, 468–472 (1985)
17. R. B. Lyubovskiĭ, R. N. Lyubovskaya, N. V. Kapustin: Sov. Phys. JETP **66**, 1063–1067 (1987)
18. R. P. Shibaeva, L. P. Rozenberg: Sov. Phys. Crystallogr. **33**, 834–837 (1988)
19. M. Kurmoo, D. R. Talham, K. L. Pritchard, P. Day, A. M. Stringer, J. A. K. Howard: Synth. Metals **27**, A177–A182 (1988)
20. R. N. Lyubovskaya, E. I. Zhilyaeva, S. I. Pesotskiĭ, R. B. Lyubovskiĭ, L. O. Atovmyan, O. A. D'yachenko, T. G. Takhirov: JETP Lett. **46**, 188–191 (1987)
21. R. N. Lyubovskaya, E. A. Zhilyaeva, A. V. Zvarykina, V. N. Laukhin, R. B. Lyubovskiĭ, S. I. Pesotskiĭ: JETP Lett. **45**, 530–533 (1987)
22. M. Jørgensen, T. Bjørnholm, K. Bechgaard: Synth. Metals **27**, A159–A163 (1988) and references therein
23. P. Day, (Personal Communication to A. M. K. August 1989)
24. A. J. Schultz, H. H. Wang, J. M. Williams, A. Filhol: J. Am. Chem. Soc. **108**, 7853–7855 (1986)
25. M.-H. Whangbo, J. M. Williams, A. J. Schultz, T. J. Emge, M. A. Beno: J. Am. Chem. Soc. **109**, 90–94 (1987)
26. P. J. Nigrey, B. Morosin, J. F. Kwak: in *Novel Superconductivity*, ed. by S. A. Wolf and V. Z. Kresin, (Plenum Press, New York 1987) pp. 171–179
27. G. C. Papavassiliou, G. A. Mousdis, J. S. Zambounis, A. Terzis, A. Hountas, B. Hilti, C. W. Mayer, J. Pfeiffer: Synth. Metals **27**, B379–B383 (1988)
28. A. M. Kini, M. A. Beno, D. Son, H. H. Wang, K. D. Carlson, L. C. Porter, U. Welp, B. A. Vogt, J. M. Williams, D. Jung, M. Evain, M.-H. Whangbo, D. L. Overmyer, J. E. Schirber: Solid State Commun. **69**, 503–507 (1989)
29. K. Kikuchi, Y. Honda, Y. Ishikawa, K. Saito, I. Ikemoto, K. Murata, H. Anzai, T. Ishiguro, K. Kobayashi: Solid State Commun. **66**, 405–408 (1988)
30. J. E. Schirber, E. L. Venturini, A. M. Kini, H. H. Wang, J. R. Whitworth, J. M. Williams: Physica C **152**, 157–158 (1988)
31. M. A. Beno, A. M. Kini, U. Geiser, H. H. Wang, K. D. Carlson, J. M. Williams: This volume.

Organic Conductors and Superconductors Based on (BEDT-TTF)-Polyiodides

R.P. Shibaeva, E.B. Yagubskii, E.E. Laukhina, and V.N. Laukhin

Institute of Chemical Physics, USSR Academy of Sciences,
SU-142432 Chernogolovka, USSR

Abstract. A series of BEDT-TTF polyiodides, in which the ratio of BEDT-TTF : I varies from 1:1.5 to 1:5, is known to have a wide spectrum of physical properties.

Since the discovery of the first BEDT-TTF (ET)-based organic superconductor at ambient pressure, β-$(ET)_2I_3$ [1], the ET-I system has been intensively studied. It turned out that there existed a whole family of ET polyiodides having a wide range of physical properties - semiconductors, metals, superconductors [2,3].

A variety of ET polyiodides is explained by the diversity of possible polyiodide anions (I_3^-, I_5^-, I_8^{2-} etc.) on one hand, and by the lability of the ET cation-radical conformation and in particular by the conformational mobility of its terminal ethylene groups, on the other. Table 1 lists ET polyiodides known so far, whose crystal structures were determined and properties studied.

The η-, ε-, λ-, and ζ- polyiodides are obtained in pure chemical oxidation; those of θ, k, γ, δ in electrochemical oxi-

Table 1. Cation-radical salts in the ET-I system

ET-polyiodide	Properties	Refs.
α-$(ET)_2I_3$	M > 130 K	[2,4,5]
β-$(ET)_2I_3$	Sc ($T_c \sim 1.4 - 8$ K)	[1,6,7]
θ-$(ET)_2I_3$	Sc ($T_c \sim 3.6$ K)	[8]
k-$(ET)_2I_3$	Sc ($T_c \sim 3.6$ K)	[9]
γ-$(ET)_3(I_3)_{2.5}$	Sc ($T_c \sim 2.5$ K)	[10]
η-$(ET)I_3$	Sm	[11]
δ-$(ET)I_3(C_2H_3Cl_3)_{0.333}$	M > 160 K	[12]
ε-$(ET)_2I_3(I_8)$	Sm	[13]
λ-$(ET)_2I_3I_5$	Sm	[14]
ζ-$(ET)_2I_2I_8$	Sm	[15]

M, Sc, Sm are metal, super-, and semiconductor, respectively.

dation; α- and β- polyiodides in the combined way. It is noteworthy, that the θ- and k-phases are usually formed in the presence of AuI_2 [8,9]. As to the θ-phase crystals (with T_c ~ 3.5 K), they were obtained (though of a poor quality) in the presence of IBr_2 in our investigations at the early stages. Recently we have been surprised to have the θ-phase crystals together with the α- and β-phases in the electrochemical reaction of $(ET + ZnI_4)$ in benzonitrile.

Firstly it should be noted, that α-, β-, θ-, and k-phases are polymorphous modifications of the cation-radical salt $(ET)_2I_3$. Their crystal structure is characterized by the presence of cation-radical layers of ET alternating with those of linear I_3 anions. However, the internal structure of these layers is different. All these modifications are two-dimensional organic metals. Three of them undergo a superconducting transition. The other polyiodides (except the γ-phase) listed in Table 1, i.e. η, δ, ε, λ, and ζ, are simple salts in their stoichiometry (cation:anion = 1:1), therefore a priori they are semiconductors. There are no isolated cation-radical layers of ET in the structure of these phases. The ζ-phase structure is characterized by the presence of strongly dimerized ET stacks and $I_{10}^{2-} = I_8^{2-} + I_2$ complex polyanions. In the ε- and λ- structures mixed cation-anion layers of the $(ET)_2I_3$ composition alternate with the layers of I_8^{2-} and I_5^- anions, respectively.

The η-, ε-, λ-, and ζ-phases are dielectrics in their initial state. However, the η-, ε-, ζ-phases rich in iodine begin losing it under certain conditions and convert to $β-(ET)_2I_3$ with T_c ~ 6-8 K [7, 16, 17].

It is noteworthy, that among all the organic superconductors known so far $β-(ET)_2I_3$ is the most puzzling, giving much to be thought about. At present it is well defined that the organic quasi-two-dimensional metal $β-(ET)_2I_3$ has two low-temperature phases, $β_L$ and $β_H$ (with the temperature of the superconducting transition of 1.5 and 7-8 K, respectively). Besides, in some crystals there are typical pretransition phenomena observed as if notifying the presence of phases with the T_c values ranging from 1.5 to 8 K [17,18]. Some of these phases were really isolated, as became known later. The reasons for $β_L$ and $β_H$ formation are not clear so far and the nature of $β_H$ is not fully understood.

The method of stable $β_H-(ET)_2I_3$ production may be divided into two methods: chemical and physical ones.

Physical methods are (1) the effect of hydrostatic pressure $β_L \longrightarrow β_H-(ET)_2I_3$ [6,19-21]; (2) the conditions promoting twinning

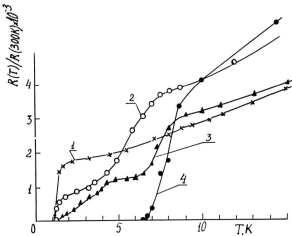

Fig. 1: Superconducting transition curves for the starting and deformed samples β-(ET)$_2$I$_3$: 1 - the starting sample; 2 - the sample after uniaxial compression at 295 K; 3 - the sample after inhomogeneous deformation at P = 15 kbar, T = 78 K; 4 - the sample after inhomogeneous deformation at P = 50 kbar, T = 295 K.

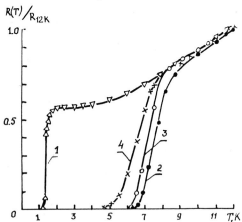

Fig. 2: Superconducting transition curves for crystals of β-(ET)$_2$I$_3$ obtained by different methods:
(1) β$_L$-(ET)$_2$I$_3$; (2) ε-(ET)$_2$I$_7$ ⟶ β$_H$-(ET)$_2$I$_3$;
(3) ζ-(ET)$_2$I$_{10}$ ⟶ β$_H$-(ET)$_2$I$_3$; (4) α-(ET)$_2$I$_3$ ⟶ β$_H$-(ET)$_2$I$_3$.

(uniaxial deformation of single crystals, nonhydrostatic compression) in single crystals β$_L$ [22] (Fig.1); (3) thermolysis α-(ET)$_2$I$_3$ ⟶ β$_H$-(ET)$_2$I$_3$ [23,24], η, ε, ζ ⟶ β$_H$-(ET)$_2$I$_3$ (Fig.2); and (4) annealing β$_L$ ⟶ β$_H$ [25].

Table 2. Structural data for β-(ET)$_2$I$_3$ under various conditions

β-(ET)$_2$I$_3$ phase	Conditions	Salient features	Refs.
β_L, $T_c = 1.5$ K	P_{amb}, 293 K	ET$_{disord}$ ($\beta + \beta^*$)	[29,30]
	P_{amb}, 125 K	ET$_{ord}$ (β,β^*) mod.str.	[31]
	P=9.5 kbar, 293 K	ET$_{ord}$ (β^*)	[32]
	P=1.5 kbar, 4.5 K	ET$_{ord}$ (β^*)	[33]
	P_{amb}, 4.5 K	ET$_{ord}$ (β,β^*) mod.str.	[33]
β_H, $T_c = 4.9$ K	P_{amb}, 293 K	ET$_{disord*}$ ($\beta + \beta^*$)	[27]
	P_{amb}, 125 K	ET$_{ord}$ (β^*)	[27]
$T_c = 6.4$ K	P_{amb}, 293 K	ET$_{disord}$ ($\beta + \beta^*$)	[34]

Chemical methods are (1) doping via the cationic part by the introduction of ET-derivatives, e.g., PT and MET, into the reaction of electrocrystallization [26,27]; (2) doping via the anionic part by the CuSCN participation (besides TBA.I$_3$) in the reaction in TCE [28].

The examination of the experimental data raised important problems to be solved:

1. Is the nature of the β_H-(ET)$_2$I$_3$ crystals the same, independent of the way of their production?
2. Why is the T$_c$ not the same for different β_H?

Unfortunately, in most cases the quality of crystals does not fit for the complete structural analysis, however, we still have some structural data for β_H to be analyzed and compared with those for β_L. These data are summarized in Table 2.

Under ambient conditions β_L is characterized by the presence of the internal disorder caused by the random orientation of one of the terminal ethylene groups of the ET cation-radical. Thus, the ET exists in the crystal with two equally probable conformations, e.g., staggered, β^*, and eclipsed, β, for the ethylene groups. The ordering of the ethylene groups at ambient pressure occurs at the temperature decrease, probably resulting in the modulated structure at T ~ 175 K [31]. However, in this case in the crystal ET again exists in two conformations, β and β^*, both at 125 and 4.5 K.

The pressure application leads to the complete positional ordering of the ethylene groups and, hence, to one conformation in ET, β^*. Such a structure is typical for β-(ET)$_2$I$_3$ at P = 9.5 kbar

($T = 293$ K) and at $P = 1.5$ kbar ($T = 4.5$ K) (see Table 2). The complete X-ray study of β_H-$(ET)_2I_3$ with $T = 4.9$ and 6.5 K showed them to have a room temperature structure with disordering of the ethylene groups. However, at 125 K as shown for β_H ($T = 4.9$ K) the modulated structure is not present and all the ET in the crystal have a β^* conformation typical for high pressures.

Thus, at first sight the $\beta_{1.5}$, $\beta_{4.9}$, and $\beta_{6.5}$ crystals at ambient pressure seem to have one and the same room-temperature structure and different structures at low temperatures.

The $\beta_{4.9}$ and $\beta_{6.5}$ crystals at room temperature are likely to differ from $\beta_{1.5}$ in the value of the β and β^* conformers contribution. If in $\beta_{1.5}$ they are equivalent, the β^* contribution is larger in β_H, therefore the temperature decrease is quite sufficient for the complete positional ordering in the ethylene groups to occur.

The analysis of these data showed that a sharp difference in T_c of β_L and β_H phases was caused by the presence of disordering and modulated structure or their absence.

The pressure increase and also the chemical doping promoting the appearance of local tension similar to the pressure effect suppress the superstructural transition.

However, such an explanation for the T_c increase is not valid for the case of the twinning effect, which is characteristic not only of β-$(ET)_2I_3$ crystals but also of isostructural crystals of β-$(ET)_2IBr_2$ and β-$(ET)_2AuI_2$, provided there are no modulated structures and all the ET are with one eclipsed conformation of the ethylene groups, β.

Besides, the samples of the β_H phase, which were obtained in the solid state reaction from the ε-phase and had a complete superconducting transition at $T_c = 6-7$ K at ambient pressure, are stable at room and higher temperatures, though far from being ideal, since they are polysynthetic twins [7].

Thus, we still have much to consider, since all the data stated above show that there are more problems than solutions regarding the organic superconductor β-$(ET)_2I_3$.

References

1. E.B. Yagubskii, I.F. Shchegolev, V.N. Laukhin, P.A. Kononovich, V.M. Kartsovnik, A.V. Zvarykina, L.I. Buravov: Pis'ma v ZhETF **39**, 12 (1984).

2. R.P. Shibaeva, V.F. Kaminskii, E.B. Yagubskii: Mol. Cryst. Liq. Cryst. **119**, 361 (1985).

3. E.B. Yagubskii, R.P. Shibaeva: J. Mol. Electron., **5**, 25 (1989).

4 K. Bender, K. Dietz, H. Endres, H.W. Helberg, I. Hennig, H.J. Keller, H.W. Schafer, D. Schweitzer: Mol. Cryst. Liq. Cryst. **107**, 45 (1984).

5 V.F. Kaminskii, V.N. Laukhin, V.A. Merzhanov, O.Ya. Neiland, Yu.V. Khodorkovskii, R.P. Shibaeva, E.B. Yagubskii: Izv. Akad. Nauk SSSR, ser. khim. 342 (1986).

6 V.N. Laukhin, E.E. Kostyuchenko, Yu.V. Sushko, I.F. Shchegolev, E.B. Yagubskii: Pis'ma v ZhETF **41**, 68 (1985).

7 A.V. Zvarykina, P.A. Kononovich, V.N. Laukhin, V.N. Molchanov, S.I. Pesotskii, V.I. Simonov, R.P. Shibaeva: Pis'ma v ZhETF **43**, 257 (1986).

8 H. Kobayashi, R. Kato, A. Kobayashi, I. Nishio, K. Kajita, W. Sasaki: Chem. Lett. 789 (1986).

9 R. Kato, H. Kobayashi, A. Kobayashi, S. Mariyama, Y. Nishio, K. Kajita, W. Sasaki: Chem. Lett. 507 (1987).

10 E.B. Yagubskii, I.F. Shchegolev, S.I. Pesotskii, V.N. Laukhin, M.V. Kartsovnik, P.A. Kononovich, A.V. Zvarykina: Pis'ma v ZhETF **39**, 275 (1984).

11 R.P. Shibaeva, P.M. Lobkovskaya, E.E. Laukhina, E.B. Yagubskii: Kristallographiya **32**, 901 (1987).

12 R.P. Shibaeva, R.M. Lobkovskaya, V.F. Kaminskii, S.V. Lindeman, E.B. Yagubskii: Kristallographiya **31**, 920 (1986).

13 R.P. Shibaeva, R.M. Lobkovskaya, E.B. Yagubskii, E.E. Kostyuchenko: Kristallographiya **31**, 455 (1986).

14 M.A. Beno, U. Geiser, K.L. Kostka, H.H. Wang, K.S. Webb, M.A. Forestone, K.D. Carlson, L. Nunez, J.M. Williams, M.H. Whangbo: Inorg. Chem. **26**, 1912 (1987).

15 R.P. Shibaeva, R.M. Lobkovskaya, E.B. Yagubskii, E.E. Kostyuchenko: Kristallographiya **31**, 1110 (1986).

16 V.A. Merzhanov, E.E. Kostyuchenko, V.N. Laukhin, R.M. Lobkovskaya, M.K. Makova, R.P. Shibaeva, I.F. Shchegolev, E.B. Yagubskii: Pis'ma v ZhETF **41**, 146 (1985).

17 E.B. Yagubskii, I.F. Shchegolev, V.N. Topnikov, S.I. Pesotskii, V.N. Laukhin, P.A. Kononovich, M.V. Kartsovnik, A.V. Zvarykina, S.G. Dedukh, L.I. Buravov: ZhETF **88**, 244 (1985).

18 L.I. Buravov, M.V. Kartsovnik, V.F. Kaminskii, P.A. Kononovich, E.E. Kostyuchenko, V.N. Laukhin, M.K. Makova, S.I. Pesotskii, I.F. Shchegolev, V.N. Topnikov, E.B. Yagubskii: Synth. Met. **11**, 207 (1985).

19 K. Murata, M. Tokumoto, H. Anzai, H. Bando, G. Saito, K. Kajimura, T. Ishiguro: J. Phys. Soc. Japan **54**, 1236 (1985).

20 V.B. Ginodman, A.V. Gudenko, I.I. Zasavitskii, E.B. Yagubskii: Pis'ma v ZhETF **42**, 384 (1985).

21 F. Creuzet, G. Creuzet, D. Jérome, D. Schweitzer, H.J. Keller: J. Phys. Lett. **46**, L1079 (1985).

22 A.V. Zvarykina, M.V. Kartsovnik, V.N. Laukhin, E.E. Laukhina, R.B. Lyubovskii, S.I. Pesotskii, R.P. Shibaeva, I.F. Shchegolev: ZhETF **94**, 277 (1988).

23 G.O. Baram, L.I. Buravov, A.S. Degtyarev, M.E. Kozlov, V.N. Laukhin, E.E. Laukhina, V.G. Onishchenko, K.I. Pokhodnya, M.K. Sheinkman, R.P. Shibaeva, E.B. Yagusbkii: Pis'ma v ZhETF **44**, 293 (1986).

24 D.Schweitzer, P. Bele, H. Brunner, E. Gogu, U. Haeberlen, I. Henning, T.Klutz, R. Swietlik, H.J. Keller: Z. Phys.B, Cond. Matter, **67**, 489 (1987).

25 S. Kagoshima, M. Hasumi, Y. Nagami, N. Kinoshita, H. Anzai, M.Tokumoto, G. Saito: Solid State Comm. (*in press*).

26 L.K. Montgomery, U. Geiser, H.H. Wang, M.A. Beno, A.J. Schultz, A.M. Kini, K.D. Carlson, J.M. Williams, J.R. Whiteworth, B.D. Gates,C.S. Cariss, C.M. Pipan, K.M. Donega, C. Wenz, W.K. Kwok, G.W. Grabtree: Synth. Met. **27**, A195 (1988).

27 M.A. Beno, A.M. Kini, L.K. Montgomery, J.R. Whiteworth, K.D. Carlson, J.M. Williams: Synth. Met. **27**, A219 (1988).

28 E.B. Yagubskii, N.D. Kushch: (*unpublished data*).

29 V.F. Kaminskii, T.G. Prokhorova, R.P. Shibaeva, E.B. Yagubskii: Pis'ma v ZhETF **39**, 15 (1984).

30 R.P. Shibaeva, V.F. Kaminskii, V.K. Bel'skii: Kristallographiya **29**, 1089 (1984).

31 P.C.W. Leung, T.J. Emge, M.A. Beno, H.H. Wang, J.M. Williams, V. Petricek, P. Coppens: J. Am. Chem. Soc. **107**, 6184 (1985).

32 V.N. Molchanov, R.P. Shibaeva, V.N. Kachinskii, E.B. Yagubskii, V.I. Simonov,B.K.Vainshtein: Dokl.Akad.Nauk SSSR **286**, 637 (1986).

33 A.J. Schultz, H.H. Wang, J.M. Williams, A. Filhol: J. Am. Chem. Soc. **108**, 7853 (1986).

34 R.P. Shibaeva (*unpublished data*).

Stoichiometry Control in Organic Metals

K. Bechgaard, K. Lerstrup, M. Jørgensen, I. Johannsen, and J. Christensen

Department of General and Organic Chemistry, H.C. Ørsted Institute,
University of Copenhagen, Universitetsparken 5, DK-2100 Copenhagen, Denmark

We discuss some simple ideas in order to obtain stoichiometry control as well as interesting Fermi surface effects in organic metals. The experimental approach is to prepare and investigate a series of new dimeric TTF's, so-called Bis-TTF's.

1. Introduction

So far the stoichiometry of new materials is obtained by serendipity, although in retrospect, when a new crystal structure is formed, it can often be "explained" by taking into account the space-filling properties of the molecules.

Especially we focus on the possibility of obtaining "2:1" stoichiometry in binary donor-acceptor materials, as this may lead to an interesting double Fermi surface. We also discuss the expected effects in simple 4:1 (TTF_4X) cation radical salts. In these materials also the effects of varying the bandfilling can be evaluated by employing mono- and divalent anions (X) respectively.

2. The concept

In a simple approximation the band structure of TTF-TCNQ and $TMTTF_2X$ can be depicted as shown in Fig. 1a and 1b. In this model the full symmetry properties of the real crystal systems are not utilized; we assume a model of uniform, independent stacks.

In TTF-TCNQ k_F is related to the charge transfer by the relation $k_F = (\pi/a)(\delta/2)$ and $k_{F(TCNQ)} = k_{F(TTF)}$. If a $2k_F$ charge density wave develops on one stack it will likely couple to the other stack and eventually a gap opens up all over the one-dimensional Fermi surface [1]. If, however, a 2:1 stoichiometry can be induced a different situation arises (Fig. 2).

The doubly degenerate TTF band contains twice as many states as the TCNQ band. This means that if a CDW-driven gap opens at $2k_F$ on the TCNQ stack, it will induce only a $4k_F$ periodicity on the TTF stack. Only for the unrealistic charge transfer 2/3 will $2k_F$ be the same on the two types of stacks.

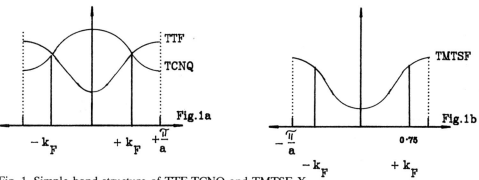

Fig. 1. Simple band structure of TTF-TCNQ and $TMTSF_2X$.

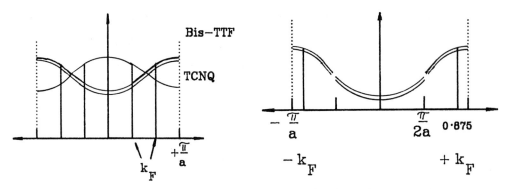

Fig. 2. Simple band structure of Bis-TTF-TCNQ. Fig. 3. Simple band stucture of "TMTTF$_4$X".

If, in contrast, the TTF stack drives a 2k$_F$ CDW transition, a gap opens all over the Fermi surface. The occupied number of states in the TCNQ band is twice that of the vacant number of states in the TTF band, and thus the normal Bragg condition k$_F$=n(π/a)(δ/2) is fulfilled for n=2, when δ denotes the average charge transfer from the TTF's.

In the TMTTF$_2$X and TMTSF$_2$X series a triclinic structure and a small dimerization opens a gap at 4k$_F$,[2]. If the basic features of this particular structural type can be conserved in a new series of Bis-TMTTF's, this will formally correspond to a stoichiometry "TMTTF$_4$X". The charge transfer in the simple model will be 0.125 and the dimerization gap at π/2a corresponds to an 8k$_F$ periodicity (Fig. 3).

If the desired material can be formed, the influence of an 8k$_F$ periodicity can be evaluated as well as the effects of bandfilling, because divalent anions are also expected to form conducting materials. In these, the bandfilling and dimerization will correspond to the 4k$_F$ situation characteristic of the TMTTF$_2$X series.

Finally we note that a "2:1" stoichiometry is expected if Bis-TTF's are intramolecularly bound to monovalent anions. If the lattice period remains a single intermolecular spacing, the band will be 3/4 filled.

3. Experimental approach

As described previously [3] dimeric TTF's are now available in reasonable quantities. Therefore inspection of stereotype stacking patterns in TTF-TCNQ type materials should help select good candidate molecules. We note that there are three typical stacking patterns: Chessboard-like (type 1,[4]) and two different sheet-like: Type 2, [5] and type 3, [6] interstack arrangements.(See fig. 4).

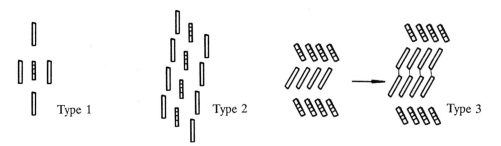

Fig. 4. Schematic stack arrangements in TTF-TCNQ type materials.

Fig. 5. Bis-(DMTTF)-ethane.

Of these we single out type 3 as the best choice because an extra layer of TTF's can be very easily introduced by connecting two TTF's end-to-end as in the Bis-DMTTF-ethane shown in fig. 5. Note that the chessboard-like structure does not support the present approach, whereas the TTF-TCNQ (type 2) structure could work if TTF's are "dimerized" side-by-side.

Also Bis-DMTTF-ethane is a good candidate for the $8k_F$ type material described above because the terminal space filling properties of the molecule mimics that of TMTTF and could induce the characteristic zig-zag stacking known from the TMTTF and TMTSF series.

4. Results

When solutions of Bis-DMTTF-ethane are treated with TCNQ or DMTCNQ black solids of composition (Bis-DMTTF-ethane)TCNQ and (Bis-DMTTF-ethane)DMTCNQ are formed i.e. the desired 1:1 stoichiometry. Both solids are conducting (compaction pellet), but only (Bis-DMTTF-ethane)DMTCNQ so far, in our hands, yielded single crystals. Fig. 6 shows the preliminary resistivity vs. T^{-1} curve for this material [7]. The behavior resembles that of a Coulomb localized material, and there is so far no direct evidence for a $2k_F$ transition.

When solutions of Bis-DMTTF-ethane in chlorobenzene/butyronitrile containing PF_6^- anions are electrochemically oxidized, black needle-like crystals are obtained. The material is semiconducting and the composition corresponds to (Bis-DMTTF-ethane)PF_6,$(C_4H_7N)_{(0.5)}$. The structure of this material has been solved and it exhibits the expected zig-zag structure, but in isolated stacks separated by solvent molecules and with no common anions for the individual stacks. In fig. 7 the structure is shown schematically [8]. The material contains twice as many

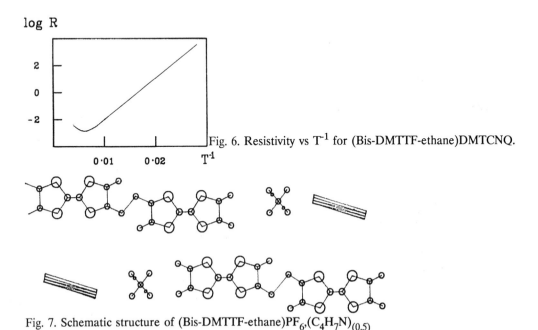

Fig. 6. Resistivity vs T^{-1} for (Bis-DMTTF-ethane)DMTCNQ.

Fig. 7. Schematic structure of (Bis-DMTTF-ethane)PF_6,$(C_4H_7N)_{(0.5)}$

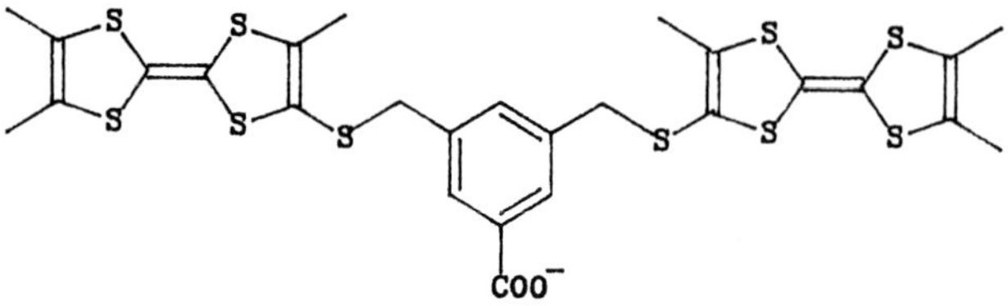

Fig. 8. Internally charge transfer controlled structure.

anions as expected and it is <u>not</u> an "8 k_F" situation; the simple band structure corresponds to generation of a "4 k_F" potential by the anions, and it is further complicated by the disordered solvent molecules. More experimental information is needed. We speculate that it is possible that the "stable" structure obtained also reflects that, if charge transfer is too small (0.125 in the expected material), the Coulombic forces are not sufficient to stabilize the structure. Instead, the present material where the charge transfer amounts to 0.25 as in TMTTF$_2$X is generated.

Finally we present a potential material where stoichiometry control is generated intramolecularly. The material shown below (Fig. 8) contains two TTF-moieties connected to a benzoate. The ion has been prepared [9], but so far we have not obtained a well defined material upon oxidation.

5. Conclusion

We have described a simple analysis of expected band structure effects, when employing a new series of dimeric TTF's. In only one case, (Bis-(DMTTF)DMTCNQ), did we obtain a solid suitable for further investigation. We emphasize that, as always, the main problem is to generate single crystals with good morphology. We expect, that further derivatization of dimeric TTF's [3] may give new materials, where the described ideas can be evaluated.

Acknowledgements

This work was supported by ESPRIT BRA contract no:3121, and The Danish Natural Science Research Council contract no:5.21.99.38

References

1. Barisic and D.Bjelis in "Theoretical aspects of band structures and electronic properties of one-dimensional solids". H.Kamimura ed., D.Reidel Publ. 1985, 49.
2. D.Jérome and H.Schulz, Adv. in Phys., <u>31</u> (1982) 299.
3. K.Lerstrup et al., this volume.
4 T.E.Phillips, T.J.Kistenmacher, A.N.Bloch and D.O.Cowan, JCS, Chem.Comm., (1976) 334.
5. T.J.Kistenmacher, T.J. Phillips and D.O.Cowan, Acta Cryst. <u>B30</u> (1978) 763.
6. J.R.Andersen, K.Bechgaard, C.S.Jacobsen, G.Rindorf, H.Soling and N.Thorup, Acta Cryst., <u>B34</u> (1978) 1901.
7. C.S.Jacobsen, private communication.
8. G.Rindorf et al., submitted to Acta. Cryst.
9. J.Christensen et al., to be published.

Unusual Molecular Systems of Organic-Inorganic Character and Increased Architectural Complexity

P. Batail, K. Boubekeur, A. Davidson, M. Fourmigué, C. Lenoir, C. Livage, and A. Pénicaud

Laboratoire de Physique des Solides Associé au CNRS,
Université de Paris-Sud, F-91405 Orsay, France

Abstract. Novel, unprecedented structure types are described and rationalized for three sets of molecular hybrid salts between TTF or BEDT-TTF and the all-inorganic or metal organic anions $Re_6Q_5Cl_9^-$, (Q = S, Se); $SiW_{12}O_{40}^{4-}$ and $V(dddt)_3^-$.

1. Introduction

This contribution focuses on the structural design of three sets of molecular hybrids. These examples demonstrate the ability [1] of three-dimensional all-inorganic or metal organic molecular ions, with low-lying essentially delocalized frontier orbitals, to serve as frameworks or templates in the construction of TTF-based charge transfer salts with novel patterns of association and solid state phenomena.

2. Modelling the Donor Patterns with Monovalent Hexanuclear Chalcohalide Re_6 Clusters

The donor patterns described in Fig. 1 are unprecedented in any TTF or BEDT-TTF salt. Despite the close proximity of the dimers in **1** (S...S distances of 3.615 and 3.643 Å for Q = S and Se), there is no interaction between the dimer spins as exemplified by Curie-like susceptibilities [2a]. This contrasts with the situation in **2** (Fig. 1) where longer (3.735 Å),

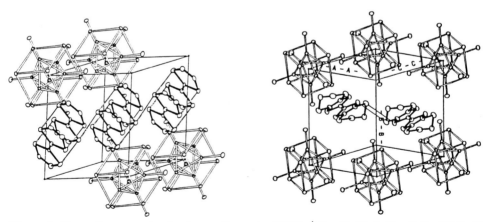

Figure 1: From discrete, mixed-valence dimers in $(TTF)_2^+ \cdot Re_6Q_5Cl_9^-$, Q = S, Se (**1**) (left); to uniform, non-interacting stacks of spins in $BEDT-TTF^+ \cdot Re_6Se_5Cl_9^-$ (**2**) (right).

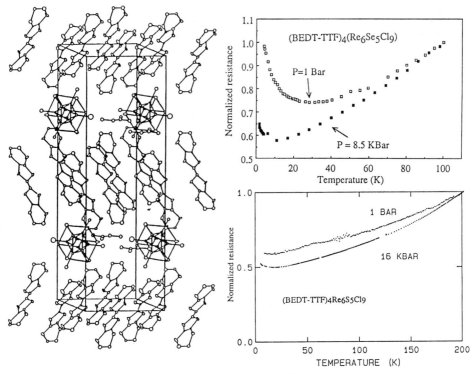

Figure 2: Structure and conductivity of $(BEDT-TTF)_4^+ \cdot Re_6Q_5Cl_9^-$, Q = S, Se (3).

albeit properly oriented, S...S contacts result in an antiferromagnetic interaction characteristic of non-interacting, very 1-D Heisenberg chains of spins with $J_{//} = 67$ K [2b].

Donor sheets are formed in 3 (Fig. 2) with a large donor/anion molar ratio and a very dilute charge on the organic monolayers, still compatible however with metallic conductivities ($\sigma_{300\ K}$ = 50 and 25 S cm^{-1} for Q = S and Se) down to low temperature [2c].

3. BEDT-TTF - Polyoxometallate Extended Acentric Sandwiches

Semiconducting sandwiches, exemplified by $(BEDT-TTF)_8SiW_{12}O_{40}$ (4) (Fig. 3), are obtained with metal oxide anions with the acentric (T_d) α-Keggin structure. Since the translational symmetry-related in-plane anions have the same orientation, the inorganic monolayer itself is non-centrosymmetrical. The mode of alternated stacking of the organic and inorganic layers on top of each other is such that this acentric character is transferred in the transverse direction to the whole solid (I2) via specific C-H...O hydrogen contacts [3].

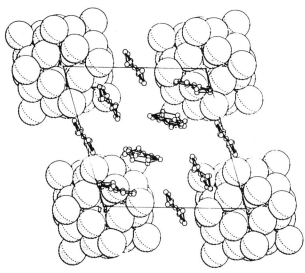

Figure 3: Relative anchoring of the organic and $SiW_{12}O_{40}^{4-}$ monolayers in **4**.

4. TTF[V(dddt)$_3$]: A Molecular Analog of the Prototypical Layered Compounds NbS$_2$ and TaS$_2$

The anticipation that trigonal prismatic transition metal dithiolene complexes such as V(dddt)$_3^-$ (dddt = 5,6-dihydro-1,4-dithiin-2,3-dithiolate) [4a] are appropriate candidates for the design of 2-D pseudo-tetragonal or hexagonal nets (Scheme 1) with three-fold (triangular) aromatics or molecular ions is verified by the analysis of the structure of TTF[V(dddt)$_3$] (P2$_1$/c, a = 15.385(6), b = 10.071(5), c = 19.114(8) Å, β = 95.17(2)°) [4b]. Indeed, the monovalent complex anions (symbolized by spheres of arbitrary radius in Fig. 4) form a pseudo-cubic face centered array in which half of the octahedral interstices are occupied by the divalent cations (TTF$^+$)$_2$. As demonstrated in Figure 4, this is accomplished by filling all the interstices between hexagonal close-packed anion layers in *alternate* layers. This arrangement, described here for the first time for a *molecular* material, is known as the cadmium hydroxide structural type [5] and is found in numerous layered materials. The compound is insulating and single-crystal ESR experiments performed on the plate-like crystals reveal two resonance lines at room temperature [4b].

Continuing collaborations with E. CANADELL, C. COULON, G. HERVE, D. JEROME and A. PERRIN are gratefully acknowledged as well as EEC Esprit Basic Research Action (MOLCOM n° 3121) for support.

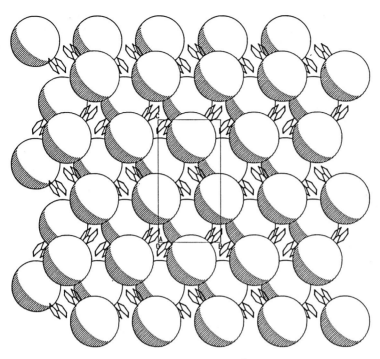

Figure 4: Projection of the [V(dddt)$_3^-$]-(TTF$^+$)$_2$-[V(dddt)$_3^-$] layer.

References

[1] P. Batail, L. Ouahab, A. Pénicaud, C. Lenoir and A. Perrin, *C. R. Hebd. Seances Acad. Sci. (Paris)*, 304 sér. II, 1111 (1987); A. Pénicaud, P. Batail, C. Coulon, E. Canadell and C. Perrin, submitted to *Chemistry of Materials*; A. Penicaud, P. Batail, P. Davidson, A-M. Levelut, C. Coulon and C. Perrin, submitted to *Chemistry of Materials*; G. Matsubayashi, K. Akiba and T. Tanaka, *Inorg. Chem.*, 27, 4744 (1988); W.E. Broderick, E.M. McGhee, M.R. Godfrey, B.M. Hoffman and J.A. Ibers, *Inorg. Chem.*, 28, 2904 (1989); A. Renault, D. Talham, J. Canceill, P. Batail, A. Collet and J. Lajzerowicz, *Angew. Chem. Int. Ed. Engl.*, (1989), in press.

[2] (a) K. Boubekeur, C. Livage, P. Batail, E. Canadell and A. Perrin, to be submitted; (b) A. Pénicaud, C. Lenoir, P. Batail, C. Coulon and A. Perrin, *Synth. Met.*, 32, 25 (1989); (c) A. Pénicaud, K. Boubekeur, P. Batail, E. Canadell and A. Perrin, to be submitted.

[3] A. Davidson, K. Boubekeur, A. Pénicaud, P. Auban, C. Lenoir, P. Batail and G. Hervé, *J. Chem. Soc. Chem. Commun.*, (1989), in press.

[4] (a) J.H. Welch, R.D. Bereman and P. Singh, *Inorg. Chem.*, 27, 2862 (1988); (b) C. Livage, P. Batail and E. Canadell, to be submitted.

[5] B.G. Hyde and S. Andersson, *'Inorganic Crystal Structures'*, J. Wiley & Sons, New York, p. 66 (1989).

Part IX

New Molecules and Materials

Salts Derived from Bis(ethylenedioxa)tetrathiafulvalene ("BO")

F. Wudl, H. Yamochi, T. Suzuki, H. Isotalo, C. Fite, K. Liou, H. Kasmai, and G. Srdanov

Institute for Polymers and Organic Solids, Departments of Physics and Chemistry, University of California, Santa Barbara, CA 93106, USA

Abstract. The oxygen analog of BEDT-TTF, BEDO-TTF (BO), which has the same color as BEDT-TTF but is much more soluble in most organic solvents and has much lower ionization potential in solution($^1E_{1/2}$ = 0.435 and $^2E_{1/2}$ = 0.699 V vs Ag/AgCl, CH$_3$CN), forms organic metals with the same anions as its sulfur analog[I$_3$, IBr$_2$, Cu(SCN)$_2$]. Its solid state structure consists of sheets of tub-shaped molecules with very close intermolecular S•••O and S•••S distances. The salt (BO)I$_{1.25}$ is a metal with the following characteristics: a resistance ratio between 300K and 20K of ~250, a Pauli magnetic susceptibility in the same temperature interval and a monotonically decreasing EPR linewidth with decreasing temperature. This combination of properties is unprecedented in organic metals. Other properties of this and other salts of the new donor are also presented.

Introduction

In the TTF family, the organic metallic state can be achieved with the unsubstituted and variously substituted sulfur, selenium, and tellurium heterocycles [1] but the superconducting state can, so far, be achieved only with derivatives akin to tetramethyl tetraselenafulvalene and bis(ethylenedithio)-tetrathiafulvalene (BEDT-TTF, a.k.a. "ET") [2] as well as their hybrids [2e]. The highest recorded superconducting transition temperature for an organic superconductor is T_c = 11.2K [3] for (d_8-BEDT-TTF)$_2$Cu(NCS)$_2$, and for very pure [3] (BEDT-TTF)$_2$Cu(NCS)$_2$ previously reported by Saito's group [4] as T_c = 10.4K for the protio sample. While these temperatures are considerably lower than those of the inorganic ceramic, copper oxide-based, superconductors (125K) [5]; the rate at which the organic superconductor transition temperatures have been rising with time is still quite impressive [(TMTSF)$_2$PF$_6$, 0.9K/25kilobar in 1980 [6] to (d_8-BEDT-TTF)$_2$Cu(SCN)$_2$, 11.2K/1bar in 1988] [3,4].

The advent of the copper oxide superconductors [7] has inspired a number of theorists to suggest that the chemical species responsible for the mixed valence in these solids is not the copper couple (Cu^{II}/Cu^{III}) but essentially oxygen radical cations [8]. Some have gone as far as suggesting that results of their calculations can be extrapolated to the design of organic polymeric superconductors [8(a),(e)]. Inspired by these results, we have been thinking about the possibility of observing metallic and perhaps superconducting behavior with charge transfer salts based on oxygen containing donors. Indeed, if the organic superconductors were BCS superconductors, the lighter the component atoms within a series of identical donors, the higher T_c is expected to be ($M^\alpha T_c$ = constant, theoretical α = 1/2). Also, assuming that the TTF core could be equated to copper, the following analogy may obtain [10]:

However, TTF derivatives with *resonance* electron donating substituents, such as sp^3 hybridized oxygen or nitrogen *directly* attached [9] to TTF did not exist until our preparation of BO [10]; the same can be said for tetraoxafulvalenes themselves [11]. In this presentation we describe the preparation and some properties of several salts derived from BO.

Results

The new donor is an air-stable, orange, crystalline solid which is soluble in most organic solvents. It reacts with the usual acceptors and produces organic metals with, for e.g. I_3, IBr_2, and $Cu(SCN)_2$, etc as shown in Table I. For the I_3 salt, the resistance ratio between 300 and 20K is *ca* 250 (see Figure 1), its magnetic susceptibility is temperature independent (300 - 20K, Figure 2), its epr signal linewidth decreases linearly with temperature (Fig. 2, inset) and its thermopower is small and tends to zero in two crystal axes directions (see Figure 3). In Figure 4 we show the temperature dependence of the conductivity of the gold diiodide salt. In Figures 5 and 6 we show the temperature dependence of the conductivity of the salts with ClO_4 and BF_4 tetrahedral anions. In Figure 7 we show the temperature dependence of the conductivity of the nitrate salt.

Table I Electrical Conductivity of BEDO-TTF Salts

BEDO-TTF salt	Conductivity (Scm^{-1})
(BO)I$_{1.25}$	100-280
(BO)$_m$(IBr$_2$)$_n$	253
(BO)$_m$[Cu(SCN)$_2$]$_n$	300
(BO)$_m$(ClO$_4$)$_n$	10
(BO)$_m$(AsF$_6$)$_n$	10
(BO)$_m$(PF$_6$)$_n$	10
(BO)$_m$(BF$_4$)$_n$	60
(BO)$_m$(NO$_3$)$_n$	50
(BO)$_2$N(CN)$_2$·2H$_2$O	80
(BO)$_m$(AuI$_2$)$_n$	160
(BO)$_m$(AuBr$_2$)$_n$	80
(BO)$_{2.5}$[(Ag(CN)$_2$]	80
(BO)$_m$(CuI$_2$)$_n$	0.19[a]

[a] Compressed powder, the value is the same as compressed powder of the Ag(CN)$_2$ salt

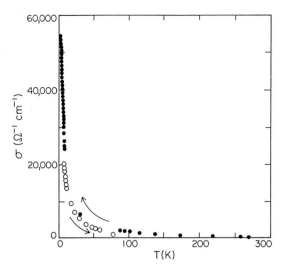

Fig. 1. Temperature dependence of the conductivity of $BOI_{1.25}$ between 300 and 1.2K. •, decreasing temperature; O, increasing temperature. In the temperature range 50-1.2K, the decreasing and increasing temperature data points overlap.

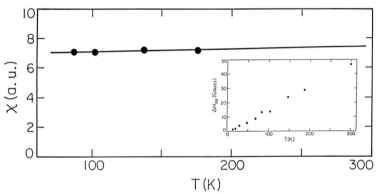

Fig. 2. Temperature dependence of the spin paramagnetic susceptibility of a single crystal of $BOI_{1.25}$ as a function of temperature. The χ scale is in arbitrary units. Considering the weight of the crystal, the "a.u." of 6.4 corresponds to a χ_M of 1.35×10^{-4} emu/mole. The inset is the temperature dependence of the EPR linewidth.

Discussion.

As can be seen from Figures 1 - 3, the iodide salt exhibits behavior typical of a classical three dimensional metal. Especially when one considers the thermopower results which are unlike any reported to date even for the more highly dimensional ET salts. The only other molecular metals, which are not strictly speaking organic because they contain transition metals, that show similar behavior are $(DMTCNDQI)_2Cu$ [12] and Ni Phthalocyanine I_n [13]. The salts of tetrahedral anions exhibit transitions in the resistivity at ca 160 K and the trigonal anion (NO_3) anion salt at ca 60K. None of the salts studied to date show a transition to the superconducting state at atmospheric pressure. However, the salts which exhibit a transition at atmospheric pressure (ClO_4, BF_4, NO_3) are the best candidates to show superconductivity under pressure.

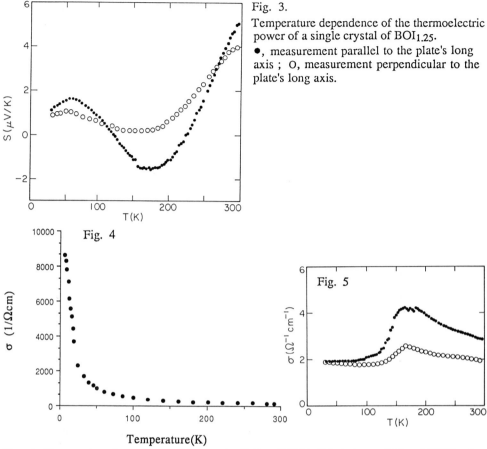

Fig. 3. Temperature dependence of the thermoelectric power of a single crystal of $BOI_{1.25}$.
●, measurement parallel to the plate's long axis ; O, measurement perpendicular to the plate's long axis.

Fig. 4. Temperature dependence of the conductivity of $BO(AuI_2)_n$ between 300 and 20K by the four probe technique.

Fig. 5. Temperature dependence of the conductivity of $BO(ClO_4)_n$ between 300 and 20K by the four probe technique.

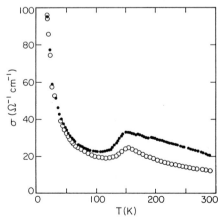

Fig. 6. Temperature dependence of the conductivity of $BO(BF_4)_n$ between 300 and 20K by the four probe technique.

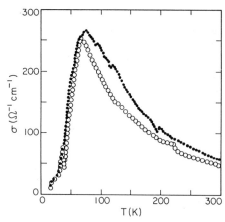

Fig. 7. Temperature dependence of the conductivity of BO(NO$_3$)$_n$ between 300 and 20K by the four probe technique.

Acknowledgment

We thank the National Science Foundation for support through Grant DMR-8820933. We also thank Nancy Keder for assistance with the x-ray structure analysis.

References

[1] D. O. Cowan, These proceedings.
[2] Proceedings of recent international conferences contain detailed accounts (a) M. Aldissi, Ed, Synth. Met. **27**, (1) - (4) (1988); **28 - 29**, (1989). (b) H. Shirakawa, T. Yamabe and K. Yoshino, Eds, Synth. Met. **17 - 19**, (1987). (c) C. Pecile, G. Zerbi, R. Bozio and A. Girlando, Eds. Mol. Cryst. Liq. Cryst. **117 - 121**, (1985). (d) R. Comes, P. Bernier, J. J. Andre and J. Rouxel, Eds. J. Phys. (Paris) Colloq. **44**, (1983), C3. (e) G. C. Papavassiliou, these proceedings
[3] D. Schweitzer, . communicated at this conference; K. Oshima, this conference, poster PB17.
[4] H. Urayama, H. Yamochi, G.Saito, K. Nozawa, T. Sugano, M. Kinoshita, S. Sato, K. Oshima, A. Kawamoto and J. Tanaka, Chem. Letters 55 (1988).
[5] S. S. P. Parkin, V. Y. Lee, E. M. Engler, A. I. Nazzal, T. C. Huang, G. Gorman, R. Savoy and R. Beyers Phys. Rev. Lett. **60**, 2539 (1988) ; S. S. P. Parkin, V. Y. Lee, A. I. Nazzal, R. Savoy , R. Beyers and S. J. La Placa Phys. Rev. Lett. **61**, 750 (1988) ; S. S. P. Parkin, V. Y. Lee, A. I. Nazzal, R. Savoy , R. Beyers and S. J. La Placa Phys. Rev. B **38**, 6531 (1988).
[6] D. Jérome, A. Mazaud, M. Ribault and K. Bechgaard J. Phys. Lett. **41**, L95 (1980).
[7] J. G. Bednorz and K. A. Müller Z. Phys. **B64**, 189 (1986).
[8] (a) K. Yamaguchi, Y. Takahara, T. Fueno and K. Nasu Jpn. J. Appl. Phys. **27**, L509 (1988); K. Yamaguchi, V. Takahara, T. Fueno, K. Nakasuji and I. Murata Jpn. J. Appl. Phys. **27**, L766 (1988). (b) V. J. Emery Phys. Rev. Lett. **58**, 2794 (1987). (c) J. E. Hirsch Ibid. **59**, 228 (1987). (d) G. Chen and W. A. III Goddard Science **239**, 899 (1988), Y. Guo, J.-M. Langlois and W. A. III Goddard ibid. **239**, 896 (1988). (e) H. Fukuyama, these proceedings

[9] W. Chen, M. P. Cava, M. A. Takassi and R. M. Metzger J. Am. Chem. Soc. **110**, 7903 (1988) prepared a pyrrole annulated TTF where the nitrogen is not directly attached to the TTF. Hsu, S.-Y. and Chiang, Y.-L. Synth. Met. **27**, B 651 (1988) prepared tetrahydrofuran annulated TTF where the oxygen is not directly attached to the TTF.

[10] T. Suzuki, H. Yamochi, G. Srdanov, K. Hinkelmann, and F. Wudl J. Am. Chem. Soc. **111**, 3108 (1989).

[11] Attempts to produce TOF (tetraoxafulvalene, i) and some derivatives failed: F. Wudl, Unpublished, 1974; H. Yamochi and F. Wudl, Unpublished, 1988. J. Ferraris, unpublished.

i

[12] H. C. Wolf, these proceedings

[13] J. Martinsen, S. M. Palmer, J. Tanaka, R. C. Greene and B. M. Hoffman Phys. Rev. B **30**, 6269 (1984). This salt does exhibit a change in resistivity as a function of temperature at ca 30K reminiscent of the quasi-one dimensional metals.

New Cation-Radical Salts (ET)$_3$CuCl$_4 \cdot$H$_2$O and (ET)$_2$CuCl$_4$ with Metallic and Semiconducting Properties

A.V. Gudenko[1], V.B. Ginodman[1], V.E. Korotkov[2], A.V. Koshelap[2], N.D. Kushch[2], V.N. Laukhin[2], L.P. Rozenberg[2], A.G. Khomenko[2], R.P. Shibaeva[2], and E.B. Yagubskii[2]

[1] Physical Institute, USSR Academy of Sciences, SU-117334 Moscow, USSR
[2] Institute of Chemical Physics, USSR Academy of Sciences, SU-142432 Chernogolovka, USSR

Abstract. Cation-radical salts (ET)$_3$CuCl$_4 \cdot$H$_2$O and (ET)$_2$CuCl$_4$ have been synthesized. Their crystal structures have been determined and their electroconducting properties studied. The (ET)$_3$CuCl$_4 \cdot$H$_2$O crystals are organic metals down to 0.5 K, those of (ET)$_2$CuCl$_4$ are semiconductors.

Earlier it has been reported, that stable organic metals based on bis(ethylenedithio)tetrathiafulvalene (ET) are obtained with tetrahedral metal complex dianions (MCl$_4$)$^{2-}$, M = Zn, Cd, Co, Mn, Cu, [1-3]. However, their exact compositions and structures are not known yet.

We obtained crystals of the new organic metal (ET)$_3$CuCl$_4$H$_2$O (A) and studied its structure and conductivity. The complex was synthesized in the ET oxidation by tetraphenylphosphonium tetrachlorocuprate (Ph$_4$P)$_2$CuCl$_4$(H$_2$O)$_x$ in benzonitrile. The crystals grew in the diffusion of the initial reagent solutions to the pure solvent at 20°C for ~ 1.5 months. Besides A crystals the process yielded B crystals of the (ET)$_2$CuCl$_4$ composition, which are semiconductors. The Table summarizes the crystal data for the A and B crystals.

The A structure is characterized by the presence of ET cation-radical layers alternating with those of centrosymmetric dimers 2[CuCl$_4 \cdot$H$_2$O] along the **c** direction (Fig. 1). The cation-radical layers are formed of ET stacks running along **a**. Crystallographically non-equivalent I-II-Iā and III-IIIi-IIIā stacks (Fig. 2) are very close in their inner structure, i.e. they have the same mode of overlapping of the neighboring ET in the stack and almost equal interplanar distances: I-II 3.74 Å, II-Ii 3.76 Å and III-IIIi 3.67 Å IIIi-IIIā 3.69 Å.

The B structure does not contain cation-radical layers, it is characterized by the presence of the ET stacks almost parallel to

Table. Crystal data for $(ET)_3CuCl_4 \cdot H_2O$ (A) and $(ET)_2CuCl_4$ (B)

Parameter	A	B
a, Å	8.994(1)	8.883(6)
b, Å	16.668(2)	12.285(7)
c, Å	16.238(2)	15.91(1)
α, °	96.78(1)	84.56(6)
β, °	90.71(1)	85.82(6)
γ, °	93.29(1)	107.36(5)
V, Å^3	2412.8	1640
space group	P$\bar{1}$	P$\bar{1}$
Z	2	2
N$_{ind.refl.}$	5012	3634
R	0.069	0.057

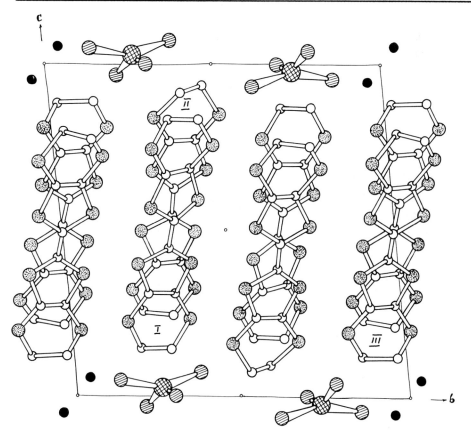

Fig. 1: $(ET)_3CuCl_4 \cdot H_2O$. The crystal structure projection along the **a** direction.

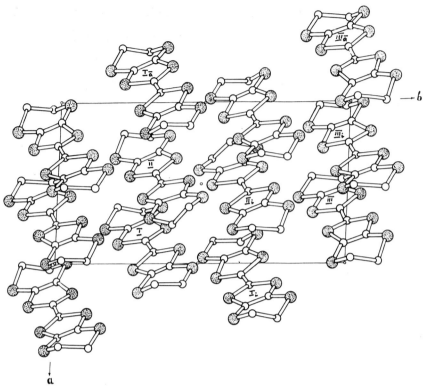

Fig. 2: $(ET)_3CuCl_4 \cdot H_2O$. The cation-radical layer projection along the **c** direction.

the **a** direction and spatially separated by the $CuCl_4$ dianions (Fig. 3).

Fig. 4 (1) presents the resistivity dependence on temperature for single crystals A in the temperature range of 295-0.5 K at ambient pressure. The complex A is seen to be a stable metal down to 0.5 K ($\sigma_{295K} \sim 10$ $ohm^{-1}cm^{-1}$; $\sigma_{0.5K} \sim 1.5 \times 10^3$ $ohm^{-1}cm^{-1}$).

We studied the pressure effect on the conductivity properties of single crystals A. Fig. 5 shows that the resistivity gradually decreases with the pressure increase up to ~ 10 kbar without phase transition at room temperature. Under the identified pressure the metallic properties retain down to ~ 1.3 K (Fig. 4). The value of residual resistivity is weakly pressure-dependent.

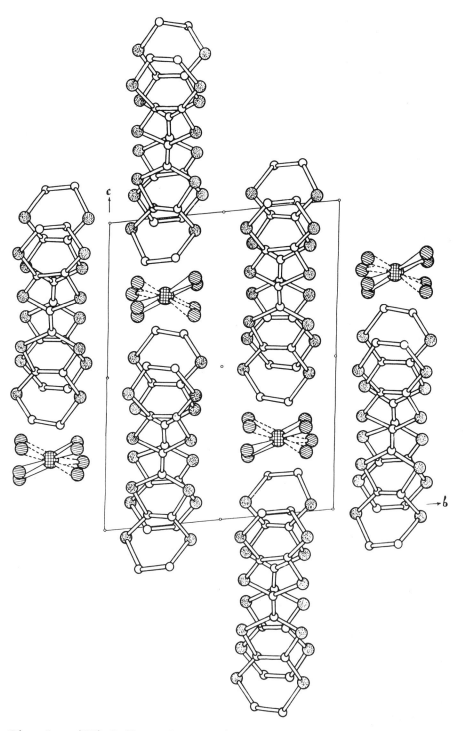

Fig. 3: (ET)$_2$CuCl$_4$. The crystal structure projection along the **a** direction.

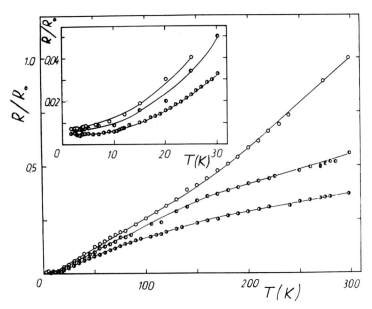

Fig. 4: The resistivity dependence of $(ET)_3CuCl_4 \cdot H_2O$ single crystals on temperature at P = 1 bar (1), 5.5 kbar (2), and 10 kbar (3). The pressure values are taken at room temperature.

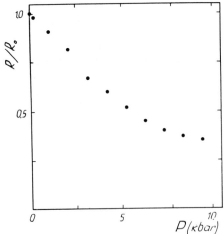

Fig. 5: The $(ET)_3CuCl_4 \cdot H_2O$ resistivity dependence on pressure.

References

1. E.B. Yagubskii, R.P. Shibaeva: J. Mol. Electr. 5, 25 (1989).
2. T. Mallah, C. Hollis, S. Bott, P. Day, M. Kurmoo: Synth. Met. 27, A381 (1988).
3. M. Lequan, R.M. Lequan, G. Maceno, P. Delhaes: J. Chem. Soc., Chem. Comm. 1 (1988).

New Organic Synthetic Metals Derived from BEDT-TTF, Ni(dsit)$_2$ and BEDO-TTF

M.A. Beno, A.M. Kini, U. Geiser, H.H. Wang, K.D. Carlson, and J.M. Williams

Chemistry and Materials Science Division, Argonne National Laboratory, Argonne, IL 60439, USA

Abstract. Three strategies have been employed by us to synthesize new organic synthetic metals and superconductors. On the basis of structure-property correlations derived for the β-(BEDT-TTF)$_2$X salts, new charge transfer salts of BEDT-TTF with large, polarizable anions have been synthesized. The occurrence of molecular dimers has been engineered into salts of the new organic acceptor molecule, Ni(dsit)$_2$ [bis (4,5 -diselenolate -1,3-dithiole-2-thione) nickelate], to synthesize salts with acceptor packing similar to the donor packing in κ-(BEDT-TTF)$_2$Cu(SCN)$_2$. Finally, two charge transfer salts of bis(ethylenedioxy)tetrathiafulvalene, BEDO-TTF, namely (BEDO-TTF)$_2$AuBr$_2$ and (BEDO-TTF)$_2$AuI$_2$ have been synthesized. The AuBr$_2^-$ salt, the first BEDO-TTF salt to be structurally characterized, is semiconducting below 263 K, while the AuI$_2^-$ salt shows metallic conductivity to low temperatures.

1. Introduction

Recent research on organic metals has produced an increasing number of organic superconductors. The ambient pressure superconducting transition temperatures (T_c) have risen steadily from T_c = 1.2 K for (TMTSF)$_2$ClO$_4$, to T_c = 1.5, 2.8, 5.0 K for the β-(BEDT-TTF)$_2$X salts where X$^-$ = I$_3$, IBr$_2$, and AuI$_2$ [1] respectively and to T_c = 10.4 K in κ-(BEDT-TTF)$_2$Cu(SCN)$_2$ [2]. In the isostructural β-phase salts, we have shown that the anion size [3] and anion–CH$_2$ hydrogen interactions [4,5] are critical factors that control the packing of the BEDT-TTF donor molecule and thus determine the transport properties of the resulting charge-transfer salts. Changes in the donor-to-anion "hydrogen bonding" interactions which accompany variations in anion size have been correlated to the lattice softness and phonon frequencies and, therefore, to the superconducting T_c values [5]. Thus, charge transfer salts of anions larger than I$_3^-$ which retain the β-phase structure should, in principle, possess higher T_c values. An alternate approach that may be used to tailor the electrical properties of synmetals is the replacement of the BEDT-TTF donor with similar organic molecules. In this paper, we will elucidate the strategies we have employed for obtaining new synthetic metals.

2. BEDT-TTF Salts with Main Group Metal Halide Complex Anions

We have attempted to synthesize new BEDT-TTF salts with anions of the type X-M-X where M is a main group metal such as Hg, Bi, or Cd and X = Br or I in order to test the correlations between anion size and increased T_c values. The results of these experiments are not new β-phase salts with linear anions but instead salts with large complex, and sometimes polymeric, anions which force the BEDT-TTF donors into new, unusual packing arrangements.

Electrocrystalization of BEDT-TTF in the presence of BiI$_3$ and (n-Bu)$_4$NI in 1,1,2-trichloroethane (TCE) produces a 1:1 salt (BEDT-TTF)BiI$_4$ [6]. The crystals are triclinic, space group P$\bar{1}$, a = 8.265(3), b = 11.118(3), c = 14.424(4)Å, α = 110.76(2), β = 96.41(2), γ = 103.57(2)°, V_c = 1176.8(7) Å^3, with Z = 2. The salt contains layers of dimerized BEDT-TTF^{+1} cations and infinite chains of BiI$_4^-$ anions. Similarly, electrocrystallization of

BEDT-TTF in the presence of CdI_2 and $(n-Bu)_4NI$ in TCE produces $(BEDT-TTF)_4Cd_2I_6$ [7]. The crystals are triclinic, space group $P\bar{1}$ $a = 8.856(7)$, $b = 11.961(9)$, $c = 18.092(11)$Å, $\alpha = 87.72(6)$, $\beta = 84.80(6)$, $\gamma = 74.04(7)°$, $V_c = 1835(2)$ Å^3, with $Z = 1$.

Two semiconducting solvated salts [7] result from the electrocrystallization of BEDT-TTF in the presence of $[(n-Bu)_4N]HgBr_3$ in TCE solvent, compound 1, $(BEDT-TTF)_2(HgBr_3)(TCE)$, monoclinic, $P2_1/c$, $a = 40.583(5)$, $b = 4.1830(7)$, $c = 22.766(3)$Å, $\beta = 104.33(1)°$, $V_c = 3744.5(9)$Å^3, $Z = 4$; and compound 2, $(BEDT-TTF)_4(Hg_2Br_6)(TCE)$, monoclinic, $P2_1/n$, $a = 19.344(5)$, $b = 13.401(3)$, $c = 29.418(10)$Å, $\beta = 103.87(2)°$, $V_c = 7404(4)$Å^3, $Z = 4$. While compound 2 contains discrete dimeric $Hg_2Br_6^{2-}$ anions, the $HgBr_3^-$ anions in compound 1 form infinite chains. In the compound 2, eight BEDT-TTF molecules repeat along the molecular stacking axis.

While structure property correlations derived for the β-phase BEDT-TTF salts predict that large polarizable anions should give superconducting salts with higher Tc values, this prediction is based on retaining the β-phase donor packing. This is clearly not the case in the semiconducting BEDT-TTF salts with main group halide complex anions studied to date.

3. Charge Transfer Salts Derived from Ni(dsit)$_2$

While the β-$(BEDT-TTF)_2X$ salts contain stacks of nearly equally spaced donor molecules linked by short interstack S⋯S interactions, the κ-phase salts such as κ-$(BEDT-TTF)_2$-$Cu(SCN)_2$ contain dimerized donor molecules which pack almost at right angles with respect to neighboring dimer pairs. The superconducting radical anion salt $Me_4N[Ni(dmit)_2]_2$ ($T_c = 3.0$ K at 3.2 kbar) [8] has an acceptor packing mode similar to the β-phase BEDT-TTF salts while the semiconducting $Me_4P[Ni(dmit)_2]_2$ salt contains dimeric acceptor molecules and exhibits κ-phase packing. The occurrence of dimeric acceptor molecules can be engineered into the Ni dithiolate acceptor by replacement of the inner S atoms of the dmit with Se atoms [9]. The central Ni atoms are pyramidally coordinated by five Se atoms so that the acceptor molecules form dimers in which the Ni(dsit)$_2$ [bis(4,5-diselenolate-1,3-dithiole-2-thione) nickelate] molecules are linked by short Ni-Se bonds. Replacement of the inner sulfur atoms in Ni(dmit)$_2$ with larger and more polarizable selenium atoms is also expected to increase the orbital overlap between molecules in the solid state and hence result in larger bandwidths. Three charge transfer salts derived from the Ni(dsit)$_2$ acceptor molecule have been synthesized: two isostructural 1:2 salts $(Me_4N)[Ni(dsit)_2]_2$ and $(Me_4P)[Ni(dsit)_2]_2$ which crystallize in the orthorhombic space group Pbnm $a = 7.409(7)$, $b = 11.875(8)$, $c = 38.32(3)$ Å, and

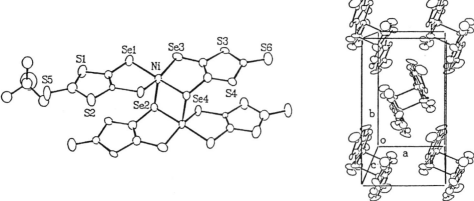

Fig. 1. Dimerization of the Ni(dsit)$_2$ acceptors (left). The κ-like packing (right) in the 1:2 salts is similar to that previously reported in $Me_4P[Ni(dmit)_2]_2$.

$V_c = 3372(4)$ Å^3 and $a = 7.418(5)$, $b = 11.942(9)$, $c = 38.5(1)$ Å, and $V_c = 3408(12)$ Å^3, respectively; and the 1:1 salt, (Et$_4$N)Ni(dsit)$_2$, which is monoclinic, P2$_1$/c, $a = 6.775(5)$, $b = 27.02(2)$, $c = 12.973(8)$ Å, $\beta = 98.09(4)°$, $V_c = 2351(3)$ Å^3. Electrical conductivity and single crystal ESR measurements show that all these salts exhibit semiconducting behavior.

4. Charge Transfer Salts Derived from BEDO-TTF

Bis(ethylenedioxy)tetrathiafulvalene, or simply BEDO-TTF, has yielded several conducting charge transfer salts [10]. We report here the conductivity and structure of a new salt, (BEDO-TTF)$_2$AuBr$_2$ which crystallizes in the monoclinic space group P2$_1$/m, Z=2, with unit cell parameters $a = 5.308(2)$, $b = 32.47(1)$, $c = 8.165(6)$ Å, $\beta = 98.47(5)°$, and $V_c = 1392(1)$ Å^3.

The unit cell contains two AuBr$_2^-$ anions located on a crystallographic mirror plane and four BEDO-TTF molecules. The two independent BEDO-TTF donor molecules, which are located with the central carbon-carbon double bond on an inversion center, are related by non-crystallographic symmetry (x,y,z+1/2). Since all of the BEDO-TTF molecules are located on inversion centers, the stacks of donor molecules are equally spaced.

Preliminary x-ray crystallographic studies on salts derived from the BEDO-TTF donor and the AuI$_2^-$ anion suggest that the poorly formed, tree-like crystals grown by electrocrystallization in TCE or THF are isostructural to the AuBr$_2^-$ salt. Four-probe

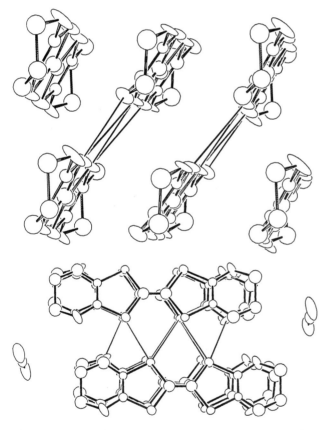

Fig. 2. The BEDO-TTF donor stacks are connected by short (less than the van der Waals sum) intermolecular S···S and S···O contacts.

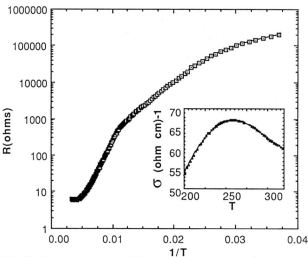

Fig. 3. Four probe resistivity measurements on single crystals of $(BEDO-TTF)_2AuBr_2$ show that this salt is metallic down ~263 K. At lower temperatures, semiconducting behavior is observed. The semiconducting gap is 0.12 eV (220 - 80K).

resistivity measurements on these twisted, twinned crystals shows that $(BEDO-TTF)_2AuI_2$ remains metallic down to 14 K. RF penetration depth measurements on polycrystalline samples of $(BEDO-TTF)_2AuI_2$ and $(BEDO-TTF)_2AuBr_2$ show that neither salt exhibits superconductivity down to 0.55 K at ambient pressure.

Acknowledgments. This work is sponsored by the US Department of Energy, Office of Basic Energy Sciences, Division of Materials Sciences, under contract W-31-109-ENG-38.

5. Literature References

1. J. M. Williams, H. H. Wang, T. J. Emge, U. Geiser, M. A. Beno, P. C. W. Leung, K. D. Carlson, R. J. Thorn, A. J. Schultz and M. H. Whangbo, in S. Lippard (ed.), Prog. Inorg. Chem., Vol. **35**, John Wiley and Sons, New York, 1987, p. p. 51-218.
2. H. Urayama, H. Yamochi, G. Saito, K. Nozawa, T. Sugano, M. Kinoshita, S. Sato, K. Oshima, A. Kawamoto and J. Tanaka, Chem. Lett. **1988**, 55 (1988).
3. T. J. Emge, P. C. W. Leung, M. A. Beno, H. H. Wang, M. A. Firestone, K. S. Webb, K. D. Carlson, J. M. Williams, E. L. Venturini, L. J. Azevedo, and J. E. Schirber, Mol. Cryst. Liq. Cryst. **132**, 363 (1986).
4. P. C. W. Leung, T. J. Emge, M. A. Beno, H. H. Wang, J. M. Williams, V. Petricek and P. Coppens, J. Amer. Chem. Soc. **107**, 6184 (1985).
5. M.-H. Whangbo, J. M. Williams, A. J. Schultz, T. J. Emge and M. A. Beno, J. Amer. Chem. Soc. **109**, 90 (1987).
6. U. Geiser, H. H. Wang, S. M. Budz, M. J. Lowry and J. M. Williams, Inorg. Chem. submitted for publication.
7. U. Geiser, H. H. Wang and J. M. Williams, in preparation.
8. A. Kobayashi, H. Kim, Y. Sasaki, H. Kobayashi, S. Moriyama, Y. Nishio, K. Kajita and W. Sasaki, Chem. Lett. **1987**, 1819 (1987).
9. P. J. Nigrey, Synthetic Met. **27**, B365 (1988).
10. T. Suzuki, H. Yamochi, G. Srdanov, K. Hinkelmann and F. Wudl, J. Amer. Chem. Soc. **111**, 3108 (1989).

Synthesis and Crystal Structures of Multi-Chalcogen TTF Derivatives and Conducting Organic Salts

T. Nogami, H. Nakano, S. Ikegawa, K. Miyawaki, Y. Shirota, S. Harada, and N. Kasai

Department of Applied Chemistry, Faculty of Engineering, Osaka University, Yamadaoka, Suita, Osaka 565, Japan

Abstract. Symmetric and unsymmetric multi-chalcogen TTF derivatives were synthesized for the study of highly conducting organic salts. They possess eight – twelve chalcogen atoms, and are expected to give two-dimensional conducting salts. Half-wave oxidation potentials of the donor molecules and the electrical conductivities of the salts were measured. X-ray crystallographic studies were made on the donor molecules and conducting salts. New transition metal salts of multi-sulfur 1,2-dithiolates were also synthesized.

1. Introduction

BEDT-TTF has given a variety of conducting ion radical salts. Some of them exhibited superconducting transition under ambient pressure. The most characteristic feature of the BEDT-TTF salts is their two-dimensional band electronic property, which is caused by the presence of eight sulfur atoms in a donor site. We have synthesized new multi-chalcogen TTF derivatives. The electrical conductivities of the salts were measured. We have also made an X-ray crystallographic work of some donor molecules and conducting salts. In order to develop conducting organic transition metal salts like $TTF[Ni(dmit)_2]_2$, we have also synthesized new transition metal salts of multi-sulfur 1,2-dithiolates.

2. Results and Discussion

The synthetic routes of the donor molecules are illustrated in Scheme 1 [1-5]. The names of the donor molecules are abbreviated as VT, DMVT, TMVT, TMTVT, TMSVT, EVT, EMVT, EDMVT, PVT, TPVT, and MVT as shown in the Scheme. All of them have vinylenedithio group at the terminal portion of the molecule. Table 1 summarizes half-wave oxidation potentials of these donor molecules in THF. These values are higher than that of BEDT-TTF (0.69, 0.82 V vs SCE) in the same solvent.

Conducting organic salts containing these donor molecules were synthesized by an electrochemical and a chemical method. Table 2 shows their electrical conductivities at room temperature and the temperature dependence of the conductivities. T_{M-I} in this table denotes a metal-insulator transition temperature [6]. Most of the complexes showed high conductivities.

X-ray crystallographic work was done for TMTVT, EVT, $(EVT)_2PF_6$, and $(EVT)_2AsF_6$. Table 3 summarizes the crystal data. Figure 1 shows the crystal structure of TMTVT. The donor molecules are stacked along c-axis to give a columnar structure. S···S contacts almost equal to the sum of the van der Waals distance (3.7 Å) were found between methylthio groups of the adjacent columns. Figure 2 shows the crystal structure of EVT. Dimeric structure of EVT molecules can be seen. This crystal structure is very close to that of BEDT-TTF. Figure 3 shows the crystal structure of EVT_2PF_6. EVT molecules are stacked along b-axis to form a columnar structure and PF_6 anions are on the center of symmetry. S···S contacts shorter than 3.7 Å were found between EVT molecules of adjacent columns. EVT_2AsF_6 gave quite similar crystal structure to EVT_2PF_6.

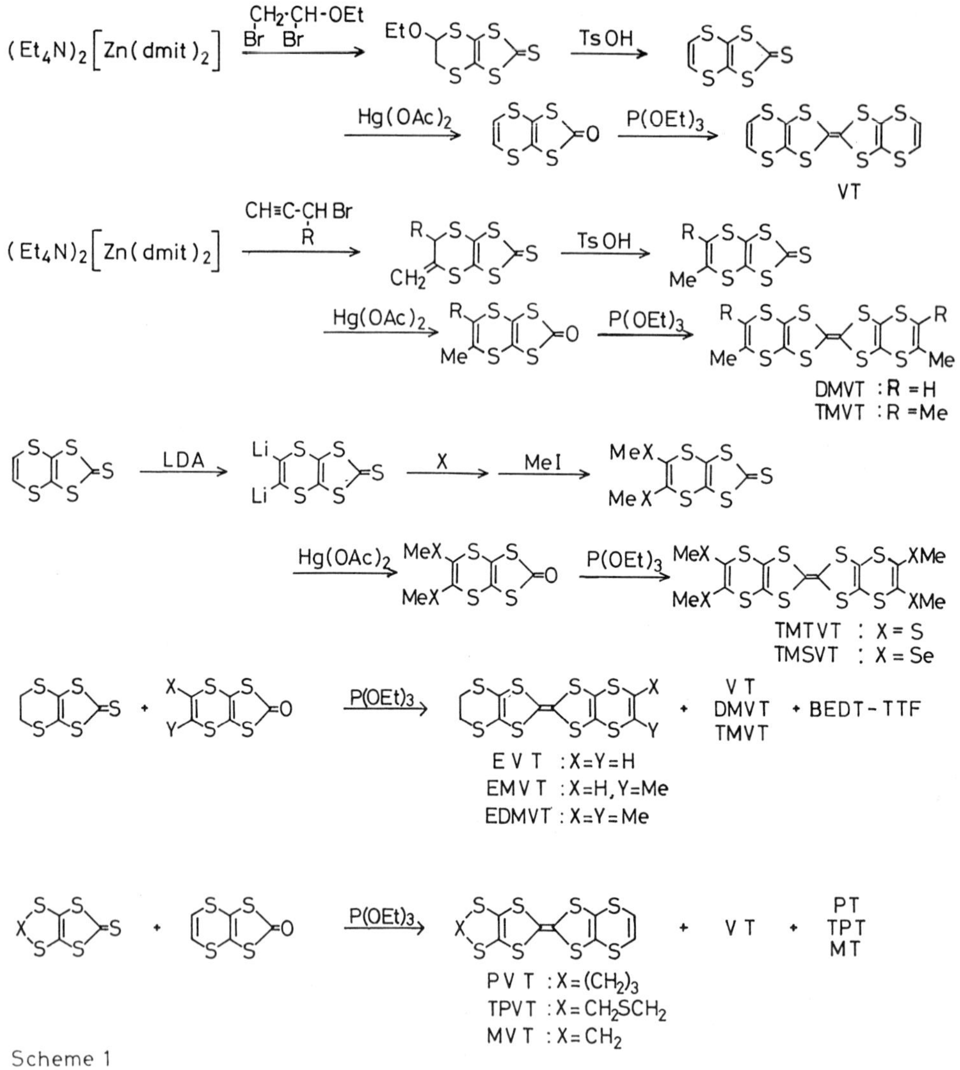

Scheme 1

Table 1. Half-wave oxidation potentials of donor molecules in THF solution

Donor	$E_{1/2}$	Donor	$E_{1/2}$
VT	0.83 V vs SCE	EVT	0.75 V vs SCE
DMVT	0.80 V	EMVT	0.76 V
TMVT	0.80 V	EDMVT	0.76 V
TMTVT	0.83 V	PVT	0.78 V
TMSVT	0.86 V	TPVT	0.80 V

Table 2. Complexes, electrical conductivities at room temperature, and temperature dependence of the conductivities

Complexes	Conductivities	Temperature dependence
VT-I$_3$ [a)]	130 S cm^{-1} [b)]	70 K (T_{M-I}) [d)]
α-VT-IBr$_2$ [a)]	170 [b)]	120 K
β-VT-IBr$_2$ [a)]	75 [b)]	250 K
α-VT-ClO$_4$ [a)]	38 [b)]	semiconductor
β-VT-ClO$_4$ [a)]	19 [b)]	semiconductor
α-VT-ReO$_4$ [a)]	8 [b)]	100 K
β-VT-ReO$_4$ [a)]	33 [b)]	180 K
VT$_2$PF$_6$	20 [b)]	180 K
VT$_2$SbF$_6$ [a)]	6 [b)]	semiconductor
VT-Ni(dmit)$_2$ [a)]	20 [b)]	160 K
EVT$_2$PF$_6$	0.8 [b)]	semiconductor
EVT$_2$AsF$_6$ [a)]	0.2 [b)]	semiconductor
EVT-I$_3$ [a)]	180 [b)]	semiconductor
EVT-IBr$_2$ [a)]	5 [b)]	semiconductor
EVT-AuBr$_2$ [a)]	2 [b)]	semiconductor
EVT-Cu(SCN)$_2$ [a)]	15 [b)]	semiconductor
VT TCNQ	insulator [b)]	
VT$_2$(F$_4$TCNQ)	6 [c)]	
VT$_2$DDQ	0.2 [c)]	
VT$_3$I$_3$	5 [b)]	
DMVT$_2$TCNQ	insulator [c)]	
DMVT$_5$(F$_4$TCNQ)$_2$ [a)]	1 [c)]	
DMVT-TCNE	0.5 [c)]	
DMVT$_3$I$_3$	1 [c)]	
TMVT$_3$TCNQ	insulator [b)]	
TMVT$_2$(F$_4$TCNQ)	3 × 10^{-3} [c)]	
TMVT$_5$(I$_3$)$_2$ [a)]	0.1 [c)]	
TMTVT-F$_4$TCNQ [a)]	1 × 10^{-4} [c)]	
TMTVT-DDQ [a)]	8 × 10^{-6} [c)]	
TMTVT-I [a)]	3 × 10^{-5} [c)]	

a) The composition of the complex was not determined.
b) Measured for a single crystal by a four probe method.
c) Measured for a compressed pellet sample by a van der Pauw method.
d) T_{M-I} denotes metal-insulator transition temperature.

Table 3 Crystallographic data of TMTVT, EVT, EVT$_2$PF$_6$, and EVT$_2$AsF$_6$

	TMTVT	EVT	EVT$_2$PF$_6$	EVT$_2$AsF$_6$
Crystal system	Monoclinic	Monoclinic	Triclinic	Triclinic
Space group	P2$_1$/n	P2$_1$/n	P$\bar{1}$	P$\bar{1}$
a / Å	7.939(1)	6.563(1)	6.47(2)	6.494(1)
b / Å	28.302(2)	13.507(8)	7.80(3)	7.760(2)
c / Å	5.067(1)	15.866(8)	15.81(5)	16.039(4)
α / °			94.8(8)	94.97(2)
β / °	95.74(1)	93.47(2)	80.2(3)	98.63(1)
γ / °			100.2(6)	79.58(2)
V / Å^3	1132.8(2)	1404(1)	772(5)	784.5(3)
Z	2	4	1	1
D$_c$ / g cm^{-3}	1.66	1.80	1.96	2.02
R	0.091	0.070	0.120	0.055

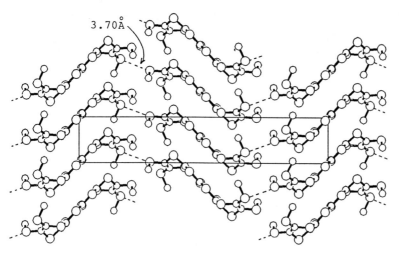

Fig. 1　Crystal structure of TMTVT.

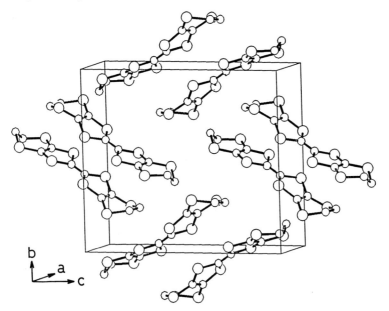

Fig. 2　Crystal structure of EVT.

Comparing the band electronic structure of EVT_2PF_6 with those of α-$(BEDT\text{-}TTF)_2PF_6$ [7] and VT_2PF_6 [8], it was found that the amount of the two-dimensional electronic nature increased in the order of VT_2PF_6 < EVT_2PF_6 < α-$(BEDT\text{-}TTF)_2PF_6$ [9].

We have also synthesized transition metal salts of 1,4-dithiin-2,3-dithiolate (ddt) [10] and 2-oxo-1,3-dithiole-4,5-dithiolate (dmio) [11] (Fig. 4). Their multi-sulfur structures are effective for suppressing Peierls transition of the conducting salts due to the multi-dimensional electronic properties.

Fig. 3   Crystal structure of EVT$_2$PF$_6$.

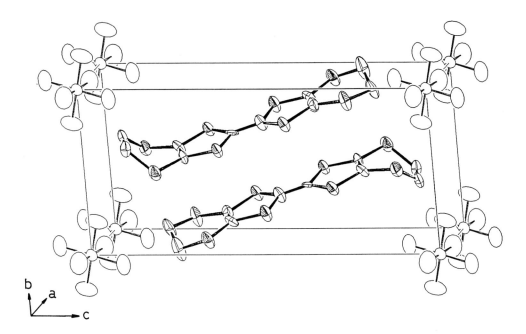

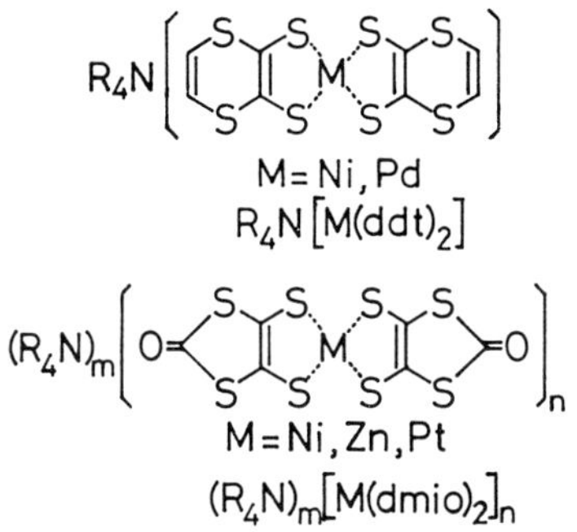

Fig. 4   Molecular structures of ddt- and dmio-salts.

References

[1] K. Inoue, Y. Tasaka, O. Yamazaki, T. Nogami, and H. Mikawa, Chem. Lett., 1986, 781.
[2] T. Nakamura, S. Iwasaka, H. Nakano, K. Inoue, T. Nogami, and H. Mikawa, Bull. Chem. Soc. Jpn., 60, 365 (1987).
[3] H. Nakano, T. Nakamura, T. Nogami, and Y. Shirota, Chem. Lett., 1987, 1317.
[4] H. Nakano, T. Nogami, and Y. Shirota, Bull. Chem. Soc. Jpn., 61, 2973 (1988).
[5] H. Nakano, K. Miyawaki, T. Nogami, Y. Shirota, S. Harada, and N. Kasai, Bull. Chem. Soc. Jpn., 62, 2604 (1989).
[6] S. Iwasaka, T. Nogami, and Y. Shirota, Synth. Metals, 26, 177 (1988).
[7] H. Kobayashi, R. Kato, T. Mori, A. Kobayashi, Y. Sasaki, G. Saito, and H. Inokuchi, Chem. Lett., 1983, 759.

[8] H. Kobayashi, A. Kobayashi, T. Nakamura, T. Nogami, and Y. Shirota, Chem. Lett., <u>1987</u>, 559.
[9] H. Nakano, K. Miyawaki, T. Nogami, Y. Shirota, S. Harada, N. Kasai, A. Kobayashi, R. Kato, and H. Kobayashi, Bull. Chem. Soc. Jpn., in contribution.
[10] T. Nakamura, T. Nogami, and Y. Shirota, Bull. Chem. Soc. Jpn., <u>60</u>, 3447 (1987).
[11] S. Ikegawa, T. Nogami, and Y. Shirota, Bull. Chem. Soc. Jpn., in press.

Preparation of Methylated BEDT-TTFs for Controlling Intermolecular S-S Contacts

A. Izuoka, S. Matsumiya, and T. Sugawara

Department of Pure and Applied Sciences, College of Arts and Sciences, University of Tokyo, Komaba 3-8-1, Meguro-ku, Tokyo 153, Japan

Abstract. Methylated BEDT-TTF(ET) derivatives, Me_2-ET and Me_4-ET were synthesized in order to control their conformations in crystals. Me_2-ET triiodide salt, which was prepared through electrocrystallization, shows semiconductive behavior down to 4K. We observed two temperature regions with different E_A's, the transition temperature being 135K. The electrical properties were discussed in terms of a modulation of transfer integrals among intermolecular S-S contacts caused by conformational flexibility of ethylene bridges.

1. Introduction

BEDT-TTF (ET, **1**) has two ethylene bridges on the periphery of the molecule. Since these ethylene bridges are out of the molecular plane, ET has two conformers called A-type and B-type (Fig.1). A and B conformers are convertible due to flip-flop motion of ethylene bridges. Organic superconductor β-$(ET)_2I_3$ contains both conformers in crystal, creating disorder at ambient temperature and pressure [1]. Under higher pressure, order appears, as all the ET molecules turn out to be A-type [2,3]. The resulting salt shows higher T_c (~8K) [3,4]. In order to study relationship between conformational flexibility of ethylene bridges and electrical property of ET salts, we synthesized methylated ETs, Me_2-ET **2** and Me_4-ET **3**.

2. What Can We Expect from Conformationally Controlled ET Salts?

The flip-flop motion of ethylene bridges of ET salts should be accompanied by a spatial change of outer sulfur orbitals, which influences intermolecular S-S transfer integrals. Thus, uniform order among S-S contacts in crystals may be disturbed by the ring-flipping of some ET molecules, and the resulting disorder in the lattice will be the cause of scattering. Therefore if one can control the conformation of ethylene bridges in ET salts without changing their electronic properties, one can get valuable information on conduction mechanism. By introducing methyl groups into ethylene bridges, conformers caused by the ring-flipping are not isoenergetic any longer, and the conformation will be fixed.

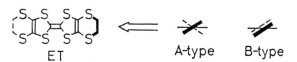

Fig. 1. Conformational isomers of an ET molecule.

3. Experimental

3.1 Synthesis of Me$_2$-ET and Me$_4$-ET

Since there are two or five isomers of Me$_2$-ET and Me$_4$-ET, respectively, it seems inevitable to synthesize the isomers stereoselectively. The trans-cyclic sulfate **4**, which can be separated from cis-isomer by a fractional distillation, was reacted with Na$_2$dmit **5** in THF to yield trans-thione **6** [5]. When cis-sulfate was used, the desired product was not formed.

After converting trans-dimethylated and unsubstituted thiones (**6** and **6'**) to ketones (**7** and **7'**) using mercuric acetate [6], equivalent amount of the two ketones were cross-coupled in triethyl phosphite to yield ET **1**, trans-Me$_2$-ET **2** and trans-trans-Me$_4$-ET **3** in 1:2:1 molar ratio. The physical properties of **1-3** are shown in Table 1. Since the oxidation potentials of Me$_2$-ET and Me$_4$-ET are almost similar to those of ET, the methyl substitution can be considered only as a sterical perturbation.

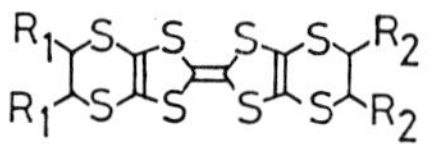

1 ; R$_1$ = R$_2$ = H
2 ; R$_1$ = CH$_3$, R$_2$ = H
3 ; R$_1$ = R$_2$ = CH$_3$

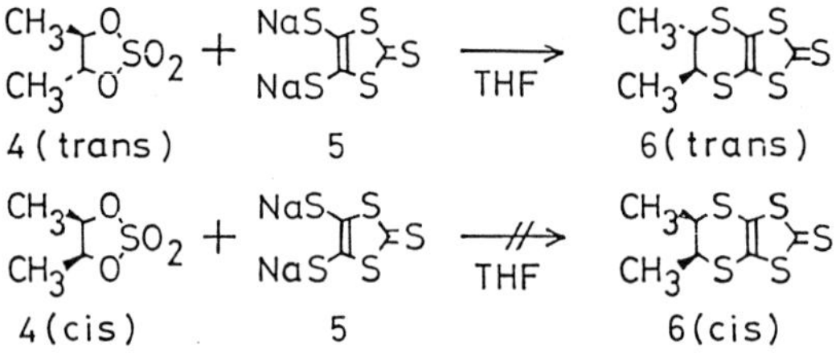

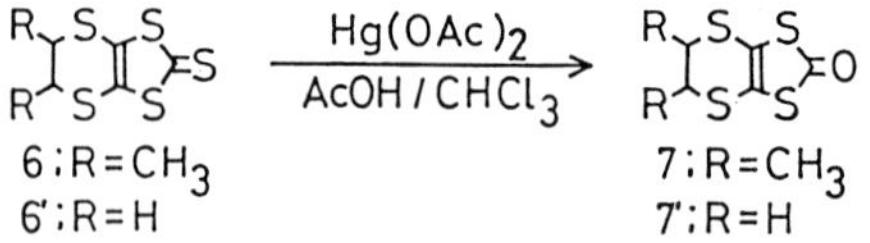

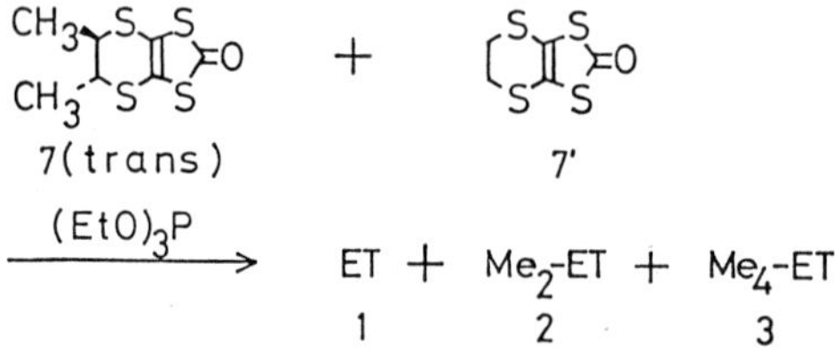

Table 1. Physical properties of ET, trans-Me$_2$-ET and trans,trans-Me$_4$-ET.

	ET 1	Me$_2$-ET 2	Me$_4$-ET 3
shape	orange needle	orange needle	orange needle
m.p. (C)	238-240 (dec.)	186-190 (dec.)	199-204 (dec.)
$E^1_{1/2}$ (V vs. SCE)	0.52	0.49	0.54
$E^2_{1/2}$ (V vs. SCE)	0.83	0.83	0.88

3.2 Me$_n$-ET Salts : Their Preparation and Properties

Crystals of trans-Me$_2$-ET triiodide were obtained by electro-crystallization, using 0.04 mmol of Me$_2$-ET and 0.16 mmol of (Bu$_4$N)I$_3$ in 1,1,2-trichloroethane or chlorobenzene with a current of 2 μA. The crystals appear as black plates with a typical size of 2x2x0.05 mm.

The electrical conductivity of Me$_2$-ET triiodide was measured by a four-probe method. The conductivity at room temperature $\sigma = 15$ Scm^{-1}, and we observed semiconductive behavior down to 4 K (Fig. 2). The temperature dependence of conductivity exhibits two temperature regions with a transition point at 135 K, which have two different activation energies, $E_A = 4.63 \times 10^{-2}$ eV (>135 K) and $E_A = 1.02 \times 10^{-2}$ eV (<135 K). X-ray structure analysis is under way.

A 1:1 salt of trans-trans-Me$_4$-ET (meso-form) and TCNQ was obtained using a diffusion method. In the crystal, Me$_4$-ET molecules form a mixed stack with TCNQ (Fig. 3). It is to be noted that the methyl groups of Me$_4$-ET are in di-equatorial conformation on one side and di-axial on the other. The preparation of Me$_4$-ET triiodide will be accomplished in the near future.

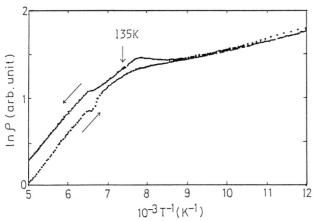

Fig. 2. Temperature dependence of resistivity of trans-Me$_2$-ET triiodide salt.

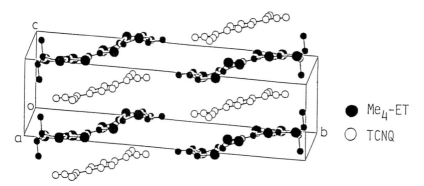

Fig. 3. Crystal structure of 1:1 salt of trans-trans-Me$_4$-ET (meso-form) and TCNQ.

4. Results and Discussion

Trans-Me_2-ET triiodide salt exhibits interesting conductivity behavior, which may originate in a structural transition. Fully elucidated crystal structure will be conclusive.

Conformation diversity found in Me_4-ET:TCNQ complex has implications for crystal design. By changing counter ions of different sizes, we will be able to alter the conformation of methylated ETs in their salts. This will result in the control of intermolecular S-S contacts. Investigation along this line is also in progress in our laboratory.

References

[1] P. C. W. Leung, T. J. Emge, M. A. Beno, H. H. Wang, J. M. Williams, V. Petricek and P. Coppens, J. Am. Chem. Soc., **107**, 6184 (1985); S. Kagoshima, Y. Nogami, M. Hasumi, H. Anzai, M. Tokumoto, G. Saito and N. Mori, Solid State Commun., **69**, 1177 (1989).
[2] V. N. Molchanov, R. P. Shibaeva, V. N. Kachinskii, E. B. Yagubskii, V. N. Simonov and B. K. Vainstein, Dokl. Akad. Sci. USSR, **286**, 637 (1986).
[3] A. J. Schultz, H. H. Wang, J. M. Williams and A. Filhol, J. Am. Chem. Soc., **108**, 7853 (1986).
[4] V. N. Laukhin, E. E. Kostyuchenko, Yu. V. Sushko, I. F. Shchegolev and E. B. Yagubskii, JETP Lett. (Engl. Transl.), **39**, 81 (1985).
[5] J. D. Wallis, A. Karrer and J. D. Dunitz, Helv. Chim. Acta, **69**, 69 (1986).
[6] K. S. Varma, A. Bury, N. J. Harris and A. E. Underhill, Synthesis, 837 (1987).

Preparation and Properties of Dimeric TTFs

K. Lerstrup, M. Jørgensen, I. Johannsen, and K. Bechgaard

Department of General and Organic Chemistry, H.C. Ørsted Institute,
University of Copenhagen, Universitetsparken 5, DK-2100 Copenhagen, Denmark

Abstract: In this paper we wish to present a number of new compounds in which two TTF units are linked together through various types of bridges. The preparation of the materials is described together with cyclic voltammetry data.

1. Introduction

In our continuous investigation of organic metals derived from the TTF family we have devoted part of our efforts to extended systems, in particular compounds formally containing two TTF units. The reasons for this work are several: In extended systems "U" could decrease, two-dimensional interaction could increase, and also in selected systems direct stoichiometry control may be possible (see also K. Bechgaard et al., this volume). The direct effect of linking TTF units depends strongly on the nature of the interconnect. Two extreme situations can be obtained: If TTF's are directly linked a new electronic structure is obtained (possibly a low "U" material). On the other hand, if the inter-connect is electronically insulating the main expectation is to obtain stoichiometry and structure control, and materials formally corresponding to TTF_2TCNQ or TTF_4X could be obtained.

2. Synthesis

The key reaction in the preparation of the dimeric TTF's is based on work reported by Ishikawa [1a] and Cava et al. [1b], but the procedure has been altered in a significant way to avoid inherent problems with that reaction. We have refined and perfected our earlier reported method [2] for the synthesis of "unsymmetrical" TTF's and expanded the scope of the reaction to include the preparation of the present class of compounds in which two TTF molecules are linked together with various carbon and/or sulfur containing chains or ring systems.

The general procedure is outlined in Scheme 1. The 4,5-dimethyl-1,3-dithiole-2-phosphonate I was chosen for most experiments because of its simple synthetic availability, but the substitution pattern for this compound does not exhibit any limitations for the general procedure. The bis-iminium compound II was prepared via the haloketone-dithiocarbamate route in the usual manner and the sulfur-carbon linked bisiminium compound III was prepared by treatment of the *meso*-ion IV [3] with a suitable dihalo compound.

The coupling reaction was carried out as follows. The phosphonate ester (10 mmol) was dissolved in ca. 25 ml dry THF, cooled to -78°C and dropwise added a solution of ca. 12 mmol potassium *tert*-butoxide in ca. 10 ml dry THF. The deprotonization is instantaneous. The bis iminium compound (4.8 mmol) was added dry and allowed to react until a uniform solution resulted (sometimes a salt precipitates out). In general, the reaction temperature was chosen as to complete the reaction within two hours and would thus vary from -40°C to +10°C depending on the solubility and the reactivity of the iminium salt. Diethyl ether was added to precipitate salts (KBr, KPF_6 unreacted iminium compound), the mixture was filtered and the filtrate concentrated *in vacuo* to a dark syrup. This material, containing the

SCHEME 1

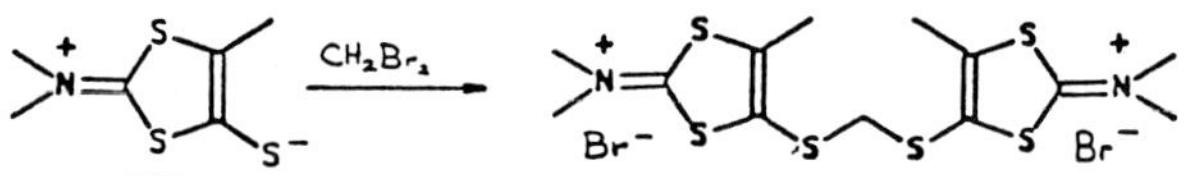

IV III

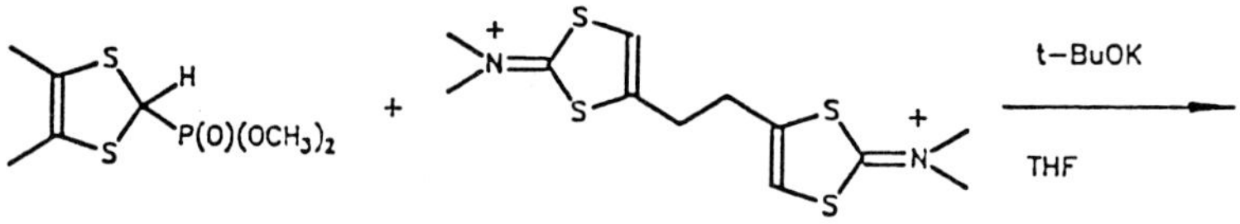

I II

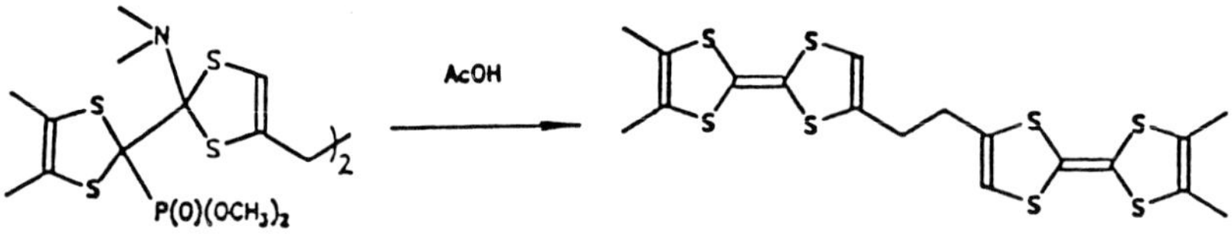

V VI

TABLE I

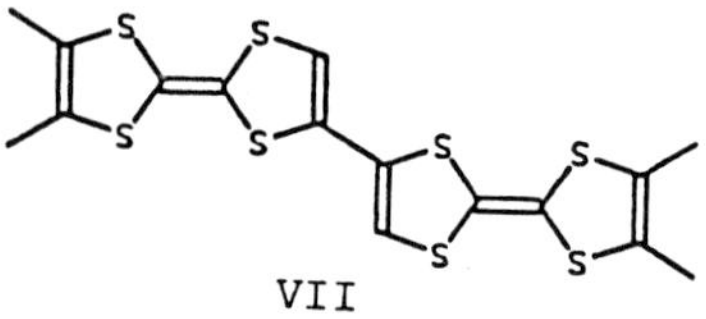

VII

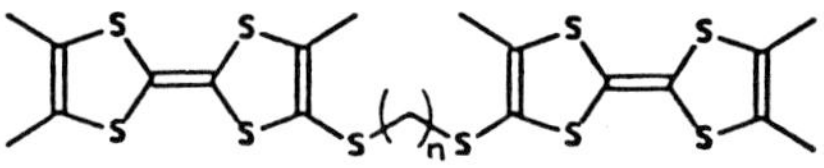

n=1: VIII
n=2: IX
n=3: X

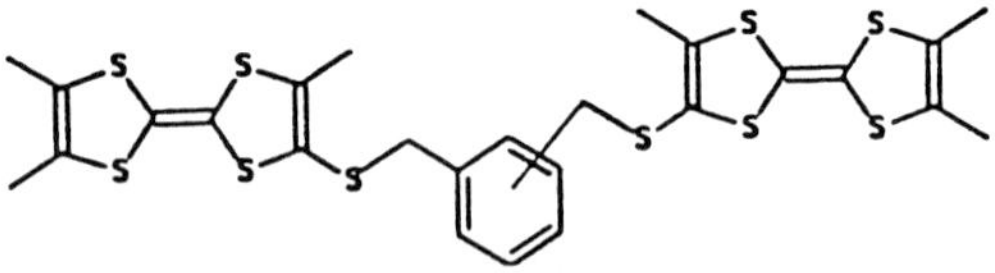

1,2: XI
1,3: XII
1,4: XIII

adduct V, was dissolved in toluene (ca. 10 ml), glacial acetic acid (5-10 ml) was added dropwise and the mixture stirred 1 to 20 hours until no more material precipitated. The product VI was filtered off and the mother liquor subjected to column chromatography to obtain further compound. The combined material was recrystallized from toluene to provide pure material in 30% yield. Other compounds (VII-XIII, Table 1) were prepared similarly in yields varying from 5 to 60%.

The synthetic procedure outlined is general and has been widely used by us and others to generate a variety of unsymmetrical TTF's including the dimeric TTF's reported here (see Table 1).

3. Electrochemistry

The dimeric TTF's (VI-X) have been investigated by cyclic voltammetry. We have investigated the series of electron transfers

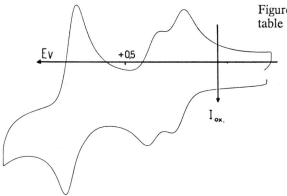

Figure 1. Cyclic voltamogram of VIII. See table 2 for details.

$$\text{Di-TTF}^0 \rightleftarrows \text{DiTTF}^{+1} \rightleftarrows \text{DiTTF}^{+2} \rightleftarrows \text{DiTTF}^{3+} \rightleftarrows \text{DiTTF}^{4+}$$

Thus, in principle 4 one-electron steps are expected but not necessarily observed (see Fig. 1). The behavior of the compounds appears more complicated.

Compound VII (the directly conjugated TTF's) exhibits two close-lying one electron oxidations followed by a wave appearing as a two electron wave. The closeness ($E = E^2_{1/2} - E^1_{1/2}$) of the $0 \rightarrow +1$ and $+1 \rightarrow +2$ oxidation waves is somewhat surprising. Normally even in large conjugated systems the waves will be well separated.

The electrochemical oxidation potentials, $E_{1/2}$, depend on several factors: Vertical Ip, solvation energy and Coulombic effects. Given the fairly limited material investigated the observed relative effects are difficult to evaluate, but we tentatively suggest that the small E for VII may arise mainly from orbital effects combined with an effect from the lack of coplanarity of the TTF units in solution.

In the compounds VI, VIII-X, where the TTF units are formally electronically isolated, they are expected to exhibit oxidative waves, where the main effect on E arises from Coulombic effect. This should decrease with increasing chain length, but no significant trend has been observed. Note, however, that when the flexible links get longer the formation of folded intramolecular dimers may influence $E_{1/2}$'s and ΔE. Thus, no direct conclusions can be drawn from the results. Eventually an extensive electrochemical and spectroscopic investigation may give some more insight.

Note also that the $+2 \rightleftarrows +3$ and $+3 \rightleftarrows +4$ waves are coalescent in all compounds and that the analysis of these waves is further complicated by precipitation on the electrodes.

Table 2 summarizes the results.

TABLE 2

Halfwave oxidation potentials of dimeric TTFs. $E_{1/2}$ vs. SCE in dichloromethane w. tetrabutylammonium hexafluorphosphate as electrolyte. Scanning rate 100 mV·s^{-1}

Compound	$E^1_{1/2}$	$E^2_{1/2}$	$E^{3,4}_{1/2}$
VII	240 (2e⁻)		740(irrev.)
VI	290	350	860
VIII	215	350	800
IX	230	360	740
X	270 (2 e⁻)		750
TMTTF	180	730	

References

1) a) K. Ishikawa, K. Akiba, N. Inamoto, Tetrahedron Lett. 41 (1976) 3695-98.
   b) N. C. Gonella, M. P. Cava, J. Org. Chem. 43 (1978) 369-70.
2) K. Lerstrup, I. Johannsen, M. Jørgensen, Synth. Metals, 27 (1988) B9-B13.
3) A. Souizi, A. Robert, Synthesis 1982, 1059-61.

Preparation and Properties of p-Quinodimethane Analogues of Tetrathiafulvalene

Y. Yamashita[1], Y. Kobayashi[2], and T. Miyashi[2]

[1]Institute for Molecular Science, Myodaiji, Okazaki 444, Japan
[2]Department of Chemistry, Faculty of Science, Tohoku University, Aramaki, Sendai 980, Japan

Abstract. The title compounds were prepared using a Wittig-Horner reaction of 2-dimethoxyphosphinyl-1,3-dithioles and a retro-Diels-Alder reaction. The oxidation potentials of the new donors with quinoid structures were much lower than those of the corresponding tetrathiafulvalenes. They gave good conducting complexes with tetracyanoquinodimethane.

Molecular design of new electron donors for organic conductors has been continuously done. One of the important strategies for the design is to extend conjugation to decrease on-site Coulomb repulsion [1]. From this viewpoint, tetrathiafulvalene (TTF) analogues containing quinodimethane structures are interesting. However, the parent compound, 2,2'-p-quinobis(1,3-dithiole) 1, was hitherto unknown and the reported examples of derivatives were limited to the dibenzo ones 2a,b [2], although some dication salts were known [3]. We report here a simple synthesis of 1 and its derivatives as well as their properties.

The donor 1 was synthesized according to the method shown in Scheme 1, in which a Wittig-Horner reaction of 2-dimethoxyphosphinyl-1,3-dithiole (4) [4] and a retro-diels-Alder reaction were used. First, a double bond of p-benzoquinone was protected by cyclopentadiene since the direct Wittig-Horner reaction of 4 with p-benzoquinone gave a complex mixture of products without formation of 1. The reaction of the cyclopentadiene adduct 3 with 4 in the presence of n-BuLi gave a bis(1,3-dithiole) 5 in 27% yield. Thermolysis of 5 at 200 °C under reduced pressure gave 1 in 54% yield. Similarly, the benzo derivatives 6a,b were prepared from the corresponding naphthoquinones. The dibenzo derivative 7 could be obtained in 74% yield by direct Wittig-Horner reaction of 4 with 9,10-anthraquinone. The dibenzo-TTF analogues 2a, 8a,b, and 9 were similarly obtained in high yields using the Wittig-Horner reaction of a benzo derivative of 4. Tetrathio derivatives 10a,b were also obtained from Wittig-Horner reagents 11a,b, respectively.

The longest absorption maxima in the electronic spectra of the bis(1,3-dithiole) donors are shown in Table 1. Comparison of the value of 1 with that of 5 shows that removal of the cyclopentadiene giving a quinoid structure makes the absorption red-shift. The oxidation potentials of the donors measured by cyclic voltammetry are also shown in Table 1. The values for the donors with quinoid structures are lower than that of TTF. This fact indicates that the introduction of quinoid structure is very effective for strengthening electron donating properties. Another interesting feature is that the difference between the first and second oxidation potential is very small and two-electron oxidation waves are observed in the other new donors. This result indicates that the donors with quinoid structures more easily become dications upon

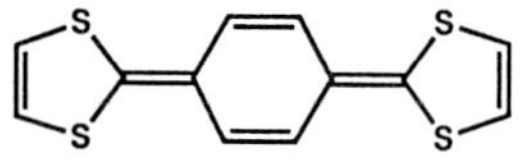

Scheme 1

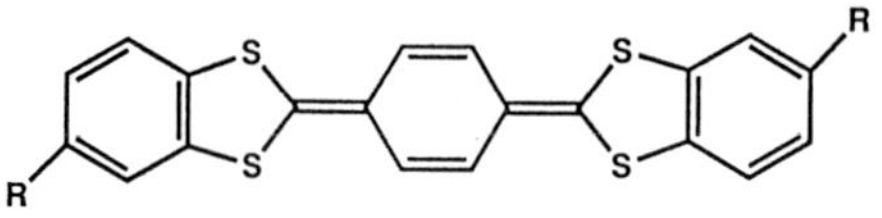

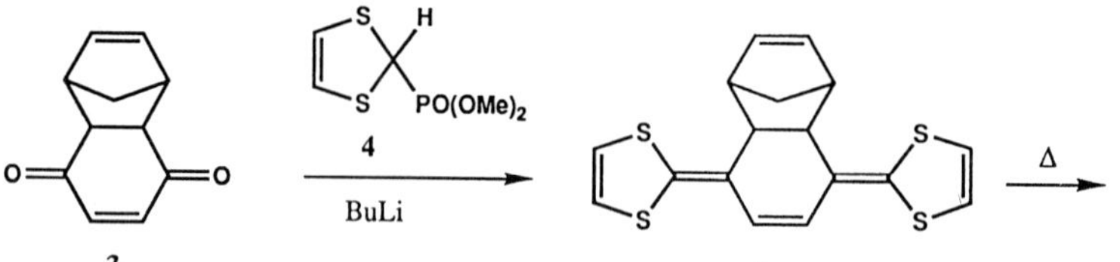

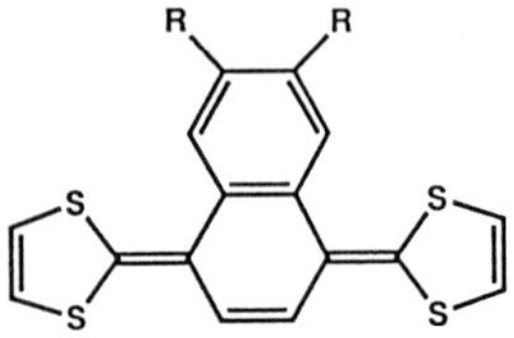

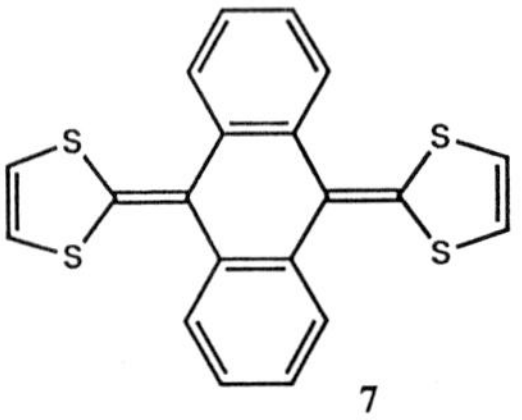

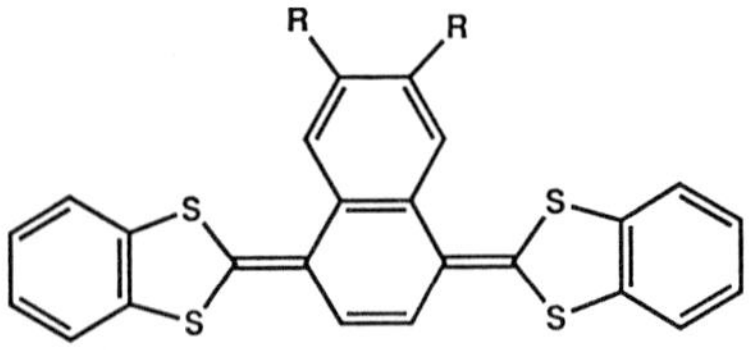

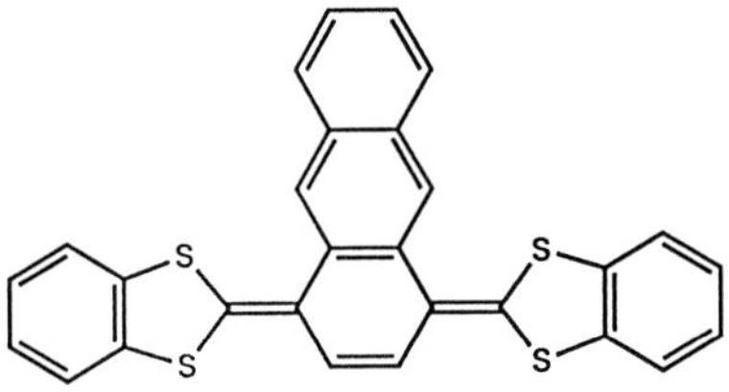

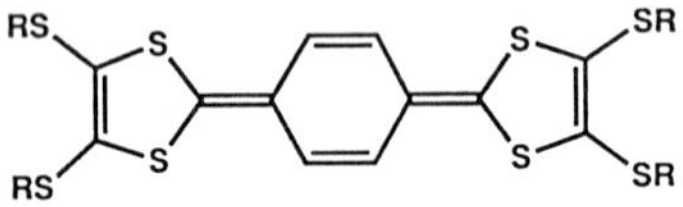

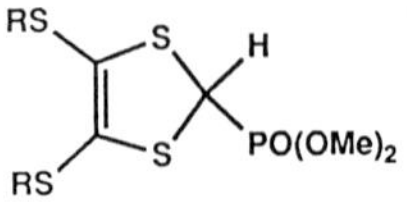

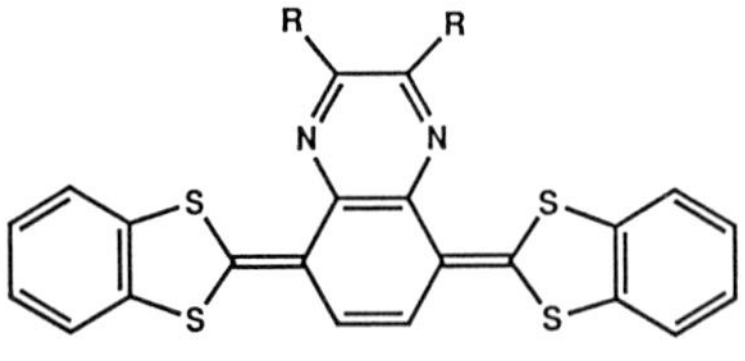

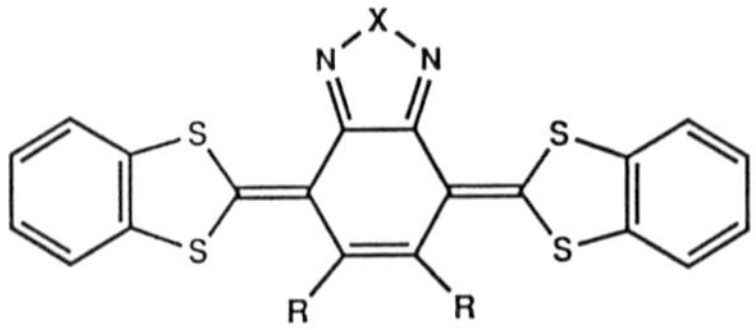

Table 1. Absorption maxima in the electronic spectra [a] and oxidation potentials [b] of donors

Donor	λmax/nm (logε)	E/V
1	495 (4.78)	-0.11, -0.04
5	440 (4.65)	+0.17
6a	479 (4.55)	0.00
6b	480sh (4.70)	-0.04
7	420 (4.40)	+0.25 [c]
TTF	320 (4.12)	+0.28, +0.64
2a	485 (4.53)	---
8a	468 (4.74)	+0.18
8b	468 (4.82)	+0.15
9	---	+0.24
12a	492 (4.52)	+0.37, +0.61
12b	489 (4.90)	+0.31, +0.56
12c	582sh (3.77)	+0.44, +0.67
13a	482	+0.55 [c]
13b	450 (4.77)	+0.52
13c	436 (4.65)	+0.54
dibenzo-TTF	320 (4.30)	+0.71, +1.13

[a] In CH_2Cl_2. [b] E/V vs. SCE, Pt electrode, scan rate 100 mV/s, 0.1 M Et_4NClO_4 in MeCN except for **12**, **13**, and dibenzo-TTF, for which 0.1 M Bu_4NClO_4 in CH_2Cl_2. [c] Irreversible. Calculated as E_{pa} (anodic peak potential) - 0.03 V.

Table 2. Properties of charge-transfer complexes with TCNQ

Donor	Molar ratio [a] D:A	ρ/Ω cm [b]
1	3:4	1.9×10^3
6a	2:3	3.4×10
6b	2:3	5.2×10
2a	1:1	2.4×10^2
8a	3:5:3 H_2O	5.4×10
8b	1:1.8:H_2O	4.5×10
12a	1:1:H_2O	1.6
12b	1:1	2.1×10

[a] Based on elemental analyses. [b] Electrical resistivities measured on compaction pellets at room temperature.

oxidation due to the decreased Coulombic repulsion. Donors **1**, **6a,b**, **2a**, and **8a,b** gave charge-transfer complexes with tetracyanoquinodimethane (TCNQ), which exhibit good conductivities as shown in Table 2. In contrast, complexes of tetrathio derivatives **10a,b** with TCNQ could not be obtained.

The new donors described above were relatively unstable and readily decomposed in solvents. Therefore, we designed pyrazine-fused quinodimethane analogues **12**, in which the electron withdrawing heterocycle was introduced to stabilize the unstable quinoid structure. The polarization resulted in interesting physical properties, and interheteroatom interactions were also expected by the introduction. The donors **12** were prepared from quinoxaline-5,8-diones by a similar method to that of **1**. Thiadiazolo and selenadiazolo derivatives **13** were also synthesized from the corresponding diones. Their absorption maxima shown in Table 1 are observed at longer wavelengths, indicating the contribution of the polarization effect caused by electron withdrawing heterocycles. Their oxidation potentials are also shown in Table 1 and the values are lower than that of dibenzo-TTF measured under the same conditions although they contain electron withdrawing heterocycles. It should be noted here that two well-defined reversible one-electron oxidation waves were observed in pyrazine derivatives **12**, indicating that the cation radical states of **12** are thermodynamically more stable. The donors **12a,b** gave highly conducting charge-transfer complexes with TCNQ as shown in Table 2 although **12c** and **13a-c** gave no complexes with TCNQ. The donors **12** and **13** gave conducting complexes with iodine, in which the highest conductivity was observed in the complex of **13a** [**13a**-I_2, ρ=4.4 Ω cm (as a compaction pellet)].

References

[1] P. J. L. Galigne, B. Liautard, S. Peytavin, G. Brun, M. Maurin, J. M. Fabre, E. Torreilles, and L. Giral, Acta Cryst., **B36**, 1109 (1980); M. R. Bryce. J. Chem. Soc., Perkin Trans. 1, **1985**, 1675.
[2] Y. Ueno, A. Nakayama, and M. Okawara, J. Chem. Soc., Chem. Commun., **1978**, 74; M. Sato, M. V. Lakshmikantham, M. P. Cava, and A. F. Garito, J. Org. Chem., **43**, 2084 (1978).
[3] Y. Ueno, M. Bahry, and M. Okawara, Tetrahedron Lett., **1977**, 4607; J. M. Fabre, E. Torreilles, and L. Giral, ibid., **1978**, 3703; M. R. Bryce, Mol. Cryst. Liq. Cryst., **120**, 305 (1985); M. R. Bryce and A. J. Moore, Tetrahedron Lett., **29**, 1075 (1988); M. R. Bryce and A. J. Moore, Synth. Metals, **27**, B557 (1988).
[4] A Wittig-Horner reaction was found to be useful for preparing bis(1,3-dithiole) donors. K. Akiba, K. Ishikawa, and N. Inamoto, Bull. Chem. Soc. Jpn., **51**, 2674 (1978); Y. Yamashita and T. Miyashi, Chem. Lett., **1988**, 661; M. R. Bryce and A. J. Moore, Synth. Metals, **25**, 203 (1988).

Syntheses and Physical Properties of Oligothiophene Charge-Transfer Complexes

S. Hotta[1],* and K. Waragai[2],*

[1]Central Research Laboratories, Matsushita Electric Industrial Co., Ltd.
Moriguchi, Osaka 570, Japan
[2]Matsushita Research Institute Tokyo, Inc., 3-10-1, Higashimita,
Tama-ku, Kawasaki 214, Japan
*Present address: The Research Development Corporation of Japan,
c/o Matsushita Research Institute Tokyo, Inc., 3-10-1, Higashimita,
Tama-ku, Kawasaki 214, Japan

Abstract. Charge-transfer complexes of oligothiophene-TCNQ were prepared and characterized by X-ray diffraction and spectroscopic analyses. These complexes have 1:1 stoichiometry between the oligothiophene and TCNQ and both the molecules form mixed stacks. The spectroscopic results show that net charge transferred between the complexes is small, which is consistent with their low conductivity.

1. Introduction

The polythiophene family, a typical class of conducting polymers, is made highly conducting upon partial oxidation with dopants [1]. Recently, a series of oligothiophenes with various molecular weights were synthesized and investigated as model compounds of polythiophenes [2]. These oligothiophenes were reported to form charge-transfer (CT) complexes with TCNQ (7,7,8,8-tetracyanoquinodimethane) [3, 4]. Such complexes, however, have not yet been adequately characterized. In the present studies we present detailed results of characterization of the oligothiophene-TCNQ complexes. Implications of molecular orbital calculations are also discussed.

2. Structural Characteristics

The oligothiophenes used include terthiophene (trimer; hereafter referred to as TT), quaterthiophene (tetramer; hereafter referred to as QtT), and quinquethiophene (pentamer). The synthesis and purification methods of these compounds as well as the preparation methods of their TCNQ complexes can be seen elsewhere [4].

Elemental analysis results show that the materials are composed of an equimolar complex of the oligothiophene and TCNQ. The two-probe conductivities (at room temperature) measured on pressed discs of the CT complexes show relatively low values, ranging from 10^{-10} to 10^{-9} S/cm.

X-ray diffraction measurements were carried out on the powder and single crystal samples. The X-ray results verify that the crystals of TT-TCNQ are monoclinic with a=10.99, b=12.71, c=7.78 Å; β=101.3°; Z=2; V=1065.5 Å^3 and D_c=1.411. The X-ray data analyses suggest that the space group is one chosen from among C2, Cm, and C2/m. The crystals of QtT-TCNQ are triclinic, space group P1, with a=7.88, b=8.89, c=10.74 Å; α=115.2, β=114.8, γ=82.1°; Z=1; V=617.2 Å^3 and D_c=1.439. The calculated densities are in excellent agreement with the specific gravities of 1.41 for TT-TCNQ and 1.44 for QtT-TCNQ determined from a flotation method (in a carbon tetrachloride/methanol mixture as solvent), respectively.

Figure 1 shows the molecular packing viewed along the a-axis of the QtT-TCNQ complex. The QtT and TCNQ molecules form mixed stacks along the c-axis in face-to-face arrangement. The sulfur atoms of QtT sit in <u>trans</u> position relative to the C-C bonds connecting the thiophene rings. Both the molecules are nearly planar and

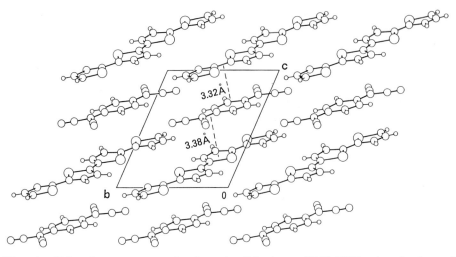

Fig. 1 Molecular arrangement of quaterthiophene (QtT)-TCNQ viewed along the a-axis. The rhomboid shows a unit cell. Note that the QtT and TCNQ molecules form mixed stacks along the c-axis.

the molecular planes are inclined to each other at the angle of 3.2°. Each pair of QtT and TCNQ is weakly dimerized; two distances between the two molecules of 3.32 and 3.38 Å are noticed from detailed analysis of the diffraction data (Fig. 1). The mixed stacks and the <u>trans</u> position arrangement of the sulfur atoms are also characteristic of other oligothiophene-TCNQ complexes [3].

3. Spectroscopic Characteristics

The oligothiophene-TCNQ complexes exhibit an overall IR spectral profile pretty close to superposition of the neutral oligothiophene and TCNQ. Figure 2, for instance, shows an IR spectrum of QtT-TCNQ taken on a KBr pressed disc as compared with neutral QtT. Four sharply resolved peaks (marked with asterisks in Fig. 2) due to the TCNQ are clearly observed in the spectrum. These peaks are assigned to the CN stretching (2215 cm^{-1}), ring stretching (B_{1u}, ν_{20}; 1536 cm^{-1}), another ring stretching (A_g, ν_4; 1441 cm^{-1}), and the CH out-of-plane (835 cm^{-1}) modes, respectively [5]. Of these, the totally-symmetric ring stretching mode (A_g, ν_4), which is IR inactive in the neutral TCNQ ($TCNQ^0$), has been definitively induced in the spectrum for the QtT-TCNQ because of symmetry breaking which results from the CT complex formation. The considerably small shift of these bands relative to $TCNQ^0$ indicates that net charge transferred between the complexes is small [6].

Electronic spectra of the CT complexes were measured in various solvents such as chloroform, methylene chloride, and acetonitrile. The dominant peaks around 400 nm in the spectra mainly arise from the $\pi-\pi^*$ transition of the neutral oligothiophene and $TCNQ^0$ moieties. A typical example is shown in Fig. 3 for TT-TCNQ. Well-resolved bands assigned to the TCNQ anion radicals ($TCNQ^-$) are observed at 750 and 840 nm (see the inset). Relatively weak absorbance for these bands again means a weak charge transfer. Thus, the spectroscopic results are consistent; these results are associated with the low conductivity of the CT complexes.

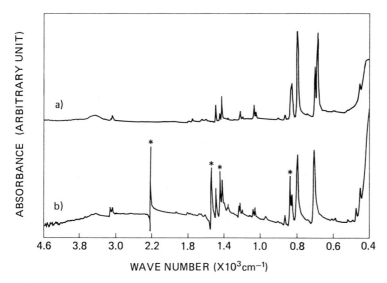

Fig. 2 IR spectra of (a) QtT and (b) QtT-TCNQ. The bands marked with asterisks for the QtT-TCNQ are assigned to the TCNQ moiety (see text).

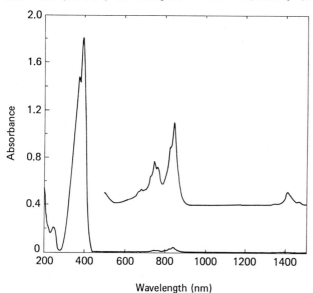

Fig. 3 Electronic spectrum of terthiophene (TT)-TCNQ complex taken in acetonitrile solution. The concentration of the complex was 5×10^{-5} M. The dominant peaks around 400 nm mainly result from the π-π^* transition of the neutral TT and TCNQ. The inset shows an enlarged profile of the bands below 600 nm.

4. Implications of Molecular Orbital Calculations

Molecular orbital calculations provide powerful means to understand and foresee the structural characteristics and physical properties of the oligothiophene-TCNQ complexes. The HOMO (Highest Occupied Molecular Orbital) of a donor (oligothio-

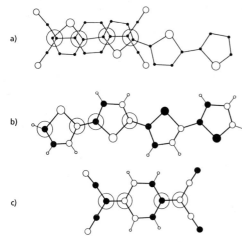

Fig. 4 (a) Molecular arrangement of QtT-TCNQ viewed along the direction of the molecular plane normal, (b) HOMO of QtT, and (c) LUMO of TCNQ. The corresponding carbons encircled are located close to each other. The open circles show one of the alternative signs of the coefficient of the atomic orbitals. The closed circles show the other.

phene) and the LUMO (Lowest Unoccupied Molecular Orbital) of an acceptor (TCNQ) play a particularly important role [7].

The molecular orbital calculations were performed using both the simple Huckel and PPP (Pariser-Parr-Pople) methods in the present studies. When an oligothiophene has an even (odd) number of thiophene rings, the oligothiophene belongs to a point group of C_{2h} (C_{2v}). In both the cases the HOMO is anti-symmetric as to the symmetry operations. An important point is that the LUMO of the TCNQ is also anti-symmetric concerning the symmetry operations of both C_{2h} and C_{2v} included as subgroups of D_{2h} which TCNQ belongs to. An example is given in Fig. 4 for QtT-TCNQ. Accordance of the sign of the two atomic orbitals for each pair of closely located carbon atoms (encircled in the figure) is evident for the molecular arrangement determined from the X-ray diffraction. Such arrangement results in significantly large overlap between the corresponding atomic orbitals, which implies that the charge transfer between the oligothiophenes and TCNQ is potentially favorable [7]. This can also be the case with polythiophene (PT) doped with TCNQ [8]; the HOMO of PT has essentially the same symmetry features as the oligothiophenes (see Fig. 4) except that the coefficients of all the sulfur atomic orbitals on PT must vanish from symmetry requirements [9].

In summary, we have prepared the oligothiophene-TCNQ complexes and characterized them through the X-ray diffraction and spectroscopic studies. The small charge transfer between the complexes confirmed from the spectroscopy is consistent with their low conductivity. Nonetheless, the molecular orbital calculations indicate the potential advantage to the charge transfer which may lead to manifestations of novel physical properties.

We would like to thank Mr. S. Kaida for making the X-ray diffraction measurements and for carefully inspecting the X-ray results.

References

1 T. Yamamoto, K. Sanechika, and A. Yamamoto, J. Polym. Sci. Polym. Lett. Ed. 18, 9 (1980).
2 F. Garnier, G. Horowitz, and D. Fichou, Synth. Met. 28, C705 (1989).
3 M. Qian, H. Fu, and Y. Cao, Jiegou Huaxue 5, 159 (1986); 5, 163 (1986).
4 S. Hotta and K. Waragai, Synth. Met. 32, 395 (1989).
5 A. Girlando and C. Pecile, Spectrochim. Acta 29A, 1859 (1973).
6 R. Bozio, A. Girlando, and C. Pecile, J. Chem. Soc. Faraday Trans. II 71, 1237 (1975).
7 R. S. Mulliken, J. Am. Chem. Soc. 74, 811 (1952).
8 S. Hotta, T. Hosaka, and W. Shimotsuma, Synth. Met. 6, 69 (1983).
9 J. L. Bredas, B. Themans, J. P. Fripiat, and J. M. Andre, Phys. Rev. B 29, 6761 (1984).

3,3' : 4,4'-Bis(thieno[2,3-b]thiophene) with an Isoelectronic Structure of Perylene

T. Otsubo[1], Y. Kono[1], H. Miyamoto[1], Y. Aso[1], F. Ogura[1], T. Tanaka[2], and M. Sawada[2]

[1]Department of Applied Chemistry, Faculty of Engineering,
Hiroshima University, Saijo, Higashi-Hiroshima 724, Japan
[2]Institute of Scientific and Industrial Research, Osaka University,
Ibaraki, Osaka 567, Japan

Abstract. The title compound, in which all of the perimetric rings consist of fused thiophenes, was synthesized by way of reductive dimerization of 3,4-dibromothieno[2,3-b]thiophene. Its molecular structure was confirmed by an X-ray crystallographic analysis. The geometry is quite planar and has a center of symmetry. The perimetric five-membered rings suffer from considerable strain in bond angles. The molecules form infinite stacking columns, which transversely interact with each other through van der Waals contact of the sulfur atoms. It, like perylene, formed a black, highly electrical conductive complex with iodine. In addition, it gave a black conductive polymer on electrolysis.

1. Introduction

Perylene **1** gained the first importance in the construction of organic solids which exhibit high electrical conductivity [1]. The iodine complex [2] and radical-cation salts of **1** [3] were regarded as low-dimensional organic metals. Wudl et al. reported an isoelectronic heterocyclic analogue of perylene, 3,4':4,3'-bis(benzo[b]thiophene) **2**, which behaved like **1** on complexation with iodine [4]. The heteroatoms incorporated in the molecule serve to not only induce high polarization but also enhance intermolecular interactions, both facilitating the formation of molecular assembly. It is empirically proposed that suitable electron-donor components for organic metals satisfy some requirements, that is, high symmetry, high planarity, high polarizability, low ionization potential, and strong intermolecular interaction [5]. In this connection, 3,3':4,4'-bis(thieno[2,3-b]thiophene) **3**, in which all of the perimetric rings consist of thiophenes, might be an isoelectronic heterocycle of perylene, which is superior to **2** as an electron donor. We here want to report the synthesis, structure, and some properties of **3**.

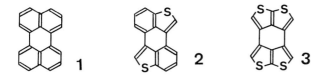

2. Synthesis

We have designed, as one of the simplest approaches to **3**, reductive coupling of 3,4-dibromothieno[2,3-b]thiophene **6**, which was prepared from thieno[2,3-b]thiophene **4** according to the literature [6]. The reductive coupling of **6** with catalytic bis(triphenylphosphine)nickel(II) chloride, excess active zinc, and tetraethylammonium iodide in refluxing benzene [7] gave only a trace of the desired product **3** and 4,4'-dibromo-3,3'-bithieno[2,3-b]thiophene **7** (maximum yield

28%) in the reaction mixture. The second coupling did not proceed so readily as the first one because of enlarged steric hindrance. However, the increasing quantity of the nickel reagent and prolonged reaction time overcame this difficulty, forming 3 in 14 % yield. The optimal conditions were found by treating a mixture of 6 (0.5 g, 1.68 mmol), bis(triphenylphosphine)nickel(II) chloride (0.54 g, 0.84 mmol), active zinc (1.42 g, 21.7 mmol), and tetraethylammonium iodide (2.76 g, 10.7 mmol) in dry benzene (15 ml) under reflux for 28 h.

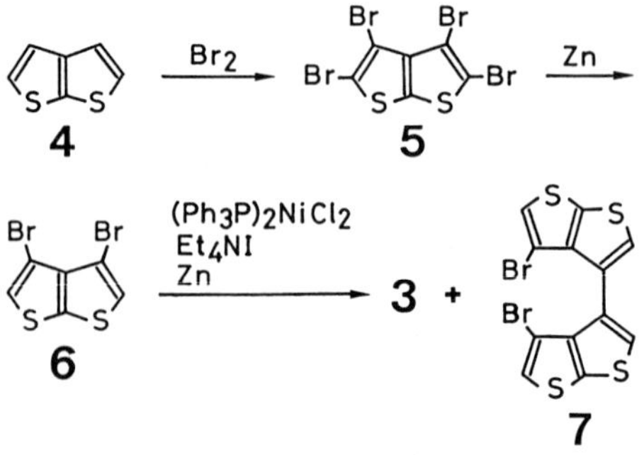

3. Structure

The molecular structure of 3 was confirmed by an X-ray crystallographic analysis. The crystal data are as follows: orthorhombic, space group $Pcab$, a=15.029(3), b=5.179(1), c=13.380(3) Å, V=1041.0 Å^3, Z=4, ρ_{calcd}=1.764 g cm^{-3}, ρ_{obsd}=1.76 g cm^{-3}, R=0.033 for 1100 unique reflections (Mo$K\alpha$, λ=0.71069Å). As shown in Fig. 1, the geometry is quite planar and has a center of symmetry in contrast to the slight bending of perylene molecule in the crystal [10]. The perimetric five-membered rings suffer from considerable strain in bond angles rather than in bond

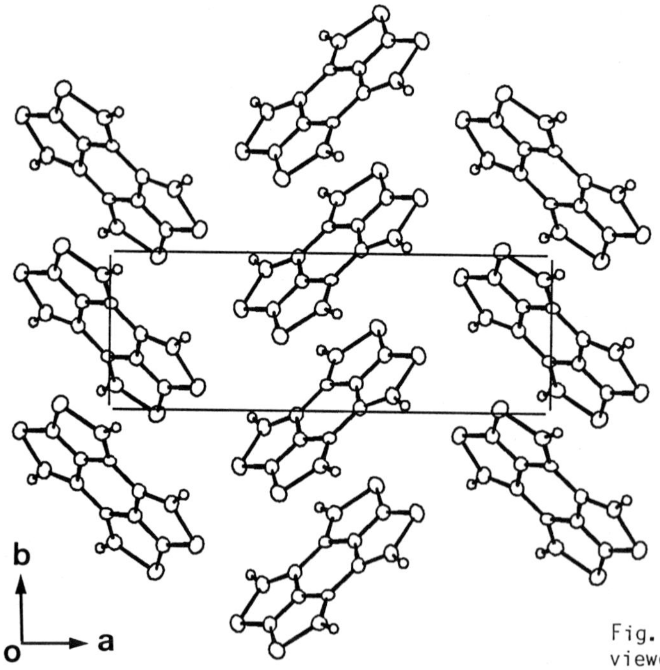

Fig. 1. Crystal structure of 3 viewed from the c-axis.

lengths as compared with the structures of thiophene [8] and thieno[3,2-b]thiophene [9]. In conformity with this view, the central six-membered ring is much deformed from a regular hexagon. The crystal structure is quite different from that of perylene which is constructed by isolated pairs of the molecules as often seen in the case of normal aromatic hydrocarbons and rather characteristic of that of its radical cation salts [3]. Thus the molecules overlap each other with slipping, forming infinite stacking columns. The face-to-face distance is 3.45 Å, just holding van der Waals approach. In addition, the stacking molecules can interact through the sulfur atoms with the other ones in the neighboring column, that is, the two sulfur-sulfur distances between the nearest molecules are 3.40 and 3.52 Å, which are shorter than sum (3.60 Å) of van der Waals radii.

4. Properties

Compound 3 was very stable and crystallized in the form of faint brown needles from carbon disulfide. It sublimed around 270°C at atmospheric pressure and did not melt up to 300°C in a sealed tube. Its highly symmetrical structure was reflected with simple patterns in the proton and carbon-13 NMR spectra: PMR (60 MHz, CS_2) δ=7.23 (s); C-13 NMR (67.8 MHz, DMSO-d_6) δ=121.95, 125.15, 132.98, 150.17. It was not so colored as perylene 1, but the electronic absorption spectrum exhibited marked bathochromic and hyperchromic shifts as compared with that of thieno[2,3-b]thiophene 4, reflecting extended π-conjugation through central peri-bond of 3; UV/VIS (tetrahydrofuran) 3: λ_{max}=252 nm (ε=46400), 254 (34200), 330 (22000); 1: λ_{max}=253.5 nm (ε=42800), 388 (12750), 410 (27880), 436.5 (36170); 4: λ_{max}=225 nm (ε=25600), 248 (5020), 255 (4510), 260 (3630).

Cyclic voltammetry of 3 was measured with Pt working and counter electrodes at scan rate 100 mV/sec in benzonitrile solution containing 0.1 M tetrabutylammonium perchlorate as supporting electrolyte at room temperature. It showed one quasi-reversible redox wave with a half-wave potential at 1.01 V vs. a Ag/AgCl standard electrode. This value is considerably lower than that (>1.5 V) of thieno[2,3-b]thiophene 4 but equal to that (1.00 V) of perylene 1. The repetitive cycling soon changed the voltammogram to two redox waves at 0.74 and 1.10 V, depositing a black polymeric material on the working electrode. The same material was also formed by potentiostatic electropolymerization of 3 at 1.05 V in benzonitrile solution containing tetrabutylammonium perchlorate as supporting electrolyte. It must take a complex network structure, because 3 has four reactive sites for polymerization, The elemental analysis (C, 45.57; H, 1.05%) suggested that it was doped with the counter ion ClO_4 (doping level 0.5). Its electrical conductivity measured on compressed pellet at room temperature was 0.06 Scm^{-1}.

Compound 3 formed charge transfer complexes with stronger π-electron acceptors such as 7,7,8,8-tetracyanoquinodimethane (TCNQ), its 2,3,5,6-tetrafluoro derivative (TCNQ-F_4), 2,3-dichloro-5,6-dicyano-p-benzoquinone (DDQ), and 1,1,2,3,4,4-hexacyanobutadiene (HCBD). All the complexes are, however, low conductive as shown in Table 1. This is probably because of less charge transfer due to the weak electron accepting ability of 3. On the other hand, 3 like 1 and

Table 1. Complexation of 3 with typical electron acceptors

Complex	Appearance	D:A	D.p./°C	Conductivity/Scm^{-1}
3·TCNQ	dark blue needles	1:1	232	2.5×10^{-9}
3·TCNQ-F_4	dark blue needles	1:1	191	1.5×10^{-8}
3·DDQ	dark green needles	1:1	>300	1.2×10^{-7}
3·HCBD	dark blue needles	1:1	177	1.6×10^{-7}
3·I_2	black powder	1:1	145	1.1×10^{-1}

2 formed fine black crystals from benzonitrile solution containing equivalent amounts of iodine.The complex had 1:1 stoichiometry of 3 to iodine and showed a relatively high electrical conductivity of 0.11 S cm^{-1}, suggesting a mixed valence state of 3 in the complex.

Acknowledgment: This work was supported by a grant-in-aid of scientific research from the Ministry of Education, Science and Culture, Japan.

References

[1] H. Akamatu, H. Inokuchi, and Y. Matsunaga, Nature, **173**, 168 (1954); Bull. Chem. Soc. Jpn., **29**, 213 (1956); M. M. Labes, R. Sehr, and M. Bose, J. Chem. Phys., **33**, 868 (1960); J. Kommandeur and F. R. Hall, ibid.. **34**, 129 (1961).
[2] H. I. Kao, M. Jones, and M. M. Labes, J. Chem. Soc., Chem. Commun., **1979**, 329.
[3] H. J. Keller, D. Nöthe, H. Pritzkow, D. Wehe, M. Werner, P. Koch, D. Schweitzer, Mol. Cryst. Liq. Cryst., **62**, 181 (1980); H. J. Keller, D. Nöthe, H. Pritzknow, D. Dehe, M. Werner, R. H. Harms, P. Koch, and D. Schweitzer, Chem. Scr., **17**, 101 (1981); V. Enkelmann, B. S. Morra, C. Kröhnke, G. Wegner, and J. Heinze, Chem. Phys., **66**, 303 (1982); D. Schweitzer, I. Hennig, K. Bender, H. Endres, and H. J. Keller, Mol. Cryst. Liq. Cryst., **120**, 213 (1985); V. Enkelmann, K. Göckelmann, G. Wieners, and M. Monkenbursch, ibid., **120**, 195 (1985).
[4] F. Wudl, R. C. Haddon, E. T. Zellers, and F. B. Bramwell, J. Org. Chem., **44**, 2491 (1979).
[5] G. Saito and J. P. Ferraris, Bull. Chem. Soc. Jpn., **53**, 2141 (1980).
[6] P. Fournari and P. Meunier, Bull. Soc. Chim. Fr. **1974**, 583.
[7] M. Iyoda, K. Sato, and M. Oda, Tetrahedron Lett., **26**, 3829 (1985) and references cited therein.
[8] B. Bak, D. Christensen, L. Hansen-Nygaard, and J. Rastrup-Andersen, J. Mol. Spectrosc., **7**, 58 (1961).
[9] E. G. Cox, R. J. J. H. Gillot, and G. A. Jeffrey, Acta Crystallogr., **2**, 356 (1949).
[10] A. Camerman and J. Trotter, Proc. R. Soc., **A 279**, 129 (1964).

Design of Organic Molecular Metals Based on New Multi-Stage Redox Systems in the Non-TTF Family: Peri-Condensed Weitz-Type Donors

K. Nakasuji

Institute for Molecular Science, Myodaiji, Okazaki 444, Japan

Abstract. New multi-stage redox systems, "peri-condensed Weitz type donors", have been designed as component molecules for organic molecular metals. All these donors showed high donor abilities comparable to that of tetrathiafulvalene. Among charge transfer complexes prepared, 2,7-MTDTPY has produced new organic molecular metals which contain neither a TTF nor a TCNQ type framework.

1. Introduction

Starting from organic semiconductors, organic superconductors have already been explored [1]. The most important skeleton in the donor molecules is tetrathiafulvalene (TTF). Furthermore, the chemical modifications of the TTF skeleton play a central role in the development of this chemistry and physics. However, exploration of new classes of donors which do not contain TTF type skeletons is also important to extend the component molecules for CT complexes and to explore new organic conductors. Recently, we have designed and synthesized *"peri-condensed Weitz type donors"* as new multi-stage redox systems belonging to the non-TTF family [2]. Furthermore, non-TTF and non-TCNQ type organic molecular metals have been realized [3]. We now describe the present stage of our investigations.

2. Correlation of Solid State Properties with Molecular Properties

In order to explore new organic molecular solids related to electrical, magnetic and optical properties, it is important to correlate the solid state properties with the molecular properties. Figure 1 shows the four schematic models for the electrical conduction processes from a chemical viewpoint. The HOMO-LUMO energy gap in (1), the intra-molecular Coulombic repulsion in (2), and Peierls transition in (4) prevent smooth movement of electrons. In model (3), there exist no such unfavorable energies, that is,

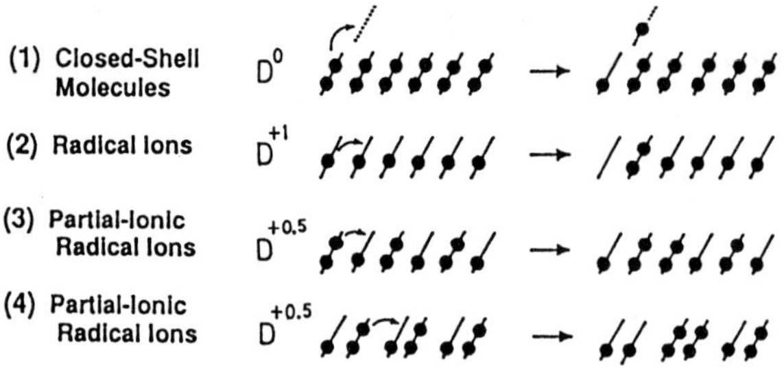
(1) Closed-Shell Molecules
D^0
(2) Radical Ions
D^{+1}
(3) Partial-Ionic Radical Ions
$D^{+0.5}$
(4) Partial-Ionic Radical Ions
$D^{+0.5}$

Fig. 1. Schematic electronic structures for molecular metals.

electron configurations before and after electron movement are isoelectronic.

In molecular level considerations, multi-stage redox type molecules are essential to reduce the intramolecular Coulombic repulsion. A multi-dimensional electronic structure is useful to suppress the Peierls transition and to construct a metallic state. Therefore, we can draw two essential molecular design strategies to realize new molecular metals: 1) construction of new multistage redox systems and 2) chemical modifications to introduce interstack interactions.

3. New Multi-stage Redox Systems, "Peri-condensed Weitz Type Donors"

Polycyclic condensed arenes, such as perylene and pyrene are an important class of component molecules for molecular conductors. These are usually weak donors. Replacing two of the sp^2 carbon atoms in such a polycyclic arene by two chalcogens produces a new heterocycle which gives a stable dication by removal of two electrons. Therefore, such heterocycles might have an increased donor ability.

We synthesized 3,9-DTPR [4], 1,7-DTPR [4], 3,10-DTPR , 1,6-DTPY, and their derivatives. Their electrochemistry measured by cyclic voltammetry shows that these heterocycles are actually two-stage redox systems and the oxidation potentials are much lower than polycyclic arenes. The donor abilities are comparable to that of TTF (0.34 eV). These new multi-stage redox systems are termed "*peri-condensed Weitz type donors* ".

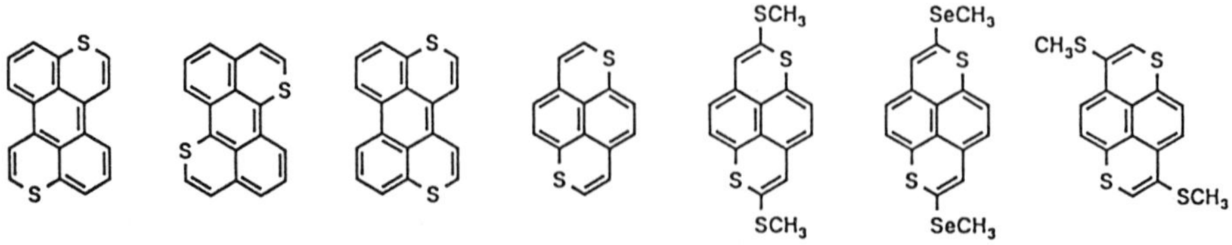
S
S
S
S
S
S
S
S
SCH3
S
S
SCH3
SeCH3
S
S
SeCH3
CH3S
S
S
SCH3

3,9-DTPR	1,7-DTPR	3,10-DTPR	1,6-DTPY	2,7-MT-DTPY	2,7-MS-DTPY	3,8-MT-DTPY

4. Realization of New Organic Molecular Metals

We prepared more than 100 charge transfer complexes of these donors containing low-conducting semiconductors, high-conducting semiconductors, and molecular metals.

An important relationship known as the V-shape correlation between $h\nu_{CT}$ and $\Delta E^{o,r}$ for the mixed stacking CT complexes is convenient for selecting the potentially highly conducting complexes [5]. Here, $h\nu_{CT}$ and $\Delta E^{o,r}$ denote the CT transition energy and the difference between the oxidation potential of the donor and the reduction potential of the acceptor, respectively. The plots of the experimental values of $h\nu_{CT}$ against the $\Delta E^{o,r}$ for our complexes are shown in Fig. 2. In order to make allowance for the experimental errors, the dotted lines are drawn on the upper and lower side from the V-shape in a range of 0.12 eV (1000 cm^{-1}). From this plot, we can classify our complexes into three groups, I, II, and III, which are characterized as neutral, ionic, and non-V-shape complexes, respectively.

The CT transition energies for group III complexes deviate from the correlation to the lower energy side. In such a case, it is highly probable to obtain high-conducting CT complexes, and even metallic complexes.

Within group III, the complexes of 2,7-MTDTPY with TCNQ, chloranil (CHL), and bromanil (BRL) and of 2,7-MSDTPY with TCNQ are found to show metallic conducting behavior. *The two p-benzoquinone complexes, 2,7-MTDTPY-CHL and 2,7-MTDTPY-BRL are the first pure organic molecular metals within the limits of component molecules having non-TTF- and non-TCNQ-type conjugated electronic systems.*

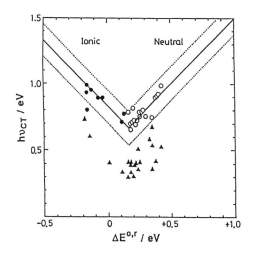

Fig. 2. V-shape correlation

5. Selected Physical Properties of New Organic Molecular Metals

The crystal structures of 2,7-MTDTPY-TCNQ and 2,7-MTDTPY-CHL show the segregated, uniform stacking of the component molecules. Shorter interstack S...S distances which are less than the sum of van der Waals radii are found between sulfur atoms in the DTPY skeleton. The ionicity (the degree of CT) is 0.6-0.7 and 0.6 for the TCNQ complex and the CHL complex, respectively. Such segregated stacking structures and their partial ionicity are consistent with the metallic conductivity of the complexes. The sharp ESR linewidth suggests one-dimensionality. Thus the origin of the metal-insulator transition can be attributed to a Peierls instability [6]. Although the electronic structures show one-dimensional nature, the slightly shorter S...S contacts found in the crystal structures might contribute to realizing the segregated stacking modes of crystal structures.

Apparently, our next step is the enhancement of the dimensionality in the electronic structures through stronger interstack interactions. In addition to enhancing the dimensionality, introduction of a third force, for example hydrogen bonding, might also be important for exploring new organic conductors.

References

1. For general reviews, see: (a) J. B. Torrance: *Acc. Chem. Res.*, **12**, 79. (1979), (b) F. Wudl: *Acc. Chem. Res.*, **17**, 227 (1984), (c) J. M. Williams, M. A. Beno, H. H. Wang, P. C. W. Leung, T. J. Emge, U. Geiser, K. D. Carlson: *Acc. Chem. Res.*, **18**, 261 (1985), (d) D. O. Cowan, F. M. Wiygul: *Chem. Eng. News*, July 21, 28 (1986)
2. K. Nakasuji, H. Kubota, T. Kotani, I. Murata, G. Saito, T. Enoki, K. Imaeda, H. Inokuchi, M. Honda, C. Katayama J. Tanaka: *J. Am. Chem. Soc.*, **108**, 3460 (1986)
3. K. Nakasuji, M. Sasaki, T. Kotani, I Murata, T. Enoki, K. Imaeda, H. Inokuchi, A. Kawamoto, J. Tanaka: *J. Am. Chem. Soc.*, **109**, 6970 (1987)
4. K. Nakasuji, A. Oda, I. Murata: J. C. S. Chem. Commun., in press
5. (a) J. B. Torrance, J. E. Vazquez, J. J. Mayerle, V. Y. Lee: *Phys. Rev. Lett.*, **46**, 253 (1981), (b) K. Nakasuji, M. Nakatsuka, H. Yamochi, I. Murata, S. Harada, N. Kasai, K. Yamamura, J. Tanaka, G. Saito, T. Enoki, H. Inokuchi: *Bull. Chem. Soc. Jpn.*, **59**, 207 (1986), (c) K. Nakasuji, I. Murata: *Synthetic Metals* **27**, B289 (1988) (d) K. Nakasuji, M. Sasaki, I. Murata, A. Kawamoto, J. Tanaka, *Bull. Chem. Soc. Jpn.*, **61**, 4461 (1988)
6. K. Imaeda, T. Enoki, T. Mori, H. Inokuchi, M. Sasaki, K. Nakasuji, I. Murata: *Bull. Chem. Soc. Jpn.*, **62**, 372 (1989)

Conjugated Heteroquinonoid Isologues of TCNQ as Novel Electron Acceptors

F. Ogura, K. Yui, H. Ishida, Y. Aso, and T. Otsubo

Department of Applied Chemistry, Faculty of Engineering, Hiroshima University, Saijo, Higashi-Hiroshima 724, Japan

Abstract. The extensively conjugated homologues (2, 3, 4, and 5) of hetero-TCNQ (1, X=O, S, Se) have been recognized as potential electron acceptors for organic electrical conductors. These compounds have the double advantage of diminution of on-situ Coulomb repulsion due to the extensively conjugated system and enhancement of intermolecular interaction due to the increasing chalcogen atoms. Furthermore, chemical modifications have been designed to improve their electron accepting abilities. Accordingly they formed various conductive complexes with many electron donors.

1. Introduction

An excellent acceptor, TCNQ, has been widely modified to improve its physical properties up to date. In this current, extension of its quinonoid conjugation has been examined to develop new acceptors with small on-site Coulomb repulsion in dianion state. TCNDQ has attracted considerable interest as one such promising candidate. However, it turned out to be a nonpersistent molecule in neutral state owing to steric repulsion between the biphenylic ortho hydrogens [1]. This was nearly verified by stable isolation of a TCNDQ derivative with replacement of the hydrogens by ethylene bridges, i.e., the TCNQ analogue of tetrahydropyrene type [2]. On the other hand, thiophene- and selenophene-TCNQs (T1 and S1) were synthesized by Gronowitz and Uppström in 1974 [3] and have been known as the first heteroquinonoid isologues of TCNQ. Though the introduced chalcogen atoms are expected to act advantageously on the formation of their molecular complexes, they have attracted little attention because of their inferior electron accepting abilities and insulating molecular complex formation.

We are interested in a possibility that more extension of heteroquinonoid conjugation in thiophene- and selenophene-TCNQs might give rise to new electron acceptors for conductive complexes. In contrast to biphenylic one, heterocyclic systems such as bithiophene, biselenophene, and bifuran can avoid the ortho hydrogen interactions by taking the trans conformation in the quinonoid conjugation. Furthermore, there is another advantage of enhancement of intermolecular interaction due to the increasing chalcogen atoms. In order to scrutinize such possibilities, we have synthesized TCNQ isologues of thiopheno-, selenopheno-, and furano-quinonoid types and studied their physical properties.

2. Results and Discussion

There are two kinds of conjugated heteroquinonoid systems; one is of a linearly conjugated type corresponding to TCNDQ and TCNTQ, and another one is of an annularly condensed type corresponding to TNAP and TANT. The structural correspondence between TCNQ and hetero-TCNQ families is illustrated in Fig. 1. We synthesized both types of acceptors in thiophene series [4,5], but only the linear type in selenophene [6] and in furan series [7]. Syntheses were carried out starting from equivalent bromides by utilizing two kinds of reactions as

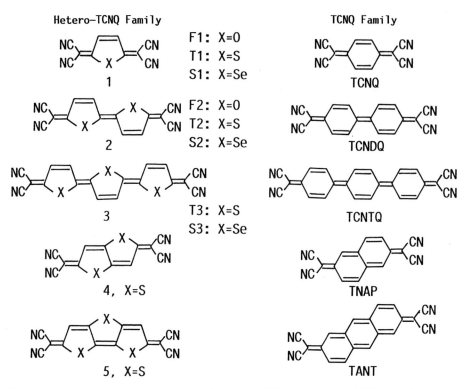

Fig. 1. Structural correspondence between TCNQ and Hetero-TCNQ families.

follows. 2,5-Dibromo-thiophene or selenophene was allowed to react with tetracyanoethylene oxide (TCNEO) in refluxing 1,2-dibromoethane to afford thiophene- or selenophene-TCNQ, respectively, and carbonyldicyanide [3]. We have found "the Gronowitz Reaction" to be applicable to both linearly conjugated and annularly condensed systems. Another kind of reaction is a substitution of the bromides with dicyanomethanide anion, catalyzed by zero-valent Pd metal complex, and following oxidation. Structures of these hetero-TCNQs were confirmed by conventional analytical methods [8].

The redox properties of these new acceptors were measured by cyclic voltammetry. As summarized in Table 1, all the compounds show two pairs of reversible redox waves. The electron affinities decrease in the order of S, Se, O, though the differences are small. The magnitude is mainly dominated by the aromaticity of the heterocyclic ring. The expected electronegative effect of oxygen is counterbalanced by the less aromaticity of furan than thiophene or selenophene. The first redox potentials of the unsubstituted conjugated hetero-TCNQs 2-5 are similar to or rather less than the ones of T1 and S1, indicating that the extension of heteroquinonoid conjugation is not capable of improving their electron accepting abilities. However, the difference between first and second wave potentials, ΔE, becomes smaller with the expansion of heteroquinonoid conjugation to demonstrate clearly the diminution of on-site Coulomb repulsion in the dianion state. Especially, both of T3 and S3 show remarkable coalescence of the two waves to minimize the electrostatic repulsion. On the other hand, introduction of electron-withdrawing halogen and sulfone substituents was designed in order to increase the electron accepting abilities of the parent compounds 2, 4, and 5. The resulting compounds T2-Br, T2-Br$_2$, T2-Cl, S2-Br, 4-Br, 5-Br, and 5-SO$_2$ except F2-Br show the first wave potentials comparable to that of

Table 1. Half-wave Redox Potentials (V) of Hetero-TCNQs.

	F1	T1	S1	F2	T2	S2	T3	S3	4	5
$E^1_{1/2}$	0.03	0.07	0.03	-0.09	-0.03	-0.05	-0.03	-0.07	0.06	0.05
$E^2_{1/2}$	-0.55	-0.54	-0.54	-0.31	-0.26	-0.25			-0.36	-0.23
ΔE	0.58	0.61	0.57	0.22	0.23	0.20	0	0	0.42	0.28

	F2-Br	T2-Br	T2-Br$_2$	T2-Cl	S2-Br	4-Br	5-Br	5-SO$_2$
$E^1_{1/2}$	0.08	0.20	0.28	0.18	0.15	0.25	0.16	0.28
$E^2_{1/2}$	-0.12	0.03	0.13	-0.06	-0.08	-0.13	-0.11	-0.05
ΔE	0.20	0.17	0.15	0.24	0.23	0.38	0.27	0.33

Measuring conditions: Pt working and counter electrodes, Ag/AgCl reference electrode, scan rate 100 mV/sec in dichloromethane solution containing 0.1 M tetrabutylammonium perchlorate as supporting electrolyte at room temperature.

F2-Br: X=O, R=Br, R'=H
T2-Br: X=S, R=Br, R'=H
T2-Br$_2$: X=S, R=R'=Br
T2-Cl: X=S, R=Cl, R'=H
S2-Br: X=Se, R=Br, R'=H

4-Br

5-Br: X=S, R=Br
5-SO$_2$: X=SO$_2$, R=H

TCNQ. These modified hetero-TCNQs also keep very small ΔE values. Thus, both features, large electron affinities and small on-site Coulomb repulsions, allow them to behave as superior electron acceptors.

These acceptors did form many molecular complexes with a variety of electron donors such as tetrathiafulvalene (TTF), tetrathiotetracene (TTT), tetraphenylbipyranylidene (TPBP), and hexamethylenetetratellurafulvalene (HMTTeF), and also give some radical anion salts with alkaline metals [4,5,6]. The selected electrical conductivity data, measured on compressed pellets at room temperature, are summarized in Table 2, together with the infrared stretching vibration data of C≡N groups. Comparison of TTF complexes of T1, T2, and T3 demonstrates a marked increase of conductivity in order of extensive conjugation, though they are all nonionic. Similar trends are observed for the other complexes of the thiophene-TCNQ series as well as other hetero-TCNQ series. These results strongly support that the diminished on-site Coulomb repulsion in the acceptors makes a significant contribution to the formation of conductive molecular complex. Acceptors 4 and 5 of condensed type, having almost the same accepting abilities as 1-3, formed more complexes with various donors than the linear isologues. Moreover, most complexes showed high conductivities. In particular, 4·HMTTeF complex recorded a very high value (140 Scm^{-1}). In this regard, the condensed type is superior as an electron acceptor to the linear type. In addition, acceptors modified by electron-withdrawing bromine or chlorine could form further complexes with various donors. Most of them were highly conductive. Specially high conductivities were observed for HMTTeF complexes of T2-Br, T2-Cl, S2-Br, and 4-Br. On the other hand, the sulfone derivative, 5-SO$_2$, though the strongest acceptor in this research, behaved differently. It complexed only just the same strong donors as did the mother compound 5. In addition, the conductivities of these complexes were not so high as those of the corresponding

Table 2. Electrical Conductivities of Charge Transfer Complexes (Scm^{-1})

	F1 (2245)	T1 (2222)	S1 (2230)	F2 (2234)	T2 (2220)	S2 (2223)	T3 (2213)	4 (2223)	5 (2219)
TTF	~10^{-10}	~10^{-8}	~10^{-9}	~10^{-10}	~10^{-4}	Non	0.003	13	2.3
	(2242)	(2221)	(2230)	(2228)	(2217)		(2215)	(2199)	(2198)
TTT	Non	Non	Non	Non	~10^{-4}	~10^{-4}	0.71	~10^{-9}	3.0
					(2219)	(2222)	(2204)	(2216)	(2172)
TPBP	Non	0.49	----	Non	Non	----	1.1	2.8	0.62
		(2184)					(2215)	(2195)	(2202)
HMTTeF	Non	~10^{-8}	~10^{-9}	Non	Non	~10^{-7}	0.37	140	20
		(2221)	(2223)			(2221)	(2209)	(2200)	(2200)
	F2-Br (2236)	T2-Br (2224)	T2-Br$_2$ (2217)	T2-Cl (2217)	S2-Br (2220)	4-Br (2224)	5-Br (2221)	5-SO$_2$ (2222)	
TTF	~10^{-8}	15	31	9	----	11	23	0.012	
	(2224)	(2203)	(2200)	(2197)		(2195)	(2205)	(2192)	
TTT	2.4	13	18	4.6	11	3.1	5.8	2.1	
	(2203)	(2203)	(2193)	(2190)	(2200)	(2194)	(2194)	(2193)	
TPBP	----	0.45	11	~10^{-4}	4.6	----	4.9	7.7	
		(2196)	(2195)	(2193)	(2195)		(2190)	(2192)	
HMTTeF	----	57	29	170	86	170	14	~10^{-5}	
		(2169)	(2182)	(2180)	(2190)	(2192)	(2192)	(2192)	

Values in parentheses indicate nitrile stretching vibration (cm^{-1}).
Non: Noncomplexation.

5-Br complexes. The steric bulkiness of the sulfone moiety might be responsible for disturbing a conductive molecular arrangement on complexation.

In conclusion, the electron accepting ability of hetero-TCNQ could be successfully improved by introduction of both extensive heteroquinonoid conjugation and electron-withdrawing substituents. Especially in the linear series, extensive conjugation was very effective for diminution of on-situ Coulomb repulsion, which is one of the important requisites for designing conductive complexes. It has proved that our acceptors possessing both small on-site Coulomb repulsion and high electron affinity are promising and reliable components for forming a variety of highly conductive molecular complexes.

Acknowledgments: This work was supported by Grant-in-Aid for Scientific Research from the Ministry of Education, Science, and Culture, Japan and Mazda Foundation's Research Grant.

References

[1] W. R. Hertler, U. S. Patent, **1964**, No. 3153658 (Chem. Abstr., **62**, 4145 (1965)); D. J. Sandman and A. F. Garito, J. Org. Chem., **39**, 1165 (1974); M. Morinaga, T. Nogami, and H. Mikawa, Bull. Chem. Soc. Jpn., **52**, 3739 (1979); A. W. Addison, N. S. Dalal, Y. Hoyano, S. Huizinga, and L. Weiler, Can. J. Chem., **52**, 3739 (1979).
[2] E. Ahalon-Schalom, J. Y. Becker, and I. Agranat, Nouv. J. Chim., **3**, 643 (1979); M. Maxfield, D. O. Cowan, A. N. Bloch, and T. O. Poehler, ibid., **3**, 643 (1979).

[3] S. Gronowitz and B. Uppström, Acta Chem. Scand., Ser. B, **28**, 981 (1974).
[4] K. Yui, Y. Aso, T. Otsubo, and F. Ogura, J. Chem. Soc., Chem. Commun., **1987**, 1816; idem, Bull. Chem. Soc. Jpn., **62**, 1539 (1989).
[5] K. Yui, H. Ishida, Y. Aso, T. Otsubo, and F. Ogura, Chem. Lett., **1987**, 2339; Y. Aso, K. Yui, H. Ishida, T. Otsubo, F. Ogura, A. Kawamoto, and J. Tanaka, Chem. Lett., **1988**, 1069; K. Yui, H. Ishida, Y. Aso, T. Otsubo, F. Ogura, A. Kawamoto, and J. Tanaka, Bull. Chem. Soc. Jpn., **62**, 1547 (1989).
[6] K. Yui, Y. Aso, T. Otsubo, and F. Ogura, Chem. Lett., **1988**, 1179.
[7] H. Ishida, K. Yui, Y. Aso, T. Otsubo, and F. Ogura, Presented at the 58th National Meeting of the Chemical Society of Japan (4IIIA09, Abstracts II, p.1395, Kyoto 1989) and at the 6th International Symposium on Novel Aromatic Compounds (Workshop II and B-20, Abstracts, p. 123, Osaka 1989).
[8] Examination of the molecular models as well as the electronic spectra of these compounds strongly suggests a planar trans structure concerning the double bond geometry of the linearly conjugated heteroquinonoid system.

Design of Two-Dimensional Stacking Structures: Twin-Type Molecules and Steric Interaction of Axial Substituents

T. Inabe[1], T. Mitsuhashi[2], and Y. Maruyama[1]

[1] Institute for Molecular Science, Myodaiji, Okazaki 444, Japan
[2] Department of Chemistry, University of Tokyo, Bunkyo-ku, Tokyo 113, Japan

Abstract. Two kinds of novel approaches to constructing a two-dimensional molecular array are presented. The first approach is based on the interleaved stacking of twin-type molecules, which are composed of two units of a donor or an acceptor. The second approach is based on the slipped stacking due to the steric interaction of axial substituents. As examples, charge transfer complexes of OCNAQ, which is a twin-TCNQ-type acceptor, and radical salts of an axially substituted phthalocyanine are presented.

1. Introduction

Two-dimensionality is recognized to be one of the important structural requisites for stabilizing the metallic state and consequently for achieving a superconducting state in organic compounds. From this point of view, molecules which can interact transversely through chalcogen-chalcogen contacts have been synthesized (Fig. 1(a)). All of the organic superconductors so far discovered are composed of this type of molecule, i.e., BEDT-TTF, TMTSF, MDT-TTF, DMET, Ni(dmit)$_2$, and Pd(dmit)$_2$. Since the chemical modifications by chalcogen atoms are rather limited, another approach to creating two-dimensional systems becomes more important to extend the varieties of organic superconductors.

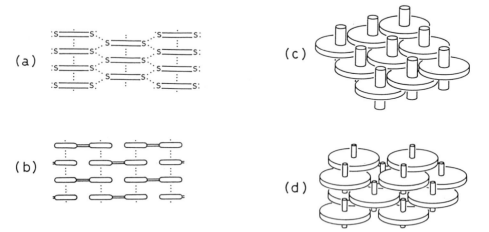

Fig. 1. Schematic representation of molecular arrangements in organic conductors; transverse interaction through S···S contacts (a), interleaved stacking structure of twin-type molecules (b), and two-dimensional (c) and three-dimensional (d) stacking structures of axially substituted molecules.

The first approach proposed is shown schematically in Fig. 1(b). If a molecule, which can be a constituent of an organic conductor, is connected with the same molecule, this twin-type molecule has the possibility of forming interleaved stacking as shown in Fig. 1(b), in which the stacking between the units keeps the pattern in the original one-dimensional conductor. As a candidate for such a molecule, the twin-TCNQ-type acceptor OCNAQ has been synthesized [1, 2], and several charge transfer complexes have been prepared [3]. The next section is devoted to describing the stacking patterns observed in these OCNAQ complexes.

The second approach is based on the slipped stacking due to the steric interaction of axial substituents of planar molecules, as shown in Fig. 1(c) and (d). These two types have been observed when the dicyanophthalocyaninatocobalt(III) anion, $[Co(Pc)(CN)_2]^-$, was electrochemically oxidized [4, 5]. In the third section, the structures obtained are presented.

2. Twin-Type Molecules: OCNAQ

TCNQ is a well-known acceptor which forms low-dimensional conductors. OCNAQ is the twin-TCNQ-type acceptor in which a pair of TCNQ nuclei are linked together through two methlidyne groups. The

important thing to realize in our proposal of interleaved stacking of twin-type molecules is that the bridging group should not influence the LUMO structure of the TCNQ unit, so that the original stacking between the TCNQ units is not disturbed. The simple Hückel MO calculation has indicated that the LUMO structure of the TCNQ unit in OCNAQ is essentially the same as that of TCNQ.

If the LUMO-LUMO interaction which is assumed to be analogous to the TCNQ conductors dominates in determining the stacking form of the molecules, then two types are expected. One is the one-dimensional type, which is observed for $Et_4N \cdot OCNAQ$ as shown in Fig. 2(a). The other is the two-dimensional type, which appears in $(TTF)_2OCNAQ$ (Fig. 2(b)). In both cases, the idealized overlaps (top figures in Fig. 2) between the TCNQ units are the same. Unfortunately, the real overlapping deviates slightly from the ideal cases, since the molecule is not planar due to the steric interaction between the inner cyano groups. This does not make the interaction in $(TTF)_2OCNAQ$ ideally two-dimensional, but rather one-dimensional. Consequently, the metal-to-insulator transition is observed at 43 K. As is well known, the planarity of the component molecule is essential to produce an effective interaction between the molecules. Another molecular design to obtain more planar twin-type molecules is being planned.

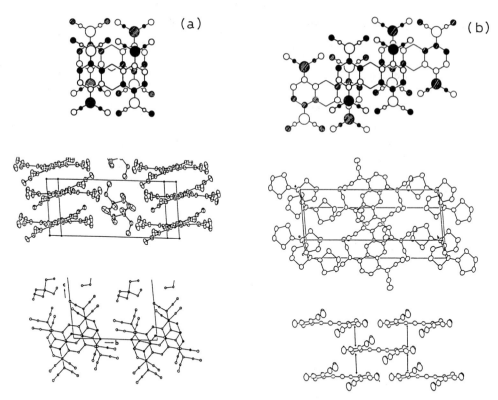

Fig. 2. Stacking structures in the OCNAQ complexes; one-dimensional stacking in Et_4N-OCNAQ (a) and two-dimensional stacking in $(TTF)_2OCNAQ$ (b).

3. Axially Substituted Phthalocyanine

Phthalocyanines are known to form low-dimensional conductors, in which the conduction occurs through the overlapped ligand π-orbitals. One of the advantages of this system is the feasibility of substitution at the axial sites. If the molecule has projections at the center (axial substituents), it cannot be stacked directly above another molecule. Taking the size of the phthalocyanine ring into account, π-orbital overlaps are expected to be sufficiently effective for electrical conduction, although the overlapping is partial.

As one such molecule, $[Co(Pc)(CN)_2]^-$ was chosen for electrochemical oxidation to obtain conducting salts. Up to now, two kinds of crystals, $K[Co(Pc)(CN)_2]_2 \cdot 5CH_3CN$ and $Co(Pc)(CN)_2 \cdot 2H_2O$, have been obtained. The crystal structures are shown in Fig. 3(a) and (b), respectively. In both compounds, the phthalocyanine rings form slipped stacks and each stack is intertwined with one another, so that the total intermolecular interaction becomes two-dimensional for $K[Co(Pc)(CN)_2]_2 \cdot 5CH_3CN$ and three-dimensional for $Co(Pc)(CN)_2 \cdot 2H_2O$. The oxidation states of the Pc rings are $Pc^{1.5-}$ and Pc^-, respectively. Thus, metallic conductivity is expected for $K[Co(Pc)(CN)_2]_2 \cdot 5CH_3CN$. Unfortunately, this crystal is unstable outside of the solution. Even when the crystal is mosaically distorted after exposure to air, the conductivity at room temperature is high, ~10 $\Omega^{-1}cm^{-1}$. Thus, the intrinsic properties are expected to

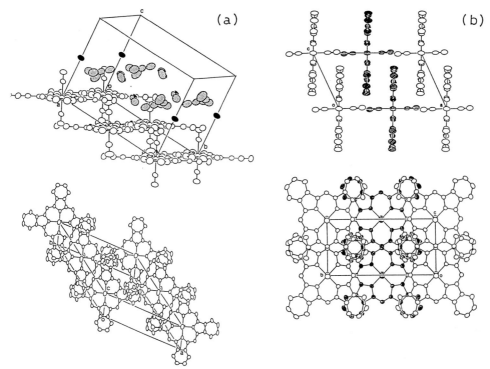

Fig. 3. Two-dimensional stacking of phthalocyanine rings in K[Co(Pc)(CN)$_2$]$_2$·5CH$_3$CN (a) and three-dimensional stacking in Co(Pc)(CN)$_2$·2H$_2$O (b).

be metallic. On the other hand, the conductivity of Co(Pc)(CN)$_2$ is ~1 Ω^{-1}cm^{-1} at room temperature, which seems to be too high for a fully oxidized salt, suggesting the existence of hydrated protons in the lattice instead of some of the water molecules. Since the crystals obtained were too small to measure the anisotropy of the conductivity by a four-probe method, the anisotropy was roughly estimated from two-probe measurements. The obtained anisotropy is 3-10, which is consistent with the three-dimensional stacking form.

Since the molecular shape is tunable by replacing the substituents, the central metal, and the π-ligand, it may be possible to construct other conducting systems for further progress.

References

1. T. Mitsuhashi, M. Goto, K. Honda, Y. Maruyama, T. Sugawara, T. Inabe, and T. Watanabe, J. Chem. Soc., Chem. Commun., 1987, 810.
2. T. Mitsuhashi, M. Goto, K. Honda, Y. Maruyama, T. Inabe, T. Sugawara, and T. Watanabe, Bull. Chem. Soc. Jpn., 61, 261 (1988).
3. T. Inabe, T. Mitsuhashi, and Y. Maruyama, Bull. Chem. Soc. Jpn., 61, 4215 (1988).
4. T. Inabe and Y. Maruyama, Chem. Lett., 1989, 55.
5. T. Inabe and Y. Maruyama, submitted to Bull. Chem. Soc. Jpn.

New Molecular Conductors Based on Metal Complex Anions

A.E. Underhill, K.S. Varma, R.A. Clark, and C.E. Wainwright

Department of Chemistry and Institute of Molecular and Biomolecular Electronics, University of Wales, Bangor, Gwynedd LL57 2UW, UK

Abstract. The influence of the ligand, central metal and cation on the structures and properties of molecular metals based on metal complex anions are discussed and the points illustrated by reference to current work on metal complexes of dmit, dmt and dmid with Group I counter cations.

Introduction

Over the past five years there has been increasing research activity in the search for new molecular metals and new molecular superconductors containing metal complex anions. There are now four low temperature molecular superconductors involving [M(dmit)$_2$] anions [1-3]. Although three of these compounds contain the organic radical cation TTF, the fourth [(CH$_3$)$_4$N][Ni(dmit)$_2$]$_2$ contains the closed-shell spectator ion [(CH$_3$)$_4$N]$^+$ as the counter ion and therefore the superconducting properties of this compound must involve only the metal complex anion [3].

Many of the metal complexes which form molecular metals and superconductors have close structural and electronic similarities to the organic donors involved in molecular metals and superconductors. Thus the [M(dmit)$_2$] anion is related to BEDT-TTF, since the [C=C]$^{2+}$ moiety is isolobal with M^{2+}. In a series of molecular metals based on an organic donor such as [BEDT-TTF]$_2$X the properties of the material may be modified by substitution within the ring system, substitution on the periphery of the ring and by changing the size and nature of the counter anion. This has been a successful strategy in the development of organic molecular metals and superconductors [4].

In the case of molecular metals based on metal complex anions a similar strategy can be adopted involving the following.
(a) Modification of the ligand. This includes changes to the ring system of the ligand as well as to changes of the substituent atoms or groups,
(b) Variation of the counter cation. As with the organic materials the size and nature of the counter ion has been shown to be very important. Large bulky anions such as (Bu$_t$)$_4$N$^+$ or small group I cations can be incorporated into the structure.
(c) Variation of the central metal atom. The presence of the metal at the centre of a complex gives an additional variable to molecular conductors based on metal complexes compared with the organic cation radical salts. The metal atom affects the nature of the anion in two important respects. Firstly, the central metal atom plays an important role in determining the structure of the anion. Up to the present time studies have concentrated on metal cations which form square coplanar complexes with chelate ligands. These complex anions under appropriate conditions readily form columnar stack structures allowing short intra- and inter-stack contacts, which, in turn, favour the formation of a molecular metal. Recently, it has been suggested that tris-chelate octahedral complexes may also form molecular metals through interactions between the anions based on packing arrangements involving their three-fold symmetry [5]. The nature of the central metal atom also plays a role through the involvement of the metal d-orbitals in the HOMO and LUMO orbitals of the ligand. Because of the different spatial extensions and energies of the 3d, 4d and 5d orbitals, there are often significant differences between, for example, the Ni(II), Pd(II) and Pt(II) complexes of the same ligand with the same counter cation [6].

To successfully molecular engineer new molecular metals and superconductors based on metal complex anions, it is necessary to have the correct combination of all the three variables described above in order to obtain the required combination of structure and electronic properties necessary for the formation of molecular metals and superconductors. The general considerations outlined above will be illustrated by reference to salts based on [M(dmit)$_2$] anions with group I counter cations.

Results and Discussion

(a) Modification of the ligand. There are many variations possible with the dmit ligand. Replacement of some or all of the sulphur atoms by selenium will clearly have a considerable effect on inter-anion interactions and some studies of the synthesis and properties of these compounds have been reported [7]. An isomer of dmit, dimercaptotrithione (dmt) exists and some molecular conductors based on the complexes of dmt have been reported [8]. For instance, TTF$_2$[Ni(dmt)$_2$] exhibits a room temperature conductivity of 0.2 S cm^{-1} [8]. However, attempts to produce molecular conductors based on [M(dmt)$_2$] (M = Ni, Pd or Pt) with group I counter cations by electrocrystallisation resulted in the formation of a thin insulating film on the anode. Pedersen and Parker have previously shown that anodic oxidation of 1,2-dithiole-3-thiones leads to the formation of dimers via disulphide linkages[9]. Thus it is thought that the film on the anode is a result of the coupling of dmt molecules via disulphide linkages. Unfortunately, insufficient material was obtained to show whether the material was the dimerised ligand or the polymer based on cross-linking of the [M(dmt)$_2$] anions.

[M(dmi-d)$_2$]

[M(dmi-t)$_2$]

[M(dmt)$_2$]

BEDT-TTF

Another alternative in the modification of the dmit ligand is to convert the thiole group to a ketone (dmid) [10]. The electrocrystallisation of the TBA$_2$[Ni(dmid)$_2$] in the presence of excess sodium or potassium perchlorate in acetonitrile solution produced small quantities of black microcrystalline material on the anode. Electrical conduction studies on compressed discs of these materials showed them to be semi-conducting with room temperature conductivities of 10 S cm^{-1} and 1.6 x 10^{-1} S cm^{-1} for the sodium and potassium salts respectively. For comparison single crystals of Na[Ni(dmit)$_2$]$_2$ and K$_{0.4}$[Ni(dmit)$_2$] both behave as molecular metals at room temperature with conductivities of 20 and 100 S cm^{-1} respectively [11,12]. Further work on these materials is in progress.

(b) Variation of the counter cation. A study has been made of the products obtained by electrocrystallising solutions of TBA[M(dmit)$_2$] (where M = Ni, Pd, Pt or Au) in the presence of a large excess of group I counter cations. The products obtained usually have a stoichiometry Z[M(dmit)$_2$]$_2$ except for the potassium and rubidium salts of [Ni(dmit)$_2$] where the ratio of cation to anion is 0.4:1. The most widely studied series of compounds are those of nickel. It is found that single crystals can only be obtained for a few of the group I counter cations whilst the remainder give microcrystalline products. The sodium and potassium salts were obtained as single crystals but unfortunately the quality of the crystals did not allow a complete X-ray structure determination. Both these salts exhibit metallic properties down to low temperatures without any sign of a metal to semiconductor transition. Only compressed pellets of the lithium and rubidium salts could be studied and these both showed semiconducting behaviour but with relatively high room temperature conductivities of 0.55 and 14 S cm^{-1} at room temperature. In the palladium series of compounds microcrystalline products were obtained for the potassium and rubidium salts and these again were examined as compressed pellets. Both behaved as semiconductors but with room temperature conductivities of 45 and 15 S cm^{-1} respectively.

(c) Variation of central metal. The results obtained so far on the group I cation salts suggest that the platinum complexes exhibit much lower conductivities than the corresponding nickel complexes. This is best illustrated by comparing the sodium salts where for the nickel complex the room temperature conductivity for a single crystal is 20 S cm^{-1} whereas a compressed pellet of the microcrystalline platinum complex only exhibits a conductivity of 10^{-5} S cm^{-1}. Of the compounds for which single crystals can be obtained, the two nickel complexes retain their metallic conductivity down to very low temperatures, as described earlier, whilst the palladium complex $Cs[Pd(dmit)_2]_2$, although exhibiting a higher room temperature conductivity than the nickel complexes, undergoes the metal to semi-conductor transition at 60 K [13]. So far, metallic behaviour has not been observed for any of the platinum complexes.

Recently we have started to investigate the group I cation salts of the $[Au(dmit)_2]$ anion. Single crystals have been obtained for the lithium salt and this exhibits a metallic conductivity around room temperature with $\sigma_{RT} = 10$ S cm^{-1} and then appears to undergo a metal to semi-conductor transition around 200 K. Although the corresponding sodium and potassium salt have only been obtained as microcrystalline products, nevertheless, compressed pellets of these salts also exhibit a metallic type of temperature dependence with room temperature conductivities of 105 and 10 S cm^{-1} respectively. Thus the gold complexes show great promise for the future.

It is clear from the above discussion that within the series of $[M(dmit)_2]$ salts of group I cations, both the cation and the central metal play an important role in determining the structure and properties of this type of material. However, studies are needed on single crystals of several salts before a full understanding of these effects can be obtained.

We would like to thank the SERC for support and Johnson Matthey plc for the loan of precious metal salts.

References

[1] Cassoux, P., Valade, L., Legros, J.P., Interrante, L., Rocs, C., (1986) *Physica B & C*, **143**, 313;
Brossard, L., Hurdequint, H., Ribault, M., Valade, L., Legros, J.P. Cassoux, P., (1988) *Synthetic Metals*, **27**, B157.
[2] Brossard, L., Ribault, M., Valade, L., Cassoux, P., (1986) *Physica B & C*, **143** (1-3), 378;
Schirber, J.E., Overmyer, D.L., Williams, J.M., Wang, H.H., Valade, L Cassoux, P., (1987) *Phys. Lett. A.*, **120**(2), 87.
[3] Kim, H., Kobayashi, A., Sasaki, Y., Kato, R., Kobayashi, H., (1987) *Chem. Lett.*, 1799;
Kobayashi, A., Kim, H., Sasaki, Y., Kato, R., Kobayashi, H., Moriyama, S., Nishio, Y., Kajita, K., Sasaki, W., (1987) *Chem. Lett.*, 1819.
[4] Williams, J.M., Wang, H.H., Emge, T.J., Geiser, U., Beno, M.A., Leung, P.C.W., Carlson, K.D., Thorn, R.J., Schultz, A.J., (1987) *Prog. in Inorg. Chem.*, **35**, 51.
[5] Broderick, W.E., McGee, E.M., Godfrey, M.R., Hoffman, B.M., Ibevs, J.A., (1989) *Inorg. Chem.*, **28**, 2902.
[6] Underhill, A.E., Clark, R.A., Varma, K.S., (1989) *Phosphorus, Sulphur and Silica*, **43**, 111.
[7] Nigrey, P.J., (1988) *Synthetic Metals*, **27**, B365.
[8] Coustumer, G. Le, Bennasser, N., Molliev, Y., (1988) *Synthetic Metals*, **27**, B523.
[9] Pederson, C.T., Parker, V.D., (1972) *Tet. Lett.*, **9**, 771.

[10] Olk, R.M., Dietzsch, W., Köhler, K., Kirmse, R., Reinhold, J., Hoyer, E., Golič, L., Olk, B., (1988) *Z. Anorg. Allg. Chem.*, 131.
[11] Clark, R.A., Underhill, A.E., Parker, I.D., Friend, R.H., (1989) *J. Chem. Soc., Chem. Commun.*, 229.
[12] Clark, R.A., Underhill, A.E., This volume.
[13] Clark, R.A., Underhill, A.E., (1988) *Synthetic Metals*, **27**, B515.

Conducting Evaporated Film of $Pt_2(CH_3CS_2)_4I$

I. Shirotani[1], *Y. Inagaki*[1], *and M. Yamashita*[2]

[1]Muroran Institute of Technology, 27-1, Mizumoto, Muroran 050, Japan
[2]College of General Education, Nagoya University, Chikusa-ku, Nagoya 464, Japan

Abstract. $Pt_2(CH_3CS_2)_4I$ is a new linear chain complex having dimeric units bridged through iodine. Evaporated thin films of the complex were prepared in high vacua. The conductivity, absorption spectrum and X-ray diffraction of the films have been studied. The conductivity of the film was about 0.2 S·cm^{-1} at room temperature, much higher than that of the powder. The film showed a semiconductive behavior with an activation energy of 0.04 eV. The absorption band of the film was observed at around 1300 nm. This strong, broad band may be assigned to the intermolecular d-d transition between the neighboring platinum dimers.

1. Introduction

New linear chain complexes having dimeric units bridged through halogen have been prepared by Bellitto et al.(1-3). Iodo tetrakis(dithioacetato)diplatinum, $Pt_2(CH_3CS_2)_4I$, is the linear chain complex with a nearly symmetrical metal-halogen-metal bridge. The structure consists of linear chains of $-I--Pt_2S_8--I--Pt_2S_8--I-$, stacking along the b-axis. The Pt-Pt distance in the dimer is 2.677 Å, about 0.1 Å shorter than the atomic distance in platinum metal. The Pt-I distances are respectively 2.975 and 2.981 Å. The formal oxidation number of Pt in $Pt_2(CH_3CS_2)_4I$ is + 2.5. Interesting physical properties have been found in the complex (3). The powder electrical conductivity is 3×10^{-3} S·cm^{-1} at 300 K. The conductivity follows an exponential temperature dependence with a very low activation energy, 0.05 eV. The complex shows a strong, broad asymmetric absorption band in the near infrared region.

Recently, the preparation and physical properties of evaporated thin films of the one-dimensional Pt complex bis(1,2-benzoquinonedi-oximato)Pt(II) have been studied (4). An anomalous spectrum is found for the evaporated film prepared on NaCl substrate.

We have tried to prepare evaporated thin films of $Pt_2(CH_3CS_2)_4I$, which shows interesting physical properties. Conducting evaporated films of the complex were successfully prepared. In this paper, the electrical and optical properties of the thin films are reported.

2. Experimental

$Pt_2(CH_3CS_2)_4I_n$ (n: 0,1,2) were synthesized according to the methods described by Bellitto and co-workers (1-3). Compositions of all materials were confirmed by elemental analysis. The purity was checked by normal physicochemical methods.

The thin films of Pt complexes with the dithioacetato ligand were prepared by evaporation onto a quartz or a glass substrate held at room temperature in a vacuum of 1.33×10^{-4} Pa (10^{-6} Torr)(4). The thickness of the film was monitored by means of a quartz-crystal oscillator. The absorption spectra, the electrical conductivities and

X-ray diffraction patterns of these films were measured at room temperature. Differential scanning colorimetry (DSC) and thermogravimetric analysis (TGA) of the powder sample were performed with a Rigaku TAS100 system.

3. Results and Discussion

Figure 1 exhibits TGA and DSC curves of $Pt_2(CH_3CS_2)_4I$. The first anomaly in DSC and TGA was found at around 290°C. This endothermic anomaly is due to a sublimation or a melting. The second anomaly was observed at 320-354°C. The magnitude of weight loss was about 27 % in weight at the temperature range. If one iodine atom from the complex is released the magnitude of the weight loss is about 15 % in weight. This corresponds to the anomaly at 333 K observed in the DTG curve, but weight is also lost by evaporation. Thus, the second anomaly mainly comes from the dissociation of the iodine atom from the complex. When the evaporated film is prepared, the iodine atoms are partially released from the complex by the experimental conditions. We have carefully controlled the rate and temperature of the evaporation to prepare good films of the complex.

Figure 2 depicts absorption spectra in evaporated films of Pt_2-$(CH_3CS_2)_4I_n$ (n: 0,1,2). An absorption band of the evaporated film of $Pt_2(CH_3CS_2)_4$ was located at around 850 nm. This band is not observed in the reflectance spectra of the complex diluted in MgO. The dithioacetato palladium(II) complex exists in two solid-state phases (form A and form B)(1). Form A consists of mononuclear and binuclear units, alternating along the a-axis of the unit cell. Form B contains a one-dimensional arrangement of dimeric molecules with CS_2 incorporated between the columns. Form B is a green crystal with a chemical composition of $Pd_2(CH_3CS_2)_4CS_2$. The color of the evaporated film of Pt_2-$(CH_3CS_2)_4$ prepared by us was green. The film corresponds to the form B of the Pd complex. As $Pt_2(CH_3CS_2)_4$ is easily decomposed, a green film including CS_2 molecules seems to be formed. Absorption bands in the films of the halogen bridged complexes were observed at around 565 nm and 1300 nm for $Pt_2(CH_3CS_2)_4I$, and at around 565 nm for Pt_2-$(CH_3CS_2)_4I_2$. These bands agree with the reflectance spectra measured by Bellitto et al.(3).

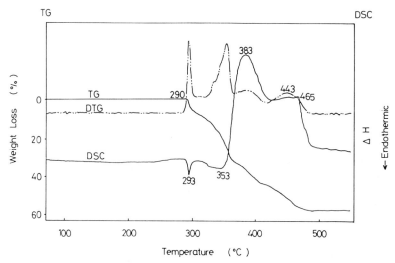

Fig. 1. DSC and TGA curves of the powder of $Pt_2(CH_3CS_2)_4I$.

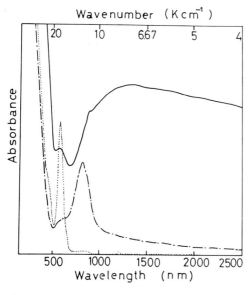

Fig. 2 Absorption spectra of thin films of $Pt_2(CH_3CS_2)_4I_n$ (n: 0,1,2).

—·— $Pt_2(CH_3CS_2)_4$
——— $Pt_2(CH_3CS_2)_4I$
- - - - $Pt_2(CH_3CS_2)_4I_2$

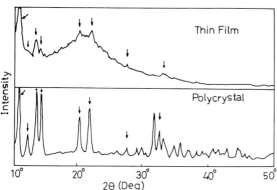

Fig. 3. X-ray diffraction profiles of thin film and polycrystal of $Pt_2(CH_3CS_2)_4I$.

The highest occupied orbital (σ^*) in $Pt_2(CH_3CS_2)_4$ is a d_{z^2}-d_{z^2} antibonding orbital. The 850 nm band in the complex may be assigned to the $5d_{z^2}$-$6p_z$ transition. For two complexes oxidized by halogen this band disappeared, but a new band was observed at around 565 nm. This is a ligand-to-metal charge transfer band (3). As the d_{z^2}-d_{z^2} antibonding orbital in $Pt_2(CH_3CS_2)_4I$ is not filled, the σ-σ^* transition is allowed. The 1300 nm band in the complex may be assigned to the transition between the two bands. But the strong absorption band at 1300 nm broadened to the infrared region. This band is absent in $Pt_2(CH_3CS_2)_4I_2$. Therefore, the band at 1300 nm may mainly arise from the intermolecular d-d transition between neighboring platinum dimers.

Figure 3 shows X-ray diffraction profiles of the evaporated film and the polycrystal of $Pt_2(CH_3CS_2)_4I$. The thickness of the film is about 800 Å. The diffraction pattern for the film was very broad, similar to an amorphous one. The diffraction lines indicated by arrows correspond to those in the polycrystal. Similar results were also found in the X-ray diffraction patterns of $Pt_2(CH_3CS_2)_4I_2$.

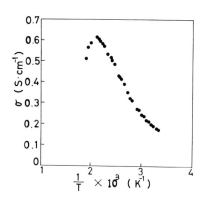

Fig. 4. Temperature dependence of electrical conductivity of $Pt_2(CH_3CS_2)_4I$ film.

Figure 4 depicts the temperature dependence of the electrical conductivity of the $Pt_2(CH_3CS_2)_4I$ film. The thickness of the film is about 500 Å. The conductivity of the film was 0.2 S.cm^{-1} at room temperature. This value is about two orders lower than that of the powder conductivity obtained by Bellitto et al.(3). The conductivity followed an exponential temperature dependence with a low activation energy, 0.04 eV. The maximum in the σ - 1/T curve was observed at 360 K. Above this temperature the conductivity decreased with increasing temperature. This may be due to the release of the iodine atoms from the film.

The electrical conductivity of the single crystal of $Pt_2(CH_3CS_2)_4I$ has been measured at high pressures (5). The conductivity along the stacking axis increased from 2 S.cm^{-1} at atmospheric pressure to about 10 S.cm^{-1} at 7 GPa. The activation energy at 7 GPa was 0.03 eV, nearly equal to the values for the polycrystal and the film. This suggests that the activation enrgy of the complex does not give an intrinsic energy gap. The electrical conduction may occur by an extrinsic band model mechanism.

4. References

1. O. Piovesana, C. Bellitto, A. Flamini and P.F. Zanazzi, Inorg. Chem., 18, 2258(1979).
2. C. Bellitto, A. Flamini, O. Piovesana and P.F. Zanazzi, Inorg. Chem., 19, 3632(1980).
3. C. Bellitto, A. Flamini, L. Gastaldi and L. Scaramuzza, Inorg. Chem., 22, 444(1983).
4. I. Shirotani, N. Minobe, Y. Ohtsuki, H. Yamochi and G. Saito, Chem. Phys. Lett., 143, 231(1988).
5. I. Shirotani, S. Kawamura, M. Yamashita, W. Utsumi and T. Yagi, Molecular Structure Symposium, September, Sapporo(1989).

Electroactive Langmuir Blodgett Films of Tetrathiafulvalene Derivatives

A.S. Dhindsa, M.R. Bryce, and M.C. Petty

Department of Chemistry and Molecular Electronics Research Group, School of Engineering and Applied Science, University of Durham, Durham, DN1 3LE, UK

Mono-substituted tetrathiafulvalene derivatives form stable, high-quality Langmuir-Blodgett (LB) films which are highly conducting after iodine doping (σ_{max} 1 S cm^{-1}). The molecular arrangements within these films have been studied by a range of techniques. Alternate-layer LB films of TTF and TCNQ derivatives, both substituted with one long chain, have been fabricated, and shown to be semi-conducting (σ_{rt} 5 x 10^{-3} S cm^{-1}).

1. Introduction

There is considerable interest in using the Langmuir Blodgett (LB) technique to organise donor and acceptor charge-transfer materials at the molecular level with the aim of producing highly conducting ultra-thin films [1]. Currently our studies are focussed on salts of TTF derivatives and high quality, electroactive multilayer films have been characterised by a range of techniques [2]. The distinctive features of the Durham materials which will be discussed in this paper are as follows.

    (i)   They are mono-substituted TTF derivatives, available in a one-step synthesis (15-40% yield) from TTF.
    (ii)   The TTF moiety is the hydrophilic part of the molecule.
    (iii) High quality, air-stable LB films are produced in as-deposited and doped forms.
    (iv) Maximum conductivity values obtained are σ_{rt} = 1 S cm^{-1}.
    (v)   The first alternate-layer TTF-TCNQ films have been fabricated.

2. LB Films of Hexadecanoyltetrathiafulvalene (HDTTF) (1)

2.1 LB Film Formation

Hexadecanoyltetrathiafulvalene (HDTTF) (1) is representative of the materials we have been studying. The surface pressure (π) versus area per molecule isotherm has been reported and the condensed nature of the isotherm is striking [2a]. Y-type deposition was achieved on hydrophobic glass substrates at a dipping pressure of 30 mNm^{-1}. Films of more than thirty monolayers have been assembled without any significant loss of quality. The u.v.-visible transmission spectra of these films show that the optical density varies linearly with the number of layers which is indicative of reproducible monolayer deposition.

HDTTF (1)   HDTTTF (2)

2.2 Conductivity Studies

The lateral d.c. conductivity of the as-deposited films of HDTTF was typically 1×10^{-5} S cm^{-1}, measured using a two-probe technique with air-drying silver paste contacts. A monolayer thickness of 3.5 nm was assumed in all conductivity calculations. On exposure to iodine vapour for 2-3 min in a sealed container the films became insulating, then, following this doping procedure, the conductivity increased with time reaching a maximum value of $\sigma_{rt} = 1 \times 10^{-2}$ S cm^{-1} ($E_a = 0.19$ eV between 300-77 K) after ca. 10 h. The conductivity then does not change after storage in air for a few weeks. The conductivity is isotropic in the film plane. Exposure of the as-deposited films to bromine vapour also gave an insulating film, but thereafter there was no increase in conductivity with time [2a,b].

2.3 Ultra violet/Visible Spectroscopy

In the uv/visible spectra of the undoped films of HDTTF (1) absorption bands are present at 284 and 490 nm which can be assigned to the π-π^* and n-π^* electronic transitions, respectively. On iodine doping, a marked change in the spectrum occurs: absorption bands at 295, 380 and 870 nm are observed, along with a CT band at 2100 nm. Apart from the 870 nm band, the changes in the spectrum parallel those observed when HDTTF in acetonitrile solution is oxidised with iodine, and can be assigned to intermolecular transitions of the HDTTF$^{+\cdot}$ radical cation. Studies using linearly polarised light indicate that in the undoped and doped states, the molecular plane of HDTTF is almost perpendicular to the substrate. Spectra obtained at different times after iodine doping show that the intensity of the CT band reaches a maximum at ca. 8 h after doping which corresponds to the time when the lateral conductivity in the layers reaches a maximum. Taken together these data suggest that immediately upon doping, a fully-oxidised (insulating) system is formed which decomposes with time giving rise to the stable, mixed-valence system HDTTF^{x+}I$_x^-$ or TTF^{x+}(I$_3^-$)$_x$, where $x < 1$, which is conducting.

2.4 Infra red Spectroscopy

The transmission i.r. spectra for 24 layers of HDTTF deposited on calcium fluoride substrate, before and after doping, have been obtained. The main absorptions in the as-deposited films are assigned as follows: ν 2916 and 2848 cm^{-1} are the asymmetric and symmetric CH$_2$ stretches of the side chain; ν 1657 cm^{-1} is the C=O stretching vibration; a series of bands at 1500-1600 cm^{-1} is assigned to exo- and endo-cyclic C=C stretches of the TTF ring. Such absorptions are known in Raman spectra of TTF and attachment of the side chain has allowed these bands to become i.r. active in HDTTF. Absorption at 1262 cm^{-1} arises from in-plane bending of the C=CH group. After iodine doping a number of changes in the spectra become apparent. A broad charge transfer band (ca. 1700 - beyond 4000 cm^{-1}) is observed; there is a marked increase in intensity for most of the bands; some of the bands shift to lower wavenumbers and a new band appears at 1320 cm^{-1}. Most of these changes can be explained by the coupling of conduction electrons to the vibrational modes of HDTTF producing frequency shifts and allowing previously inactive i.r. bands to become active. Data obtained from time-dependent RAIRS and ATR spectra provide information concerning the orientation of the molecules within the film. The carbonyl dipole is aligned approximately parallel to the substrate (strong absorption in ATR, weak in RAIRS spectra) and the relative strengths of the molecular

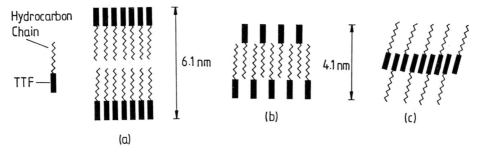

Figure 1. Schematic representation showing proposed structure of HDTTF (1) LB layers: (a) as-deposited; (b), (c) alternative models for LB layers after doping.

electron vibronic coupling band in the ATR and RAIRS spectra, demonstrate that electron motion is parallel to the substrate surface.

2.5 X-Ray Diffraction Studies

From X-ray diffraction studies, the films of HDTTF before doping, were found to consist mainly (> 90% by volume) of a single phase possessing a d-spacing of 6.1 ± 0.1 nm; this corresponds to ca. twice the length of HDTTF, consistent with the observed Y-type (head-to-head) deposition for the layers. Prolonged doping with either Br_2 or I_2 produces a single-phase structure with a d-spacing of 4.1-4.2 nm. This difference in d-spacing corresponds to the length of the hydrocarbon chain of HDTTF, implying that either the chains or the head groups are interdigitated in the doped multilayers. Schematic packing arrangements for the as-deposited and doped multilayers are shown in Fig. 1. Our data do not distinguish between the two possible arangements shown in Fig. 1b and 1c. We favour the structure in Fig. 1c which allows closer packing of the TTF rings, and hence relatively high lateral conductivity along the TTF stacks which is found for the bulk material.

3. LB Films of O-Hexadecylthiocarboxytetrathiafulvalene (HDTTTF) (2)

Preliminary work has established that LB films of compound (2) exhibit significantly higher conductivity than films of compound (1). The lateral d.c. conductivity of as-deposited films of HDTTTF (2) was typically 7×10^{-2} S cm^{-1}; for iodine doped films a maximum value $\sigma_{rt} = 1.0$ S cm^{-1} was obtained ($E_a = 0.09$ eV between 300-77 K). Detailed studies on these films will be reported in due course.

4. Alternate-layer LB Films of Octadecanoyl-TTF and Octadecyl-TCNQ

We have used a specially-designed alternate-layer trough to form alternate-layers of a long-chain TTF and a long-chain TCNQ derivative [2c]. This technique provides a novel way of enforcing segregated stacking upon a charge-transfer complex. Figure 2 shows a schematic diagram of the alternate layer structure. To improve monolayer transfer, each material was mixed with 10% stearic acid (in chloroform) before being spread onto the two areas of the alternate-layer trough. The uv-visible absorption spectra obtained

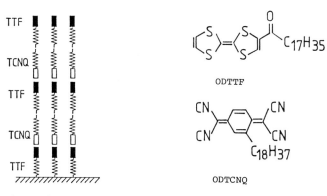

Figure 2. Schematic representation of alternate layer LB film structure of ODTTF and ODTCNQ [2c].

for different numbers of bilayers deposited onto calcium fluoride have been obtained. The optical density scales almost linearly with film thickness, indicating reproducible bilayer deposition. Bands characteristic of TCNQ (365 nm) and TTF (285 nm) are observed, the former component dominating the spectrum in agreement with the high absorption coefficient characteristic of the TCNQ system. The room temperature conductivity of the alternate-layer films was found to be $(5\pm 1) \times 10^{-3}$ S cm^{-1}, assuming a bilayer thickness of 6.0 nm. The activation energy for conductivity over the temperature range 300-100 K was 0.26 ± 0.01 eV.

Acknowledgement

We thank SERC for funding this work.

References

[1] (a) A. Barraud, A. Ruaudel-Teixier, M. Vandevyver and P. Lesieur, Nouv. J. Chim., 9, 365 (1985); (b) M. Matsumoto, T. Nakamura, F. Takei, M. Tanaka, T. Sekiguchi, M. Mizuno, E. Manda and Y. Kawabata, Synth. Met., 19, 675 (1987); (c) A.S. Dhindsa, M.R. Bryce, J.P. Lloyd and M.C. Petty, Synth. Met., 22, 185 (1987); (d) Y. Kawabata, T. Nakamura, M. Matsumoto, M. Tanaka, T. Sekiguchi, H. Komizu, E. Manda and G. Saito, Synth. Met., 19, 663 (1987); (e) J. Richard, M. Vandevyver, A. Barraud, J.P. Morand, R. Lapouyade, P. Delhaes, J.F. Jacquinot and M. Roulliay, J.C.S. Chem. Commun., 754 (1988); (f) T. Nakamura, H. Tanaka, M. Matsumoto, H. Tachibana, E. Manda and Y. Kawabata, Chem. Lett., 1667 (1988).

[2] (a) A.S. Dhindsa, M.R. Bryce, J.P. Lloyd and M.C. Petty, Thin Solid Films, 165, L97 (1988); (b) A.S. Dhindsa, M.R. Bryce, J.P. Lloyd and M.C. Petty, Synth. Met., 27, B563 (1988); (c) C. Pearson, A.S. Dhindsa, M.R. Bryce and M.C. Petty, Synth. Met., 31, 275 (1989); [d] A.S. Dhindsa, C. Pearson, M.R. Bryce and M.C. Petty, J. Phys. D, (1989) in press.

Physical Properties of Conductive Langmuir-Blodgett Films of Tridecylmethylammonium-Au(dmit)$_2$ and Its Derivatives

T. Nakamura, Y. Miura, M. Matsumoto, H. Tachibana, M. Tanaka, and Y. Kawabata

National Chemical Laboratory for Industry, Tsukuba, Ibaraki 305, Japan

Abstract. Highly conductive Langmuir-Blodgett (LB) films of tridecylmethylammonium-Au(dmit)$_2$ (3C10-Au) and related compounds are described. The conductivity of the LB films of 3C10-Au was 30 - 50 S/cm after electrochemical oxidation at room temperature, which showed the metallic temperature dependence down to around 150 K. At low temperature, the temperature dependence of the conductivity was explained by the VRH mechanism.

1. Introduction

The conductive LB films so far reported are made mostly of long chain derivatives of typical low-dimensional organic materials — radical salts, charge transfer complexes or conductive polymers [1]. Owing to the one-dimensional nature of conduction stacks, these materials are sensitive to defects and disorders and have intrinsic instability. To suppress these instabilities and achieve high and metallic conductivity, we introduced the metal(dmit)$_2$ (H_2dmit = 4,5-dimercapto-1,3-dithiol-2-thione) complexes as a film-forming material [2-4]. This dithiolene complex has provided several kinds of molecular superconductor [5-7]. By the introduction of the alkylammonium group as a counter cation, the complexes become amphiphilic and suitable for the construction of LB films.

Among others, the LB film of a gold complex 3C10-Au (Fig. 1, m = 3, n = 10) showed high conductivity at room temperature [3,4]. In this paper, we describe the electrical transport properties of the LB films of 3C10-Au and related compounds.

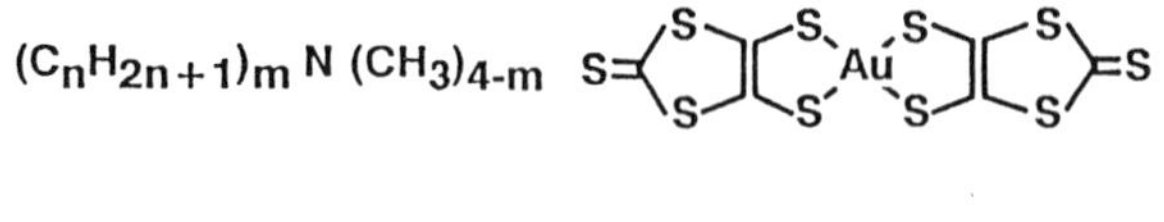
(CnH2n+1)m N (CH3)4-m
S
S
S
S
S
Au
S
S
S
S
S

mCn - Au

Fig. 1 Molecular structure of alkylammonium-Au(dmit)$_2$.

Springer Proceedings in Physics, Vol. 51
The Physics and Chemistry of Organic Superconductors
Editors: G. Saito · S. Kagoshima

2. Preparation of LB Films

The amphiphilic Au(dmit)$_2$ complexes with several ammonium cations were synthesized according to the procedures described previously [2]. The 1:1 mixture with icosanoic acid formed monolayers which were deposited on hydrophobized solid substrates by the horizontal lifting method. The gold electrodes (gap distance: 0.1 - 0.5 mm) were formed by vacuum deposition before the transfer of monolayers. The LB films were led to the conductors by chemical or electrochemical oxidation [2]. The electrochemical oxidation of the LB films was achieved by a constant current method in 0.1 mol/l LiClO$_4$ aqueous solution. The conductivity of the films was measured for a 20-layered sample by a dc 2-probe or 4-probe method.

3. Electrical Transport Properties of the LB Films

3.1 Conductivity of the LB Films of Au(dmit)$_2$ Complexes

The bulk conductivities of the oxidized LB films of Au(dmit)$_2$ complexes at room temperature are summarized in Table I, assuming a monolayer thickness of 3 nm, the thickness of the matrix acid. The results of the chemical oxidation by bromine gas [2] are also shown in the table. These values are the highest observed by a dc 2-probe method. The LB films of 3Cn-Au with shorter (n = 10, 14) alkyl chains showed fairly high conductivity (20 - 30 S/cm) after electrochemical oxidation, in contrast to those of 2C10-Au. The effect of the number of side alkyl chains in the ammonium moiety is not clear at present, however, in both cases, the alkyl-chain length strongly affects the conductivity of the films.

Table I. Conductivity of the LB films of Au(dmit)$_2$ complexes

Material	Bulk conductivity (S/cm)	
	Bromine oxidation	Electrochemical oxidation
3C10-Au	15	33
3C14-Au	5.4	19
3C16-Au	2.6	0.46
3C18-Au	1.4	0.12
2C10-Au	0.12 [a]	1.4
2C14-Au	0.15 [a]	
2C18-Au	0.005 [a]	

a) Silver paste was used as the electrode.

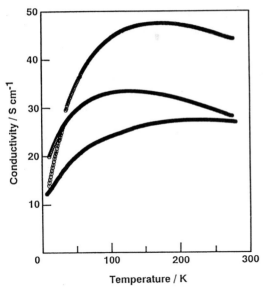

Fig. 2 Temperature dependence of the conductivity of three different samples of 3C10-Au LB film after electrochemical oxidation. The measurement was carried out by a dc 4-probe method in vacuum. The cooling rate was 0.5 K/min.

3.2 Transport Properties of the LB Films of Tridecylmethylammonium-Au(dmit)$_2$

Figure 2 shows the temperature dependence of the conductivity of three different samples of 3C10-Au LB films after electrochemical oxidation in the range of 280 - 10 K. The film showed metallic temperature dependence of conductivity down to around 200 - 130 K, then the film showed semiconducting behavior. The activation energy estimated from the Arrhenius plot around 100 K was at about 2 meV. At present, the following three explanations can be offered for the metallic nature of the films.

(a) The film is a metal which undergoes a metal-insulator transition at low temperature.

(b) The film is a narrow-gap semiconductor. At higher temperatures, the metallic temperature dependence of the conductivity is caused by an increase in the mobility with a decrease in temperature due to the small value of the activation energy.

(c) The film is made of metallic polycrystals, which act as a narrow-gap semiconductor due to the presence of the domain boundary between crystals.

As seen in Fig. 2, the temperature at which the conductivity reaches a maximum varies with the sample. The fact that the transition from the metallic to the insulating region is influenced by the film-forming process strongly suggests the existence of disorder and defects arising from poly-crystalline structure of the film (possibility (c)). The possibility (a), however, cannot be eliminated because such disorders will affect the behavior of the metal-insulator transition. Further studies on the thermoelectric power and the Hall effect are now in progress.

At low temperature, a distinct deviation from the Arrhenius plot was observed. The conductivity is approximately proportional to $\exp(-(T_0/T)^{-1/4})$ below 50 K, which suggests the predominance of a VRH mechanism in this temperature region.

The conductivity of the film decreased by one to two orders of magnitude in vacuum over 1 week. The aged film did not show a metallic temperature dependence of the conductivity. The activation energy was 0.02 eV in the range 300 - 100 K. At lower temperatures, again a large deviation from the Arrhenius plot was observed.

4. Conclusion

In this paper, highly conductive LB films of the tridecylmethylammonium-Au(dmit)$_2$ complex and related compounds are described. The conductivity of 3C10-Au LB film after electrochemical oxidation was 30 - 50 S/cm, which showed a metallic temperature dependence of the conductivity down to around 150 K. The metallic nature of the film is not clear at present. At low temperature, the temperature dependence of the conductivity approximately followed $T^{-1/4}$. The film showed semiconducting behavior in the range 300 - 10 K after aging in vacuum.

References

1. T. Nakamura and Y. Kawabata, Techno Japan, **22**, 8 (1989).
2. T. Nakamura, H. Tanaka, M. Matsumoto, H. Tachibana, E. Manda, and Y. Kawabata, Chem. Lett., 1667 (1988); Synth. Met., **27**, B601 (1988).
3. T. Nakamura, K. Kojima, M. Matsumoto, H. Tachibana, M. Tanaka, E. Manda and Y. Kawabata, Chem. Lett., 367 (1989).
4. T. Nakamura, H. Tanaka, K. Kojima, M. Matsumoto, H. Tachibana, M. Tanaka, and Y. Kawabata, Thin Solid Films, **179**, 183 (1989).
5. L. Brassard, M. Ribault, M. Bousseau, L. Valade, and P. Cassoux, Physica, **143B**, 378 (1986).
6. L. Brassard, H. Hurdequint, M. Ribault, L. Valade, J.P. Legros, and P. Cassoux, Synth. Met., **27**, (1988) B157.
7. A. Kobayashi, H. Kim, Y. Sasaki, R. Kato, H. Kobayashi, S. Moriyama, Y. Nishio, K. Kajita, and W. Sasaki, Chem. Lett., 1819 (1987).

Polymerization of Diacetylenes in Liquid Crystal Phases and Its Application to the Preparation of High Spin Polydiacetylenes

A. Izuoka, T. Okuno, and T. Sugawara

Department of Pure and Applied Sciences, University of Tokyo, Komaba 3-8-1 Meguro-ku, Tokyo 153, Japan

Abstract. A hydroxymethyl-substituted diacetylene (1) with benzylidene aniline moiety as a mesogenic core was polymerized at 130° C in the nematic phase. The degree of polymerization turned out to be over 50 units after heating for 25 h. A hydroxybiphenyl-substituted diacetylene (2) with the same mesogenic core was prepared and chemically oxidized to give a stable phenoxy radical (3). The stability of 3 both in benzene solution and in the solid state turned out to be reasonably high. The degree of spin delocalization in 3 was investigated through the hfs in ESR spectra. Polymerization of 2 was also undertaken at 185° C, and the obtained pentamers were isolated by gel permeation chromatography. Oxidation of pentamers gave the polyphenoxy radical; its stability and the structure were investigated by ESR spectroscopy.

1. Introduction

Polydiacetylene, which can be constructed by solid state polymerization of diacetylenes, is expected to have various physical properties. This is to be realized by modulating the π conjugated system of polydiacetylene through an electronic perturbation of its hanging groups [1,2]. However, the polymerization reactivity of diacetylenes depends heavily on the spatial arrangement of monomers in crystals [3]. In the case of monomers containing interesting substituents, polymerization often does not occur. In order to overcome such difficulties, we have designed diacetylenes with a functional substituent on one end and a mesogenic core on the other, and examined polymerization reactivity in liquid crystal phases [4]. Since liquid crystal phases are structurally more flexible than crystals, they are expected to have an advantage as reaction media. Choosing a stable radical or its precursor as the functional substituent, we applied this methodology to the preparation of high spin polydiacetylenes, which are key materials for achieving organic ferromagnetism.

2. Results and Discussion

2.1 Polymerization of the hydroxymethyl-diacetylene derivative (1) in the nematic phase

Hydroxymethyl diacetylene (1) with benzylidene aniline as a mesogenic core was prepared and found to have a nematic phase above 124° C according to the DSC diagram and observation of dynamic scattering under polarized light [5]. Monomer 1 was polymerized at 130° C in the nematic phase, and the decay was observed through the decrease of the $\nu_{C\equiv C}$ band in the IR spectrum. A characteristic induction time (about 50 h) was observed in the crystal phase,

$C_8H_{17}O-\langle\text{Ph}\rangle-N=\overset{H}{C}-\langle\text{Ph}\rangle-C\equiv C-C\equiv C-CH_2OH$

1

whereas the decay proceeds exponentially in the nematic case. The distributions of the molecular weight distribution of the products in both phases were analyzed by Gel Permeation Chromatography (GPC) and exhibited similar patterns. The degree of polymerization for polymers in solution turned out to be over 50 units.

2.2 Preparation of the hydroxybiphenyl diacetylene derivative (2) and ESR study of the phenoxy radical generated from (2)

Based on the successful nematic polymerization of model monomer 1, a diacetylene derivative substituted by a stable radical precursor was designed; the mesogenic part of the monomer was kept the same.
Para-ethynyl phenyl lithium was added to the benzoquinone derivative in ether solution. The product was reduced by aluminium hydride to give hydroxybiphenyl acetylene. After removing the trimethylsilyl group, hydroxybiphenyl acetylene was cross-coupled with acetylene derivative substituted by N-benzylidene ethylamine. The ethyl group in the unsymmetrical diacetylene was replaced by para-octyloxy phenyl to give the final product (2).
The monomer 2 was chemically oxidized in benzene solution and the ESR spectrum of the phenoxy radical (3) was recorded (Fig.1a). The hfs of the radical was reproduced by the following hfs constants: a(2H)=0.9G, a(4H)=1.7G (Fig.1b). This result shows that the spin is delocalized over the second phenyl ring. The radical persists over a few days at room temperature, and is also stable in solid state at temperatures lower than 100°C. (Polymerization of the radical was not observed when heated at 100°C.)

$$\text{2,6-di-t-butylbenzoquinone} \xrightarrow[\text{(2) NH}_4\text{Cl aq}]{\text{(1) TMS}-\equiv-\langle\text{Ph}\rangle-\text{Li} / \text{Et}_2\text{O}} \xrightarrow{\text{AlH}_3 / \text{Et}_2\text{O}} \text{TMS}-\equiv-\langle\text{Ph}\rangle-\langle\text{Ph(t-Bu)}_2\rangle-\text{OH}$$

$$\xrightarrow[\text{(2) H}-\equiv-\langle\text{Ph}\rangle-\overset{H}{\text{C}}=\text{NEt}/\text{CuCl}/\text{pyridine}]{\text{(1) Na}_2\text{B}_4\text{O}_7\cdot 10\text{H}_2\text{O}/\text{EtOH}} \text{EtN}=\overset{H}{C}-\langle\text{Ph}\rangle-C\equiv C-C\equiv C-\langle\text{Ph}\rangle-\langle\text{Ph(t-Bu)}_2\rangle-\text{OH}$$

$$\xrightarrow[\text{(2) }C_8H_{17}O-\langle\text{Ph}\rangle-\text{NH}_2/\text{pyridine}]{\text{(1) 1N HCl/EtOH}} C_8H_{17}O-\langle\text{Ph}\rangle-N=\overset{H}{C}-\langle\text{Ph}\rangle-C\equiv C-C\equiv C-\langle\text{Ph}\rangle-\langle\text{Ph(t-Bu)}_2\rangle-\text{OH}$$

2

$C_8H_{17}O-\langle\text{Ph}\rangle-N=\overset{H}{C}-\langle\text{Ph}\rangle-C\equiv C-C\equiv C-\langle\text{Ph}\rangle-\langle\text{Ph(t-Bu)}_2\rangle-O\cdot$

3

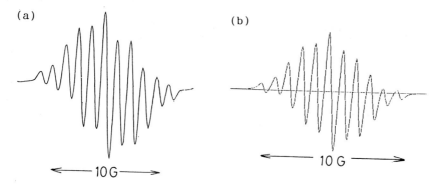

Fig.1.(a)ESR spectrum of 3.(b)Its simulation spectrum.

2.3 Polymerization of hydroxybiphenyl-diacetylene derivative (2) and generation of phenoxyradicals from diacetylene pentamers

According to DSC measurement and macroscopic observation, the monomer 2 doesn't exhibit liquid crystal phases. But it polymerizes on being heated for 11h at 185°C. Molecular weight distribution of the soluble oligomers is shown in Fig.2. The pentamer fraction was isolated by GPC, and oxidized in benzene solution. A broad signal around the g=2 region is observed by ESR of oxidized products (Fig.3), suggesting intramolecular exchange interaction among electron spins. The multiplicity of phenoxy radicals is not determined at the present stage.

The control of magnetic interaction among high spin oligomers is crucial in realizing macroscopic ferromagnetism. The methodology can be applicable to constructing conductive polydiacetylenes by using the liquid crystalline monomer with a donor group at one end. The polydiacetylene may have dual electronic interactions; one is along

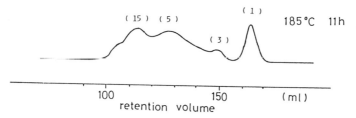

Fig.2. Molecular weight distribution of oligomers obtained from 2.

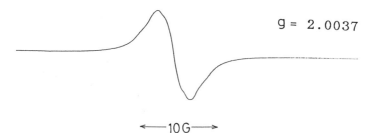

Fig.3. ESR spectrum of the oxidized pentamer.

stacking of partially oxidized donors and the other is through the conjugated π bond. These coexisting interactions are expected to manifest a unique elctronic property. Study along this line is also in progress in these laboratories.

References.

[1] R.R.Chance and R.H.Baughman, J. Chem. Phys. **4**, 3889 (1976).
[2] C.Sauteret, J.-P.Hermann, R.Frey, F.Pradere, J.Ducuing, R.H.Baughman and R.R.Chance, Phys. Rev. Lett. **36**, 956 (1976).
[3] V.Enkelmann, Advances in Polymer Science, edited by H.-J.Cantow, (Springer-Verlag, Berlin,Heidelberg,1984), Vol.63.
[4] A.F.Garito, C.C.Teng, K.Y.Wong and O.Z.Khamiri, Mol. Cryst. Liq. Cryst. **106**, 219 (1984).
[5] A.Izuoka, S.Kamei, K.Tohyama, T.Ito, N.Sato and T.Sugawara, to be published.

Part X

Theory

Novel Superconductivity from an Insulator

Y. Takada and M. Kohmoto

Institute for Solid State Physics, University of Tokyo,
7-22-1 Roppongi, Minato-ku, Tokyo 106, Japan

Abstract. We consider a new pairing Hamiltonian in which a filled valence band is separated from an empty conduction band by an energy gap. In the normal state, the system is insulating, but when a large enough attractive interaction acts between states in the two bands, it goes into a superconducting phase at T_c of the order of the energy gap. Although we obtain the Meissner effect and the infinite dc-conductivity similar to the BCS superconductors, there are many unusual properties due to a different gap equation. Possible applications to CuCl and double-chain organic materials are discussed.

1. Introduction

In spite of its simplified model Hamiltonian and the mean-field treatment, the BCS theory [1] is successful in explaining all the basic properties of usual superconductors. The core of the theory is the Cooper instability. Namely the normal metallic state is unstable against an infinitesimal attractive interaction between electrons near the Fermi surface. Thus we do not expect superconductivity in a system without a Fermi surface, like an insulator.

In recent years, however, novel types of superconductors have been found near the insulating phases in organic materials and copper oxides. Besides, large diamagnetic anomalies have been observed at temperatures as high as 200 K in some specially prepared samples of CuCl under high pressures [2]. If we interpret the anomalies as the occurrence of superconductivity, we have a very unique situation in which a semiconducting phase showing an electronic resistivity with a finite activation energy enters directly into a superconducting phase through not a second- but a first-order phase transition at such a high temperature.

Motivated by these experimental facts, we investigate the possibility of a new superconducting instability which could take place even in an insulator in which no Fermi surface exists.

2. Model

Consider a system composed of a filled valence band and an empty conduction band. The Fermi level μ lies in the middle of the gap of the one-particle spectrum. The fermion annihilation operators for the valence band

and the conduction band are denoted by $a_{k\sigma}$ and $b_{k\sigma}$, respectively. If the gap G is larger than the Debye cutoff energy ω_c, the normal state is stable against the Cooper pairing $<a_{k\uparrow}a_{-k\downarrow}>$ because of the absence of states in phase space for multiple scatterings to form the pair. This is the case even if the attractive phonon-mediated potential is very deep. Note, however, that the same attractive potential works for a pair of electrons in different bands. When electrons in the valence band are partially promoted to the conduction band to make a pair $<b_{-k\downarrow}a_{k\uparrow}>$, there is a plenty of room for multiple scattering even for $\omega_c < G$. Thus we may have superconductivity originating from a new type of pairing $<b_{-k\downarrow}a_{k\uparrow}>$ rather than the usual Cooper one.

A simple model Hamiltonian to examine the instability against an attractive interaction between a valence electron and a conduction electron is

$$H = \sum_k \{ \varepsilon_a(k) a_k^\dagger a_k + \varepsilon_b(k) b_k^\dagger b_k \} - U \sum_{k \neq k'}{}' a_{k'}^\dagger b_{-k'}^\dagger b_{-k} a_k , \quad (1)$$

where $\varepsilon_a(k)$ and $\varepsilon_b(k)$ are the energy dispersions minus μ of the valence band and the conduction band, respectively. The spin degrees of freedom are neglected for simplicity. The primed sum means that the attractive interaction works only for $|\varepsilon_a(k) - \varepsilon_a(0)| < \omega_c$ and $|\varepsilon_b(k) - \varepsilon_b(0)| < \omega_c$ with a cutoff energy ω_c.

Expecting $<b_{-k}a_k> \neq 0$ [3], we solve the Hamiltonian (1) in the mean-field approximation to obtain a gap equation

$$|\Delta| = U \sum_k{}' \frac{|\Delta|}{2E_k} [f\{\frac{\varepsilon_a(k) - \varepsilon_b(-k)}{2} - E_k\} - f\{\frac{\varepsilon_a(k) - \varepsilon_b(-k)}{2} + E_k\}] , \quad (2)$$

where $E_k = [|\Delta|^2 + \{\varepsilon_a(k) + \varepsilon_b(-k)\}^2/4]^{1/2}$ and $f(\varepsilon) = 1/\{\exp(\beta\varepsilon) + 1\}$ is a Fermi distribution function.

So far the theory does not significantly deviate from the BCS theory. The crucial difference arises when the valence band and the conduction band are explicitly considered. For simplicity, we assume symmetric bands $\varepsilon_a(k) = -k^2/2m - G/2 - \mu$ and $\varepsilon_b(k) = k^2/2m + G/2 - \mu$. The chemical potential μ is zero irrespective of temperature or U due to the symmetry between the two bands. Then one has $E_k = |\Delta|$ which is independent of k. The gap equation is reduced to

$$|\Delta| = U \sum_k{}' [f\{\varepsilon_a(k) - |\Delta|\} + f\{\varepsilon_b(k) - |\Delta|\} - 1]/2 . \quad (3)$$

Note that the energy denominator is absent in (3). So the attractive potential U works in its full strength in the entire k sum in the present pairing theory. Thus we do not need a very strong U to have a non-zero solution of (3).

3. Calculated Results and Discussion

We can solve (3) analytically at T=0. Nonzero Δ exists if

$$\lambda > d(2\omega_c / G)^{1-d/2} + d/(2\omega_c / G)^{d/2}, \tag{4}$$

where λ is a dimensionless coupling constant, defined by $\lambda = UN(G)$ with $N(G)$ the density-of-states at energy G (*i.e.*, $G/2$ from the bottom of the conduction band) and d is the dimensionality of the system. The total energy of the superconducting state ($\Delta \neq 0$) becomes lower than that of the normal state ($\Delta = 0$) if

$$\lambda > 2d^2(2\omega_c / G)^{1-d/2}/(d+2) + 2d/(2\omega_c / G)^{d/2}. \tag{5}$$

In two and three dimensions, (5) gives a more restrictive condition than (4) and there are no quasi-particle states at μ. In one dimension, this is also true for $\omega_c < 1.5\, G$, but (4) provides a stronger condition for $\omega_c > 1.5\, G$. In the latter case, we have a totally exotic situation in which quasi-particle states exist at μ in the superconducting phase. (We have confirmed the existence of the Meissner effect and the infinite conductivity in the present pairing model.)

The order parameter Δ depends on λ linearly. This is quite distinct from the BCS theory in which Δ depends on λ exponentially. This difference stems from the absence of the energy denominator in our gap equation (3). Thus the present model may lead to a high-temperature superconducting transition for a large λ. At the same time we need a *finite* strength of λ, namely, of the order of 1, to produce the superconductivity (note that the Cooper instability of a metal takes place with an infinitesimal attractive interaction).

We solve (3) numerically at finite temperatures. In contrast to the BCS theory, $\Delta(T)/\Delta(0)$ and $H_c(T)/H_c(0)$ are not universal functions of T/T_c. Also $\Delta(0)/T_c$ is not universal and approaches 2 in the large-λ limit, which is different from the BCS universal value 1.76. In most cases we obtain a second-order phase transition. However, when λ is fixed and ω_c is near the critical value determined by (5), we have a first-order phase transition as shown in Fig. 1 in which T_c is given as a function of ω_c and λ. Since T_c is about G or less in this first-order transition region, there are only a few thermally excited carriers in the normal state near T_c. Thus we have, for the first time, an unusual *insulator(or semiconductor)-superconductor first-order phase transition*. This is consistent with the experimental situation in CuCl.

Since superconductivity takes place when U is of the order of G in our model, the theory should treat the strong correlation in a non-perturbative way and the band picture may not be appropriate. We have employed the mean-field treatment, but its validity must be examined. This question is already addressed in the one-dimensional system in which the exact result by the Bethe ansatz method is available [4].

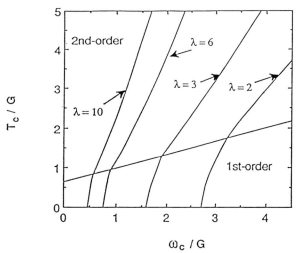

Fig.1 Calculated T_c / G as a function of ω_c / G for $\lambda = 2, 3, 6$, and 10. The straight line divides the regions of the first-order and the second-order phase transitions.

In applying the present theory to real solids, we need to include the effects of a weak non-symmetry in $\varepsilon_a(k)$ and $\varepsilon_b(k)$. However, it can be shown that this does not lead to a drastic change in the predictions of the theory. Perhaps the most important question is the origin of a strong attractive interaction which gives $\lambda \geq 2$. One of the simplest candidates is the exchange of phonons as in the BCS theory. The possibility of having such a large λ from optical phonons exists in some insulators in which a strong electron-phonon interaction is not screened at all. Another candidate for an attractive interaction is an electronic origin, either charge fluctuations like excitons or spin fluctuations. We have already proposed a new exciton mechanism in a specially synthesized double-chain organic system with polarizable materials surrounding the chains [4]. Whatever the origin of the attractive force is, we have to treat a completely new many-body problem for superconductivity, namely, an *instability without a Fermi surface*.

References

[1] J. Bardeen, L.N Cooper, and J.R. Schrieffer, Phys. Rev. **108**, 1175 (1957).
[2] C.W. Chu, A.P. Rusakov, S. Huang, S. Early, T.H. Geballe, and C.Y. Huang, Phys. Rev. **B18**, 2116 (1978); I. Lefkowitz, J.S. Manning, and P.E. Bloomfield, Phys. Rev. **B20**, 4506 (1979).
[3] It should not be confused with the excitonic insulator in which $\langle b_k^+ a_k \rangle \neq 0$.
[4] Y. Takada and M. Kohmoto, to appear in Phys. Rev. **B41** (1990).

Bethe Ansatz Wavefunction, Momentum Distribution and Spin Correlation in the One-Dimensional Strongly Correlated Hubbard Model

*M. Ogata and H. Shiba**

Institute for Solid State Physics, University of Tokyo, Roppongi, Minato-ku, Tokyo 106, Japan

Abstract. The momentum distribution function $n(k)$ and spin correlation function $S(k)$ are determined for one-dimensional large-U Hubbard model with various electron densities. The Bethe Ansatz wavefunction is used, which has a simple form in the large-U limit. The singularity of $n(k)$ at k_F is analyzed from the system size dependence. In addition to the k_F-singularity $n(k)$ has a weak singularity at $3k_F$; however no detectable singularity is present at $2k_F$. The singularity of $S(k)$ at $2k_F$ is also examined in detail.

I. Introduction

The Hubbard model is the simplest Hamiltonian containing the essence of strong correlation. Notwithstanding its apparent simplicity, our understanding of physics of the Hubbard model is still limited even in the one-dimensional case.[1] In fact, although its thermodynamics was clarified, various important quantities such as momentum distribution $n(k)$ and spin correlation function $S(k)$, which require an explicit form of the wavefunction, have not been explored yet.

The purpose of this paper is to present $n(k)$ and $S(k)$ in the large-U limit of the one-dimensional Hubbard model based on the exact Bethe Ansatz wavefunction.[2] To this end we first observe that Lieb and Wu's wavefunction[1] has a very simple form in the large-U limit owing to a decoupling of charge and spin degrees of freedom. This observation enables us to calculate exactly $n(k)$ and $S(k)$ in relatively large systems, which are not available in other methods.

II. Bethe Ansatz Wavefunction in the Large U Limit

The one-dimensional (1D) Hubbard model is given by

$$H = -t \sum_{(i,j)\sigma} (c_{i\sigma}^\dagger c_{j\sigma} + \text{h.c.}) + U \sum_i n_{i\uparrow} n_{i\downarrow}. \qquad (2.1)$$

The system consists of N electrons on N_A sites.

---

* Present address: Department of Physics, Tokyo Institute of Technology, Oh-okayama, Tokyo 152

Let $f(x_1, \cdots, x_N)$ be the amplitude in the wavefunction when down spins are located at the sites $x_1, \cdots, x_M$, and up spins at $x_{M+1}, \cdots, x_N$. f has been given explicitly under the Bethe Ansatz.[1] In the limit of $U/t \to \infty$, we can show that f is rewritten as[2]

$$f(x_1 \cdots x_N) = (-1)^Q \det\left[\exp(ik_i x_{Q_j})\right] \Phi(y_1 \cdots y_M), \qquad (2.2)$$

within each region of $x_{Q1} < x_{Q2} < \cdots < x_{QN}$, where $Q = (Q1, Q2, \cdots, QN)$ is a permutation of $(1, 2, \cdots, N)$ and $(y_1 \cdots y_M)$ are the "coordinates" of M down spins along the chain with vacant sites omitted. The determinant depends only on the sites of particles $(x_{Q1} < \cdots < x_{QN})$ and not on their spins. Thus it is the same as the Slater determinant of spinless fermions with momenta k_j's. Furthermore we can show that $\Phi(y_1, \cdots, y_M)$ is just the same as the Bethe's exact solution of 1D $S = \frac{1}{2}$ Heisenberg spin system. This notable feature holds for any electron density. Equation (2.2) means a complete decoupling between the charge- and spin degrees of freedom for $U/t \to \infty$.

Note that the wavefunction in Eq. (2.2) represents a singlet wavefunction which is connected naturally to the ground state in the whole region of $0 < U < \infty$, although all the spin configurations become degenerate at $U = \infty$.

III. Momentum Distribution Function

In practice we construct the ground state wavefunction as follows: We determine $\Phi(y_1, \cdots, y_M)$ by diagonalizing the 1D Heisenberg model. The Slater determinant is calculated by means of the Vandermonde determinant.

The momentum distribution $n(k)$ is given by

$$n(k) = <c_{k\sigma}^\dagger c_{k\sigma}> = \frac{1}{N_A} \sum_{j\ell} <c_{j\sigma}^\dagger c_{\ell\sigma}> e^{ik(r_j - r_\ell)}. \qquad (3.1)$$

Since the wavefunction contains the Slater determinant of spinless fermions, one may think that $n(k)$ must have a jump at $k = 2k_F$. However this is not the case. When an electron is transferred from the ℓ-th site to the j-th site as in the r.h.s. of Eq. (3.1), the spin configuration $\Phi(y_1, \cdots, y_M)$ changes in general. This change in Φ leads to an essential difference from the spinless fermion case.

We present the results of momentum distribution in Fig. 1. As a typical case, the quarter-filled system (i.e. $N/N_A = 1/2$) is chosen here to study the size-dependence systematically. It is readily seen that there is no detectable singularity at $k = 2k_F$. Instead one notices a strong singularity at k_F and a weak singularity at $3k_F$. As shown in Figure 2, there always exist noticeable singularities at k_F and $3k_F$. Note that, when $3k_F$ becomes larger than π, it is folded back into the Brillouin zone and appears at $2\pi - 3k_F$.

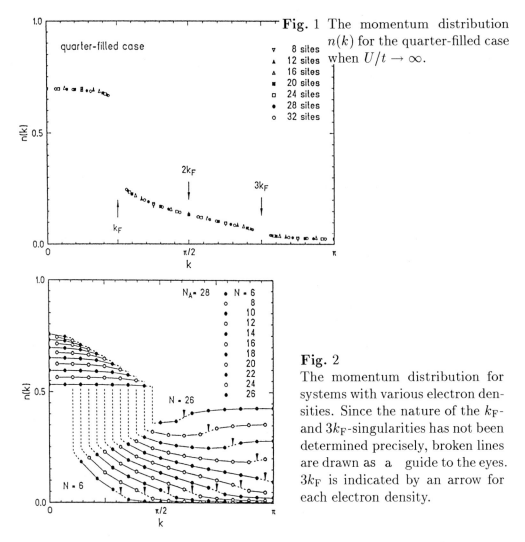

Fig. 1 The momentum distribution $n(k)$ for the quarter-filled case when $U/t \to \infty$.

Fig. 2 The momentum distribution for systems with various electron densities. Since the nature of the k_F- and $3k_F$-singularities has not been determined precisely, broken lines are drawn as a guide to the eyes. $3k_F$ is indicated by an arrow for each electron density.

Let us investigate the singularity at $k = k_F$. A power-law singularity

$$n(k) = n_{kF} - C|k - k_F|^\alpha \mathrm{sgn}(k - k_F), \qquad n_{kF} = 1/2, \qquad (3.2)$$

is expected from the g-ology[3] in the *weakly correlated regime*. Although the present results are in the *strongly correlated regime*, we analyze the data in the vicinity of k_F having the formula (3.2) in mind. From the two points closest to k_F, we obtain $\alpha = 0.136$ for $k < k_F$ and $\alpha = 0.147$ for $k > k_F$. We conclude from the size-dependence that $0.136 < \alpha < 0.147$, i.e. $\alpha \simeq 0.14$.

The $3k_F$-singularity is due to a pair of electron-hole excitations as suggested from the perturbation calculation. It is natural then to think that a progressively weaker singularity caused by multi-pairs of electron-hole excitations must be present, in principle, also at $k = 5k_F, 7k_F, \cdots$.

IV. Spin Correlation Function

The Fourier transform of the spin correlation function is defined by

$$S(k) = \frac{1}{N_A} \sum_{j,\ell} <S_j^z S_\ell^z> e^{ik(r_j - r_\ell)}. \qquad (4.1)$$

Although the spin degrees of freedom in (2.2) are completely described by that of the Heisenberg model, $S(k)$ for less-than-half-filled systems is, of course, different because of the presence of holes: $|j - \ell|$ is in general different from the distance in the Heisenberg model.

We calculate $S(k)$ for various electron densities, examples of which are presented in Fig. 3 with $S_H(k)$ (i.e. $S(k)$ in the Heisenberg model). Clearly a peak shows up at $k = 2k_F$, which is incommensurate when the system is away from half filling. Its dependence on the electron density is consistent with the perturbation calculation.

To examine the singularity at $2k_F$, we have studied $S(k)$ for the quarter-filled case in detail. It is readily seen from Fig. 4 that the size-dependence is evident especially at $k = 2k_F = \pi/2$ suggesting convincingly a singularity at this point. We find that the peak $S(2k_F)$ is drastically reduced compared to $S_H(\pi)$ in the Heisenberg model and it has a weaker size-dependence than $\ln N_A$, while $S_H(\pi)$ has a stronger size-dependence than $\ln N_A$.[4]

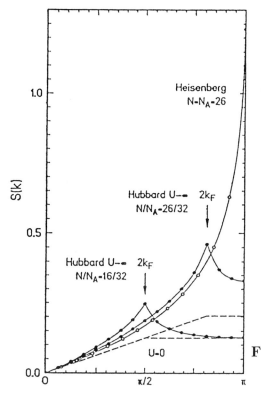

Fig. 3 The spin correlation functions $S(k)$ when $U/t \to \infty$.

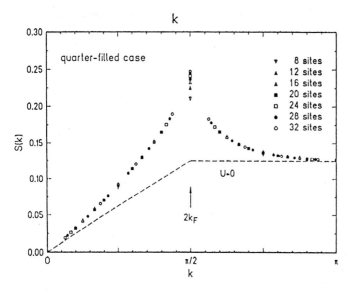

Fig. 4 The spin correlation function for $U/t \to \infty$ in the quarter-filled case.

This size-dependence of $S(2k_F)$ indicates that $<S_j^z S_\ell^z>$ decays faster than $1/|r_j - r_\ell|$. From the size-dependence of $<S_j^z S_\ell^z>$ for the largest $(r_j - r_\ell)$, we obtain $<S_j^z S_j^\ell> \sim |r_j - r_\ell|^{-\eta}$ with $1.29 < \eta < 1.44$.

V. Summary and Discussion

We have studied $n(k)$ and $S(k)$ in the large-U limit of the 1D Hubbard model. It is based on the Bethe Ansatz wavefunction of finite, but fairly large systems.

Generally, the propagation of electrons in strongly correlated systems is greatly affected by rearrangements of surrounding spins. In the present case, this interplay shows up in a simple way. Namely, the ground state wavefunction can be expressed as a product of a Slater determinant of spinless fermions and the spin wavefunction of the 1D Heisenberg model. The rearrangement of spin configuration plays an essential role in $n(k)$.

Recently Sorella et al.[5] have proposed a new algorithm of Monte Carlo simulation and studied $n(k)$ in 1D Hubbard model. It would be interesting to compare our results with what their Monte Carlo method tells us about the critical exponent.

Details of the present work will be reported separately.[2]

[1] E.H. Lieb and F.Y. Wu: Phys. Rev. Lett. **20**, 1445 (1968); C.N. Yang: Phys. Rev. Lett. **19**, 1312 (1967).
[2] M. Ogata and H. Shiba: preprint.
[3] J. Sólyom: Adv. Phys. **28**, 201 (1979).
[4] K. Kubo, T.A. Kaplan and J.R. Borysowicz: Phys. Rev. B **38**, 11550 (1988).
[5] S. Sorella, E. Tosatti, S. Baroni, R. Car and M. Parrinello: Progress in High Temperature Superconductivity (World Scientific) **14**, 457 (1988); M. Imada and Y. Hatsugai: preprint.

New High-Temperature Cooper-Pairing Phase for Vibronic Superconductivity

A. Tachibana[1,2], S. Ishikawa[2], T. Tada[2], and T. Yamabe[1,2,3];

[1]Department of Hydrocarbon Chemistry, Faculty of Engineering,
Kyoto University, Kyoto 606, Japan
[2]Division of Molecular Engineering, Graduate School of Engineering,
Kyoto University, Kyoto 606, Japan
[3]Institute for Fundamental Chemistry, 34-4, Nishihiraki-cho,
Takano, Sakyo-ku, Kyoto 606, Japan

Abstract. A new vibronic mechanism for real-space electron pairing is developed using an exact vibronic Hamiltonian. The attractive force of pairing is demonstrated locally on the sites where the derivatives of the LCAO MO coefficients are sufficiently large. At a higher temperature than that of the conventional BCS phase transition, a new Cooper-pairing phase appears by strong vibronic interaction for electrons remote from the Fermi level, which may lead to high-Tc superconductivity.

1. Introduction

The recent discovery of copper-oxide superconductors stimulated us to propose new mechanisms for high-Tc superconductivity. There is a general consensus that the driving force for superconductivity is the formation of Cooper pairs, or time-reversal pairs in general. In this connection, we have been performing ab initio quantum chemical calculations for a hidden vibronic attractive force between a pair of electrons which may lead to superconductivity [1–5].

Conventional BCS theory predicts that the superconducting phase transition is brought about at the same time as electron pairing. But pairing and condensation of pairs do not necessarily occur at the same time. A strong vibronic interaction for electrons remote from the Fermi level brings about the BCS vacuum polarization [2,6]. Then, we will see two steps for the superconducting phase transition. The high-temperature phase corresponds to the "glass" phase or "liquid" phase of Cooper pairs due to the BCS vacuum polarization, and the low-temperature phase corresponds to the BCS superconducting phase, or the "solid" phase [6].

In this paper, we examine the oxygen-containing low-dimensional model polymer $(CHO)_n$, and calculate the vibronic interaction for the polymer.

2. Theory

The effective interaction energy of the time-reversal pair associated with $i \to j$ ($i \neq j$) electron scattering process is written as [1]

$$V_{eff}^{ij} = V_{Coul}^{ij} - \Delta_{ij}, \tag{1}$$

where V_{Coul}^{ij} and Δ_{ij} denote Coulombic repulsion energy and vibronic attraction energy, respectively. Δ_{ij} is given as

$$\Delta_{ij} = \sum_n \Delta_n^{ij}, \tag{2.a}$$

$$\Delta_n^{ij} = |\langle \psi_i | \frac{\partial \psi_j}{\partial Q_n} \rangle|^2 = |f_n^{ij}|^2 / (\varepsilon_i - \varepsilon_j)^2, \tag{2.b}$$

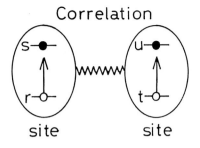

Fig.1 Intrasite excitation and intersite correlation for the Cooper pair formation.

$$f_n^{ij} = \langle \psi_i | \frac{\partial f}{\partial Q_n} | \psi_j \rangle, \quad (2.c)$$

where f, Q_n, and f_n^{ij} denote the dynamic Fock operator for vibronic orbitals $\{\psi_i\}$ and their orbital energies $\{\varepsilon_i\}$ [1], the n-th normal coordinate, and the extended orbital vibronic constant (EOVC) [4], respectively. V_{Coul}^{ij} decreases in inverse proportion to the size of the polymer, and Δ_n^{ij} increases in proportion to that. Therefore, over a critical number N^* giving the size of the polymer, attractive energy for V_{eff}^{ij} is obtained [1-3].

Using the LCAO approximation for ψ_i, we have found that the MO coefficient derivatives are the dominant factor for Δ_{ij}. Then, Δ_n^{ij} is approximately written as

$$\Delta_n^{ij} \simeq \frac{1}{2} \sum_{rstu} (c_{ri}^* c_{ui} \frac{\partial c_{tj}^*}{\partial Q_n} \frac{\partial c_{sj}}{\partial Q_n} \langle r|s \rangle \langle t|u \rangle + c_{rj}^* c_{uj} \frac{\partial c_{ti}^*}{\partial Q_n} \frac{\partial c_{si}}{\partial Q_n} \langle r|s \rangle \langle t|u \rangle), \quad (3)$$

where the symmetry in the i → j vibronic scattering process is explicitly shown. It should be noted that the major contributions in Δ_n^{ij} are 1) local excitation with respect to r → s and t → u, and 2) non-local correlation between these local sites. This is schematically shown in Fig.1.

3. Results and Discussion

Ab initio calculations were performed with the 3-21G basis set employing the GAUSSIAN 80 and GAUSSIAN 82 programs [8] with suitable modifications. We have taken $(CH_2)_2O$ as a structural unit for the $(CHO)_n$ polymer. The N^* is here equal to the critical number of the unit cell. The structure of the unit cell is optimized. As shown in Fig.2, (1) and (2) are the optimized structures in C_{2v} and D_{2h} symmetry, respectively. We have also calculated geometry (3), which has a longer C-O bond length than that of (2). The total energy becomes unstable in this order: (1)<(2)<(3).

In this unit, we pick up three combinations of orbitals which are characteristic of the vibronic scattering processes in the vicinity of the Fermi level. For (2) as a key structure, the orbital patterns and the vibrational modes are shown in Fig.3, and the interaction energies are shown in Table 1.

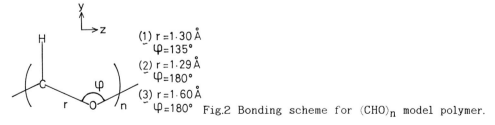

Fig.2 Bonding scheme for $(CHO)_n$ model polymer.

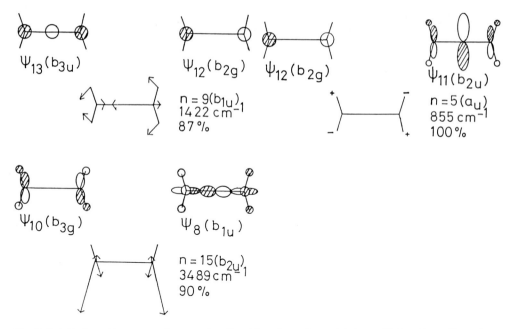

Fig.3 Orbital patterns and normal vibrational modes which mediate the vibronic scattering processes for (2), the contribution of mediation being shown in %.

Table 1. Effective interaction energies in au.

	i–j	V^{ij}_{Coul}	Δ^{ij}_n	$|f^{ij}_n|^2$	$(\varepsilon_i-\varepsilon_j)^2$	N^*
(2)	12–13	1.24×10^{-1}	2.66×10^{-5}	3.72×10^{-6}	1.40×10^{-1}	130
	11–12	4.70×10^{-3}	6.64×10^{-5}	3.07×10^{-6}	4.62×10^{-2}	17
	8–10	4.90×10^{-2}	5.34×10^{-4}	6.12×10^{-6}	1.15×10^{-2}	18
(3)	9–8	2.05×10^{-2}	1.10×10^{0}	2.42×10^{-6}	2.19×10^{-6}	2

HOMO(ψ_{12}) and LUMO(ψ_{13}) are both out-of-plane π orbitals, so that the vibronic scattering process is mediated by longitudinal vibrational mode 9 as shown in Fig.3. The N^* is 130 as shown in Table 1.

HOMO–1(ψ_{11}) is an in-plane π orbital, and hence the atomic orbitals have no overlaps with HOMO. This makes the Coulombic interaction for the 11→12 scattering process small, while the out-of-plane vibrational mode 5 can mediate the vibronic interaction as shown in Fig.3. The N^* becomes 20, smaller than the HOMO→LUMO case.

A similar situation is observed for the 8→10 scattering process, and remains unchanged for the structural relaxation (2)→(1) or structural deformation (2)→(3). Especially for (3), the 8→10 vibronic interaction is enhanced because of the near degeneracy of the orbitals as shown in Table 1: the orbital energy of ψ_{10} becomes lower than ψ_8, and the orbital indices are renumbered from 10,8 to 8,9. The EOVC f^{ij}_n remains almost constant during the deformation. The change of Δ^{ij}_n is due to a decreasing orbital energy difference. A small orbital energy difference brings about large MO coefficient derivatives. The MO coefficient derivatives are dominant for large vibronic interactions. In Fig.4, the MO coefficients for ψ_8 and its derivatives are schematically shown. The pattern of the MO derivatives of ψ_8 coincides with ψ_9. Transversal vibration polarizes ψ_8 and induces a $2p_z$ orbital of oxygen, which ψ_8 originally does not

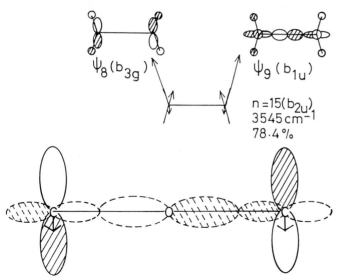

Fig.4 Orbital patterns of ψ_8 (solid line) and its derivative (broken line) for (3).

Table 2. MO coefficients of ψ_9 and MO coefficient derivatives of ψ_8 in amu.

	O 2s	2p$_z$	C 2s	2p$_z$	H 1s
MO of ψ_9	0.00	$-$0.43	$-$0.07	0.14	$-$0.06
MO deriv. of ψ_8	0.00	$-$19.32	$-$2.80	6.06	$-$2.83

have. In Table 2, we show the MO derivatives of ψ_8 and the MO coefficients of ψ_9. Only outer valence orbitals are shown. The contribution of O_{2p_z} to Δ_n^{ij} is 10 %, while that of C_{2p_z} and H_{1s} is less than 1%.

To conclude, the attractive force of pairing is demonstrated locally on the sites where the MO coefficient derivatives are large. The site itself may not necessarily be vibrating as in the case of oxygen in this model. This local excitation may be correlated with other oxygen sites as shown in Fig.1, which may lead to Cooper pair condensation. If the BCS vacuum polarization is allowed, those orbitals which have small N^* in the vicinity of the Fermi level bring about the glass phase or the liquid phase of the Cooper pairs [2,6].

Small V_{Coul}^{ij} and large local MO coefficient derivatives have also been found in the structural unit CuO_4 for 2-dimensional sheets of novel copper-oxide superconductors: this is mainly due to the 2-dimensional character of the electron scattering processes [5,8].

Acknowledgement

This work was supported by a Grant-in-Aid for Scientific Research from the Ministry of Education, Science and Culture of Japan, for which the authors express their gratitude. We are also grateful to the Data Processing Center of Kyoto University and Computer Center of the Institute for Molecular Science for their generous permission to use the FACOM M-780 and VP-400E, and HITAC M-680H and S-820 computer systems, respectively.

References

[1] A.Tachibana, Phys.Rev.**A35**,18(1987); Synth.Met.**19**,105(1987).
[2] A.Tachibana in "High-Temperature Superconducting Materials" ed.by W.E.Hatfield and J.H.Miller, Jr. (Marcel Dekker, New York, 1988), p.99.
[3] A.Tachibana, K.Hori and T.Yamabe, Chem.Phys.Lett.**112**,279(1984); A.Tachibana, T.Inoue, H.Fueno, T.Yamabe and K.Hori, Synth.Met.**19**,99(1987); A.Tachibana, T.Inoue, T.Yamabe and K.Hori, Int.J.Quantum Chem.**30**,575(1986).
[4] A.Tachibana, S.Ishikawa, T.Inoue and T.Yamabe, Chem.Phys.Lett.**154**,403(1989).
[5] A.Tachibana, H.Fueno, S.Ishikawa, and T.Yamabe, Chem.Phys.Lett.**160**,353(1989).
[6] A.Tachibana, this volume.
[7] J.S.Binkley, R.A.Whiteside, R.Krishnan, R.Seeger, D.J.DeFrees, H.B.Schlegel, S.Topiol, L.R.Kahn and J.A.Pople, GAUSSIAN 80, QCPE 13, No.406(1981); J.S.Binkley, M.J.Frisch, D.J.Defrees, K.Raghavachri, R.A.Whiteside, H.B.Schlegel, E.M.Fluder and J.A.Pople, GAUSSIAN 82 release A version, An ab initio molecular orbital program (Carnegie-Mellon University, Pittsburgh, PA, September 1983).
[8] Paper in preparation.

Examination of Pairing via Effective van der Waals Interaction in High-Temperature Superconductivity

K. Tanaka[1], Y. Yamaguchi[1,2], and T. Yamabe[1,2]

[1]Department of Hydrocarbon Chemistry and Division of Molecular Engineering, Faculty of Engineering, Kyoto University, Sakyo-ku, Kyoto 606, Japan
[2]Institute for Fundamental Chemistry, 34-4 Nishihiraki-cho, Takano, Sakyo-ku, Kyoto 606, Japan

Abstract. We examine a mechanism for the pairing of holes via "effective" van der Waals interaction in the superconducting state of the high-T_c ceramic materials, which is also applicable to the excitonic organic superconductors. Using the conventional perturbation technique, the fluctuation of the effective oxygen charge felt by the hole and the inter-oxygen distance has been estimated for a definitely attractive coupling interaction to appear between two holes.

1. Introduction

As is well known, the van der Waals force is a weak attractive interaction which depends on the Coulombic force between atoms or neutral molecules. Some interesting aspects have been reported as follows:

(1) The superconducting carriers in these materials are holes [1,2] except for $Nd_{2-x}Ce_xCuO_{4-y}$ [3]. These holes are mainly created in the occupied oxygen p-orbitals experimentally [4,5]. Then, it is expected that the hole-hole interaction is induced by the polarization of oxygen atoms. The nearest interatomic distance of the pair of oxygen atoms concerned is about 3.8Å [6] to which the scheme of the long-range interaction is already applicable.

(2) The coherence lengths are very small [7,8]. The orders of these are ξ_c=3-7Å for the direction of the c-axis and ξ_{ab}=20-30Å in the a-b plane [7,8]. Thus, the coherence lengths are of similar order to the lattice constants. Therefore, it is presumed that the pairing of two holes reflects the local circumstances in the perovskite structure.

(3) A very small isotope effect has been observed experimentally [9].

2. The Effective van der Waals Interaction

We wish to examine the possibility of the pairing of two holes via an "effective" van der Waals interaction in superconducting materials by matching the present numerical results to the order of the coherence length ξ_{ab} [7,8] and the coupling interaction energy, $\sim$100 meV [10].

We set up a simple system consisting of two holes and two negative ions of high-T_c superconductors with an instantaneous long-range interaction as illustrated in Fig. 1, where the two negative ions are oxygen atoms with a certain separation both residing on the Cu-O plane or at the apical sites perpendicular to this plane [11,12]. The itinerant motion of each hole in the Cu-O plane is assumed to be described by the Bloch-type function for the unperturbed system in the ground state based on an excitonic picture. The fluctuation of the effective negative charge on the oxygen seen by the hole is set to be $-\rho e$, taking account of the polarization effect.

Taking up to the dipole-dipole interaction in this system, the perturbed Hamiltonian H' as well as the unperturbed Hamiltonian H_0 can be written

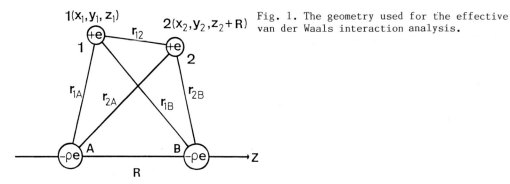

Fig. 1. The geometry used for the effective van der Waals interaction analysis.

$$H_0 = -(\hbar^2/2|m_0|)(\nabla_1^2 + \nabla_2^2)$$
$$+ [e^2/(4\pi\varepsilon_0)](-\rho/r_{1A} - \rho/r_{2B} + \rho^2/R^2) \qquad (1)$$

$$H' = [e^2/(4\pi\varepsilon_0)](1/r_{12} - \rho/r_{1B} - \rho/r_{2A})$$
$$= [e^2/(4\pi\varepsilon_0)][\{(1-2\rho) - (z_1-z_2)(\rho-1)\}/R$$
$$+ \{(\rho-1)(x_1^2 + x_2^2 + y_1^2 + y_2^2 - 2z_1^2 - 2z_2^2)/2$$
$$+ (x_1x_2 + y_1y_2 - 2z_1z_2)\}/R^3] \qquad (2)$$

with usual multipole expansion, where m_0 and ε_0 signify the mass of a free electron and the static dielectric constant of the vacuum, respectively, and other variables are indicated in Fig.1. The internuclear repulsion between oxygen atoms is included in H_0 since this term is supposed to be a part of the lattice energy.

The perturbation energy up to second order in the Unsöld approximation [13] is obtained as follows for each type of atomic orbital for the tight-binding basis:

(i) 1s-1s type (just for trial)

$$E(R) = [e^2/(4\pi\varepsilon_0)][(1-2\rho)/R - 8a_B^3(1-\rho)^2/(3\rho R^4)$$
$$- \{14a_B^5(1-\rho^2) + 8a_B^5\}/(\rho R^6)] \qquad (3)$$

(ii) $2p\pi$-$2p\pi$ type

$$E(R) = [e^2/(4\pi\varepsilon_0)][(1-2\rho)/R - 12a_B^2(1-\rho)/R^3$$
$$- 432a_B^5(1-\rho)(2-3\rho)/(5\rho R^4)$$
$$- 36a_B^5\{468(1-\rho)^2 + 504\}/(5\rho R^6)] \qquad (4)$$

(iii) $2p\sigma$-$2p\sigma$ type

$$E(R) = [e^2/(4\pi\varepsilon_0)][(1-2\rho)/R + 24a_B^2(1-\rho)/R^3$$
$$- 432a_B^5(1-\rho^2)/(5\rho R^4)$$
$$- 36a_B^5\{900(1-\rho)^2 + 1368\}/(5\rho R^6)] \qquad (5)$$

$$(a_B = 4\pi\varepsilon_0\hbar^2/(|m_0|e^2\rho))$$

Table 1. Effective attraction energy (negative quantities of E(R)) between two holes (in eV) [a]

Type	ρ	$C(1/R)$	$C(1/R^3)$	$C(1/R^4)$	$C(1/R^6)$	Total
1s-1s						
	0.10	1.15	0	-4.61	-11.56	-15.02
	0.12	1.09	0	-2.13	-3.77	-4.81
R=10 Å	0.14	1.04	0	-1.10	-1.45	-1.51
	0.15	1.01	0	-0.81	-0.95	-0.75
	0.16	0.98	0	-0.61	-0.63	-0.26
	0.17	0.95	0	-0.47	-0.44	0.04
	0.05	0.65	0	-5.14	-12.33	-16.82
	0.06	0.64	0	-2.42	-4.08	-5.86
R=20 Å	0.07	0.62	0	-1.28	-1.59	-2.25
	0.08	0.60	0	-0.73	-0.70	-0.83
	0.09	0.59	0	-0.45	-0.34	-0.20
	0.10	0.58	0	-0.29	-0.18	0.11
2pπ -2pπ						
	0.30	0.58	-0.38	-1.75	-4.33	-5.88
R=10 Å	0.50	~0	-0.10	-0.07	-0.17	-0.34
	1.0	-1.44	~0	~0	~0	-1.44
R=20 Å	1.0	-0.72	~0	~0	~0	-0.72
R=30 Å	1.0	-0.48	~0	~0	~0	-0.48
2pσ -2pσ						
	0.30	0.58	0.75	-2.07	-10.68	-11.42
R=10 Å	0.50	~0	0.19	-0.22	-0.44	-0.47
	1.0	-1.44	~0	~0	-0.01	-1.45
R=20 Å	1.0	-0.72	~0	~0	~0	-0.72
R=30 Å	1.0	-0.48	~0	~0	~0	-0.48

[a] $C(1/R^n)$ signifies the contribution of the $1/R^n$ term.

The van der Waals force is a long-range interaction which is usually understood to act between atoms or molecules. Here we regard that this perturbation energy E(R) giving an attractive interaction contributes in some portion to couple the two holes through oxygen ions in the material. Substituting R=10, 20 and 30 Å corresponding to the coherence length ξ_{ab} into Eqs.(3)-(5), one obtains the data of the attraction energy from E(R) as listed in Table 1. An attractive coupling interaction of the order of 100 meV for the 1s-1s type appears when ρ =0.16-0.15 for R=10Å and 0.09-0.10 for R=20Å. On the other hand, for the 2pπ -2pπ and the 2pσ -2pσ types, which are more realistic, an attractive interaction of this order appears around ρ =0.5-1.0 for all the R-values. The order of the corresponding Bohr radius a_B becomes 1-10Å for these ρ, as would be expected from the usual excitonic point of view.

It is seen that the contribution from the R^{-4} and R^{-6} terms strongly depends on the oxygen charge parameter ρ. This suggests a subtle role of a fluctuation effect working in the high-Tc materials. In any case, all of these terms are independent of nuclear mass, consistent with a very small isotope effect in the high-Tc superconductors. Although it may seem that superconductivity is then predicted in any material using this model, it should be borne in mind that it becomes possible only in a <u>stable</u> lattice framework such as a perovskite structure which involves strongly polarized oxygen atoms.

Acknowledgements - We are grateful to Prof. K. Fukui for valuable discussion. This work was partially supported by a Grant-in-Aid for Scientific Research from the Ministry of Education, Science and Culture of Japan.

REFERENCES

1. S. Uchida, H. Takagi, H. Ishii, H. Eisaki, T. Yabe, S. Tajima, and S. Tanaka, Jpn. J. Appl. Phys., 26, L440(1987).
2. S. W. Cheong, S. E. Brown, Z. Fisk, R. S. Kwok, J. D. Thompson, E. Zirngiebl, G. Gruner, D. E. Peterson, G. L. Welles, R. B. Schwarz, and J. R. Cooper, Phys. Rev., B36, 3913(1987).
3. Y. Tokura, H. Takagi, and S. Uchida, Nature, 337, 345(1989).
4. A. Bianconi, A. Congiu-Castellano, M. De Santis, P. Delogu, A. Gargano, and R. Giorgi, Solid State Commun., 63, 1135(1987).
5. A. Fujimori, E. Takayama-Muromachi, Y. Uchida, and B. Okai, Phys. Rev., B35, 8814(1987).
6. See, for example, J. E. Greedan, A. H. O'Reilly, and C. V. Stager, Phys. Rev., B35, 8770(1987).
7. T. K. Worthington, W. J. Gallagher, and T. R. Dinger, Phys. Rev. Lett., 59, 1160(1987).
8. A. Kapitulnic, M. R. Beasley, C. Castellani, and C. Di Castro, Phys. Rev., B37, 537(1988).
9. K. J. Leary, H. C. zur Loye, S. W. Keller, T. A. Faltens, W. K. Ham, J. N. Michaels, and A. M. Stacy, Phys. Rev. Lett., 59, 1236(1987).
10. Our estimation from the experimental data.
11. A. Bianconi, M. De Santis, A. M. Flank, A. Fontaine, P. Lagarde, A. Marcelli, H. Katayama-Yoshida, and A. Kotani, Physica C 153, 1760(1988).
12. N. Nücker, J. Fink, J. C. Fuggle, P. J. Durham, and W. M. Temmerman, Physica C 153, 119(1988).
13. See, for example, L. I. Schiff, **Quantum Mechanics (3rd ed.)**, McGraw, New York(1968), Chap. 8.

Reassessment of the Excitonic Mechanism of Little's Superconductivity

K. Tanaka[1], M. Okada[1], Y. Huang[1], and T. Yamabe[1,2]

[1] Department of Hydrocarbon Chemistry and Division of Molecular Engineering, Faculty of Engineering, Kyoto University, Sakyo-ku, Kyoto 606, Japan
[2] Institute for Fundamental Chemistry, 34-4 Nishihiraki-cho, Takano, Sakyo-ku, Kyoto 606, Japan

Abstract. The excitonic mechanism of Little's superconductivity has been reassessed with consideration of the time-dependent behavior of the charge oscillation in the side chain attached to a *trans*-polyacetylene chain. Emphasis has been put on finding the possibility of the appearance of effective attractive interaction between two π electrons. Conventional time-dependent perturbation theory has been employed to check the energy variation of the π-electron system.

1. Introduction

In 1964 Little [1] proposed a possible excitonic mechanism for superconductivity, in which the effective attractive interaction between π electrons moving in an organic polymer chain originates from the virtual oscillation of charge in the side chain attached to the polymer. Although this idea is of interest, no experimental evidence for the existence of a superconductor utilizing this mechanism has been published. Also, over the past two decades, quite a few arguments have appeared on the excitonic mechanism [2-9].

The time-dependence of the oscillation, however, has not been explicitly discussed hitherto. Moreover, recent experimental progress in high-Tc ceramic superconductors seems to require re-consideration of the role of the excitonic mechanism [10].

In this article, we wish to reassess the excitonic mechanism by paying attention to the time-dependent behavior of the charge oscillation in the side chain modelled by a point charge. The time factor is explicitly included in the Hamiltonian. Our main effort will be towards finding an effective attractive interaction between two π electrons. A *trans*-polyacetylene chain is employed for the medium for the π electron motion after Little. Two kinds of chain, that is, the non-bond-alternating and the bond-alternating cases, are to be treated.

2. Numerical Estimation

The model system to be dealt with is shown in Fig. 1. When the polyacetylene has a non-bond-alternating skeleton (i.e., undimerized), $\vec{b}$ is equal to $\vec{a}/2$. The polarizing side chain is simplified by a point charge [3] changing its value with a time-dependent sinusoidal function. The charge oscillation includes a phase difference ξ between each point charge.

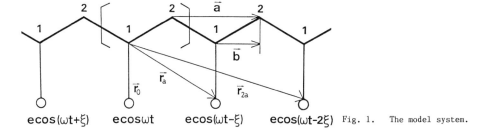

Fig. 1. The model system.

The effective interaction between two π electrons in the polymer chain can be evaluated by

$$\hat{V}_{eff} = (1/2) \sum_{k,k'} \{V_c(\xi/a) - (2|V(\xi)|^2/W)\} a_k^\dagger a_{k-\xi/a}^\dagger a_{k-k'-\xi/a} a_{k'}, \quad (1)$$

which becomes attractive when the value in the brace becomes negative. W represents the energy required for the polarization; $a_k^\dagger$ and a_k are the conventional fermion operators for the π electrons.

We pick up the atomic orbital (AO) integrals of the one- and the two-center types in the same way as in the zero differential overlap approximation [11]. In the non-bond-alternating case, $b=a/2$, the matrix element $V_c(\xi/a)$ standing for the repulsive interaction between two π electrons in the polymer chain is

$$V_c(\xi/a) = (1/2\,N)\{<\chi_1(j)\chi_1(j)|\hat{V}|\chi_1(j)\chi_1(j)> \\
+ 2<\chi_1(j)\chi_1(j)|\hat{V}|\chi_2(j)\chi_2(j)> \cos(\xi/2)\}, \quad (2)$$

where $\chi_m(j)$ stands for the $p\pi$ AO on the carbon m in the j-th cell and V is expressed in Eq.(5). The matrix element $V(\xi)$ standing for the attractive interaction between a π electron and a polarized charge is

$$V(\xi) = (1/2\sqrt{N})\{<\chi_1(j)|V_j|\chi_1(j)> \\
+ 2\cos(\xi/2)<\chi_2(j)|V_j|\chi_2(j)> \\
+ 2\cos\xi <\chi_1(j+1)|V_j|\chi_1(j+1)>\}, \quad (3)$$

where V_j is the Coulomb interaction between a π electron in the polymer chain at r and a polarized charge at r_{ja} in the side chain in the j-th cell and is explicitly described by

$$V_j \equiv -Ce^2/|\vec{r} - \vec{r}_{ja}| \quad (C: const.). \quad (4)$$

The total Coulomb interaction becomes

$$V = \sum_j V_j\{\exp[i(\omega t+j\xi)]d_j + \exp[-i(\omega t+j\xi)]d_j^\dagger\} \quad (5)$$

$$(\omega = W/\hbar).$$

The maximum value of W (W_M) under which the <u>attractive</u> interaction between two π electrons is guaranteed is plotted in Fig. 2.

Let us examine the energy variation of the π-electron system due to the introduction of the oscillating point charges. Applying conventional time-dependent perturbation theory [12], the time average of the variation of the energy of the π-electron system per unit cell becomes

$$\overline{\Delta E_{el}} = \lim_{T\to\infty}(1/T)\int_0^T \Delta E_{el}(t)dt = [32\beta/(\pi W^2)]F(\xi), \quad (6)$$

where β is a negative energy constant. $F(\xi)$ can be obtained from

$$F(\xi) = (1/4)\{ [<\chi_1(j)|V_j|\chi_1(j)> + 2\cos\xi <\chi_1(j+1)|V_j|\chi_1(j+1)> \\
+ 2\cos(\xi/2)<\chi_2(j)|V_j|\chi_2(j)>]^2[\cos(\xi/2) - 1] \\
- [<\chi_1(j)|V_j|\chi_1(j)> + 2\cos\xi <\chi_1(j+1)|V_j|\chi_1(j+1)> \\
- 2\cos(\xi/2)<\chi_2(j)|V_j|\chi_2(j)>]^2[\cos(\xi/2) + 1] \}. \quad (7)$$

The minimum value of $\overline{\Delta E_{el}}$ under the condition of the appearance of the effective attractive interaction is estimated as listed in Table I.

Table I. Time average of variation of the π-electronic system energy, $\overline{\Delta E_{el}}$, (in eV) at various ξ.

ξ	$\overline{\Delta E_{el}}$	
	Non-bond-alternating	Bond-alternating
0	2.08	2.12
$\pi/4$	3.22	3.60
$\pi/3$	4.46	4.73
$\pi/2$	1.01×10^1	1.14×10^1
$2\pi/3$	4.49×10^1	4.78×10^1
$3\pi/4$	1.24×10^2	1.37×10^2
π	1.54×10^4	1.57×10^4

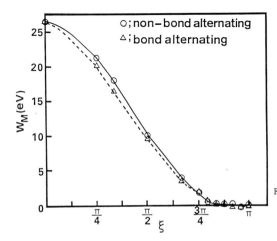

Fig. 2. Maximum W (W_M) for the attractive $\hat{V}_{eff}$ versus ξ in Eq.(1).

3. Conclusion

The instability of the π-electron system becomes minimum at $\xi = 0$ as seen in Table I. The reason for increasing $\overline{\Delta E_{el}}$ at $\xi = \pi$ stems from the very small value of W_M there as seen in Fig. 2. Thus it is understood that the phase difference ξ also affects the energy variation of the π-electronic system.

Three major conclusions are:

(1) There are upper limits to the energy required for the charge polarization (W_M) for each phase difference (ξ) to guarantee the effective attractive interaction for the possible formation of the Cooper pair.

(2) Introduction of the oscillating point charge brings about an energy destabilization of the π-electronic system on average. There should be a certain optimum combination of ξ and W for the molecular design of Little superconductors.

(3) No significant differences in the result are seen when considering the non-bond-alternating and the bond-alternating polyacetylene chains as the medium.

Acknowledgments

This work was partially supported by a Grant-in-Aid for Scientific Research from the Ministry of Education, Science and Culture of Japan. Numerical calculations were carried out at the Data Processing Center of Kyoto University.

References

[1] W. A. Little, Phys. Rev., 134, A1416(1964); H. Gutfreund and W. A. Little, in Highly Conducting One-Dimensional Solids (ed. by J. T. Devreese, R. P. Evrard, and V. E. van Doren, Plenum, New York, 1979), Chap.7.
[2] K. F. G. Paulus, Mol. Phys., 10, 381(1966).
[3] L. Salem, Mol. Phys., 11, 499(1966).
[4] V. L. Ginzburg, Contemp. Phys., 9, 335(1968).
[5] J. Ladik, G. Biczo, and J. Redley, Phys. Rev., 188, 710(1969).
[6] D. Allender, J. Bray, and J. Bardeen, Phys. Rev., B7, 1020(1973).
[7] J. Yoshida, K. Nishikawa, and S. Aono, Prog. Theoret. Phys., 50, 830(1973).
[8] D. Davis, H. Gutfreund, and W. A. Little, Phys. Rev., B13, 4766(1976).
[9] J. E. Hirsch and D. J. Scalapino, Phys. Rev., B32, 117(1985).
[10] W. A. Little, Science, 242, 1390(1988)
[11] J. A. Pople and D. L. Beveridge, Approximate Molecular Orbital Theory (McGraw, New York, 1970), Chap. 3.
[12] See, for instance, L. I. Schiff, Quantum Mechanics, 3rd ed. (McGraw, New York, 1968), Chap. 8.

Multi-Valence Resonance-Condensation Model: A Possible Novel and Universal Origin of Superconductivity

A. Nakamura

Japan Atomic Energy Research Institute, Chemistry Division,
Tokai-mura, Naka-gun, Ibaraki 319-11, Japan

Abstract. The Multi-Valence Resonance-Condensation Model (MVRC Model) is invoked as a possible novel and universal origin of superconductivity and is successfully applied to various high T_c oxide superconductors, A-15 type intermetallic Nb_3Ge, etc. The MVRC state of superconductivity is a Bose-condensate of the microscopic resonance hybrids shown in Fig. 1, which is realized macroscopically through the self-organizational process (valence fluctuation → valence ordering → resonance-condensation) of the electron system as the fundamental chemical bonding force. This is demonstrated to provide a possible unified picture of the superconducting state of all these systems in a straightforward way for both the so-called 'hole' and 'electron' type superconductors in a symmetrical manner. A possible expression for T_c derived on this basis can also give a fair account of the isotope effects of these superconductors in a systematic manner. Thus, the model is expected to be useful as a chemical and structural principle in both the search for new superconductors and the construction of a new microscopic theory of superconductivity.

1. Introduction

In this paper, a brief account is given of the MVRC model recently proposed by the present author as a possible novel and universal origin of electron pairing and superconductivity in high T_c oxide superconductors and A-15 type Nb_3Ge [1,2].

In contrast to the conventional phonon-mediated BCS mechanism of superconductivity [3], in the MVRC model, the origin of superconductivity is sought in the more fundamental and universal nature of matter, i.e., the chemical bond, more specifically, the microscopic quantum chemical resonances inside the tri-cation crystal molecule of the appropriate chemical species Cu (Bi, Nb, etc.) shown in Fig. 1. Adopting the formal valence representation for various cations [4,5], two modes of these resonance hybrids are in principle possible, having either higher or lower average cation valence than the intermediate valence state (↑) and being denoted ORH and RRH, respectively. Of course, in oxides, though not shown in this figure, the Cu-Cu (Bi-Bi) bond is an indirect one via the oxide ion (O^{2-}). So, depending on the nature of the hybrid MO bond between the cations and the oxygen, holes (or electrons) in the ORH (or RRH) type normal state conductors should exhibit a variable character ranging from the cation d(s)- to the oxygen 2p orbitals' character. But, the essential point here seems that such hybrid MO bonds exist and indeed the charge transfer process $\underline{Cu^{3+}} + O^{2-} \rightarrow Cu^{2+} + O^{-}$ takes place between them.

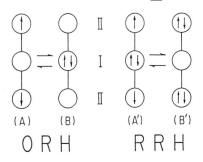

Fig. 1. Oxidized and Reduced type Resonance-Hybrids (ORH and RRH) inside the tri-cation crystal molecule.

Spin state (electron number):

◯(0), ↑(1), ↑↓(2).

These microscopic resonances(ORH and $\overline{RRH}$) of Fig. 1 are attained by the symmetrical and simultaneous paired motion of two electrons of opposite spin in opposite directions. So, if certain chemical and structural conditions are satisfied and these microscopic resonances indeed exist there and extend throughout the whole crystal, the systems are expected to be in the macroscopically coherent quantum chemically resonating state with a real-space electron pair, that is, the MVRC state of superconductivity. This is demonstrated briefly in the next section for almost all the oxide superconductors reported so far and the former high T_c record holder Nb_3Ge.

2. MVRC State of Superconductivity

The 90K high T_c superconductor $YBa_2Cu_3O_7$ [6] with a tri-layered perovskite structure and average Cu valence of 2.33+ offers the best example of the present model: It is immediately understood that the (b-c) plane of this system shown in Fig. 2 is nothing but the interconnected one-dimensional (1D') array of ORH of Fig. 1. Thus, the microscopic intra-molecular resonance($A \rightleftarrows B$), provided with the intermolecular resonance, now manifests itself as the macroscopic MVRC state of superconductivity (ABAB$\cdot\cdot \rightleftarrows$ BABA$\cdot\cdot$), in which every electron moves as half of a real-space electron pair, i.e. two electrons of opposite spin moving in opposite directions. We may regard the MVRC state as a Bose-condensate of two bosons (A and B) strongly interacting in real space.

An important aspect of this Cu-O bond is its non-cubic symmetry around Cu: Square-planar and pyramidal configurations of oxygen around Cu in the middle chain and the upper and lower 2D square planes, respectively, are essential to lift the degeneracy of Cu 3d orbitals (Hund's rule), so that this multi-valence state of Cu occurs inside the split single Cu-O HOMO bond with the designated spin degeneracy. This appears to be the definitive reason why in d (or, more properly, other than s) electron systems, superconductivity is found only in the anisotropic 1-, 2- and 3D systems, and not in the isotropic 3D systems, whereas in the s electron system (l=0) isotropic 3D superconductivity is possible as in $Ba(K)BiO_3$.

Results of the present analysis are concisely summarized in Table 1 in order of increasing dimensionality of the system. Table 1 clearly shows that 1-, 2- and 3D extensions of ORH and RRH of Fig. 1 indeed provide straightforwardly the most optimum valence states (chemical compositions) of various superconductors with at least nine kinds of multi-valent type 'soft valence' cations including also a non-oxide system, i.e. the A-15 type intermetallic Nb_3Ge with true 1D Nb-Nb chain type direct d-d non-polar HOMO bonds. For full details, see Refs.[1,2].

Application of the present model to other types of superconductors such as conventional metal systems and organic superconductors might also be possible in view of the following experimental facts: (a) Even in pure metal systems, the relatively high T_c superconductors are often found in the multi-valent type 'soft valence' elements such as Nb, Pb, Bi, Hg, Sn, and so on. (b) As for the organic superconductors, the completely intermingled quantum chemical resonance bond of A (A') and B (B') of Fig. 1 does not at all rely on whether its original normal

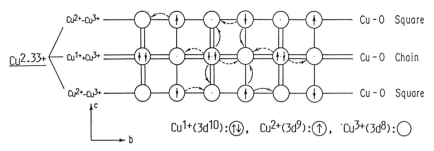

Fig. 2. (b-c) plane of $YBa_2Cu_3O_7$; tri-layered type quasi-one-dimensional (1D') superconductivity (ORH).

Table 1 Brief Summary of the Present Analysis on Various Superconductors

Dimension	Systems (Electron Conf.)	T_c(max) (K)	Cation orbital	Cation Valence (ORH or RRH)	Valence State ⇅ ↑ ○
1D Chain	$\underline{Nb}_3Ge(4d^45s^1)$	23.7	$d_{z^2}\ \sigma$	2.25+(ORH)	$Nb^{1,2,3+}(4d^{4,3,2})$
1D' Chain	$YBa_2\underline{Cu}_3O_7(3d^{10}4s^1)$	92	$3d_{x^2-y^2}\ \sigma^*$	2.33+(")	$Cu^{1,2,3+}(3d^{10,9,8})$
2D Square Plane	$La_{1.75}Sr_{0.25}\underline{Cu}O_4$	37	"	2.25+(")	" (")
	$Tl(Bi)-\underline{Cu}-O$ Systems	130	"	" (")	" (")
	$Nd_{1.85}Ce_{0.15}\underline{Cu}O_{3.95}$	20	"	1.75+(RRH)	" (")
2- to 3D' Network	$Li_{0.8}\underline{Ti}_2O_4(3d^24p^2)$	12.2	" σ	3.25+(ORH)	$Ti^{2,3,4+}(3d^{2,1,0})$
	$K_{0.75}\underline{Mo}O_3(4d^55s^1)$#	4.3	$4d_{x^2-y^2}\ \sigma$	5.25+(")	$Mo^{4,5,6+}(4d^{2,1,0})$
	$K_{0.75}\underline{W}O_3(5d^46s^2)$#	1.5	5d " "	" (")	$W^{4,5,6+}(5d^{2,1,0})$
	$K_{0.25}\underline{Re}O_3(5d^56s^2)$#	3.6	5d " "	5.75+(RRH)	$Re^{5,6,7+}(\ "\ \ )$
	$Ag_7O_8NO_3(4d^{10}5s^1)$#	1.0	4d " σ^*	2.25+(ORH)	$Ag^{1,2,3+}(4d^{10,9,8})$
	$Ba\underline{Pb}_{3/4}Bi_{1/4}O_{3.935}(6s^26p^2)$	13	6s σ	3.25+(")	$Pb^{2,3,4+}(6s^2,1,0)$
3D Network	$Ba_{3/4}K_{1/4}\underline{Bi}O_3(6s^26p^3)$	30	"	4.25+(")	$Bi^{3,4,5+}(\ "\ \ )$

<u>Underlined</u> cationic species denote those responsible for their superconductivity.
#: For these systems, theoretically most optimized compositions(cation valence) shown in this table are not realized experimentally.

state was of cationic or anionic origin, or of inorganic or organic origin, etc. So, it seems reasonable to suppose that the present model is basically applicable to such organic superconductors, in which their superconductivity appears to be mostly due to the partially filled single HOMO bond of the chalcogen (S, Se, Te) 1- to 2D network planes [7,8], a charge transfer complex type normal state conductors in a sense similar to the above inorganic superconductors.

3. Nature and Mechanism of the Superconductive Transition

As is apparent from Fig. 1 and Table 1, the onset of superconductivity is optimized at average valence states of ±0.33 and ±0.25 from the intermediate valence state ↑ for tri-layered type 1D' systems and for other 1-, 2- and 3D systems, respectively. Here, + and - refer to the ORH and RRH type superconductors.

The system may approach this particular composition either from the magnetic insulator side (La_2CuO_4, $YBa_2Cu_3O_6$) or from the metallic oxide side ($BaPbO_3$, ReO_3). Thereby, their electronic states would be significantly destabilized, driven just to the middle of the metal/insulator transition region and totally perturbed from their initial stable chemical state in either the magnetic order or the covalent metallic bond. In this sense, we could regard the electron system in the normal state of these superconductors as metastable or even unstable, so that it is destined to seek an alternative stable chemical bond in a self-organizing manner due to its intrinsic nature as the chemical bonding force.

Referring to Figs. 1 and 2, to restabilize the system, the electron system first undergoes a dynamic valence exchange: A(A') → B(B') or vice versa and generates the multi-valence state. Then, using the induced electron ⇅ - hole ○ attractive interactions inside and between the chains(planes)[we designate the intra- and inter-chain(plane) ones as E_a (exciton) and E_a (CDW), respectively] as the ordering energy of the process, this high temperature disordered multi-valence state with lots of unsymmetrical electron (spin) configurations will start to order gradually with decreasing temperature, competing against the thermal agitation of the system, i.e. E_N (phonon). As the temperature is further lowered and the thermal agitation of the system becomes sufficiently weak, the ordering reaches a critical level around which the system can no longer sustain its A(A')-B(B') type compressed state. The system drops almost spontaneously into the most stable macroscopically coherent MVRC state of superconductivity, simultaneously changing its chemical bond from basically the electrostatically bound excitonic and CDW type compressed one to the completely uniform strong resonance bond.

From the above scenario, it follows that the T_c of the system may be given by

$$T_c = 1.14|E_N|\exp(-|E_N|/|E_S-E_N|) = 1.14|E_N|\exp(-1/\lambda) . \qquad (1)$$

Here $E_{N,S}$ designate the energy density of the normal and the superconducting states at 0K, and are given by, $E_N=E_a(\text{exciton})+E_a(\text{CDW})+E_N(\text{phonon})$, $E_S=E_S(\text{MVRC})+E_S(\text{phonon})$, $J=E_S-E_N$ and $\lambda=|J|/|E_N|$, all at 0K. The completely dissociated multi-valence state is the energy reference state (E=0). The preexponential term of Eq. (1), $T_c'=1.14|E_N|$, is the critical temperature of the ordering in Fig. 2 and the exponential term $\exp(-1/\lambda)$ is the relative stability factor, which means that when $\lambda=|J|/|E_N| \to 0$ or ∞, $T_c \to 0$ or T_c'. According to Eq.(1), theoretically, in principle, there seems to exist no upper limit of T_c in agreement with our instinct in Figs. 1 and 2 that the T_c of the system could be raised in principle infinitely if the strength of the microscopic resonance is further increased. This also means that to get a higher T_c the energy difference between A(A') and B(B') in Fig. 1 should be minimized and also the covalency of the HOMO bond should be increased.

Equation (1) was applied for the isotope-effect analyses of various superconductors [1,2]. The results are concisely described in the following expression:

$$T_c'M'^{1/2}/T_cM^{1/2} = 1-\{(M'/M)^{1/2}-1\}\{1-(n+1)\lambda\}/n\lambda \qquad (2)$$

where T_c' is the T_c of the isotope (M': the mass) substituted system, and n is a parameter found by putting $|E_N|=n\hbar w_D$ (w_D: Debye frequency). For low T_c systems in which n=1-4 and $\lambda\sim0.25$, a $T_c'M'^{1/2}=T_cM^{1/2}=$constant relationship is obtained, whereas, when $n \to \infty$, $T_c' \to T_c$ (no isotope effect). Thus, we can clearly interpret the oxygen isotope effects of several high T_c oxide superconductors which exhibit the clear tendency of the higher the T_c of the system, the smaller the isotope effect $|\Delta T_c|/T_c$.

The present model implies that superconductivity, in a word, might be the quantum chemical resonance-condensation process of the electron system in Fig. 1. Then, it seems reasonable enough to suppose that through the symmetrical paired motion of two electrons as a real space electron pair, the electron system escapes and decouples almost perfectly simultaneously from the electron scatterer (phonon) and the energy destabilizer (magnetic field) of the system, giving rise to the remarkable zero resistivity and the Meissner effect.

References

[1] A.Nakamura, Proceedings of Tsukuba Seminar on High T_c Superconductivity, Tsukuba, Japan, 1989, edited by K.Masuda et al. (Tsukuba Univ., Tsukuba, 1989)pp85-90.
 A.Nakamura, Jpn.J.Appl.Phys.28, 2468(1989).
[2] A.Nakamura, Studies of High Temperature Superconductors, Vol.4, edited by A.V. Narlikar (Nova Science Publishers, New York, 1990) in press.
[3] J.Bardeen, L.N.Cooper and J.R.Schrieffer, Phys.Rev.108, 1175(1957).
[4] A.W.Sleight, Science 242, 1519(1988).
[5] J.A.Wilson, J.Phys.C21, 2067(1988).
[6] M.K.Wu, J.R.Ashburn, C.J.Torg, P.H.Hor, R.L.Gao, Z.J.Huang, Y.Q.Wang and C.W.Chu, Phys.Rev.Lett.58, 908(1987).
[7] K.Bechgaard and D.Jérome, Betsusatsu Science 84, 105(1987).
[8] G.Saito, Bull.Ceram.Soc.Jpn.7, 574(1987).

Density Functional Theory for the New High-Temperature Superconducting Phase Transition

A. Tachibana

Department of Hydrocarbon Chemistry, and Division of Molecular Engineering, Faculty of Engineering, Kyoto University, Kyoto 606, Japan

Abstract. The density functional theory is used, with no recourse to any model of the Cooper pairing force, for the description of the superconducting phase transition. Thereby we relax the BCS vacuum for polarization of the Cooper pair. Cooper pairs as "two-electron molecules" can form the "liquid phase" at higher temperature Tb above the well-defined critical temperature Tc at which the unpolarized BCS vacuum is formed as the "solid phase". Under certain conditions, the negative jump of the electronic specific heat occurs at Tb in contrast to the conventional positive jump at Tc.

1. Introduction

The BCS theory [1], which is supposed to provide an essentially correct picture of organic superconductivity [2], assumes three fundamental features for the statistical mechanical framework: (I) the amplitude $<\hat{a}_{(-\sigma)}{}^+ \hat{a}_{(\sigma)}{}^+>$ carries nonzero value in the superconducting phase, where $\hat{a}_{(\sigma)}{}^+$ is the creation operator of the quasi-particle with spin σ, constituting the superconducting pair, and where $<>$ denotes the trace with respect to the density matrix, (II) the entropy is evaluated by enumeration of the pairing states, and (III) the amplitude $<\hat{c}_{(-\sigma)}{}^+ \hat{c}_{(\sigma)}{}^+>$ is set equal to zero at every temperature, where $\hat{c}_{(\sigma)}{}^+$ is the creation operator of the pairing state obtained by the Bogoliubov transformation.

The first item (I) is essential. We follow this. But the items (II) and (III) are achieved by careful choice of the projected states for the thermal equilibrium ensemble, hence the model Hamiltonian [1,3]. So, we loosen the items (II) and (III) in such a way that the residual entropy appears for the pair-breaking states and the amplitude $<\hat{c}_{(-\sigma)}{}^+ \hat{c}_{(\sigma)}{}^+>$ carries a nonzero value even in the zero-temperature limit, leading to a polarization of the BCS vacuum. This generalization is performed using the density functional theory [4-6].

2. Density Functional Theory for Superconducting Phase Transition

The BCS vacuum is constructed by complete pairing of quasi-particles, dictating the following Bogoliubov transformation from $\hat{a}_{(\sigma)}{}^+$ to $\hat{c}_{(\sigma)}{}^+$:

$$\hat{a}_{i\sigma}{}^+ = \{\hat{c}_{i\sigma}{}^+ + g_{i\sigma}\hat{c}_{i-\sigma}\}/(1 + |g_i|^2)^{1/2} \; ; \; g_{i\alpha} = -g_{i\beta} = g_i . \tag{1}$$

The thermodynamic potential for the superconducting state is an exact functional of the electron charge density $\rho(r)$ [4-6] and may be represented as

$$\Omega[\rho] = E_p[\rho] - \mu_G \Sigma_\sigma \Sigma_i f_{i\sigma} - TS_p , \tag{2}$$

$$S_p = -k_B \Sigma_\sigma \Sigma_i \{d_{i\sigma} \ln d_{i\sigma} + (1-d_{i\sigma})\ln(1-d_{i\sigma})\} , \qquad (3)$$

$$f_{i\sigma} = \{1/(1+|g_i|^2)\}\{d_{i\sigma} + g_{i\sigma} <\hat{c}_{i\sigma}^+ \hat{c}_{i-\sigma}^+ > + g_{i\sigma}^* <\hat{c}_{i-\sigma} \hat{c}_{i\sigma}^+ > + |g_i|^2(1-d_{i-\sigma})\} , \qquad (4)$$

where μ_G is the Gibbs chemical potential. The diagonal element $f_{i\sigma}$ of the matrix $\|f_{i\sigma j\sigma}\|$ is the occupation number of the quasi-particle with spin σ. The diagonal element $d_{i\sigma}$ of the matrix $\|d_{i\sigma j\sigma}\|$ is the occupation number in the pairing state. S_p is the entropy for the pairing states. $E_p[\rho]$ is the effective energy for the pairing states:

$$E_p[\rho] = \Sigma_\sigma \Sigma_{ij} f_{i\sigma j\sigma} <\phi_{i\sigma}|-(1/2)\Delta|\phi_{j\sigma}> + F_p[\rho] , \qquad (5)$$

$$\rho(r) = \Sigma_\sigma \rho_\sigma(r) ; \quad \rho_\sigma(r) = \Sigma_{ij} f_{i\sigma j\sigma} \phi_{i\sigma}^*(r)\phi_{j\sigma}(r) , \qquad (6)$$

where $\phi_{i\sigma}(r)$ and $\rho_\sigma(r)$ denote the wave function and electron density, respectively, of the quasi-particle with spin σ. Other parts such as the kinetic energy of nuclear vibration, together with the residual entropy terms for the broken-pair quasi-particles are hidden in $F_p[\rho]$ [4,6]. The $\phi_{i\sigma}(r)$ is given by the following secular equation:

$$L_\sigma(r)\phi_{i\sigma}(r) = e_{i\sigma}\phi_{i\sigma}(r) ; \quad <\phi_{i\sigma}|\phi_{j\sigma'}> = \delta_{ij}\delta_{\sigma\sigma'} , \qquad (7)$$

$$L_\sigma(r) = -(1/2)\Delta(r) + \int [\partial F/\partial v(r')] \chi_{v\rho_\sigma}(r',r) dr' , \qquad (8)$$

$$\chi_{v\rho_\sigma}(r',r) = \delta v(r')/\delta\rho_\sigma(r) . \qquad (9)$$

Here $v(r)$ is the Lagrange multiplier adopted in the "apparatus" density functional theory [6] for the variational constraint of fixed electron charge density, and where $\chi_{v\rho_\sigma}(r',r)$ is the response function of $v(r')$ with respect to $\rho_\sigma(r)$. $e_{i\sigma}$ is the energy of the quasi-particle.

We use the "apparatus" operators involved in the density matrix [6] and make the amplitude $<\hat{a}_{i\beta}^+ \hat{a}_{i\alpha}^+>$ fixed in the variational procedure. This fixation stems from the item (I). We shall minimize $\Omega[\rho]$ with respect to the variational parameters $\|d_{i\sigma j\sigma}\|$ and g_i involved in $\rho(r)$. First, $<\hat{c}_{i\beta}^+ \hat{c}_{i\alpha}^+>$ is assumed to be fixed or "optimized", i.e. $\delta<\hat{c}_{i\beta}^+ \hat{c}_{i\alpha}^+>/\delta z = 0$, with respect to the variational parameters $z = \|d_{i\sigma j\sigma}\|$ and g_i, which will be loosened finally in Eq.(15). We have obtained

$$d_{i\sigma} = 1/[\exp(\beta E_{i\sigma}) + 1] ; \quad E_{i\sigma} = E^\circ_{i\sigma} - \mu^\circ G ; \qquad (10)$$

$$E^\circ_{i\sigma} = \{1/(1+x)\} e_{i\sigma} - \{x/(1+x)\} e_{i-\sigma} , \quad \mu^\circ G = [(1-x)/(1+x)]\mu_G ; \quad x = |g_i|^2 , \qquad (11)$$

where $E^\circ_{i\sigma}$ and $\mu^\circ G$ denote the modified energy and chemical potential for the pairing state, respectively. The amplitude $<\hat{c}_{i\beta}^+ \hat{c}_{i\alpha}^+>$ may be written as

$$<\hat{c}_{i\beta}^+ \hat{c}_{i\alpha}^+> = (1/2)w<\hat{a}_{i\beta}^+ \hat{a}_{i\alpha}^+> , \qquad (12)$$

where w is a real weight function defined by this equation. The nonzero w demonstrates the "BCS vacuum polarization". Especially for a Cooper pair, it can be expressed as

$$w = 2(k-x)/(k+x) , \qquad (13)$$

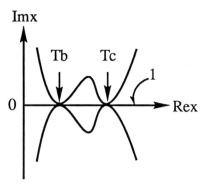

Fig. 1. Distribution of zeroes of w in the complex x-plane.

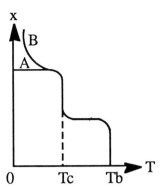

Fig. 2. Strength of x vs T.

where k is an implicit function of x. The k is the ratio of the pairing state mixing:

$$f_{i\sigma} = \{di\sigma + k(1 - di\text{-}\sigma)\}/(1 + k) , \qquad (14)$$

and the deviation of k from x measures the degree of the BCS vacuum polarization. The behavior of the zeroes of w is very interesting. In general the zeroes of w are complex if we perform analytical continuation of w in the complex x-plane and are distributed symmetrically with respect to the real x-axis as shown in Fig.1. For the BCS vacuum, the zeroes of w are on the line 0<x<1 in the real x-axis. The superconducting phase transition may be characterized by the point where the zeroes of w approach the real x-axis asymptotically. Two such points are shown in Fig.1. Corresponding behavior of x may be shown in Fig. 2. For higher temperature region where the pairing disappears, the x should vanish. As we cool the system, nonzero x appears. First a new phase transition is observed at Tb, and second at Tc. Here, the phase transitions are assumed to be second order.

The $E_{i\sigma}$ in Eq. (10) is identified with the energy of the quasi-particle in the pairing state relative to the modified Fermi level $\mu°G$. If the pairing state has relatively high energy as shown in Fig.3, then at Tb a quasi-bound or resonant state may be formed, in contrast to the bound state at Tc, the well-defined critical temperature. The consequence is that, as shown in Fig.4, the electronic specific heat C from the pairing states brings about the negative jump at Tb, in contrast to the conventional positive jump at Tc. The new phase at Tb may be called a "liquid phase" or "glass phase" of the Cooper pairs, whose anomalous contribution to the electronic specific heat may be observed in quite unusual circumstances, such as high-Tc superconductors for which electrons remote from the Fermi level participate in the pairing [4,7]. If the pairing is formed only in the very vicinity of the Fermi level, then there may not be room for quasi-bound state to be formed, only the bound state for "solid phase" should be observed. This is the conventional situation.

The mirror image, with respect to E=0, of the curves in Fig. 3 are for holes, $e_{\iota\sigma}\text{-}\mu G < 0$, where ι denotes the mirror image of i. There emerges a strong configuration interaction at Tc. This interaction is introduced here mathematically through the loosening of the variational constraint mentioned just before Eq.(10). The interaction brings about the renormalization of the energy for the quasi-particles. The secular equation is then written as

$$\begin{vmatrix} E_{i\sigma} - E & V_{i\iota} \\ V_{i\iota}{}^* & E_{\iota\sigma} - E \end{vmatrix} = 0 , \qquad (15)$$

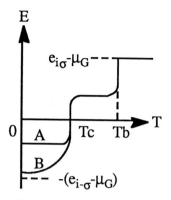
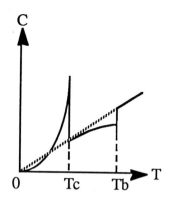

Fig. 3. Energy spectrum of quasi-particle in the pairing state.

Fig. 4. Electronic specific heat shows negative jump at Tb and positive jump at Tc.

where V_{it} is the matrix element of the interaction. The renormalized energy is obtained as

$$E = (\varepsilon^2 + \Delta^2)^{1/2} , \qquad (16)$$

where we have replaced $E_{i\sigma}$ by the same energy ε as a reference energy, and where we have assumed $|V_{it}|$ to be a constant Δ for every i. Eq.(16) is exactly the BCS energy.

Regional chemical potentials and hardnesses for general acids and bases that are useful concepts for electron or charge transfer in the regional density functional theory are used to predict the liquid-solid phase transition [6]. A possible mechanism for the new superconducting phase transition may be given by the hidden vibronic superconductivity [2,8].

Acknowledgement. This work was supported by a Grant-in-Aid for Scientific Research from the Ministry of Education of Japan, for which the author expresses his gratitude.

References

[1] J. Bardeen, L.N. Cooper and J.R. Schrieffer, Phys. Rev.**108**,1175(1957).
[2] A.Tachibana, Phys.Rev.A**35**,18(1987).
[3] J.G.Valatin,Nuovo Cimento **7**,843(1958).
[4] A.Tachibana,in:High-Temperature Superconducting Materials,eds.W.E.Hatfield and J.H.Miller,Jr. (Marcel Dekker, New York,1988),p.99.
[5] P.Hohenberg and W.Kohn,Phys.Rev.**136**,B864(1964).
[6] A.Tachibana, Int.J.Quant.Chem.S.**21**,181(1987); Int.J.Quant.Chem.**34**,309(1988);**35**, 361(1989) ;A.Tachibana and R.G.Parr,paper in preparation.
[7] S.E.Inderhees, M.B.Salamon,N.Goldenfeld,J.P.Rice,B.G.Pazol,D.M.Ginsberg,J.Z.Liu and G.W.Crabtree, Phys.Rev.Lett.**60**,1178(1988).
[8] A.Tachibana,S.Ishikawa,T.Tada and T.Yamabe, paper in this volume.

Possible Role of Two-Dimensionality for the Enhancement of Superconducting T_c

K. Fukushima[1] and H. Sato[2]

[1] Toshiba Research and Development Center, 4-1, Ukishima-cho,
Kawasaki-ku, Kawasaki 210, Japan
[2] Department of Physics, Hyogo University of Education,
Yashiro-cho, Kato-gun, Hyogo 673-14, Japan

Abstract. The authors investigate the effect of two-dimensionality (2D) on the electron-electron interaction mediated by 3D longitudinal optical (LO) phonons of long wavelength in a quasi-2D system, where electrons are confined completely within an x-y plane. It is shown that, in this system, fermion confinement occurs wherein the electric field produced by LO phonons is squeezed within the x-y directions. The effective electron-electron interaction potential in the second-order perturbation for this system is logarithmic, which is quite different from the one widely used. This potential possibly leads to the high-T_c superconductivity even within the BCS scheme.

1. Introduction

Recently, a large amount of research effort has been exerted to understand electronic properties for quasi-two-dimensional (2D) materials such as organic superconductors and high-T_c copper oxides. With regard to copper oxides, carrier holes have been shown to exist in a Cu-O plane from electron energy-loss spectroscopy [1], and the Kosterlitz-Thouless transition has been observed [2], which is direct evidence that the carriers are confined within the 2D plane. It has also been shown experimentally that the superconducting T_c decreases when electron systems are 3D [3]. It is therefore useful to study the possible 2D effect on the electron-electron interaction for more enhancement of the superconducting T_c for organic superconductors. Here, the authors treat a quasi-2D system, where electrons are confined completely within an x-y plane and interact with 3D longitudinal optical (LO) phonons of long wavelength. In 3D electron systems, the electron-electron interaction potential is proportional to r^{-1} with r being the interelectronic distance, since phonons are propagated in the x-y-z directions. It is shown that, in our quasi-2D system, the exchanged phonon momentum is oriented in the x-y directions because of the momentum conservation law. The electric field produced by LO phonons is then squeezed within the x-y directions and the effective electron-electron interaction potential in the second-order perturbation for this system is proportional to $\ln(r)$. This 2D effect is called fermion confinement in field theory [4-6].

In the superconducting state, a Cooper pair is broken easily, when the attractive interaction potential is proportional to r^{-1}, because the electrons of the pair are separated, if they obtain a larger energy than the binding energy. When the electron-electron interaction potential is proportional to $\ln(r)$, the pair is not broken easily, since the binding energy is infinitely large. It can then be expected that enhancement of the superconducting T_c will occur in the quasi-2D system with weak screening. The case of copper oxides is discussed in detail elsewhere [7].

Section 2 derives the electron-electron interaction potential in the present quasi-2D system. Section 3 discusses the possibility for enhancing superconducting T_c.

2. Electron-Electron Interaction Potential in a Quasi-2D System

When electrons are itinerant, the Bloch function is a good approximation for them. When electrons are localized, the electronic correlation becomes important and an atomic-orbital-like wave function is suitable to describe electrons, as discussed in detail by Yamashita and Kurosawa [8]. Therefore, we approximate the electron wave function as $\psi(\vec{x}) = \exp[i(k_x x + k_y y)]\phi(z)$. Here, $\phi(z)$ is a localized atomic-orbital-like function and is expanded as $\phi(z) = \Sigma_{k_z}\phi_{k_z}\exp(ik_z z)$. Since the total momentum in the z-direction is zero, $\phi_{k_z} = \phi_{-k_z}$. In the present context, the field operator for electrons is expanded as $\Psi(\vec{x}) = \Sigma_{\vec{k}} c_{\vec{k}} \exp(i\vec{k}\cdot\vec{x})$ in the second-quantized form to satisfy the momentum conservation law.

The Fröhlich interaction is denoted as

$$H' = -i4\pi eF\Sigma_{\vec{q}} \int d^3x \Psi^+ |\vec{q}|^{-1}[b_{\vec{q}}\exp(i\vec{q}\cdot\vec{x}) - b_{\vec{q}}^+\exp(-i\vec{q}\cdot\vec{x})]\Psi$$
$$= -i4\pi eF\Sigma_{\vec{k}\vec{q}} |\vec{q}|^{-1}(b_{\vec{q}} - b_{-\vec{q}}^+) c_{\vec{k}+\vec{q}}^+ c_{\vec{k}} \quad , \tag{1}$$

where $b_{\vec{q}}^+$ and $b_{\vec{q}}$ represent the creation and annihilation operators for phonons with momentum $\vec{q}$, and $c_{\vec{k}}^+$ and $c_{\vec{k}}$ are corresponding operators for electrons with momentum $\vec{k}$, respectively. Note that, in Eq. (1), $\delta^3(-\vec{k}'+\vec{k}+\vec{q})$ appears with $\vec{k}$ and $\vec{k}'$ being the initial and final momenta, respectively, which expresses the momentum conservation law. The second-order electron-electron interaction is

$$H'' = \Sigma_{\vec{q}}\Sigma_{\vec{k}\vec{k}'} \frac{(4\pi eF)^2}{\vec{q}^2 + k_s^2} \frac{2\omega_D}{(\varepsilon_{\vec{k}} - \varepsilon_{\vec{k}-\vec{q}})^2 - \omega_D^2} c_{\vec{k}'+\vec{q}}^+ c_{\vec{k}'} c_{\vec{k}-\vec{q}}^+ c_{\vec{k}} \quad , \tag{2}$$

in units with $\hbar=1$, where $\varepsilon_{\vec{k}}$ is the energy of an electron energy measured from the Fermi energy ε_F, ω_D is the Debye energy and k_s is a cutoff parameter to remove the infrared divergence. The electron-electron interaction potential is approximated as

$$U_{\vec{q}} = (\frac{1}{\varepsilon_0} - \frac{1}{\varepsilon_\infty}) \frac{4\pi e^2}{\vec{q}^2 + k_s^2} \quad , \tag{3}$$

where ε_0 and ε_∞ are the static and optical dielectric constants, respectively.

The z-dependence for the state vector is discussed next. When operating H'' to $|\phi(z)\rangle$, $c_{\vec{k}-\vec{q}}^+ c_{\vec{k}} |\phi(z)\rangle$ becomes $\Sigma_{k_z}\phi_{k_z}|k_z - q_z\rangle = \exp(-iq_z z)|\phi(z)\rangle$. The state of $\exp(-iq_z z)\phi(z)$ has momentum $-q_z$, so that $q_z = 0$ is required to keep the 2D motion of electrons. When the summation in Eq. (2) is carried out using Eq. (3) over q_z associated with the weight $P(q_z)$, which is the scattering matrix element between the initial and final states, the z-dependence for the state vector should be unchanged:

$$\frac{1}{2\pi}\int dq_z P(q_z)\exp(-iq_z z)\phi(z) = \phi(z) \quad . \tag{4}$$

Then, we get

$$\frac{1}{2\pi}\int dp_z P(q_z)\exp(-iq_z z) = 1 \quad \text{for all } z \quad . \tag{5}$$

Therefore, $P(q_z)$ must be proportional to $\delta(q_z)$, which implies $q_z = 0$.

When the $q_z = 0$ selection rule is applied, the electron-electron interaction becomes

$$H_I'' = \rho_z \int dq_z H''\delta(q_z) \quad , \tag{6}$$

in the integral representation. Here, $\rho = \rho_x \rho_y \rho_z$ is the density of states for LO phonons. The Fourier inverse of the interaction potential $\rho_z\int dq_z U_{\vec{q}}\delta(q_z)$ in Eq. (6) is proportional to $\rho_z \ln(r)$ for $k_s = 0$, where r is the interelectronic distance. The electron-electron interaction is therefore much more attractive for long distances in the quasi-2D system than in the 3D one.

When acoustic phonons are emitted simultaneously with LO phonons, LO phonons can be emitted in the z-direction. Denoting the coupling constant between electrons and LO phonons by g_0, the corresponding quantity between electrons and acoustic phonons by g_A, the z-component of the momentum of acoustic phonons by Q_z and the z-component of the center of mass momentum for the crystal by P_z, it has been shown that, for $Q_z \neq 0$, acoustic phonons with $Q_z = -q_z$ are emitted simultaneously, while, for $Q_z = 0$, acoustic phonons of the uniform mode are emitted and lattice atoms are translated uniformly, meaning $P_z = -q_z$ [9]. These two processes, however, are higher-order compared to those where only LO phonons are emitted, because the coupling constant is $g_0 g_A$ in the former processes, while it is g_0 in the latter. Thus, the processes where acoustic phonons are emitted simultaneously with LO phonons can be neglected.

3. Superconducting Transition Temperature

In the present context, the mean field theory may be used for calculating superconducting T_c, because the electric 3D long-range force treated here suppresses the fluctuation in the quasi-2D system. The contribution from the electron-phonon interaction is included into m^*. The following BCS gap equation is then no longer the simple weak-coupling one:

$$1 = \int_{-\omega_D}^{\omega_D} d\varepsilon \frac{g}{2\varepsilon} \tanh \frac{\varepsilon}{2T_c} , \qquad (7)$$

where $g = -(1/\varepsilon_0 - 1/\varepsilon_\infty)\varepsilon_\infty \mu - \mu^*$ with $\mu^* = \mu/[1 + \mu \ln(\varepsilon_F/\omega_D)]$. Here,

$$\mu = \frac{\xi}{\varepsilon_\infty} \frac{m^*}{k_F} (4\pi e^2 \eta) , \qquad (8)$$

$$\eta_{k'k} = \int_{-\pi}^{\pi} d\phi \frac{1}{q^2 + k_s^2} = \frac{2\pi}{[(k^2 + k'^2 + k_s^2)^2 - (2kk')^2]^{1/2}} . \qquad (9)$$

Quantities ξ and η are calculated at the Fermi momentum and $\xi = (1/2\pi)^2 \rho_z k_F$. For the Coulomb pseudopotential μ^*, we use the same value to ρ_z for photons as for phonons. k_s is set to the inverse of the Thomas-Fermi screening length in the random phase approximation. For the quasi 2D system

$$k_s^2 = \frac{4\pi e^2 n_z}{\varepsilon_\infty} \Sigma_{k_x k_y} \delta(\varepsilon - \varepsilon_F) = \frac{4\pi e^2 n}{\varepsilon_\infty \varepsilon_F} , \qquad (10)$$

where $n = n_x n_y n_z$ is the carrier density.

T_c was calculated for a set of parameters; $n = 10^{21}$ cm^{-3}, $\omega_D = 600$ K, $\varepsilon_\infty = 4$, $\rho_z = 0.4$ Å^{-1}, $n_z = 0.15$ Å^{-1} and $m^* = 0.5$ m_0, where m_0 is the mass of a free electron. Figure 1 shows the ε_0 dependence of T_c in the quasi-2D system. With $\varepsilon_0 = 20$, T_c

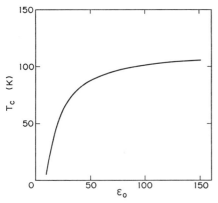

Fig. 1 Superconducting transition temperature T_c in the quasi-2D system. ε_0 refers to the static dielectric constant.

is below 1 K in 3D, while T_c is 49 K in the quasi-2D system. It is seen that, because of the weak screening, the two-dimensionality enhances T_c more for the larger values of ρ_z and $|1/\varepsilon_0 - 1/\varepsilon_\infty|$, which means the larger density of states of LO phonons and the stronger electron-phonon interaction, respectively. When these values are small, a high-T_c is not achieved even in the quasi-2D system.

4. Conclusion

The authors have shown that, in the present quasi-2D system, the effective electron-electron interaction potential via 3D LO phonons of long wavelength becomes $\rho_z \ln(r)$, which is quite different from the one widely used. In this context, it is possible to expect high-T_c superconductivity even within the BCS scheme.

Acknowledgment

The authors are grateful to Dr. Y. Gomei for stimulating discussions.

References

1. N. Nücker, H. Romberg, X. X. Xi, J. Fink, B. Gegenheimer and Z. X. Zhao, Phys. Rev. B39, 6619 (1989).
2. P. C. E. Stamp, L. Forro and C. Ayache, Phys. Rev. B38, 2847 (1988).
3. F. Herman, R. V. Kasowski and W. Y. Hsu, Phys. Rev. B37, 2309 (1988).
4. K. G. Wilson, Phys. Rev. D10, 2445 (1974).
5. J. B. Kogut, Rev. Mod. Phys. 55, 775 (1983).
6. K. Fukushima, Phys. Rev. D30, 1251 (1984).
7. K. Fukushima and H. Sato, phys. status solidi (b) 153, K141 (1989).
8. J. Yamashita and T. Kurosawa, J. Phys. Chem. Solids 5 34 (1958).
9. C. Kittel, Introduction to Solid State Physics, 6th edit. (Wiley, New York, 1986), p. 92.

Part XI

Summary

New Developments That Emerged in the ISSP Symposium

T. Ishiguro

Physics Department, Kyoto University, Kyoto 606, Japan

Abstract. Reports and related discussions on the organic superconductors during the ISSP International Symposium are reviewed from the viewpoint of physics.

In the last three years the field of condensed matter physics has been vitalized by the high T_c oxide superconductors. Historically, high T_c superconductors were first proposed 25 years ago by Little in the field of organic polymeric materials. Although this proposal has not yet been substantiated even today, it did stimulate the development of organic conductors and superconductors. Superconductivity was first observed in the organic charge transfer salts $(TMTSF)_2PF_6$ under pressure in 1979. Thus, we meet on the 10th anniversary of the birth of the organic superconductors in 1989. Until the discovery of superconductivity in $\beta-(ET)_2I_3$ in 1984, the $(TMTSF)_2X$ salts were the only family exhibiting superconductivity. They have been studied intensively and their characteristic features as exotic superconductors, such as low-dimensionality and the proximity of the superconducting phase to the SDW phase, have been revealed. The low-dimensionality in these salts has undoubtedly opened a new research field in solid-state physics. Interestingly, recent new attractive materials also exhibit restricted dimensionality to a greater or lesser extent. In fact, the oxide superconductors are of two-dimensional nature and the physics constructed through the study of organic superconductors has served as a basis for their proper characterization.

In this ISSP symposium, the superconductivity of $(TMTSF)_2X$ was hardly discussed, since experimental investigations have already revealed its general properties. One advance in a related area, reported in this symposium, is the phase diagram. The proximity of the SDW to the superconducting phases is a characteristic feature of $(TMTSF)_2X$. However, this has not been proved for $(TMTSF)_2ReO_4$ and $(TMTSF)_2FSO_3$, which exhibit obscure phase boundaries, called the glassy state, between the insulating and superconducting phases. By using a pressure-temperature cycling method, Tomic et al. showed that $(TMTSF)_2ReO_4$ exhibits a similar proximity of the SDW to the superconductivity as for $(TMTSF)_2PF_6$. This may serve to construct a unified picture of the phase diagrams of these salts.

The dimensionality of the TMTSF salts lies between one-dimensionality (1-D) and two-dimensionality (2-D). The role of the 1-D has been the source of much argument. In this symposium, the nature of the 1-D has been revisited through theoretical work. Based on NMR data, Bourbonnais et al. argued for the crucial role played by magnetic excitation of one-dimensional paramagnon-type and anti-ferromagnetic fluctuations in the condensation of instabilities such as spin-Peierls, SDW and superconducting pairing of the phase diagram of $(TMTSF)_2X$ and their sulfur analogs. Further, the interplay of the dimensionalities is studied in a newly synthesized mixed sulfur-selenium material (TMDTDSF)$_2$PF$_6$ by Jérome et al. Also, the effect of the long-range antiferromagnetic spin fluctuation on the phase diagram was discussed by Shimabara based on a quasi one-dimensional Hubbard model. The effect of nonmagnetic impurities on the transport properties of quasi one-dimensional superconductors was presented by Suzumura.

In recent years, the main subject in the physics of (TMTSF)$_2$X has been the field-induced (FI) SDW effect. By applying high magnetic fields up to 30T, Chaikin et al. discovered that (TMTSF)$_2$ClO$_4$ exhibits a reentrant metallic phase, which cannot be understood on the basis of the standard FISDW theory. Furthermore, it was reported in this symposium that the low-temperature transport phenomena in a high magnetic field in (TMTSF)$_2$PF$_6$ under pressure yield a clear quantized Hall effect (QHE) and also the fractional QHE. Originally the possibility of the QHE was pointed out for (TMTSF)$_2$ClO$_4$, but it was not confirmed because the size of the steps is not well characterized with the quantized value, presumably due to the effect of the resistance jumps appearing irregularly in the cooling process. For (TMTSF)$_2$PF$_6$ under pressure, which can be cooled without meeting the resistance jump, it is found that the sizes of the resistance steps are characterized to be either $h/4e^2$ (by Chaikin et al.) or $h/2e^2$ (by Jérome et al.), instead of h/e^2 for MOS. Further, a sudden reversal of the sign of the Hall voltage appears in a limited region of magnetic field and pressure in the new results, also under limited conditions. The difference in the observed quantized units and the Hall reversals need further investigation experimentally and theoretically.

For the SDW and FISDW phases, the electrical conduction due to the sliding motion and the sound attenuation were discussed by Maki in terms of the quasi one-dimensional Hubbard model with random impurities. Further, the magneto-thermodynamics and magneto-transport in (TMTSF)$_2$ClO$_4$ were presented by Montambaux and Garoche. In particular, Garoche et al. reported fractal-like fine structure near the FISDW phase boundaries observed in the specific heat measurements in magnetic fields and discussed the phase boundaries in terms of the fractional quantization of the nesting in relation to the pockets of unpaired carriers or higher-order nesting due to the anharmonicity of the energy dispersion near the Fermi level.

(ET)$_2$X salts are the 2nd generation salts of the organic superconductors. The main features of these superconductors had been previously reported in earlier conferences, e.g. the International Conference on Synthetic Metals held at Santa Fe in 1988. For these salts, rather high T_c superconductors, up to 11.4 K, have been realized with deuterated (ET)$_2$Cu(NCS)$_2$. Together with a review presentation on the nature of the salt, recent results on the effect of purification of this superconductor were presented by Mori (Urayama) et al. They found that careful purification makes the transition characteristics sharper and raises the T_c to 11.6 K. A T_c of 13 K in (ET)$_2$Cu(NCS)$_2$ was reported by Oshima et al., but the reproducibility of the material and its general characteristics need to be checked.

Empirically it is found that the T_c of organic superconductors decreases with a decrease in the separation between constituent molecules, e.g. ET molecules: the rapid decrease of T_c with pressure and with substitution of smaller anions for the original ones can be interpreted in this scheme. Hence, if one increases the distance between molecules, T_c is expected to increase. Based on this idea, Kusuhara et al. effectively applied a tensile strain to (ET)$_2$Cu(NSC)$_2$ within the conducting planes by sticking thin crystals on a Cu block etc., so that the extraordinarily large thermal contraction of the salt is suppressed, resulting in tensile strain. According to the band calculation for the strained crystal and the superconducting model based on the electron pairing via intramolecular vibration, the increase of T_c by this tensile strain is expected to reach 3-4 K. The experimental results exhibit a T_c increase of around 1 K. The difference indicates that the interlayer interaction, which is not taken into account by the calculation, plays a significant role in the determination of T_c.

Concerning the nature of the superconductivity, Kanoda et al. argued a possible presence of the zero-gap region for the superconducting order parameter for $(ET)_2Cu(NCS)_2$, through the temperature dependence of the penetration depth for the magnetic flux. The optical properties, including reflectance and electron-molecular vibrational spectra, and scanning tunneling images of the salt were reported. Further, an energy gap of $2\Delta=4.8\pm1.1$ meV was observed by a tunneling spectroscopy experiment (Bando et al.).

The resistance maximum appearing in $(ET)_2Cu(NCS)_2$ near 90K reminiscent of the resistivity maximum for the heavy-fermion system has been one of the most interesting subjects in recent organic conductors. In this symposium a few mechanisms were proposed for this phenomenon, such as the structural phase transition associated with phase locking (Friend et al.), the nonmetal-metal transition due to anisotropic thermal contractions (Andres et al.) and the polaron effect (Kusuhara et al.). A similar resistance maximum has been reported in other organic superconductors, such as $(DMET)_2AuBr_2$ (Kikuchi et al.) and DCNQI salts (Wolf et al., Kato et al.). For $(DMeDCNQI)_2Cu$, the resistance maximum associated with a variation in the magnetic susceptibility can probably be interpreted in terms of the dense Kondo effect caused by the spins of Cu^{2+}.

$(ET)_3Cl_2 \cdot 2H_2O$ is unique as an organic superconductor due to its 2/3-band-filling electronic structure and to the transition into the superconducting state from the nonmetallic state. Friend et al. reported that there exists a gap at the Fermi surface in the nonmetallic region.

As a novel feature for $\beta-(ET)_2I_3$ exhibiting incommensurate lattice modulation at ambient pressure, Kagoshima et al. reported that the wave number is varied by annealing. In accordance with this change the superconducting temperature is raised a little, to 2 K, indicating the appearance of a somewhat new superconducting state. It should be borne in mind that there have been articles reporting the presence of a few T_c states, e.g. starting at $\sim$4 K, $\sim$7 K and so on, for this material. This diffusivity in the transition characteristics needs further investigation of the role and nature of the incommensurate superstructure in this material.

As one of the focal points of the symposium, we should point out the arguments about the Fermiology of $(ET)_2X$, exhibiting a two-dimensional Fermi surface, based on measurements of the Shubnikov-de Haas effect and related phenomena. The magnetoresistance of the $(ET)_2X$ superconductors and conductors, such as $\beta-(ET)_2I_3$, $\beta-(ET)_2IBr_2$, $\beta-(ET)_2AuBr_2$, $\kappa-(ET)_2Cu(NCS)_2$, $\theta-(ET)_2I_3$, and $(ET)_2KHg(SCN)_4$, has been studied in many places throughout the world over the last two years. This symposium has provided a timely forum for the exchange of data and interpretations of this subject.

An amazing feature of the Shubnikov-de Haas effect, a giant magneto-oscillation in resistivity of $\beta_H-(ET)_2I_3$, is reported by Jérome et al., who ascribed the giant oscillation to the slightly warped cylindrical Fermi surface and the pureness of the crystal. Meanwhile, the controversy concerning the Fermi surfaces derived from the extended Hückel method (Mori et al.) and the local density functional approximation and the augmented-spherical-wave algorithm

(Kübler et al.) was almost reconciled by noting that the latter can also give similar results to the former.

The angular dependence of the magnetoresistance for β-$(ET)_2IBr_2$ and θ-$(ET)_2I_3$ shows oscillatory behavior for rotation of the crystal with respect to the magnetic field (Laukhin et al. and Kajita et al.). This oscillation is interpreted in terms of a nearly complete quantization of energy levels occurring for a corrugated cylinder-like Fermi surface (Yamaji). Also, it is reported by Pratt et al. that the period of the oscillation varies with a change of the current direction for measurement, which may be interpreted in terms of the domain structure formed in the crystal. The correlation between the cyclotron masses and the Dingle temperatures which are derived from the Shubnikov de-Haas oscillations in various kinds of $(ET)_2X$ salts has been interpreted by Toyota et al. on the basis of the renormalization of the many-body effect for which the on-site Coulomb interactions in the Hubbard model might play an important role.

As new superconducting materials appearing in the last three years, $(DMET)_2X$, $(MDT-TTF)_2AuBr_2$, and M-$(dmit)_2$ salts were also reviewed together with a presentation on some new aspects. For α-$TTF[Pd(dmit)_2]_2$, Cassoux et al. suggested that the same electron-phonon interaction is responsible for both the metal-insulator and superconducting transitions. In the meantime, Poujet et al. have reported diffuse lines and spots in both $Ni(dmit)_2$ and $Pd(dmit)_2$ compounds, showing typical CDW transitions in quasi one-dimensional conductors. From the form factor analysis, they asserted that the periodic lattice distortions are on $M(dmit)_2$ chains.

Concerning the mechanisms of superconductivity, various kinds of models were referred to in this symposium: electron-phonon interaction via intramolecular vibrations, libron-electron coupling, magnetic coupling, electron-electron interaction, and mechanisms in relation to the ceramic superconductors. The roles of the interlayer interactions and the effect of non-magnetic impurities are also pointed out based on experimental results. Taking account of these investigations, we should say that the arguments on the mechanism are making steady progress : theory and experiment have been interplaying rather directly in many phases. There are definite signs of progress being mode.

I have tried to summarize the reports on the physical aspects of the organic superconductors that were presented in this ISSP symposium. These subjects themselves have attracted a great deal of interest. Finally, I would like to emphasize that a huge number of organic compounds have been developed, and they are not yet the focus of our attention simply because they are not superconducting.However, if we look more closely, we see that each of them can offer a unique nature which may enrich contemporary condensed matter physics and materials science.

Index of Contributors

Akiba, K. 134,159
Allan, M. 28,290
Alvarez, S. 262
Andres, K. 195
Anzai, H. 126,130,138,142, 167,191,224,267
Ara, N. 280
Aso, Y. 395,403
Auban, P. 2
Awaga, K. 329

Bair, M. 41
Bando, H. 167
Batail, P. 68,272,353
Bechgaard, K. 68,349,383
Beno, M.A. 262,334,369
Biberacher, W. 195
Boubekeur, K. 353
Bourbonnais, C. 68
Bravic, G. 290
Brooks, J.S. 81
Brossard, L. 22
Bryce, M.R. 420

Canadell, E. 252
Carlson, K.D. 334,369
Cassoux, P. 22
Chaikin, P.M. 81,276
Chang, K.Y. 267
Chasseau, D. 290
Chiang, L.Y. 81
Christensen, J. 349
Clark, R.A. 28,412
Cooper, J.R. 2,111
Creuzet, F. 68
Cruz-Vázquez, C. 58

Davidson, A. 353
Day, P. 8,181,200,272,290
Dhindsa, A.S. 420
Dresselhaus, G. 324

Enoki, T. 32,294,298
Erk, P. 41
Evain, M. 262

Fenton, E.W. 177
Fernando, Q. 58

Ferraro, J.R. 262,334
Fite, C. 358
Fourmigué, M. 353
Friend, R.H. 28,181,272,290
Fukase, T. 191
Fukushima, K. 465
Fukuyama, H. 15

Garoche, P. 87
Gärtner, S. 146
Gaultier, J. 290
Geiser, U. 334,369
Ginodman, V.B. 122,364
Gogu, E. 146
Grimm, H. 146
Gudenko, A.V. 122,364

Hannahs, S.T. 81
Harada, S. 373
Harada, T. 107
Hasumi, M. 126,142
Hayes, W. 181,200,290
Heidmann, C.-P. 195
Héritier, M. 87
Hilti, B. 247
Honda, Y. 224,234
Hotta, S. 391
Hountas, A. 247
Huang, Y. 453
Hünig, S. 41

Ida, T. 49,54,311
Ikegawa, S. 373
Ikemoto, I. 230,234,238,242
Imaeda, K. 298
Inabe, T. 163,329,408
Inagaki, Y. 416
Inokuchi, H. 204,298
Inoue, M. 58
Inoue, M.B. 58
Ishibashi, M. 224
Ishida, H. 403
Ishiguro, T. 171,470
Ishikawa, S. 444
Isotalo, H. 358
Iwasa, Y. 319
Iwasawa, N. 319
Iye, Y. 212,324
Izuoka, A. 32,379,428

Jérome, D. 2,64,68,111
Johannsen, I. 349,383
Jung, D. 262
Jørgensen, M. 349,383

Kageshima, M. 280
Kagoshima, S. 126,130,142, 220
Kahlich, S. 146
Kaji, M. 171
Kajimura, K. 167
Kajita, K. 45,212
Kakoussis, V. 247
Kang, W. 2,81,111
Kanoda, K. 107,134,155, 159,242
Kartsovnik, M.V. 186
Kasai, N. 373
Kashiwaya, S. 167
Kasmai, H. 358
Kato, R. 36,45,49,212,302
Kawabata, Y. 424
Kawamoto, A. 298
Kawazu, A. 280
Keller, H.J. 146
Khomenko, A.G. 364
Kikuchi, K. 230,234,238,242
Kini, A.M. 262,334,369
Kinoshita, M. 315
Kinoshita, N. 126,138,142, 167,224,267
Kobayashi, A. 28,36,45,49, 212,302
Kobayashi, H. 28,36,45,49, 212,302
Kobayashi, K. 230,234,238, 242
Kobayashi, Y. 387
Koda, T. 319
Kohmoto, M. 102,434
Komazaki, T. 224,234
Kono, Y. 395
Kononovich, P.A. 122,186
Korotkov, V.E. 306,364
Koshelap, A.V. 364
Koshihara, S. 319
Kübler, J. 208
Kubota, T. 267

Kurihara, T. 267
Kurmoo, M. 181,200,272, 290
Kuroda, H. 49,54
Kushch, N.D. 306,364
Kusuhara, H. 171,284

Langohr, U. 41
Laukhin, V.N. 122,186,342, 364
Laukhina, E.E. 342
Legros, J.-P. 22
Lenoir, C. 272,353
Lerf, A. 195
Lerstrup, K. 349,383
Liou, K. 358
Livage, C. 353

Maki, K. 91
Mallah, T. 290
Marsden, I. 28,181,290
Maruyama, Y. 163,329,408
Masuda, H. 54
Matsumiya, S. 379
Matsumoto, M. 424
Mayer, C.W. 247
Meixner, H. 41
Mishima, T. 284
Mitsuhashi, T. 408
Miura, N. 220
Miura, Y. 424
Miyamoto, A. 45
Miyamoto, H. 395
Miyashi, T. 387
Miyawaki, K. 373
Miyazaki, A. 32
Montambaux, G. 81,97
Mori, H. 150,155,163,257, 280
Mori, N. 130,324
Mori, T. 204
Mota, F. 262
Mousdis, G. 247
Müller, H. 195
Murata, K. 107,167,191,224, 230,234
Murayama, C. 324

Nagashima, U. 329
Nakamura, A. 457
Nakamura, T. 424
Nakano, H. 373
Nakano, Y. 294
Nakasuji, K. 399
Nebesny, K.W. 58
Nicholls, J.T. 324
Nishio, Y. 45,212
Nogami, T. 373
Nogami, Y. 126,130
Novoa, J.J. 262

Obertelli, S.D. 181
Ogata, M. 438
Ogura, F. 395,403
Ohyama, T. 107
Ojima, G. 49
Okada, M. 453
Okui, S. 242
Okuno, T. 428
Osada, T. 142,220
Oshima, K. 150,257,276
Oshima, M. 220,257,280
Otsubo, T. 395,403

Papavassiliou, G.C. 247
Parker, I.D. 272
Pénicaud, A. 353
Pesotskii, S.I. 186
Pesty, F. 87
Petty, M.C. 420
Pouget, J.P. 252
Pratt, F.L. 181,200,290

Ravy, S. 252
Ren, J. 262
Rosseinsky, M.J. 181
Rozenberg, L.P. 364

Saito, G. 107,126,130,134, 150,155,159,163,220,257, 276,280,294,319
Saito, K. 230,234,238,242
Sakao, K. 155
Sakata, Y. 171
Sasaki, T. 142,177,191
Sasaki, W. 45,212
Sato, H. 465
Sato, N. 32
Sawada, M. 395
Schegolev, I.F. 122,186
Schultz, A.J. 334
von Schütz, J.U. 41
Schweitzer, D. 146
Shiba, H. 438
Shibaeva, R.P. 306,342,364
Shigekawa, H. 280
Shimahara, H. 73
Shioda, R. 280
Shirota, Y. 373
Shirotani, I. 416
Sieburger, R. 195
Singleton, J. 200
Sommers, C.B. 208
Spermon, S.J.R.M. 200
Srdanov, G. 358
Sugano, T. 315
Sugawara, T. 32,379,428
Suzuki, K. 134,294,298
Suzuki, T. 358
Suzumura, Y. 77

Tachibana, A. 444,461
Tachibana, H. 424
Tachiki, M. 177

Tada, K. 171,284
Tada, T. 444
Tajima, H. 49
Takada, Y. 434
Takahashi, T. 107,134,155, 159,212,242
Takeuchi, H. 298
Tanaka, J. 298
Tanaka, K. 449,453
Tanaka, M. 298,424
Tanaka, S. 150
Tanaka, T. 395
Tanigawa, S. 267
Tejel, C. 22
Terzis, A. 247
Tokumoto, M. 116,126,130, 138,142,167,191,224,267
Tokumoto, T. 167
Tokura, Y. 319
Tomić, S. 64,111
Tomomatsu, I. 294
Toyota, N. 142,177,191
Tseng, P.K. 267

Ueba, Y. 171,284
Ugawa, A. 49,54,311
Ulmet, J.-P. 22
Underhill, A.E. 28,412
Urayama, H. 276

Valade, L. 22
Varma, K.S. 412

Wainwright, C.E. 412
Wang, H.H. 262,334,369
Waragai, K. 391
Watabe, M. 155
Watanabe, K. 267
Watanabe, N. 319
Whangbo, M.-H. 262,334
Williams, J.M. 262,334,369
Wolf, H.C. 41
Wudl, F. 358
Wzietek, P. 68

Yagi, R. 220
Yagubskii, E.B. 306,342,364
Yakushi, K. 49,54,311
Yamabe, T. 444,449,453
Yamaguchi, Y. 138,449
Yamaji, K. 216
Yamakado, H. 54,311
Yamashita, M. 416
Yamashita, Y. 387
Yamochi, H. 150,163,276, 280,358
Yomo, S. 324
Yoshimura, M. 280
Yu, R.C. 276
Yui, K. 403

Zamboni, R. 146
Zambounis, J.S. 247